W0043734

TREATMENT
OF CEREBRAL
INFARCTION

JIRO SUZUKI

TREATMENT
OF CEREBRAL
INFARCTION

EXPERIMENTAL
AND CLINICAL STUDY

JIRO SUZUKI

SPRINGER-VERLAG
WIEN NEW YORK

Professor Jiro Suzuki, M.D.

Division of Neurosurgery, Institute for Brain Diseases,
Tohoku University School of Medicine, Sendai, Japan

This work is subject to copyright.
All rights are reserved, whether the whole or part of the material is concerned, specifically those of translation, reprinting, re-use of illustrations, broadcasting, reproduction by photocopying machine or similar means, and storage in data banks.

© 1987 by Springer-Verlag/Wien
Softcover reprint of the hardcover 1st edition 1987

The use of registered names, trademarks, etc. in the publication does not imply, even in the absence of a specific statement, that such names are exempt from the relevant protective laws and regulations and therefore free for general use.

Product Liability: The publisher can give no guarantee for information about drug dosage and application thereof contained in this book. In every individual case the respective user must check its accuracy by consulting pharmaceutical literature.

With 240 partly colored Figures

Library of Congress Cataloging-in-Publication Data. Suzuki, Jiro, 1924– . Treatment of cerebral infarction. Bibliography: p. Includes index. 1. Brain-Infarction. I. Title. [DNLM: 1. Cerebral Infarction. 2. Cerebral Ischemia. WL 355 S968t]. RC394.I5S89 1987. 616.8'1. 86-29758

ISBN 978-3-7091-8863-7   ISBN 978-3-7091-8861-3 (eBook)
DOI 10.1007/978-3-7091-8861-3

PREFACE

It is a great honor and pleasure for me to have Springer-Verlag publish this volume entitled "Treatment of Cerebral Infarction". I am much indebted to my colleagues for my success in publishing this book.

I have engaged in clinical work in the field of neurosurgery for these few decades and I have performed more than 5,000 major operations of intracranial surgery. Throughout this time, it has been my privilege to conduct a 2-hour-morning research seminar in our department every Wednesday and to supervise a great deal of research. At these seminars my fellow research workers and I have exchanged many ideas about the study of neurosurgery and we have designed many animal experiments. The results of the research performed in the previous week have always been reported at such seminars and research workers have benefited from the advice and criticisms given there.

In 1969 I found that the permissible occlusion time for cerebral blood flow could be prolonged by mannitol. In that year Dr. Takashi Yoshimoto started a difficult series of animal experiments to prove my hypothesis. Since then, many researchers have joined our department and many research programs and experiments on cerebral infarction have been carried out. Dr. Yoshimoto hoped that the results of research done in our department over the last sixteen years concerning cerebral infarction could be published. Moved by his enthusiasm, my colleagues took their share in writing each article. I suggested to them that this volume should cover primarily the results of our own research, but that at the same time it should also include the results of related work done by neurosurgical experts throughout the world. Therefore, I hope that this book will interest many investigators who have devoted themselves to research in the field of neuroscience.

My colleagues acknowledge that our research has been stimulated and promoted by our weekly "think tank" sessions and that any achievement we may have attained is due largely to those fruitful discussions. This is the reason why this book is of my authorship. I am very happy and honored that my own contribution to our study of neurosurgery is thus recognized, but I am very conscious that this book was completed thanks to the cooperation of the clinical neurosurgeons who have studied and engaged in clinical neurosurgery with me in Sendai. I must therefore mention those who took part in writing this book: Dr. Takashi Yoshimoto (Chapter 1), Dr. Takao Watanabe and Dr. Michiyasu Suzuki (Chapter 2), Dr. Hirobumi Seki (Chapter 3), Dr. Takamasa Kayama (Chapter 4), Dr. Shigeki Imaizumi (Chapter 5), Dr. Kazuo Mizoi (Chapter 6), Dr. Akira Ogawa and Dr. Tetsuo Kogure (Chapter 7), Dr. Satoru Fujiwara (Chapter 8), Dr. Takehide Onuma and Dr. Yoshihide Nagamine (Chapter 9A), Dr. Motonobu Kameyama (Chapter 9B), Dr. Yoshiharu Sakurai (Chapter 9C), Dr. Hiroshi Niizuma (Chapter 10), Dr. Ryuichi Katakura and Dr. Ryuichi Konda (Chapter 11), Dr. Akira Takahashi (Chapter 12), Dr. Taisuke Otsuki and all neurosurgeons who have studied with me in Sendai. I am also grateful to Dr. D. H. Waterbury, who helped us by reading the manuscript and by giving valuable advice on the linguistic problems of English. Especially without the genuine help of Mr. N. D. Cook, as a translator, this book would have never appeared. Finally, I heartily thank Dr. S. Weingärtner and Mr. R. Wieder of the Editorial Department of Springer-Verlag who encouraged us to publish this volume.

Jiro Suzuki

Nagamachi, Sendai
November, 1986

CONTENTS

Part III
Appendixes

INTRODUCTION

Among the developed nations, cerebrovascular disease (CVD) ranks as one of the top three causes of human death and must therefore be considered a major health hazard for mankind. Due to the elucidation of the risk factors involved, there has been a gradual decrease in the incidence of hemorrhagic CVD, but together with the resultant increases in longevity there has also been a gradual increase in ischemic CVD—a trend which is likely to become a worldwide phenomenon. For this reason, the development of techniques for the prevention and treatment of ischemic diseases of the brain is an issue of extreme importance for all of mankind.

Not only is normal brain function crucial to the individual, but it is worth emphasizing that the brains of world leaders play pivotal roles in current and future world events. The appearance of CVD, the concomitant loss of functions, and the decrease or complete halt in productivity thus have wide-ranging implications for the individual, for his family and for society at large. By the same token, the development of methods to prevent and treat this disease has importance not only for individual patients and their families, but also for the nations within which they work.

The essential nature of ischemic cerebrovascular disease can be described as necrosis of brain tissue due to decreased cerebral blood flow caused by stenosis or occlusion of cervical and/or intracranial arteries. Necrosis produces functional deficits in those parts of the brain and leads either to survival is an impaired condition or to death. While this pathology of CVD is, of course, common knowledge in medicine, it remains true that there is a considerable degree of uncertainty concerning the nature of the gradual intracerebral changes which occur following an ischemic attack. Among several fundamental questions which remain unanswered, the following should be mentioned: What level of reduction of cerebral blood flow and how prolonged a duration of occlusion will result in damage to brain tissue? What differences in these parameters exist in cases of focal ischemia and in cases of global ischemia? With regard to recirculation following ischemia, it must be said that, with the exception of morphological findings using the electron microscope, current research is little more than groping in the dark concerning neurophysiological and neurochemical results.

In the light of these uncertainties, it is evident that accurate evaluation of the therapeutic effects of various drugs is simply not possible. Moreover, there are significant individual differences in the capacity for developing collateral pathways—a fact which makes evaluation of the prognosis following onset equally difficult. Due to further problems posed by changes in various components of circulating blood, age, blood pressure, associated disorders of the cardiac, pulmonary and respiratory systems, diabetes, and varying degrees of arteriosclerosis, it is evident that the effect of ischemia on brain cells is an extremely complex issue.

What kinds of medical therapy are currently in use for conditions of cerebral ischemia? First of all, with regard to prevention, gradually more favorable results have been obtained by means of dietary measures and the administration of drugs such as aspirin and wafarin to prevent arteriosclerosis and to prevent and/or treat hypertension.

Unfortunately, with regard to the therapeutic steps taken during the acute stage following the onset of ischemic CVD, whatever measures have been employed, the results have been little better than those following the natural course of the disease. Logically, it is easy to imagine that, by means of surgical treatment or the administration of hemolytic agents, vascular occlusions could be removed and efforts then made to increase cerebral blood flow. Such therapy, however, has been found to cause an increase in cerebral edema and cerebral hemorrhage, and for this reason attempts to induce vascular recirculation during the acute stage are now thought to be inappropriate.

Surgical therapy is therefore also thought to be fruitless during the acute stage.

With regard to treatments currently in use for ischemic CVD, only two courses of action are open—neither of which is likely to lead to full recovery. Either the brain tissue which has been affected in the acute stage is considered beyond recovery and the patient is sent for rehabilitation with the neuronal deficits or, afternatively, by-pass surgery can be performed in order to allow for some slight recovery of the brain tissue in the penumbra of the focus.

Serious thought, as a matter of human compassion, should cirtainly be given to the problems of prevention of ischemic CVD and functional recovery through rehabilitation, but the most important questions concern the development of therapeutic techniques for the acute stage of the disease. Specifically, in order to prevent the rapid deterioration of the patient's cerebral condition and reduce the number of acute deaths, which are known to be a function of the interval from onset, it is essential to develop some form of therapy that can be to instituted as early as possible in the acute stage. That therapy, whatever it may be, would be of extreme importance and would constitute the first positive step forward in the treatment of ischemic cerebrovascular disease. During that period, while rapid deterioration of brain cells is being prevented, it would then be possible either to undertake vascular reconstruction or to allow for spontaneous increase in cerebral blood flow due to the emergence of collateral pathways.

The most difficult and frustrating task for the neurosurgeon is to be forced simply to observe the progression of cerebral infarction following onset and to be unable to take positive action. Similar to dealing with a house on fire, the most important step would be to take preventive measures at a very early stage in the event. From our own animal studies we have found that when cerebral flow is reduced to 40% of its normal level for a period of 3 hours, the morphology of brain cells is drastically changed and the cells are phagocytosed by leucocytes 24 hours later. Needless to say, despite the fact that there is total occlusion of capillaries and small vessels, leucocytes manage to destroy the infarctic brain tissue. Once the brain has entered such a condition, it is already too late to take therapeutic steps.

My own interest in and study of the ischemic brain began with an attempt to prolong the permissible time of temporary vascular occlusion for radical surgery on ruptured cerebral aneurysms—and indeed this quest has become a major lifework for me. At that time, an energetic young man, my classmate Akira Watanabe, entered the department of surgery of our college and began investigating hypothermic anesthesia in dogs. I turned my attention to the question of the permissible time for cerebral vascular occlusion using various degrees of hypothermia and soon learned how very difficult it is to reduce cerebral blood flow to zero. Eventually, I found that cerebral flow could be halted by means of throacotomy and occlusion of all the ascending arteries to the neck, but this experimental model proved to be extremely laborious!

Finding that 30 minutes of vascular occlusion could be done using hypothermic anesthesia at 27 °C, I performed my first aneurysm surgery using hypothermia on May 27th, 1961 and over the following years until 1969, I operated on many such cases. Considerable time however was needed for the lowering and subsequent raising of the patient's body temperature and, moreover, during that procedure complications such as cardiac arrest and skin burns were not uncommon. Throughout this period I pondered in the back of my mind whether or not there might be alternative means for prolonging the permissible occlusion time of cerebral vessels.

A fundamental turning point in our surgical technique came in 1969 when I experienced a case of ruptured cerebral aneurysm with severe cardiac complications. The anesthesiologist suggested that hypothermia alone would be fatal and since I had operated on some 300 aneurysm cases by then, I was confident that a successful operation could be done at normothermia. Craniotomy and dissection of the aneurysm were begun, but before actually reaching the aneurysm, there was a major re-rupture and intracerebral structures could no longer be distinguished or the aneurysm treated. Temporary occlusion of feeding and draining arteries of the aneurysm therefore became inevitable and the total occlusion time exceeded 50 minutes. Since cerebral necrosis occurs in about three minutes in medical common sense, there was no reason to expect favorable results after such a prolonged occlusion. Expecting severe postoperative sequelae, I went to the patient's bed on the following morning and called her name. To my great surprise, we found her to be in a normal state of consciousness and to be without paresis.

My first thought was that this was surely a case of divine intervention, but the scientist in me soon returned and the entire staff was convened to examine what kinds of treatment the patient had received before, during and after the operation. At this point, I recalled that, in order to reduce the cerebral volume and enlarge the surgical field, 1,000 cc of 20% hypertonic mannitol solution had been administered by *i.v.* drip in such a manner that it was completed near the end of craniotomy. It was thus apparent that the mannitol may have been the factor which allowed for the prolonged occlusion of cerebral vessels without producing sequelae.

I asked Assistant Professor Takashi Yoshimoto, who was then a young M.D., if he would investigate the apparent capability of mannitol to suppress cerebral infarction, and in this way a long series of animal experiments were initiated. In order to clarify the effects of drugs such as mannitol, it was first necessary to establish an experimental animal model in which infarctic foci of the same size could be produced consistently at the same site in the brain. Furthermore, the severity of these foci would have to be controllable. Although research into the effects of mannitol was to begin with the development of such an ideal infarction model, animal models using various

species and countless techniques did meet our expectations. Eventually, Dr. Yoshimoto went to Taiwan, where native Taiwanese monkeys were inexpensive enough to pursue such research to develop a model, but he returned despondent without success.

Upon hearing his bad news, I concluded that rigorous animal studies on the effects of drugs on brain infarction might in fact be impossible and I regretfully called a halt to the research project. But just two months after having abandoned hope, Dr. Yoshimoto strode into my office with four formalin-fixed dog brains in hand and announced: "I've done it!" When I asked what he had done, he proceeded to show me infarctic foci of similar size in the anterior thalamus of all four brains cut coronally through the optic chiasma.

As delighted as we both were, it was soon found that even when the four trunk arteries were visible when a unilateral temporal approach was used without doing damage of the temporal lobe, the incidence of anterior thalamic infarction was considerably lower than the initial indications of Dr. Yoshimoto's first four dogs. At that point, Dr. Tetsuya Sakamoto suggested that a deep electrode stereotaxically inserted into the anterior thalamus for the purpose of monitoring brain electrical activity would allow for confirmation of the infarction produced by means of occlusion of the four trunk arteries. It was then found experimentally that an ideal focus of infarction could be produced in virtually 100% of the animals using that technique. Subsequent work was then devoted to the development of related models of brain ischemia, infarction and edema and to clear demonstration of the beneficial effects on the brain of mannitol administration.

Once we were certain that mannitol had protective effects, we began to use it clinically and gradually the number of patients subjected to hypothermic anesthesia has decreased. As we have often indicated at conferences in Japan and abroad, there is no doubt that mannitol is effective, but when asked why mannitol has such protective effects, we have long been in the uncomfortable position of only responding that it does indeed have beneficial effects, so please try it! For this reason, while the clinical effectiveness of mannitol can be empirically demonstrated, the theoretical argument for its usage has remained weak.

While pondering the problem of how mannitol may work in ischemia, I read that it acts a scavenger of HO radicals. Then Demopolas reported in 1972 that, when the free radical reaction is generated in the infarctic brain, neurons are likely to be destroyed. It was therefore evident that if mannitol suppresses the pathological changes in cerebral infarction by acting as a free radical scavenger, then many other scavengers should also be capable of suppressing infarction.

In ensuing experimental work making use of our canine infarction model in which cerebral blood flow is under the complete control of an infusion pump, we have indeed discovered other scavengers with similar effects. Notable among the other scavengers are vitamin E and dexamethasone. In light of these findings, it is now certain that mannitol acts as a scavenger of the free radical reaction and thereby suppresses the development of cerebral infarction. Moreover, making use of the chemiluminescence technique, we have confirmed that the free radical reaction takes place in the ischemic brain and we have demonstrated that mannitol, vitamin E and steroids suppress that chemiluminescence.

Subsequent to such experimental work, we have published several reports on the use of these drugs in surgery on cases of cerebral aneurysms and arteriovenous malformations (AVMs) by administering them over a period of 30 minutes just prior to temporary vascular occlusion using a temporary clip. Specifically, the feeding and drainer vessels of the lesion are dissected and the above-mentioned occlusion procedure is employed. The aneurysm or AVM is then treated. In the case of deep-seated feeders, such as the basilar artery, a balloon catheter is inserted into the feeder artery and inflated to achieve the vascular occlusion. Using these techniques, we have been able to achieve a bloodless surgical field for a wide variety of neurosurgical operations.

Further findings concerning the protective effects produced by other drugs have led to the next step. Specifically, the anticonvulsant drug, Aleviatin, has recently been shown to have remarkable protective effects which are brought about, not by means of free radical scavenger action, but by stabilization of the cellular membrane. As a consequence, clinically, we currently use a combination of three drugs, mannitol, vitamin E and aleviatin, without employing steroids.

Following the lead of Dr. Kazuo Mizoi of our department, we have also investigated the effects of Ca antagonists. Although not producing a strong effect, such drugs have also been found beneficial in the case of cerebral infarction. They are, however, incapable of suppressing cerebral edema. Using a variety of experimental techniques, we are currently in the midst of investigating the mechanisms of action of a combination of above-mentioned four drugs, which include Ca antagonists. The combination of drugs originally used for protecting the brain from infarction was labeled the "Sendai Cocktail" by young members of our department.

We have only begun the clinical use of newer versions of the Sendai Cocktail in cases of cerebral infarction—administering it at as early a stage as possible following onset and thus delaying the pathological changes brought about by infarction. During that period the occluded vessels are identified by cerebral angiography, blood substitutes are administered and recirculation is established in the acute stage of cerebral infarction.

While significant improvements in the treatment of acute stage cerebral infarction have been achieved, we are still but at the beginning of such research. Therefore, several important problems remain to be tackled—such as determining more precisely the indication for such treatment, the permissible time until treatment and the causes of hemorrhagic infarction.

In summary, the present volume contains the story of a struggle over a period of 15 years by the members of our department to develop a means for safely prolonging temporary vascular occlusion for the treatment of ruptured cerebral aneurysms, a struggle which has thus far led—amidst blood, sweat and tears—to what we believe is a significant advance in acute stage treatment, the Sendai Cocktail. In addition to reviewing our own efforts, the chapters which follow also describe much work done by many researchers throughout the world. The present volume is thus our attempt to describe the most recent experimental and clinical developments concerning cerebral infarction, including both our own work and that of others. The critical evaluation of colleagues throughout the world is eagerly awaited.

Part I
Experimental Study

1. EXPERIMENTAL MODELS

1.1 Introduction

A large number of experiments using various methods and various animal species have been reported in the development of animal models of cerebral infarction. Various aspects of the pathophysiology of cerebral ischemia have thus been clarified and progress has been made in the clinical treatment of patients with cerebral infarction. Further progress is still needed, however, and even in the realm of experimental models alone, there are still many unsolved problems—a fact which is clearly reflected in the large volume of research on such models and the continuing development of new animal models.

The nature of cerebral infarctions encountered in the clinic is extremely complex, and even when a vascular occlusion occurs in the internal carotid artery (ICA) or the middle cerebral artery (MCA), the clinical picture varies widely with each patient—ranging from virtually normal to fatal. It is undoubtedly the case that the more varied the clinical picture, the greater is the need for an experimental model which can produce a "pure" pathology, which can then be studied rigorously. In brief, an ideal model for cerebral infarction would be one in which a constant level of ischemia could be produced, resulting in a defined focus of infarction and necrosis at that site. Currently, such an ideal model is not available. Although it is thought that an experimental model with the greatest reliability would be one using subhuman primates, a monkey model meeting the requirements for an ideal infarction model has not been achieved.

It has therefore become apparent that a uniform level of infarction for experimental investigation simply cannot be produced using the same animal species and the same experimental method. In light of this situation, in recent years it has become popular to measure and record the level of ischemia in experimental models of brain infarction and to note in detail the conditions under which the ischemia was obtained—particularly the residual blood flow and/or the degree of reduction in rCBF. By this means, it is thought possible to evaluate the similarities and differences among diverse models. Unfortunately, not all technical problems in the measurement of rCBF have been resolved. As described below, using a thalamic infarction model in dogs, we have found in our experimental work that the rCBF at the center of an ischemic focus and that in its immediately surrounding area shows different values depending upon the site of the focus. Moreover, there are regions within which drastic changes in the level of rCBF occur during ischemia.

Due to these and related problems, most researchers have resorted to using large numbers of animals in each experimental group and to drawing conclusions based upon statistically significant differences among the groups when studying the spread of infarction, its pathophysiology or the effects of drugs.

Other desirable features of an experimental model for brain infarction include the ready availability of the animal species and a relatively simple method for producing the ischemia. It should also be possible to maintain the animal's systemic condition in a stable state. Finally, the means for sampling the brain tissue should be precise, recirculation should be possible and a chronic preparation should also be feasible.

In the present chapter, these points will be considered and representative experimental models for cerebral infarction which are currently in use will be discussed. The experimental model, which is frequently referred to in this monograph and which has been developed in our department over more than a decade, will also be described in some detail

1.2 Cerebral Infarction Models

1.2.1 Experimental Models Using the Monkey

The use of subhuman primate, such as the monkey, in experiments on cerebral infarction is thought to be ideal. Thus far, primarily occlusion of the MCA has been studied using monkey models. In the 1950s, Meyer et al.[19] performed occlusion of the MCA at craniotomy and found that more than 50 minutes of occlusion was required in order to produce severe neurological deficits. Since then, subsequent to the introduction of the surgical microscope, many experiments have been performed and a transorbital approach[1, 10, 32] in which pressure on the brain itself, incision of the dura mater and damage to the dura mater, can be largely avoided, has come into use.

Using the hydrogen clearance method, Morawetz et al.[22] measured rCBF and found a mean decrease of 56% following MCA occlusion. After 2–3 hours, the tissue near those sites where rCBF measurements fell to below 12 cc/100 g/min were found to be infarctic. It was, however, impossible to obtain a constant level of blood flow by means of occlusion of the MCA. Sundt et al. for example[33] found that among 10 animals undergoing 3 hours vascular occlusion, 4 died whereas 6 had only mild symptoms. In an experiment by Crowell et al.[2], 10 animals underwent 4 hours of occlusion, but not one showed severe or fatal neurological symptoms.

For these reasons, various other

techniques have been employed in order to obtain more severe infarctic foci. Among these methods, one has involved the selective and permanent occlusion of the lenticulostriate artery, which leads to the production of foci of infarction at the basal ganglia or the internal capsule[41]. Another technique has involved temporary occlusion of the brachiocephalic trunk artery and subclavian artery with while simultaneous reduction of systolic blood pressure to 80 mmHg[7].

1.2.2 Experimental Models Using the Dog

Canine models are representative of large animal experimental models, unclear sounds grammatically wrong. Generally speaking, reliable experimental data can be obtained using these models, because dogs can withstand surgical intervention, including the effects of the anesthesia, blood sampling is simple, large changes in blood pressure are uncommon, and physiologically stable conditions can be maintained throughout the course of the experiment. Unfortunately, collateral vascular pathways are abundant and it has been difficult to produce constant levels of ischemia. For this reason, canine models have generally been thought to be unsuitable for experimental studies on brain infarction. In the recent reports of Nakagawa et al.[23] as well, occlusion of the MCA alone has been shown to produce a decrease in rCBF of 20–25%, but carbon perfusion in such cases does not reveal a region of carbon defect and histological analysis did not show clear signs of infarctic foci.

Using the dog to produce various experimental models, we have found—by trial and error—that it can be used for the reliable experimental study of brain infarction and we have succeeded in developing various canine models with various, welldefined properties. Details will be provided below, but it is worth noting here that in recent years a wide variety of canine experimental models have been reported. These have included: the combined occlusion of the MCA and cervical ICA[3], reduction of blood pressure to 50 mmHg by means of hypovolemia following occlusion of the MCA[30], the production of "embolic occlusion" of the MCA[17, 21, 24] or of the basilar artery using silicon beads, temporary ligation of the ascending aorta[31] and occlusion of the aorta using a balloon[20].

1.2.3 Experimental Models Using the Cat

Being less expensive than the monkey and having less collateral circulation than the dog, the cat has been thought suitable as an experimental animal for many years. A large number of studies have been reported on a cerebral infarction model in which the MCA is occluded. Unfortunately, the size of the infarctic lesion produced by MCA occlusion varies considerably and there have been reports of foci covering between 30 and 100% of the region fed by the MCA[8]. Moreover, two hour occlusion has been found to lead to "severe cortical damage" in 1/3 of the animals and "little or no cortical damage" in the remaining 2/3[35]. For this reason, various techniques have been devised to produce infarctic foci more reliably. Among these methods are: bilateral occlusion of the ICA and basilar artery, together with induced hypotension[6]; bilateral occlusion of the common carotid artery, together with hypovolemic hypotension to produce a blood pressure of 30 mmHg[26, 38], and temporary cardiac arrest by means of ventricular fibrillation[37]. In using these techniques to produce brain ischemia, however, relatively large effects on the systemic condition of the cat are induced and organ function (kidneys, lungs, liver, etc.) altered[9]. In such cases, evaluation of the brain ischemia itself becomes problematical.

1.2.4 Experimental Models Using the Rat

Rat models have various merits, including the ready availability of this species, the possibility of using large numbers of animals in each experiment, and the ease and rapidity of freezing the brain following occlusion or other experimental manipulations. Moreover, because of the theoretical similarities between positron CT scanning (used with increasing frequency in the clinic) and autoradiography (frequently used in animal experiments), the merits of using small animals such as the rat have recently attained much greater appeal. For these and other reasons, we also have devised a rat model for brain infarction, which will be described below.

Previously, various rat models have been developed. For example, brain metabolism during various forms of brain ischemia has been studied, following microsphere injection to a unilateral common carotid artery[15]. Instead of microspheres, Kudo *et al.*[16] have used homologous blood clots and have reported the possibility of obtaining a higher frequency of embolic occlusion. An infarction model in which the bilateral common carotid arteries are ligated has been used for many years[4] and it has been reported that a higher frequency of ischemia in the SHR rat (developed by Okamoto and Yamori[25]) can be obtained than in the normal rat[5, 39].

Rat models which have frequently been used in recent years include: the regional cerebral ischemia model of Tamura *et al.*[36] and the global ischemia model of Pulsinelli and Brierley[27]. Since selective occlusion of the rat MCA was thought to be impossible, it was not attempted, but Tamura succeeded in doing so, thus making possible a variety of chronic experimental models, which are just now beginning to be explored. Pulsinelli and Brierley developed a 4-vessel occlusion model in which the bilateral vertebral arteries and bilateral common carotid arteries are occluded. That is, the bilateral vertebral arteries are occluded by coagulation at the second cervical vertebra and, 24 hours later, clamps are placed on the bilateral common carotid arteries for 10–30 minutes. Using this technique, the surgical procedure is simple and the incidence of post-ischemia convulsions is low. As judged by EEG recordings and clinical symptoms, a constant level of ischemia is obtained and has been used as a model for "forebrain ischemia"[18].

1.2.5 Experimental Models Using the Gerbil

Making use of the fact that some gerbils have incomplete development aplasia of the circle of Willis—specifically, no ACA and PComA, it is possible to produce an ischemia model in a unilateral cerebral hemisphere by means of unilateral ligation of the common carotid artery. The gerbil also has the merit of being a small animal like the rat, and therefore suitable for autoradiography, etc. Although it is possible to perform experiments under constant conditions for evaluating the ischemic symptoms following vascular occlusion, the success rate is said to be only 25–50%[11]. In addition, although a higher rate of infarctic foci can be obtained using bilateral rather than unilateral common carotid artery ligation, effects due to respiratory difficulties are common and some believe the gerbil is therefore unsuitable as an animal model for ischemia[12]. It is noteworthy, however, that there have recently been reports of reliable regional cerebral ischemia using the gerbil, by means of selective occlusion of the MCA and PComA[42] or the basilar artery[40].

1.3 The Cerebral Infarction Model Developed in Sendai

1.3.1 Production of Various Cerebral Infarction Models in the Dog by Means of Occlusion of Intracranial Trunk Arteries

In this model, the dog—which had previously been thought unsuitable because abundant collateral pathways prevented production of a constant level of ischemia—was used. Employing a unilateral subtemporal approach and a surgical microscope, all of the bilateral trunk arteries entering and leaving the circle of Willis can be exposed and dissected. By occluding various combinations of these vessels, a variety of cerebral infarction models can be produced[34] (**Fig. 1-1**). In the following sections, the characteristics of various experimental models will be discussed.

1.3.1.1 Thalamic infarction model

Four ipsilateral sites are occluded in this model: the ICA, ACA, MCA and PComA. By means of such occlusion, a small infarctic focus of about 5 mm diameter and 4 mm in the anterior-

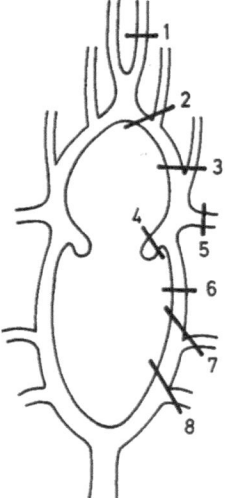

Fig. 1-1. Relationship between occluded arteries in the brain of the dog and various kinds of cerebral infarction models. Occluded arteries: *1*, A2 portion of ACA, *2*, ACA bifurcation ethmoidal, *3*, ACA bifurcation opthalmic, *4*, internal carotid, *5*, MCA, *6*, PComA, *7*, PCA bifurcation PComA, *8*, anterior cerebellar. Cerebral infarction models and the occluded arteries: Anterior thalamic infarction[3, 4, 5, 6]. Extensive thalamic infarction[3,4,5,7]. Cerebral mantle infarction[1,2,3,5,7,8]. Complete cerebral hemisphere infarction[1, 2, 3, 4, 5, 7, 8]. Incomplete cerebral hemisphere infarction[1, 2, 4, 7, 8]. Bilateral cerebral infarction (combinations of the above—mentioned arteries bilaterally)

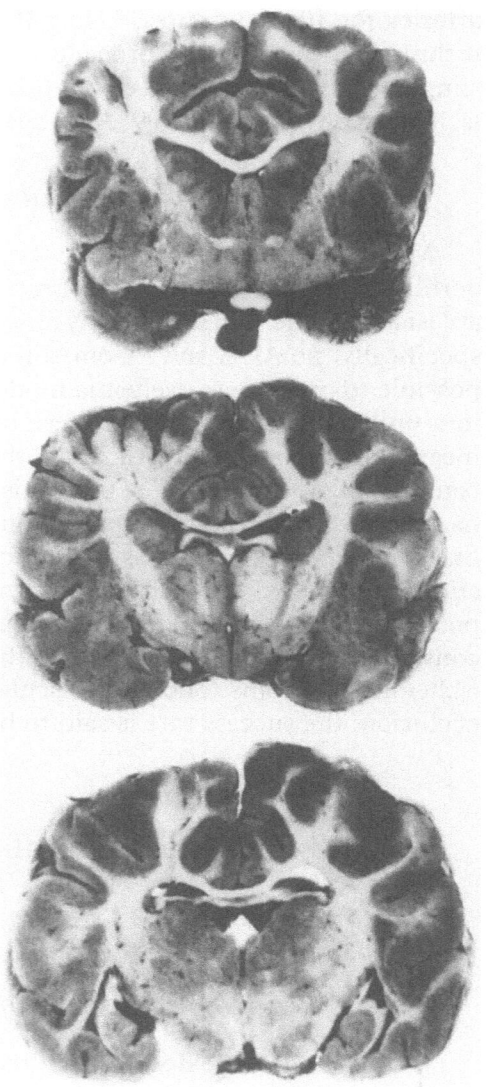

Fig. 1-2. The non-perfused area (arrow) confined to the anterior portion of the thalamus was revealed by carbon perfusion. This was obtained from the frontal slice at a point 5 mm behind the optic chiasma (middle)

posterior direction is produced in the nucleus ventralis of the anterior half of the canine thalamus (**Fig. 1-2**). When occlusion exceeds 60 minutes, such foci of infarction are produced in 2/3 of dogs[43]. Moreover, since the site of the infarction is constant in this model, it is possible to insert an electrode prior to surgery for vascular occlusion and the changes in thalamic electrical activity following 4-vessel occlusion can then be monitored (**Fig. 1-3**). In this way, infarctic foci can be obtained in nearly 100% of dogs[28]. The surgical intervention in this model is minimum and dogs can be maintained under spontaneous respiration. Furthermore, chronic preparations of this kind can also be made.

L-Th

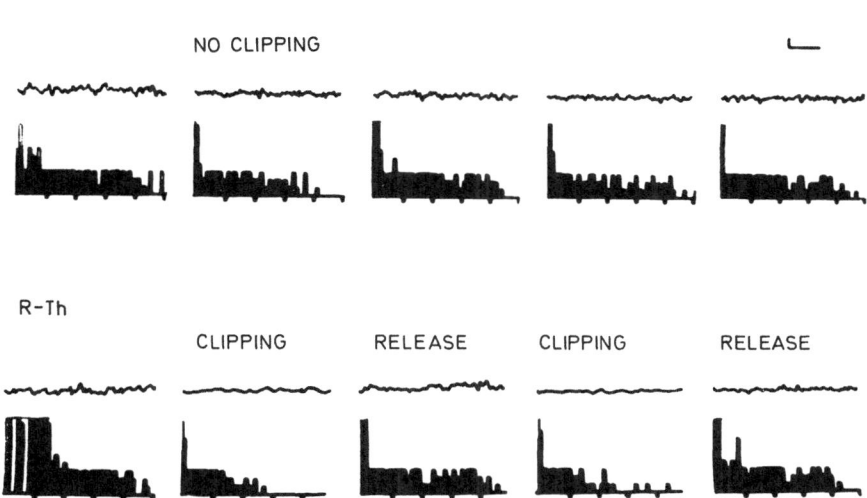

R-Th

Fig. 1-3. Changes in the thalamic depth EEG pattern and power spectrum following repeated temporary occlusions of cerebravascular occlusion. LTH: Depth EEG from non-occluded side (upper) and its power spectrum (lower). RTh: Depth EEG from occluded side (upper) and its power spectrum (lower). Calibration of EEG: 1 second, 50 μV. Abscissa of power spectrum indicates frequency, graduated at 2-Hz intervals

1.3.1.2 Cerebral Mantle Infarction Model

In this model, vascular occlusion is done simultaneously at the following 6 sites: the ICA, A1 portion of the ACA at the bifurcation of the ophthalmic artery, the ACA bifurcation of the ethmoidal artery, the A2 portion, the PCA and the anterior cerebellar artery. Infarctic foci are produced on the side of the occlusion at the cerebral cortex, over the entire region of subcortical white matter, at the basal ganglia and at the arterior portion of the internal capsule. The thalamus and hypothalamus, however, do not become infarctic (**Fig. 1-4**). A bilateral cerebral mantle infarction can also be prepared by means of bilateral occlusion of the same arteries.

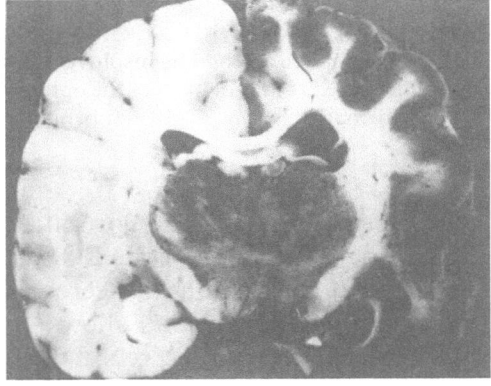

Fig. 1-4. Cerebral mantle infarction model

1.3.1.3 Complete Cerebral Hemisphere Infarction Model

In this model, 7 locations on trunk arteries are occluded unilaterally. They are: the ICA, the bifurcation of the A1 portion of the ACA at the ophthalmic artery and the ACA bifurcation at the

1.3.1.4 Incomplete Cerebral Hemisphere Infarction Model

This model is based upon the perfusion of a cerebral hemisphere, which is normally fed by the external carotid artery through the ophthalmic artery. This is achieved by means of unilateral

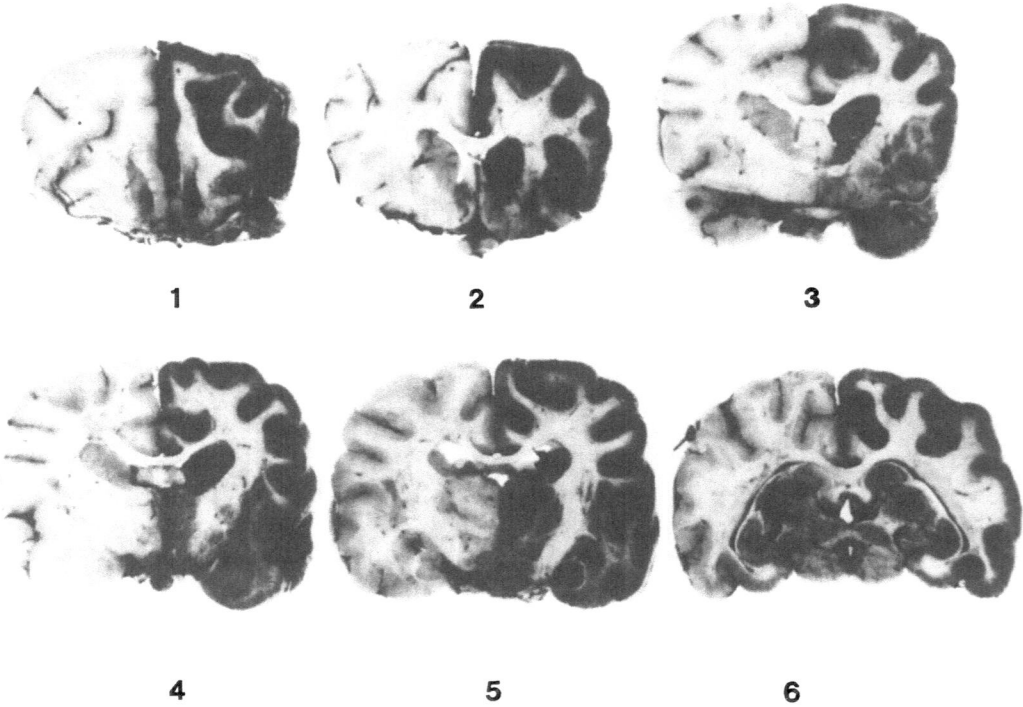

Fig. 1-5. Complete cerebral hemisphere infarction model—frontal sections *1*, 10 mm in front of the optic chiasma, *2*, 5 mm in front of the optic chiasma, *3*, passing through the optic chiasma, *4*, 5 mm behind the optic chiasma, *5*, 10 mm behind the optic chiasma, *6*, 15 mm behind the optic chiasma

ethmoidal artery, the A2 portion, the MCA, the bifurcation of the PCA and the PComA and the anterior cerebellar artery. Such occlusion allows for ischemia of an entire cerebral hemisphere (**Fig. 1-5**), and also makes possible models 4 and 5, described below.

occlusion of 5 sites: the ICA, the A2 portion of the ACA at the bifurcation of the ethmoidal artery, the A2 portion, the bifurcation of the PCA and PComA and the anterior cerebellar artery. Brain regions fed by the A1 portion of the ACA, the MCA and the PComA are then fed via the ophthalmic artery. By performing carbon perfusion, it has been found that, unlike in the complete cerebral hemisphere infarction model, there are no regions of carbon defect, but the carbon concentration in the entire occluded cerebral hemisphere is notably decreased (**Fig. 1-6**).

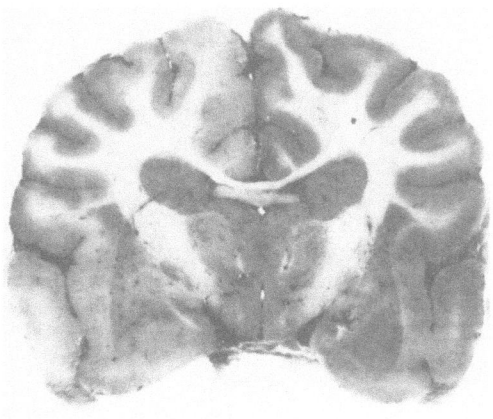

Fig. 1-6. Widespread, incomplete cerebral hemisphere infarction model

placed immediately posterior of the bifurcation of the ICA and the right MCA. At a site approximately 1 cm peripheral from the clip, the vessel is cut and the peripheral end is cauterized. The 600 µm tube is inserted approximately 1 cm into the central end toward the origin of the MCA and immobilized there. The aneurysm clip is then released.

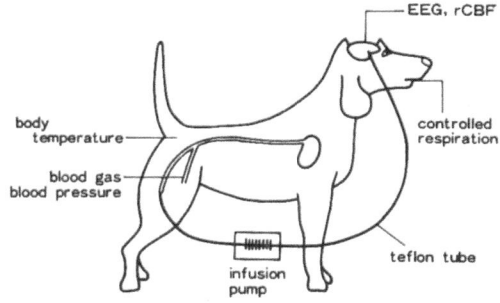

Fig. 1-7. Experimental system of the model: Cerebral perfused blood via the rt MCA cannulated tube is induced from femoral artery

1.3.1.5 Complete Ischemic Brain Regulated with the Perfusion Method

After right-sided craniotomy, the base of the brain is approached and the trunk arteries entering and leaving the circle of Willis are identified. Next, a 3 mm diameter polyethylene tube is inserted into the femoral artery and let to the abdominal aorta via a peristaltic infusion pump. The polyethylene tube, made with a 600 µm tip, is inserted retrogradely into the right MCA and the tube leading from the infusion pump is then connected (**Fig. 1-7**). Next, a temporary occlusion clip is

The right ophthalmic artery, right ethmoidal artery and the left ethmoidal artery are then occluded by cauterization in that sequence and the right optic tract is cut. Next, the left ophthalmic artery, right PCA and right anterior cerebellar artery are cauterized in that sequence. When these procedures are completed, the infusion pump is switched on and the infusion of autologous blood to the brain is slowly begun. Next, the basilar artery is occluded and the infusion is regulated to 4 ml/min. Then the right ICA is occluded and infusion is increased to 8 ml/min. Finaly, the left ICA is occluded and infusion is increased to

16 ml/min. By means of this procedure, while maintaining the venous system intact, circulation to the left cerebral hemisphere is made completely independent of the systemic circulation and is regulated by the infusion pump (**Fig. 1-8**). Using this model, it is possible to control instantaneously the flow of blood at will and easily to reduce blood flow to the brain to 0%, 10%, 20%, 30% or other levels[14]. In **Fig. 1-9**, the relationship between infusion blood volume (IBV) in this model and cortical electrical activity is illustrated and in **Fig. 1-10** the relationship between the infusion blood volume and cerebral blood flow at the bilateral frontal lobes is shown.

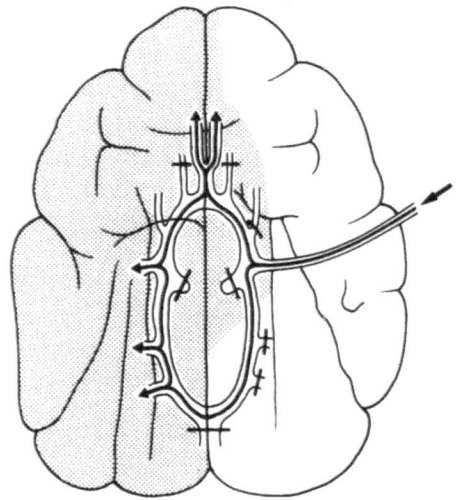

Fig. 1-8. This schema shows the main arteries of the dog brain with occluded points and the blood distribution with blood perfusion via a cannulated tube into the rt MCA retrogradely

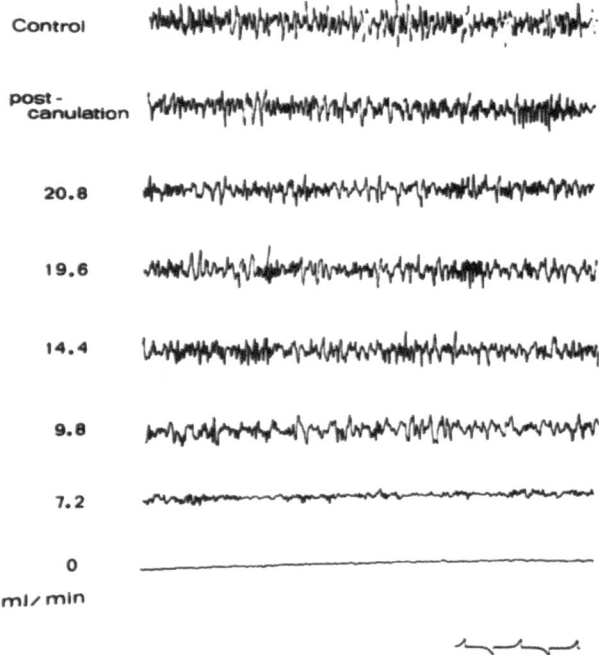

Control

post-canulation

20.8

19.6

14.4

9.8

7.2

0

ml/min

Fig. 1-9. Correlation between the IBV and the cortical EEG on the lt hemisphere. Between 20.8 ml/100 g · min of the IBV and 14.4 ml/100 g · min of the IBV no difference in shown on the cortical EEG. But below 9.8 ml/100 g · min, low voltage and slow waves appear on the cortical EEG. At 0 ml/100 g · min of the IBV the cortical EEG becomes completely flat

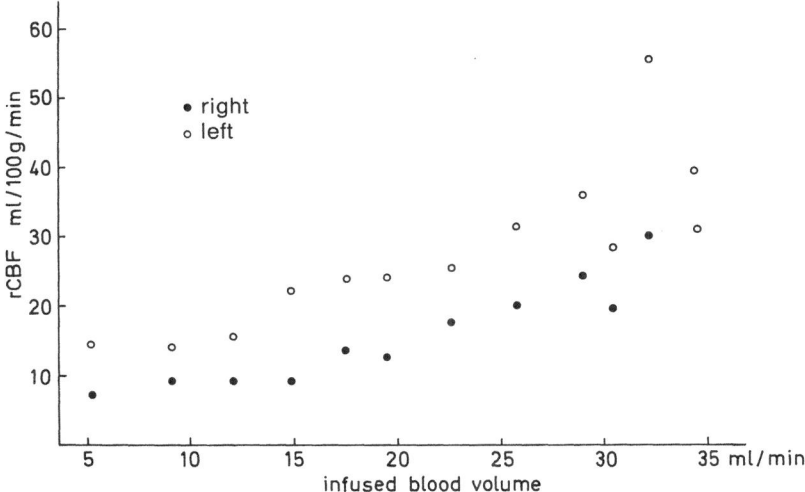

Fig. 1-10. Correlation between the infused blood volume and r CBF on the territories of bilateral anterior cerebral arteries

1.3.2 A 3-Vessel Occlusion Model for Bilateral Hemispheric Infarction Using the Rat

As mentioned earlier, due to the clinical application of positron CT scanning, the usefulness of a brain ischemia model using the rat has been reconsidered recently. In light of the fact that a more precise, small animal model that would allow for the reliable production of a defined level of ischemia was required, we developed a new model for bilateral hemispheric ischemia[13]. Below, we will describe the production of this model and make comparisons between it and the Pulsinelli and Brierley model, which is currently the most widely used rat infarction model.

1.3.2.1 Production of the Ischemia-Model

Male Wister rats weighing 200–250 grams each were used. Under the ad-

ministration of light halothane anesthesia, a 3 cm midline skin incision was made at the anterior cervical region and, proceeding bluntly using a surgical microscope, a 3×3 mm bone window was opened, thus exposing the basilar arteries. These vessels were then cauterized and cut. On the following day, again under the administration of light halothane anesthesia, an intratracheal tube was inserted, the animal was immobilized with pancronium bromide and regulated respiration was instituted. Then, the bilateral common carotid arteries were clipped, thus completing the ischemia model (**Fig. 1-11**).

For the sake of comparison, the ischemia model of Pulsinelli and Brierley was also produced. In this model, the bilateral vertebral arteries are cauterized, and on the following day the bilateral common carotid arteries are also cauterized.

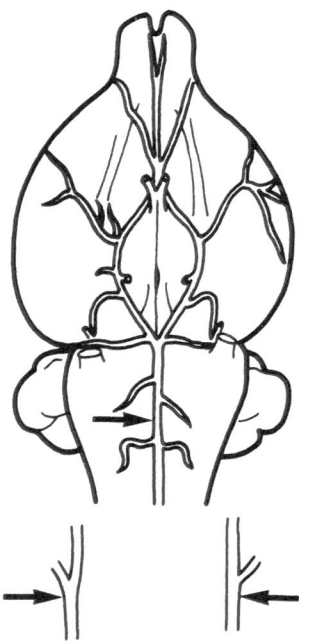

Fig. 1-11. Occluded arteries (arrows)

1.3.2.2 EEG Studies

Extradural electrodes were placed over the frontal and parietal regions and the changes in EEG occurring 1 minute

after complete occlusion were classified into 4 groups: (i) no change, in which the EEG remained normal, (ii) moderate change, in which the voltage and frequency of electrical activity decreased, (iii) severe change, in which the amplitude of the EEG was markedly reduced but no flat, and (iv) flat, in which there was complete attenuation of brain electrical activity (**Fig. 1-12**). It was found that 19 of the 20 animals in the 3-vessel occlusion model had flat EEGs, whereas a variety of EEG states were produced in the Pulsinelli and Brierley model. Statistically, the difference between the groups was highly significant (**Table 1-1**).

1.3.2.3 Studies of the Cerebral Blood Flow Using Autoradiography

Eleven animals were prepared in the 3-vessel occlusion model and 11 in the Pulsinelli and Brierley model. One minute after bilateral occlusion of the common carotid arteries, 100 μCi/kg of 14-C-iodo-antipyrine (IAP) was injected into the femoral vein, following the method of Sakurada[29]. Immediately

control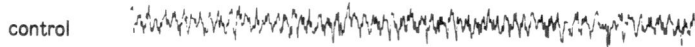

Patterns of EEG Changes Following Ischemia

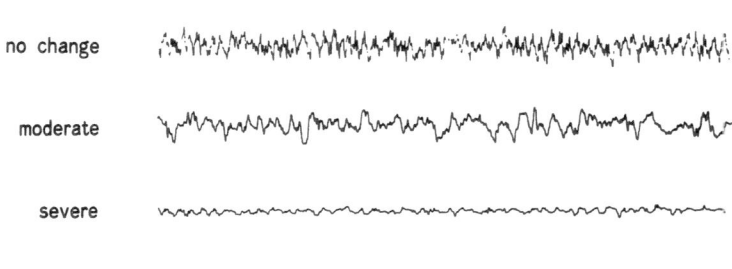

Fig. 1-12. Patterns of EEG changes

Table 1-1. EEG changes one minute after occlusion

EEG Changes / Rat model	no change	moderate	severe	flat
Pulsinelli's model (n=20)	1	7	7	5
Three-Vessel Occlusion model (n=20)	0	0	1	19

following completion of the venous injection, the animals were decapitated and 20 µm frozen brain sections were obtained using a cryostat. The sections were placed on X-ray film and 14-C-standard film for autoradiography (Amersham) to obtain images of the CBF. In order to make further comparisons of the degree of cerebral blood flow in each group, the optical density of the X-ray films was measured using a densitometer (Sakura PDA-15). After calculating the radioactivity in various regions from the 14-C-standard on the film, the relative cerebral blood flow (being the ration of that at the top cervical vertebra of the spinal cord to that of the cortex of the fronto-parietal

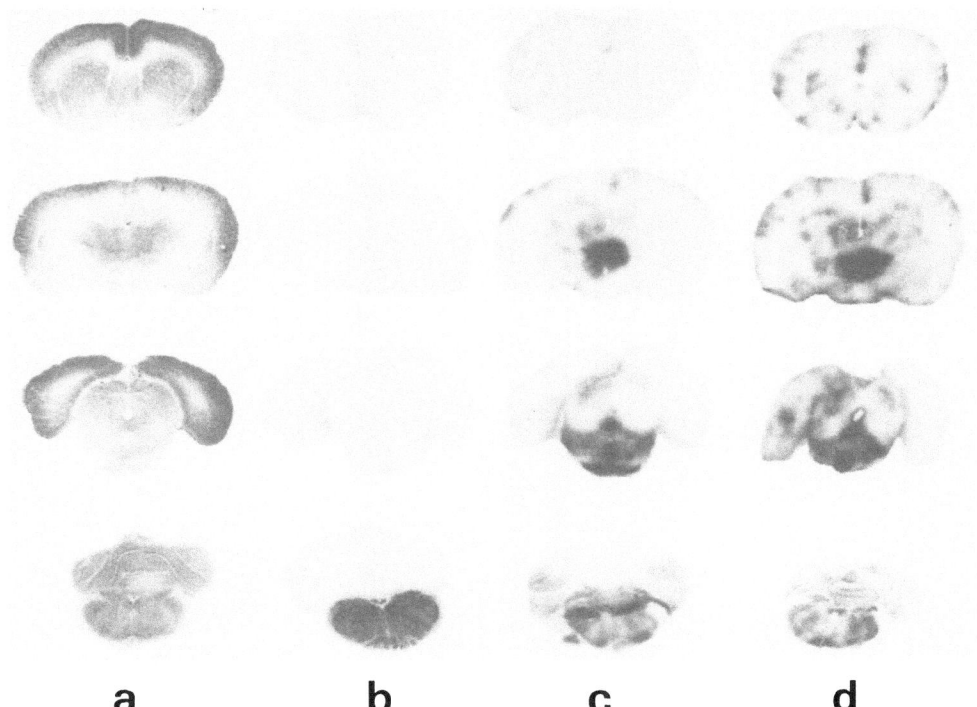

a   b   c   d

Fig. 1-13. 14C-IAP autoradiography. *a*, Control, *b*, three vessel occlusion model, *c, d*, Pulsinelli & Brierley model of high grade ischemia with flat EEG

region) in each animal was also calculated. The results showed that, among the 11 animals in the 3-vessel occlusion model, 9 showed virtually no 14C-IAP in the cerebral hemispheres except for the brainstem and cerebellum—indicating a condition of severe ischemia. The remaining 2 animals showed mild or moderate 14C-IAP in the cerebellum and cerebrum. In contrast, among the 11 animals in the Pulsinelli and Brierley model, results varied widely. Some animals showed no notable differences between the experimental group and the control group (which had undergone a sham operation involving the opening of a bone window at the base of the brain and exposure of the basilar arteries), whereas some showed relatively severe brain ischemia. Even in the two animals which showed flat EEG records, there was clear indication of blood flow to both the cerebellum and the cerebrum. Particularly in the vicinity of the thalamus, characteristic 14C-IAP accumulation was in evidence (**Fig. 1-13**). In the relative cerebral blood flow measures as well, the control group had a value of $118.3 \pm 20.1\%$, the 3-vessel occlusion had a value of $6.4 \pm 14.9\%$ and the Pulsinelli and Brierley group had a value of $61.6 \pm 38.6\%$. Statistically, the difference between the 3-vessel occlusion group and the Pulsinelli and Brierley group was highly significant.

2. HISTOLOGICAL STUDY

2.1 Introduction

Detailed research on the histopathological changes due to cerebral ischemia date from the work of Spielmeyer[58] in the early part of this century. In 1920 he reported that, following cerebral ischemia, there is shrinkage of neurons, in which the cytoplasm becomes acidophilic and the nucleus basophilic and the Nissl bodies disappear. Spielmeyer called such changes "ischemic cell changes".

Thereafter through the 1950s, various animal experiments were undertaken and cerebral ischemic pathology was investigated primarily using the light microscope[18, 19, 26, 36, 44, 60]. In the 1960s Levine[34] developed an anoxic-ischemic rat model in which a unilateral common carotid artery (CCA) was ligated and the animal was made to breath nitrogen gas. Using this experimental model, both light microscopic[3,4] and electron microscopic[5, 6, 8, 19, 43] studies of histological changes and of changes in cerebral vessels were carried out.

Brown and Brierley later reported that neurons in animals prepared using this anoxic-ischemic model underwent "microvacuolation" followed by "ischemic cell changes" and eventually "ischemic cell changes with incrustation"[3—6]. Unfortunately, this anoxic-ischemic model was found to be heavily influenced by the level of anoxia and cannot be considered to be an experimental model of pure cerebral ischemia.

In 1966, Levine and Payan reported that they had produced foci of cerebral ischemia by means of unilateral or bilateral CCA occlusion in the Mongolian gerbil with aplasia of the circle of Willis[35]. Since this experimental model is easily prepared, it was widely used throughout the 1970s and contributed significantly to our understanding of brain ischemia.

Another important finding was that of Ito et al.[23], who reported that recirculation following 15 minutes of unilateral CCA occlusion in the gerbil characteristically results in central chromatolysis and nuclear displacement in neurons of the hippocampus. They labeled such pathology "reactive change".

In 1968 Ames et al.[32] reported morphological changes, such as bleb formation. They considered the "no-reflow phenomenon"[1] to be an important factor in the development of cerebral infarction. In 1977, however, they investigated the same experimental model and showed that the bleb formation and stenosis of the lumen of extremely small capillaries were likely to be due to an artifact[13].

Hossmann *et al.* reported a similar complete ischemia model using the cat in 1970[20, 21]. They found that even after lengthy complete ischemia of 60 minutes duration, there was recovery of the functions of cerebral tissue—a finding which received considerable attention and was pursued histologically by Hossmann and colleagues[2]. During the mid-1970s animal models of focal cerebral ischemia were developed, which more closely mimic the pathology of clinical cases.

Using a unilateral occlusion of the middle cerebral artery (MCA) in the squirrel monkey and studying tissue changes under the light microscope, Garcia *et al.* (1974) found that there are four stages of pathology[15]: (i) a stage of ischemic injury (10–15 minutes after the occlusion), (ii) a stage of circulatory abnormalities (4 hours after the occlusion), (iii) an inflammatory stage (from approximately 18 hours after the occlusion), and (iv) a healing stage (after about one week). Electron microscopical study of the neuronal changes in the stage of ischemic injury[16, 17] showed that swelling of mitochondria occurs about 15 minutes after unilateral occlusion of the MCA of the cat, and after one hour there is the start of neuronal shrinkage. They subsequently did various cerebral ischemia experiments and have demonstrated that the cerebral pathology of regional ischemia and global ischemia differs[16]. Little *et al.* did an electron microscopical study of brain tissue using MCA occlusion in the squirrel monkey. They found shrinkage of cortical neurons 90 minutes after occlusion, and thereafter a gradual increase in shrunken neurons[39]. They also reported changes in synapses[40], blood vessels[41] and neuronal lysosomes[38]. In 1979, Pulsinelli *et al.* developed a 4-vessel occlusion model in the rat[53] and reported on the histological effects of the same[54]. In this experimental model, ischemic pathology was found in neurons of the hippocampus, paramedian cortex and striatum—suggesting a selective vulnerability of these structures of ischemia. Similar results have been reported in clinical cases[56] and in other cerebral ischemia models[25]. In the same year, Jenkins *et al.*[24] reported that the earliest pathology to appear in the peracute stage of cerebral ischemia is chromatin clumping and nucleolar condensation.

Petito later demonstrated the formation of platelet thrombi within severely ischemic foci[51]. Many other reports were made by Dodson *et al.*[11, 12, 42, 55]. In histological research on cerebral ischemia, however, it has been found that various artifacts, such as "dark neurons"[3], are produced, and the need for particularly careful studies has been repeatedly noted[7, 10, 37]. In the 1980s, attention has been drawn to several kinds of pathologic characteristics of cerebral ischemia—including delayed neuronal death[30], contralateral degeneration and repair[28, 29] and early proliferative change in astrocytes[32]. It has also been shown that lactic acidosis aggravates the histological pathology in foci of cerebral ischemia[27]. In recent years, various new techniques showing considerable promise have been developed—including the freeze fracture technique[59], immuno-histochemical techniques[21, 22, 46, 64, 65, 66] and multi-tracer autoradiography[14, 33, 45].

The main themes of research on the pathology of cerebral infarction have been briefly summarized above, and since 1978 we also have undertaken various histological studies of brain ischemia using various models developed by us[31, 47—50, 59, 61—63].

Based upon the results of fundamental research, we will now summarize our views concerning the sequential pathology of cerebral ischemia and the effects on ischemic pathology of recirculation in the acute period.

2.2 Sequential Changes in Cerebral Ischemia

2.2.1 Sequential Changes in Focal Ischemia

Using the canine thalamic infarction model in which rCBF is reduced to 10 ml/100 g/min (approximately 1/3 of the preocclusion level) due to vascular occlusion[57], a study was made of the histopathology at the center of the ischemic focus[61, 62] (**Table 2-1**).

(**Fig. 2-2**) to severe (**Fig. 2-3**). After 6–12 hours of occlusion, cells which had shown slight shrinkage were found to have become swollen (**Fig. 2-4**) and after 12–24 hours the cell membrane ruptured (cell disruption, **Fig. 2-5**). In contrast, neurons originally showing

Table 2-1. Number of dogs used in the experiment

Occlusion time	0	1/4	1/2	1	3/2	2	3	6	12	24 (hrs)
Number of dogs	2	2	2	2	2	2	2	2	2	2

2.2.1.1 Acute Stage Changes in an Ischemic Focus

a) Neuronal Changes

A normal neuronal morphology was maintained for 15 minutes following occlusion, but after 30 minutes microvascuoles were seen in the neuronal cytoplasm (microvacuolation, **Fig. 2-1**). After 1–3 hours of occlusion, shrinkage of the cellular cytoplasm or the nucleus was seen. Shrinkage varied from slight

severe shrinkage of the cytoplasm and nucleus presented disrupted cell bodies and fragmentation (**Fig. 2-6**).

These findings are summarized in **Fig. 2-7**.

b) Changes in Glial Cells

After 1 hour of occlusion, the cell body of astrocytes was found to be swollen, but oligodendroglia were normal or showed only slight swelling (**Fig. 2-8**). After 2 hours of occlusion

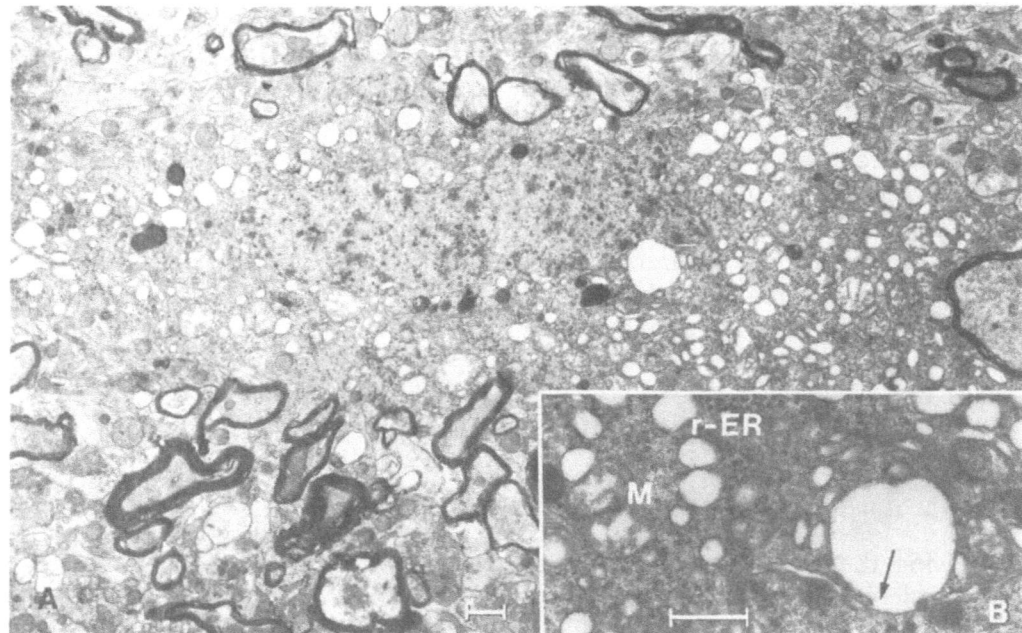

Fig. 2-1. Microvacuolation of nerve cell (30 minutes occlusion). A) The nerve cell has many microvacuoles in the cytoplasm. B) These microvacuoles are swollen mitochondria (*M*) and rough endoplasmic reticulum (*r-ER*). The nuclear membrane is swollen in a portion (arrow). (Bar = 1 μm)

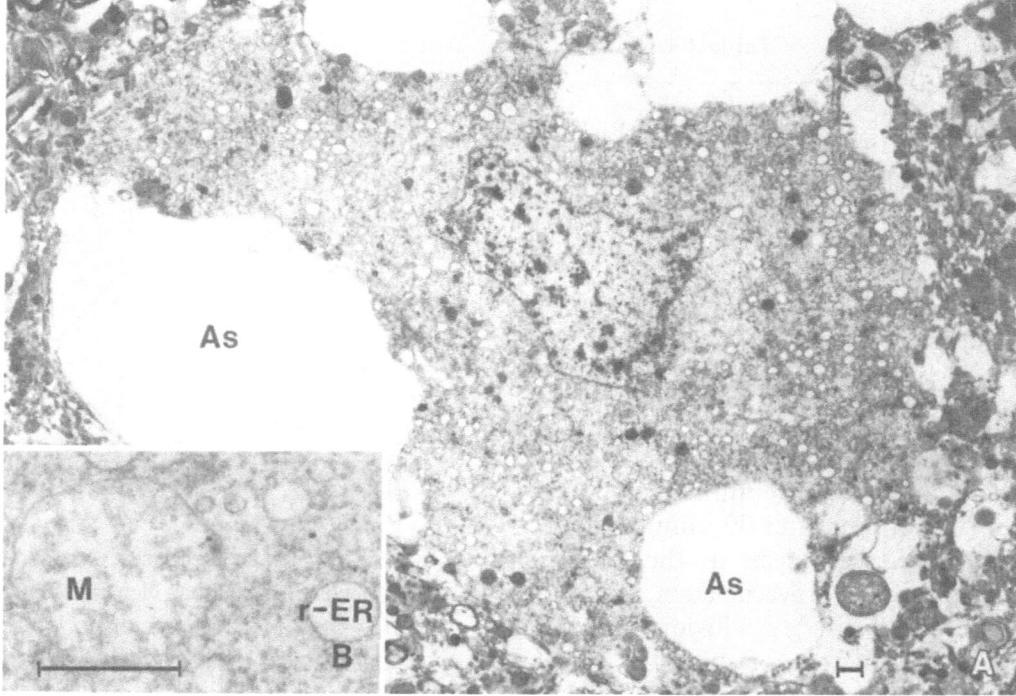

Fig. 2-2. Slight shrinkage of nerve cell (120 minutes occlusion). A) The slightly shrunken nerve cell is surrounded by swollen astrocytic processes (*As*). B) Its mitochondria (*M*) and rough endoplasmic reticulum (*r-ER*) are swollen, and polysomal rosettes have decreased in number slightly.

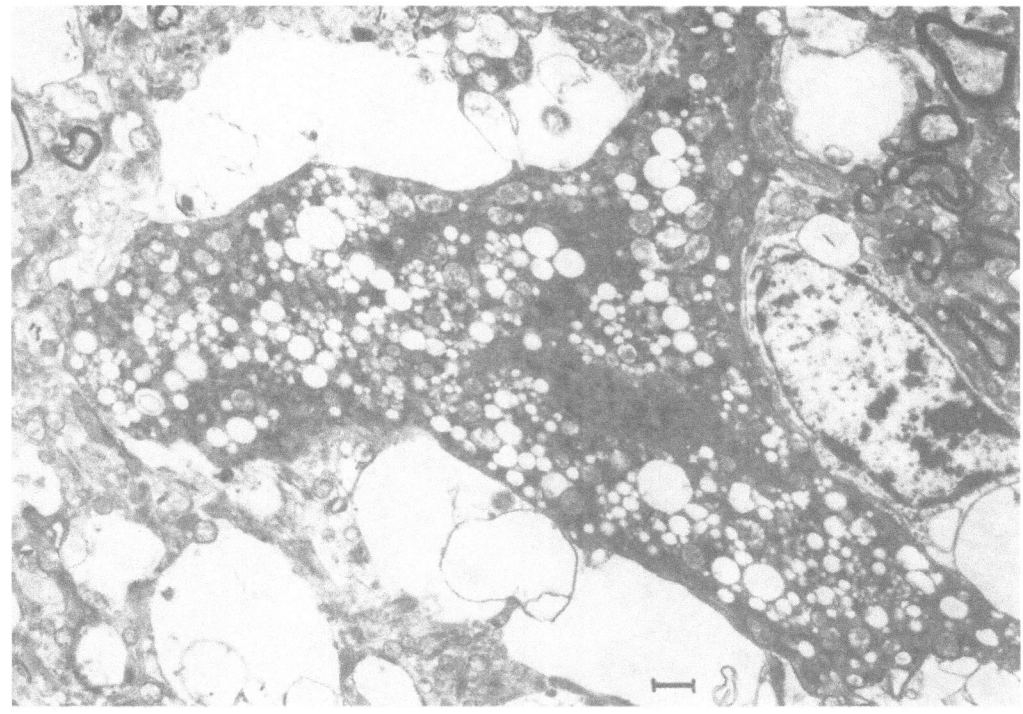

Fig. 2-3. Severe shrinkage of nerve cell (120 minutes occlusion). (Bar = 1 μm)

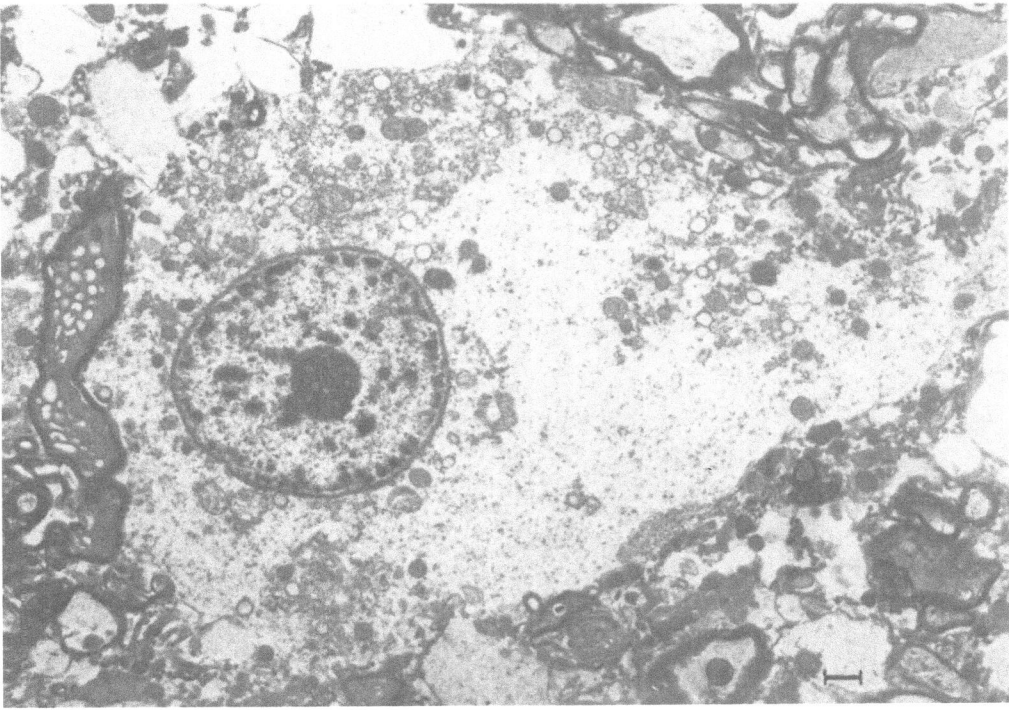

Fig. 2-4. Swelling of nerve cell (12 hours occlusion). The cytoplasm of the nerve cell is swollen and the nucleus is degenerated. (Bar = 1 μm)

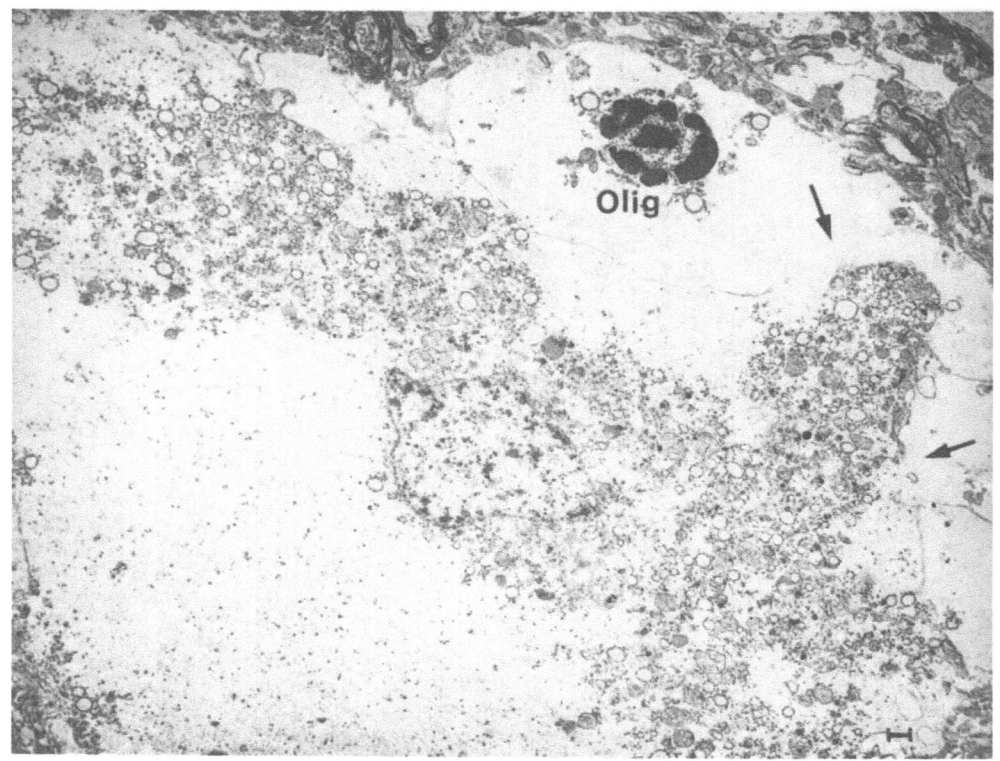

Fig. 2-5. Disruption of nerve cell (12 hours occlusion). The cytoplasmic membrane is broken (arrow). The oligodendroglia (*Olig*) is necrotic. (Bar = 1 μm)

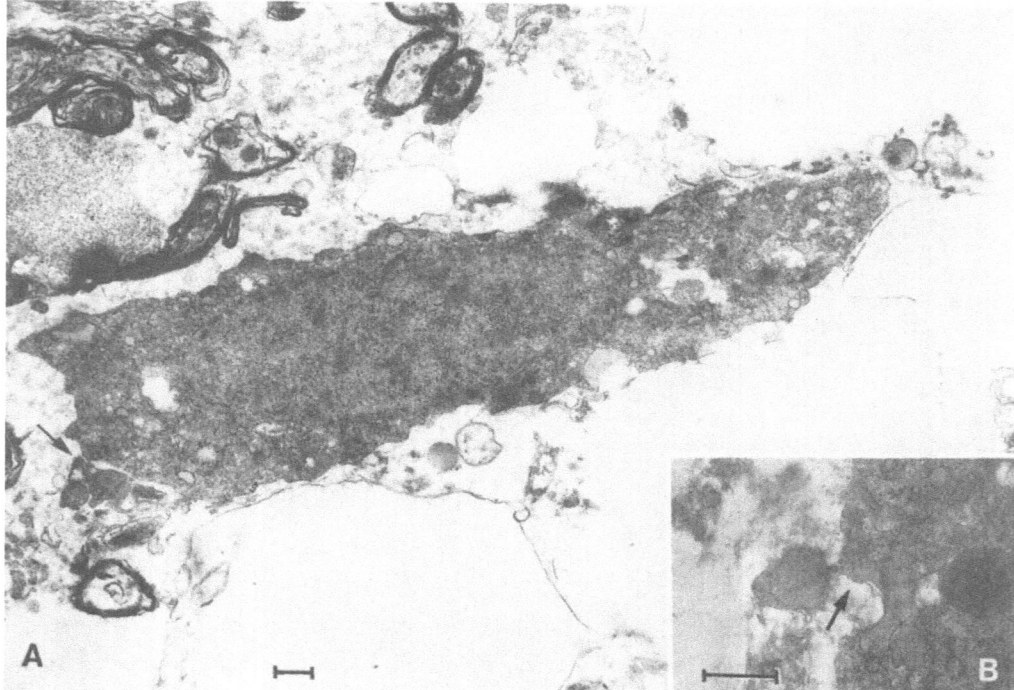

Fig. 2-6. Fragmentation of nerve cell (24 hours occlusion). A) Fragmentation of severely shrunken neuron (arrow). B) A fragment of degenerated cytoplasm is broken off from the severely shrunken

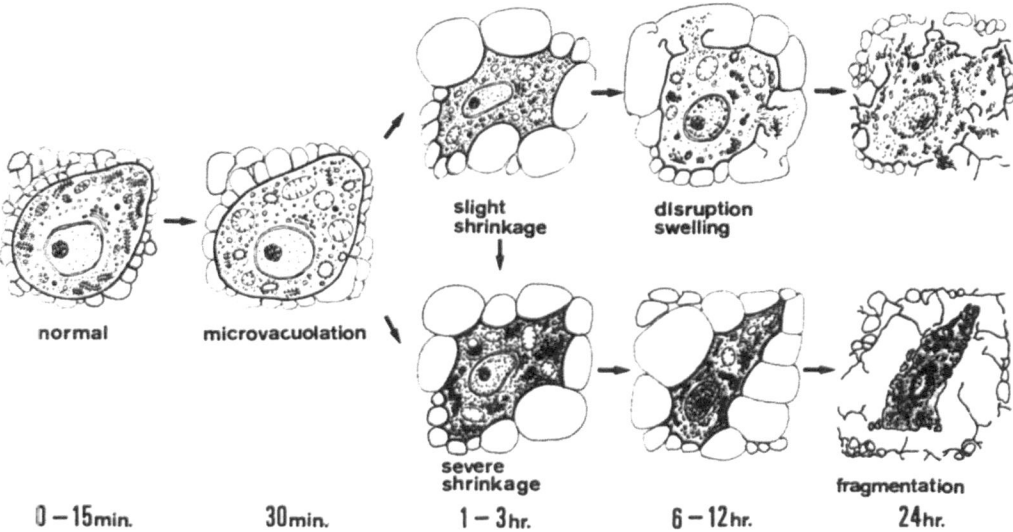

Fig. 2-7. Diagram of neuronal alterations after occlusion of cerebral arteries. (Thalamus infarction model)

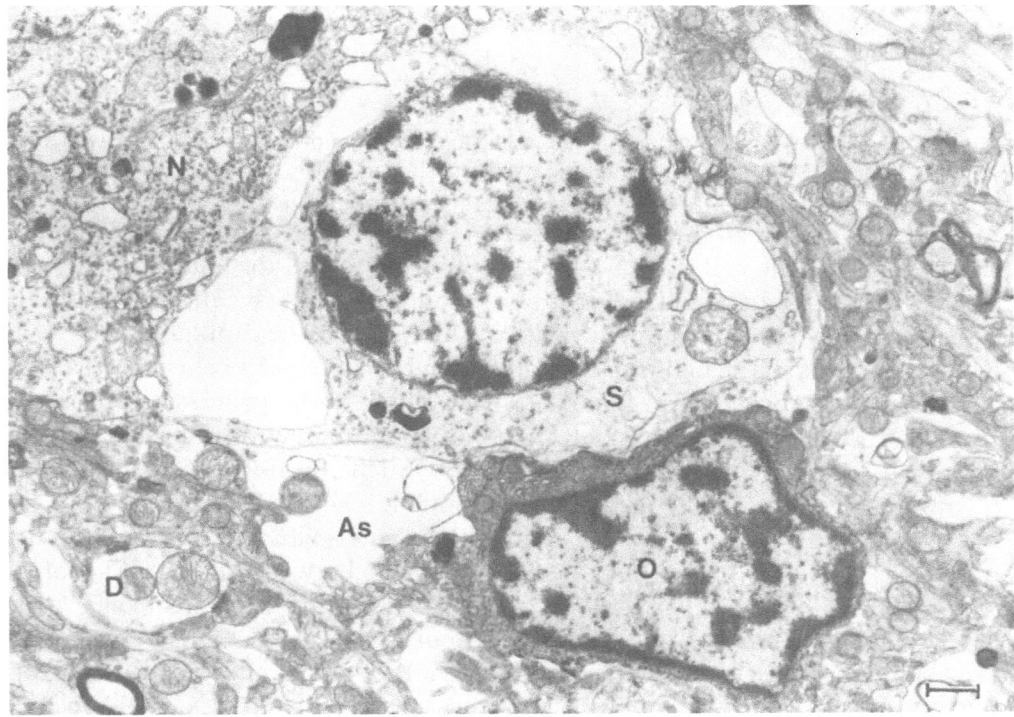

Fig. 2-8. Ischemic changes in glial cells (1 hour occlusion). A satellite cell (*S*) is swollen with swollen endoplasmic reticulum and mitochondria of a nerve cell (*N*) are swollen. An astrocytic process (*As*) and a dendritic process (*D*) are swollen in the neuropile. (Bar = 1 μm)

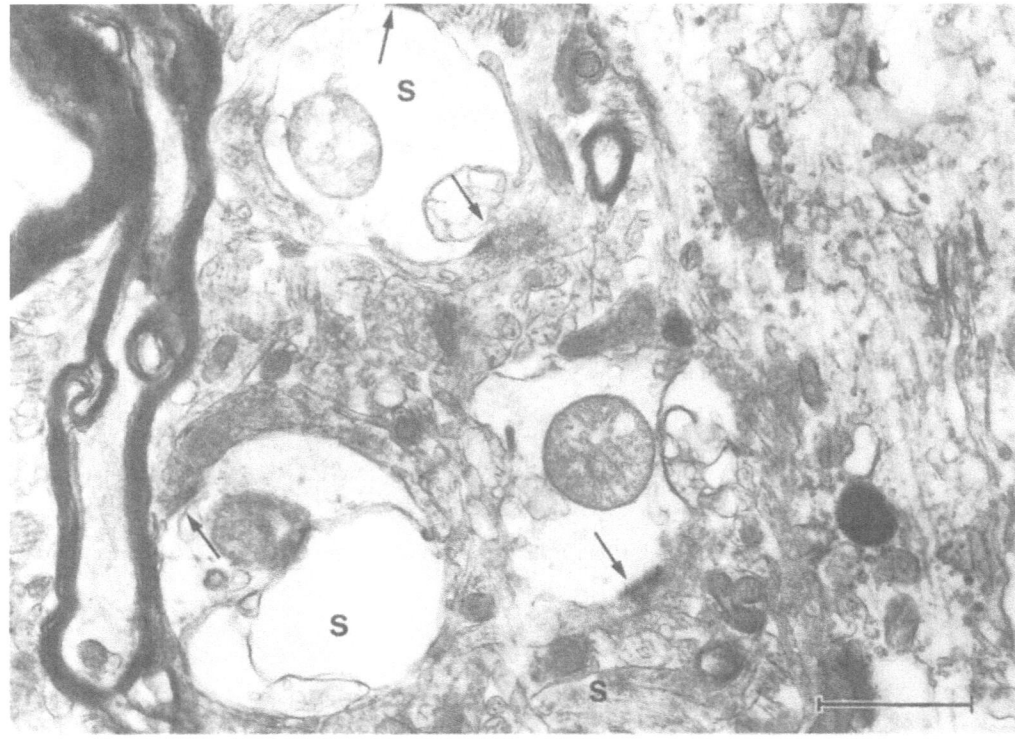

Fig. 2-9. Spongiosis of neuropile (30 minute occlusion). Postsynaptic buttons (*S*) are swollen. The arrows show synapses. (Bar = 1 µm)

there were widespread nuclear changes and swelling of the cell bodies of both astrocytes and oligodendroglia. Between 3 and 6 hours, the frequency with which glial cells with such nuclear and cytoplasmic changes were found increased with time. Between 12 and 24 hours, cells with ruptured cellular membranes were found (**Fig. 2-5**).

c) Changes in Neuropile

After 15–30 minutes of occlusion, mild spongiosis of the neuropile due to swelling of the processes of dendrites and astrocytes was observed (**Fig. 2-9**). The swelling of the dendritic processes was particularly notable at the post-

synaptic button. From about 30 minutes after the start of occlusion, could be seen swelling of astrocytes surrounding the capillaries (**Fig. 2-10**). After 1 hour of occlusion, there was swelling of the processes of astrocytes surrounding neurons (**Figs. 2-2** and **2-3**). Such findings were particularly notable around neurons showing shrinkage—and suggest a close relationship between the swelling of glial processes and the shrinkage of the neurons they surround. Between 2 and 6 hours, the swelling of the neuropile was found to increase with occlusion time, and after 12 hours there was rupture of the cellular membrane at these processes.

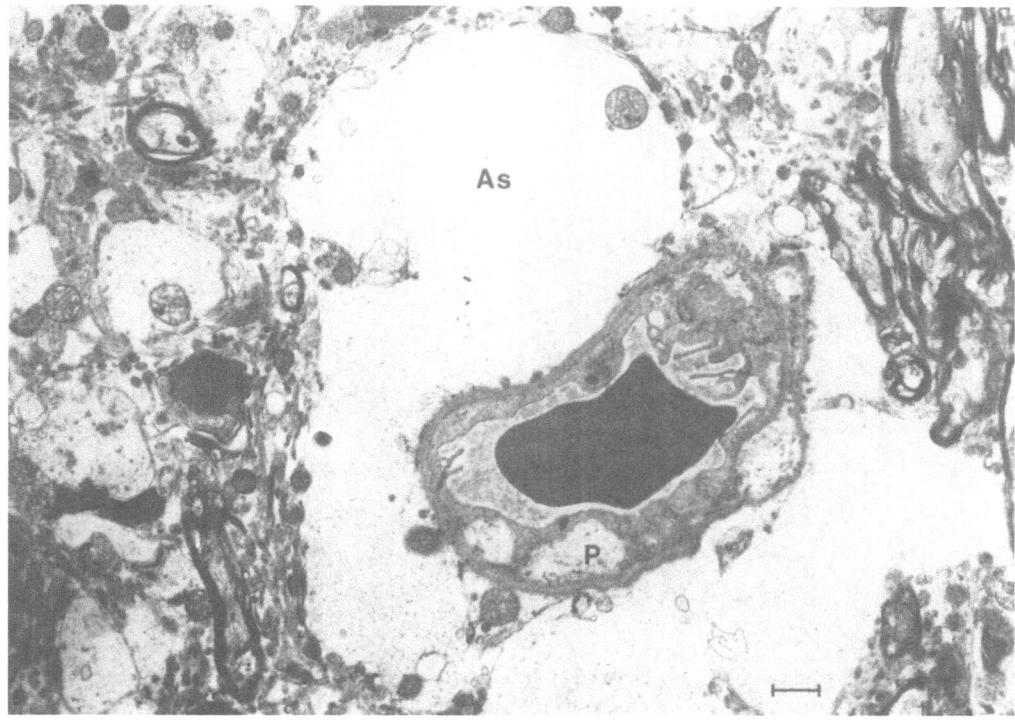

Fig. 2-10. Marked swelling of perivascular astrocytic processes (*As*) and the pericyte (*P*) of a capillary after 2 hours occlusion. (Bar = 1 μm)

d) Capillary Changes

After 30 minutes of occlusion, there was swelling of the astrocyte processes surrounding capillaries, and after 1 hour there was swelling of capillary pericytes (**Fig. 2-10**). From about two hours after the start of occlusion, there was slight swelling of endothelial cells and after 3 hours blebs were sometimes seen within the capillaries (**Fig. 2-11**). Between 6 and 12 hours, the changes in capillaries and venules became more pronounced and tight junctions were found to be opening up in necrotic vessels (**Fig. 2-12**). Exudation was seen in some of these capillaries and venules (**Fig. 2-13**). After 24 hours, exudation of neutrophils within the ischemic focus was seen (**Fig. 2-14**).

2.2.1.2 Changes in the Ischemic Focus Following Acute Stage Recirculation

a) Recirculation after 30, 60, and 120 Minutes of Occlusion

Foci of cerebral ischemia were produced with the canine thalamic infarction model and then recirculation was allowed for 1, 2, 6, 12 or 24 hours after 30, 60 or 120 minutes of occlusion. An autopsy study was then made of the histological changes in each group[63] (**Table 2-2**).

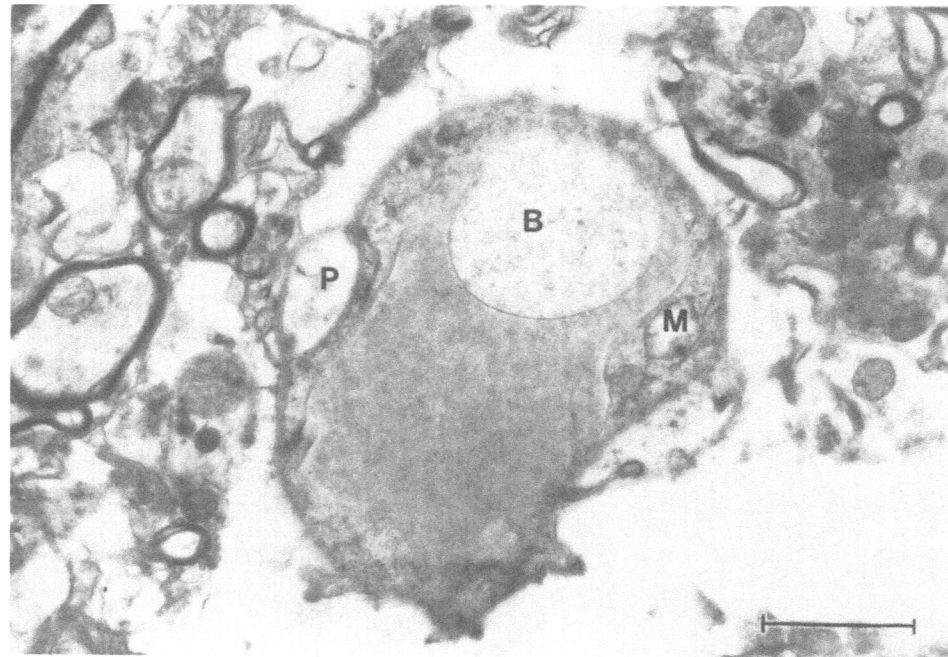

Fig. 2-11. Bleb formation in a capillary after 3 hours occlusion. A bleb (*B*) was seen in the capillary. A swollen mitochondrion (*M*) was seen in an endothelial cell. Pericytes (*P*) were moderately swollen. (Bar = 1 µm)

Table 2-2. Number of dogs used in the experiment

Occlusion time (hours)	Recirculation time (hours)					
	0	1	2	6	12	24
1/2	2	2	6	2	2	2
1	2	2	2	2	2	2
2	2	2	6	2	2	2

In the group undergoing 30 minutes of occlusion, spongiosis of the neuropile and microvacuolation of neurons was seen prior to recirculation, following recirculation such pathology continued for several hours in some cases, but normal histology was evident after 24 hours (**Fig. 2-15**). In the group undergoing 2 hour recirculation, some neurons showed the characteristic morphology of microvacuolation. In such neurons, there were large vacuoles in the cytoplasm, but normal morphology of the mitochondria and other organelles was seen in other cells. We presume that these neurons presented

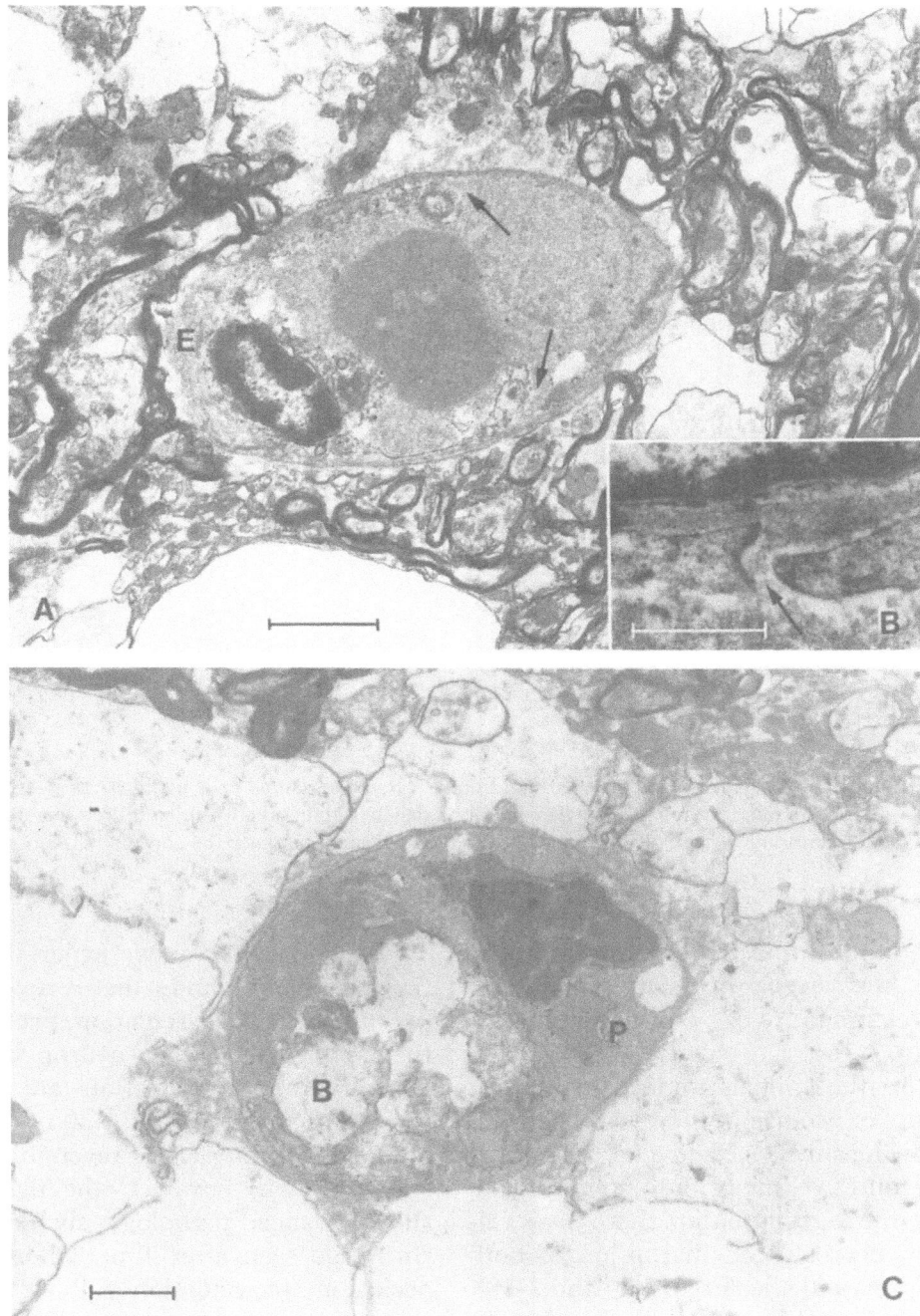

Fig. 2-12. Degeneration of capillary. A) Degeneration of an endothelial cell (*E*) of a capillary after 12 hours occlusion. The nucleus was shrunken. The cytoplasm was separated from the basement membrane and tight junctions were broken (arrow). (Bar = 1 µm). B) Dehiscence of a tight junction in a capillary after 6 hours occlusion (arrow). C) Degeneration of the pericyte (*P*) of a capillary after 12 hours occlusion. Many blebs (*B*) were seen in the lumen. (Bar = 1 µm)

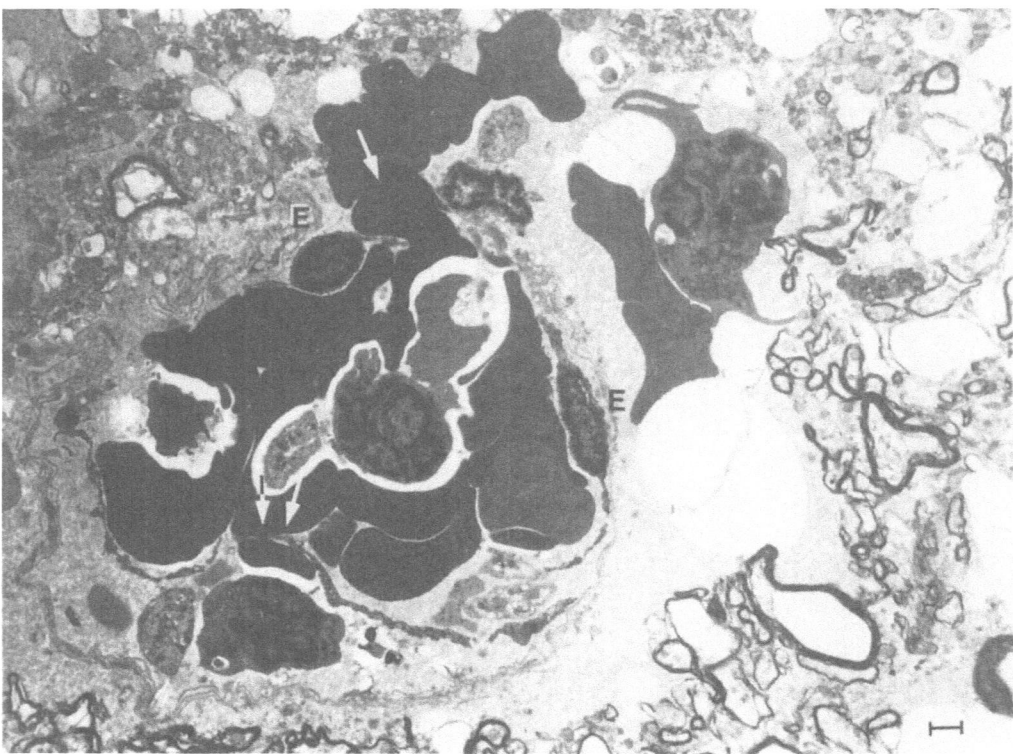

Fig. 2-13. Diapedesis from a venuole after 12 hours occlusion. Blood cells had leaked out from the opened junction of endothelial cells (*E*) which had degenerated and shrunk, and then from broken basement membrane (arrow). Blood cells also invaded the space between the endothelial cells and the basement membrane (double arrows). (Bar = 1 µm)

one kind of morphology found during the recovery process following the development of microvacuolation (**Fig. 2-16**).

In the animals undergoing 1 or 2 hours of occlusion, there were neurons showing shrinkage and glial cells showing mild swelling prior to recirculation, but due to recirculation there was swelling, necrosis and disruption of both neurons and glia (**Figs. 2-17 and 2-18**), marked swelling of capillaries (**Fig. 2-19**), signs of necrosis of endothelial cells and leakage of blood cells (**Fig. 2-15C and D**).

In light of the above experimental findings, it is thought that recovery is possible due to recirculation, provided that spongiosis of the neuropile and neuronal microvacuolation are at a stage similar to that in the 30 minutes' occlusion group (*i.e.*, reversible lesions). When, however, the neurons show advanced pathology, such as the shrinkage seen after 1 or 2 hours of occlusion, recirculation will result in neuronal disruption in the early stages following recirculation, the ischemic focus will show a severer pathology and ultimately necrosis will set in. For this

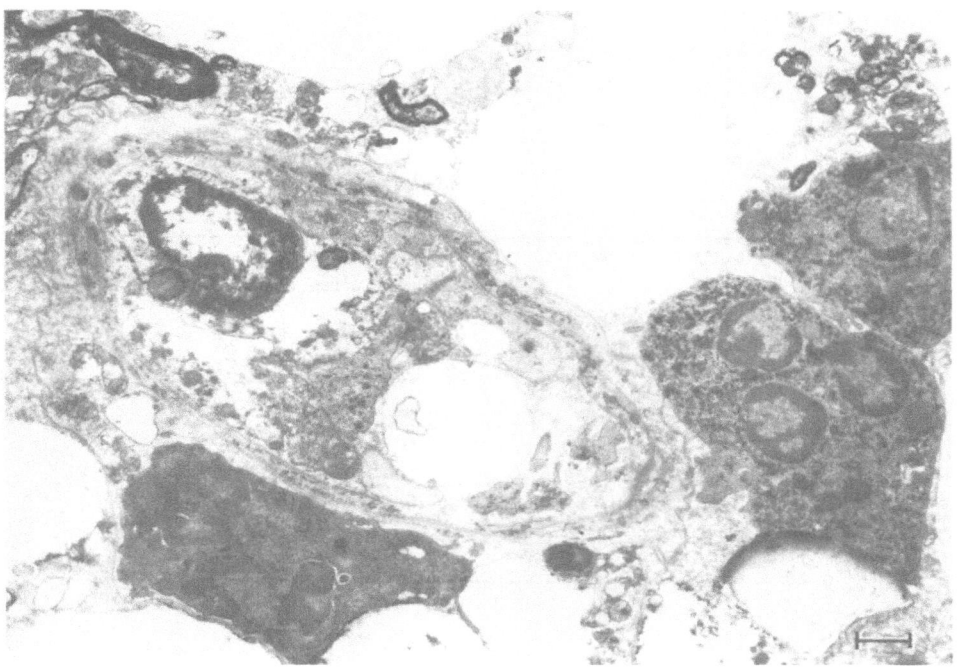

Fig. 2-14. Infiltration of neutrophiles into softened areas after 24 hours occlusion. Two neutrophiles were found in the perivascular space (right side). One neutrophile was found between widely separated layers of the basement membrane (lower side). (Bar = 1 μm)

reason, neuronal shrinkage is thought to be an irreversible lesion.

b) Hemorrhagic Infarction After Recirculation

Using the canine thalamic infarction model and performing vascular occlusion for 2–24 hours, an autopsy was done after 5–60 minutes of recirculation and the incidence of hemorrhagic infarction was studied[31] (**Table 2-3**). In the aniamals undergoing 2 hours of occlusion and then recirculation, no hemorrhage was seen, but after 3 hours of occlusion and 1 hour of recirculation, mild hemorrhage was apparent under a light microscope. When occlusion was for 6–12 hours,

Table 2-3. Number of dogs used in the experiment

Occlusion time	Recirculation time 5 min.	60 min.
2 hrs	2	2
3	1	4
6	3	1
9	1	1
12	1	1
24	2	3

hemorrhagic infarction was seen in all cases (**Fig. 2-20**). In the group undergoing 24 hours of occlusion, hemorrhagic infarction was seen in only 40% of the animals (**Fig. 2-21**).

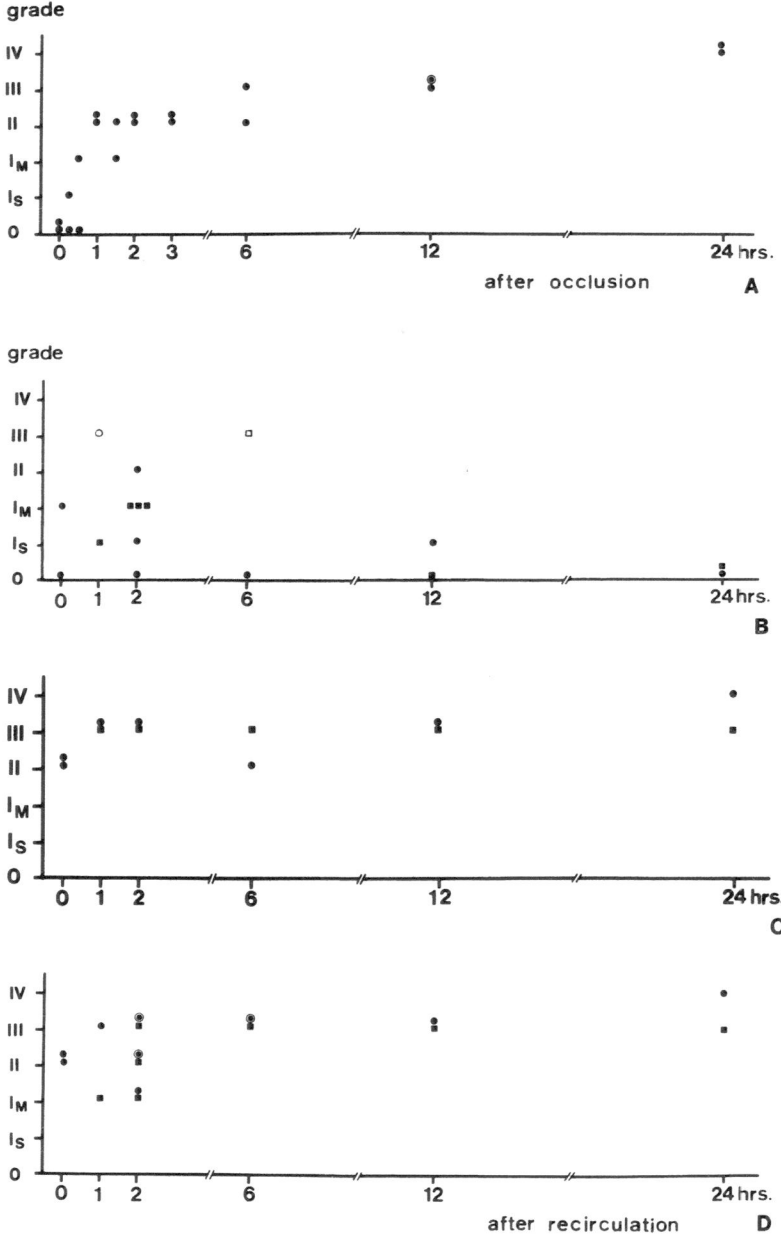

Fig. 2-15. Time course of ischemic changes in thalamic infarction model. A) Permanent occlusion, B) 30 minutes temporary occlusion, C) 1 hour temporary occlusion, D) 2 hours temporary occlusion. Gradings of ischemic changes: Grade 0: normal; Grade Is: spongiosis of neuropil, Grade Im: microvacuolation of nerve cells; Grade II: shrinkage and/or swelling of nerve cells; Grade III: fragmentation and/or disruption of nerve cells; Grade IV: infiltration of neutrophilic leucocytes

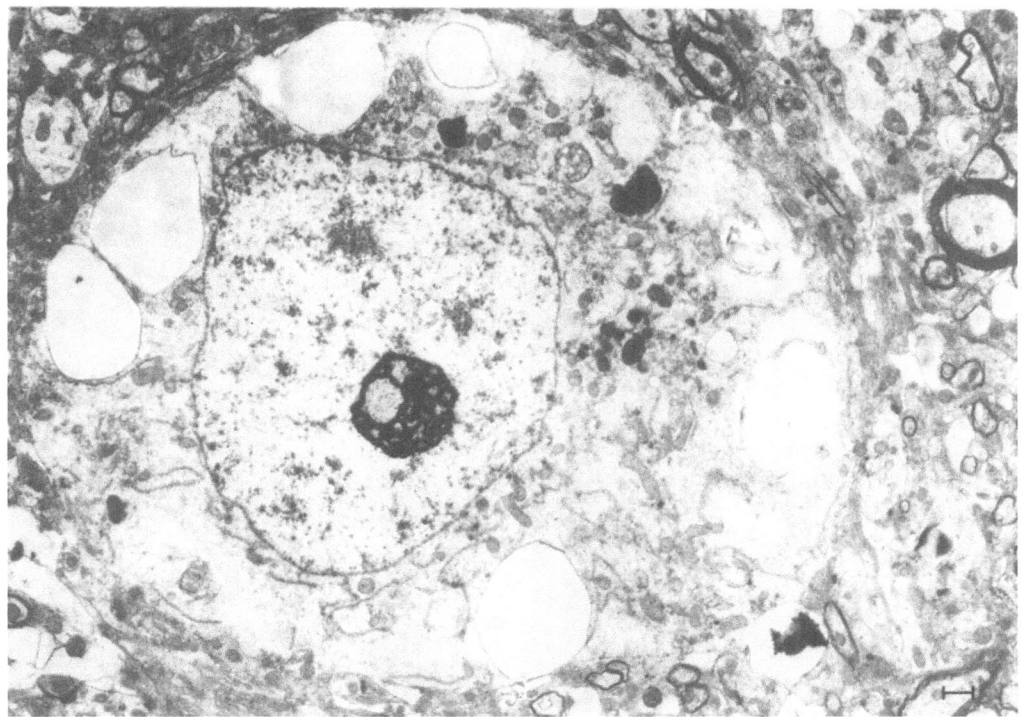

Fig. 2-16. Special type of nerve cell with microvacuolation (30 minutes occlusion—2 hours recirculation). Several large vacuoles are found in the cytoplasm, although mitochondria and other organella are preserved with normal appearance. (Bar = 1 μm)

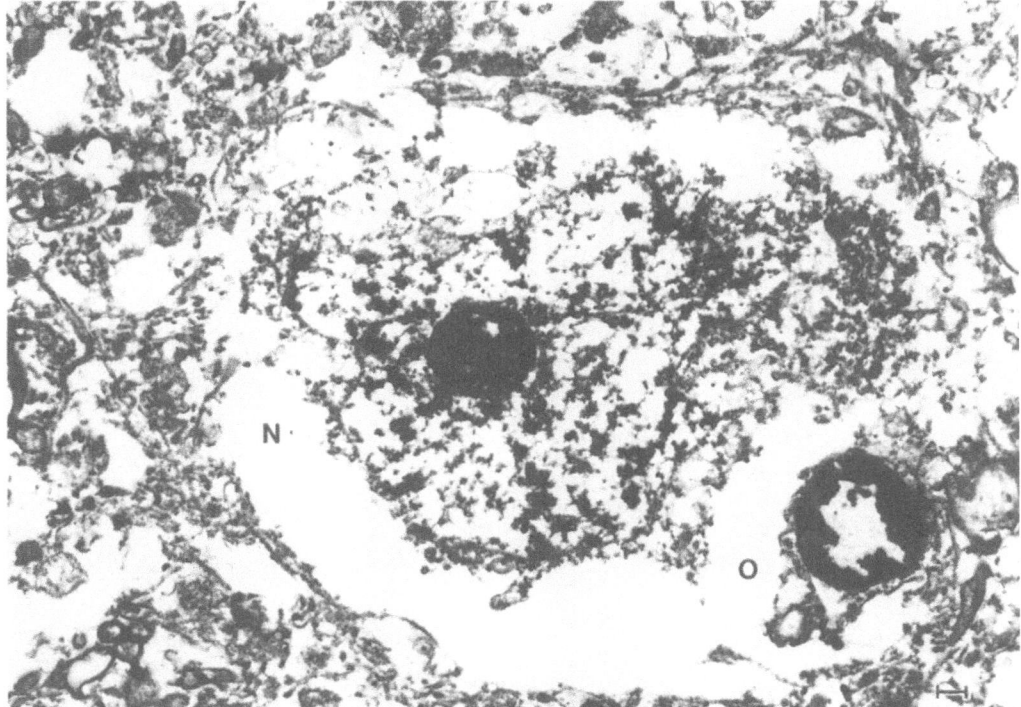

Fig. 2-17. Disruption of a nerve cell (2 hours occlusion—2 hours recirculation). The plasma membrane of a nerve cell (*N*) is disrupted and communicated to a disrupted oligodendroglia (*O*). (Bar = 1 μm)

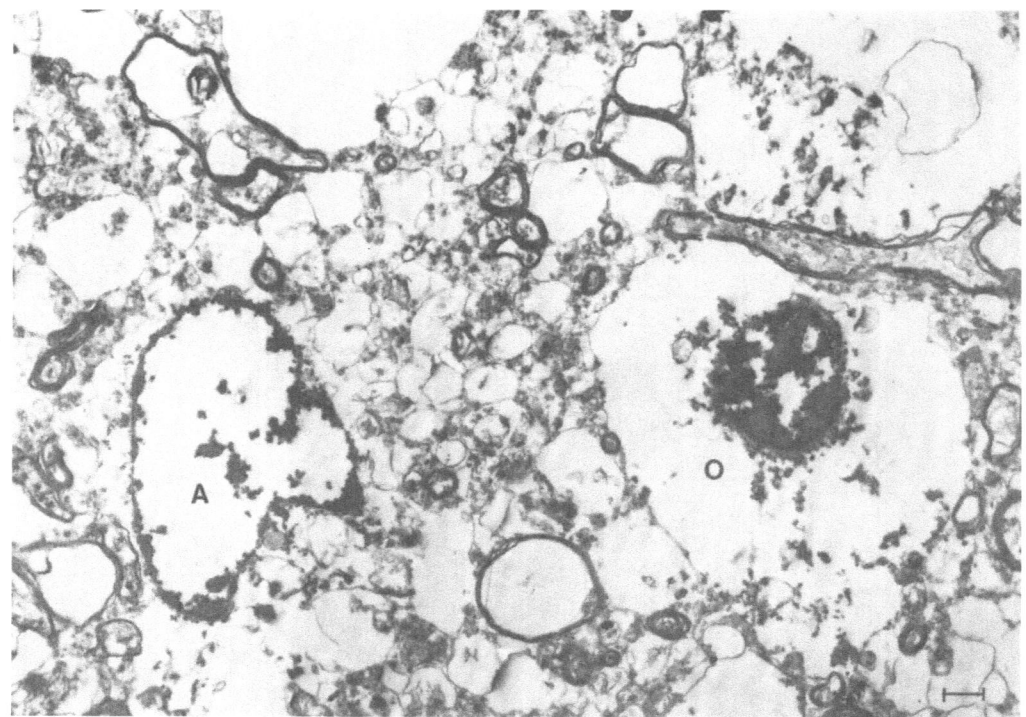

Fig. 2-18. Ischemic changes in glial cells (1 hour occlusion—1 hour recirculation). An oligodendroglia (*O*) and an astrocyte (*A*) are swollen and degenerated. Swelling and disruption of the axonal, dendritic and glial processes. (Bar = 1 μm)

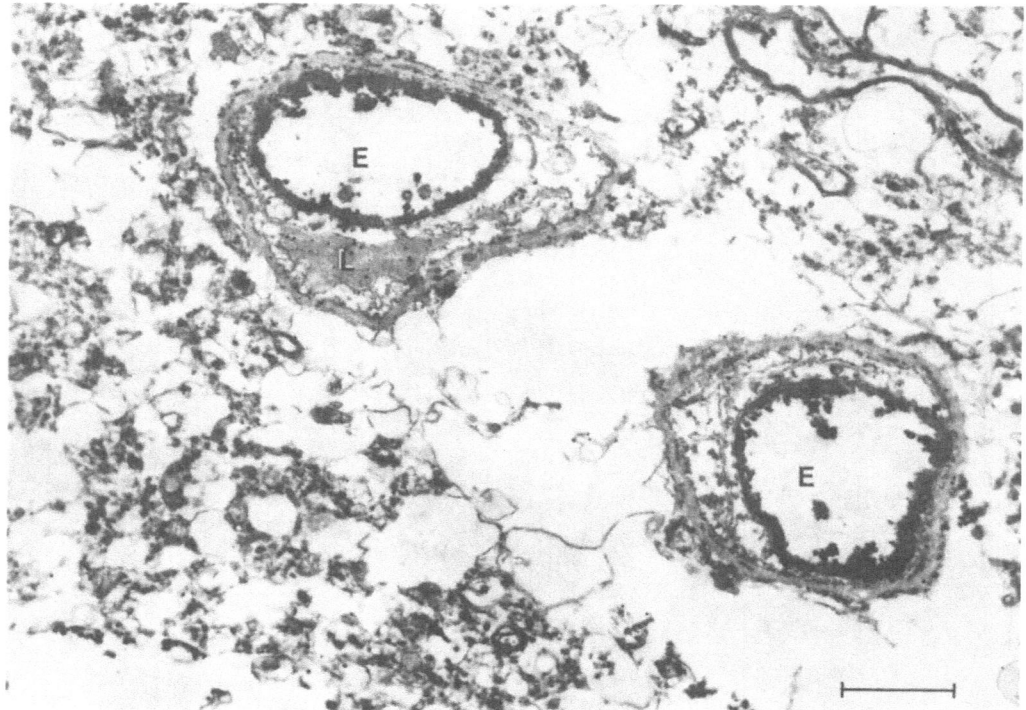

Fig. 2-19. Ischemic changes in the capillary (1 hour occlusion—1 hour recirculation). Swelling of endothelial cells (*E*) and narrowing of capillary lumen (*L*) are found. The axonal, dendritic and glial processes are swollen and disrupted. (Bar = 1 μm)

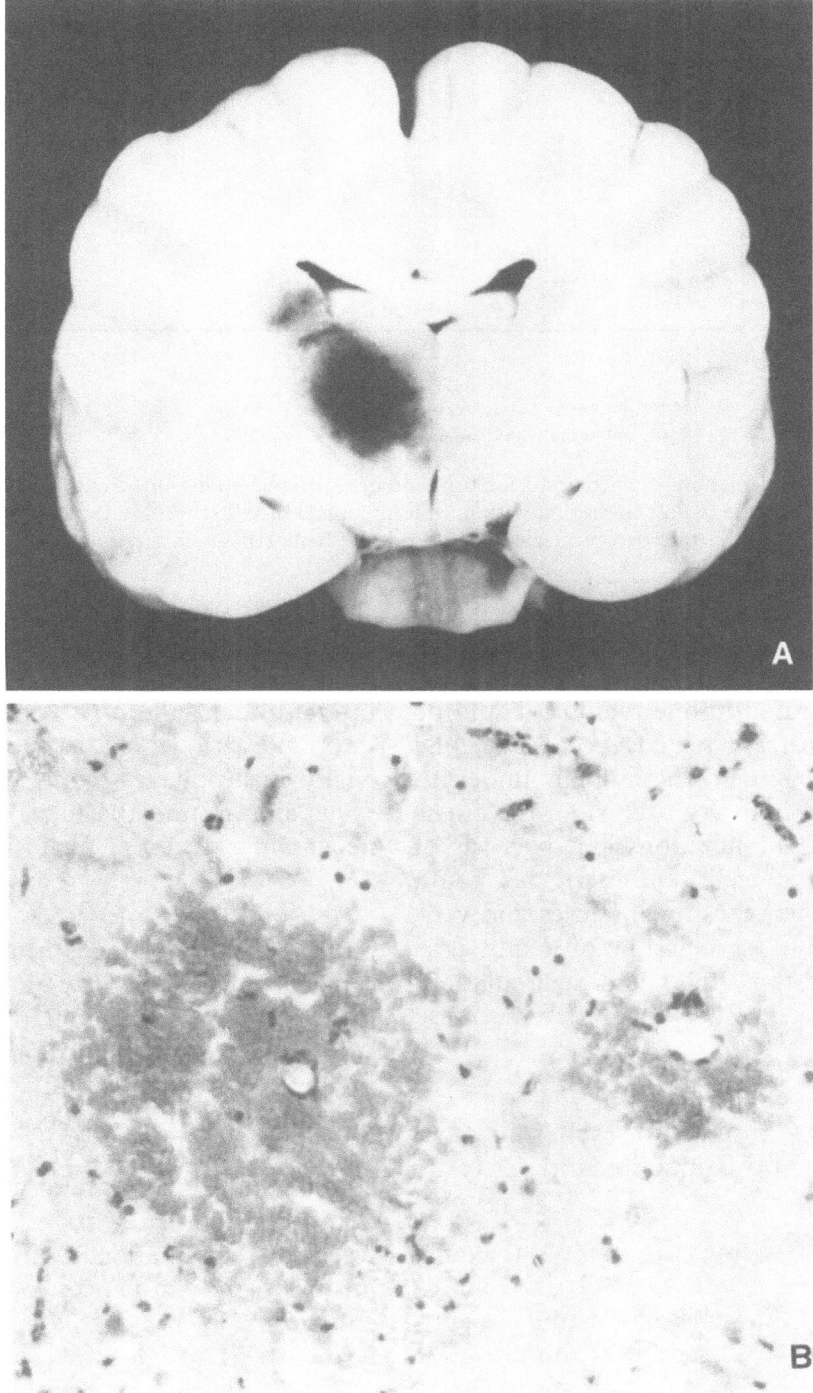

Fig. 2-20. Hemorrhagic infarction (6 hours occlusion—2 hours recirculation). A) Autopsied brain. B) Microscopical findings. Diapedesis from capillaries and small veins. (H-E stain × 200)

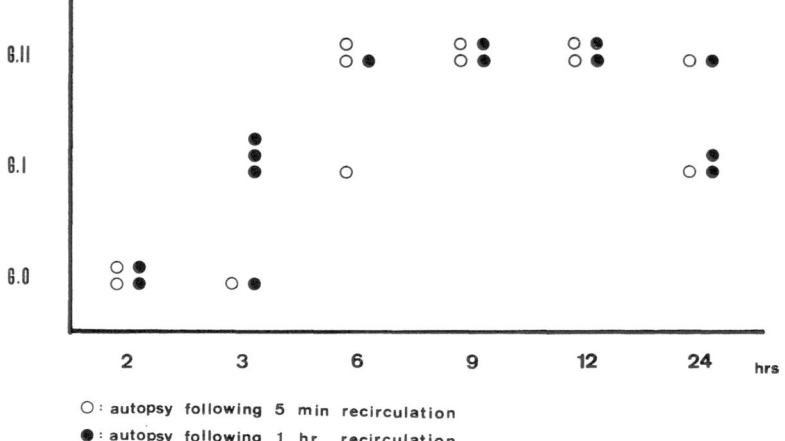

O : autopsy following 5 min recirculation
● : autopsy following 1 hr recirculation

Fig. 2-21. Correlation between occlusion time and hemorrhagic infarction. Grading of hemorrhagic infarction: Grade 0: pale infarction without microscopical bleeding; Grade I: pale infarction with microscopical bleeding; Grade II: hemorrhagic infarction

The hemorrhage in these cases was primarily exudation from capillaries or venuoles and when such exudation was concentrated in one region it could be seen with the naked eye as petechial hemorrhage. In the animals undergoing 6–12 hours of occlusion, the opening of tight junctions and necrosis of endothelial cells and pericytes could already be seen in capillaries and venules prior to recirculation[62] and it is thought that, following recirculation, hemorrhage occurred from these vessels.

2.2.1.3 Changes in Ischemic Foci During the Subacute and Chronic Stages

Again using the canine thalamic infarction model (without EEG monitoring of the ischemia), vascular occlusion of 30, 60, and 120 minutes was followed by recirculation and cerebral tissue was studied after one week survival[67] (**Table 2-4**). Among the animals undergoing 60–120 minutes of occlusion, approximately 2/3 showed infarctic foci. The foci were "pale infarction" with necrosis at their centers. Many macrophages were seen in the focus and around the focus there was neovascularization and gliosis (**Fig. 2-22**).

Next, in order to study the long-term sequential changes in the foci of cerebral infarction, the canine hemorrhagic infarction model was used. While monitoring the brain by means of both CT scans and the EEG, thalamic infarction of 6 hours duration was produced and cerebral tissue was studied under a light microscope after 2 hours, 2, 5, 7 or 9 days, or, 2, 3, 4, 5 or 6 weeks of recirculation. Study was also made of the influences on contrast enhancement at the focus of infarction[49] (**Table 2-5**).

Two hours after recirculation, neurons showing ischemic changes

Table 2-4. Correlation between thalamus infarction and length of occlusion in dogs

Thalamus infarction	Grade 0	Grade 1	Grade 2	Grade 3	Total
Occlusion time					
Sham operation	3	0	0	0	3
30 minutes	7	3	0	0	10
60 minutes	4	0	2	4	10
120 minutes	4	3	5	8	20

Grade 0 : no infarction in thalamus.
Grade 1 : small infarctic foci in thalamus.
Grade 2 : infarction affecting about half of thalamus.
Grade 3 : infarction affecting more than two thirds of thalamus

Table 2-5. Number of dogs used in the experiment

Recirculation time	2h,	2d,	5d,	7d,	9d,	2w,	3w,	4w,	5w,	6w
Number of dogs	2,	2,	2,	2,	2,	2,	2,	2,	2,	2

were found dispersed throughout the infarctic focus, glial cells were swollen and nerve fibers were poorly stained. Many sites of leakage of red blood cells could be seen around capillaries and venuoles (**Fig. 2-22**). On the second day necrosis within the infarctic focus had progressed and there was exudation of neutrophils. By the fifth day, necrosis at the center of the focus had become more severe and pronounced. In the area surrounding the focus, many macrophages could be seen and there was a small amount of neovascularization. Between the 7th and 14th days, macrophages were seen at the center of the focus and neovascularization extended throughout the focus (**Fig. 2-23A**).

Between the third and fourth week, phagocytosis of necrotic tissue by the macrophages continued and there was the beginning of formation of small vacuoles and the disappearance of new vasculature in the marginal zone of the ischemic foci (**Fig. 2-23B**). Between the fifth and sixth weeks the vacuolization extended to the center of the lesion and only small banded objects remained at its center (**Fig. 2-23C**). The walls of the vacuoles were due to thin gliosis caused by the reactive formation of astrocytes. **Figure 2-24** schematizes these sequential changes.

With regard to the disappearance of neovascularization, the effects at regions within the focus and surrounding it were studied separately. The

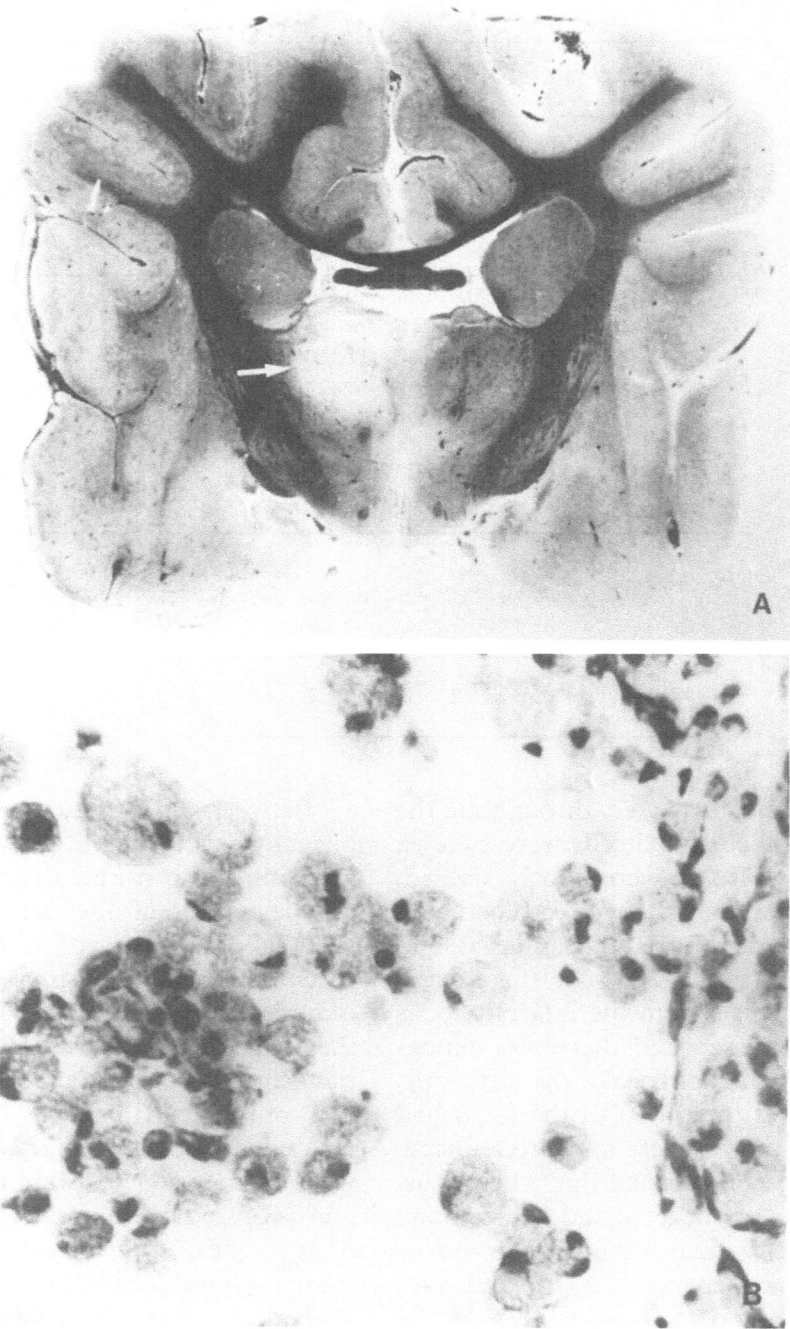

Fig. 2-22. Thalamus infarction in the dog. A) Dog's brain autopsied 7 days after 2 hours temporary occlusion. Infarction foci were found in the right thalamus (arrow). B) Microscopical findings. Numerous macrophages were found in the infarction foci. (H-E stain × 400)

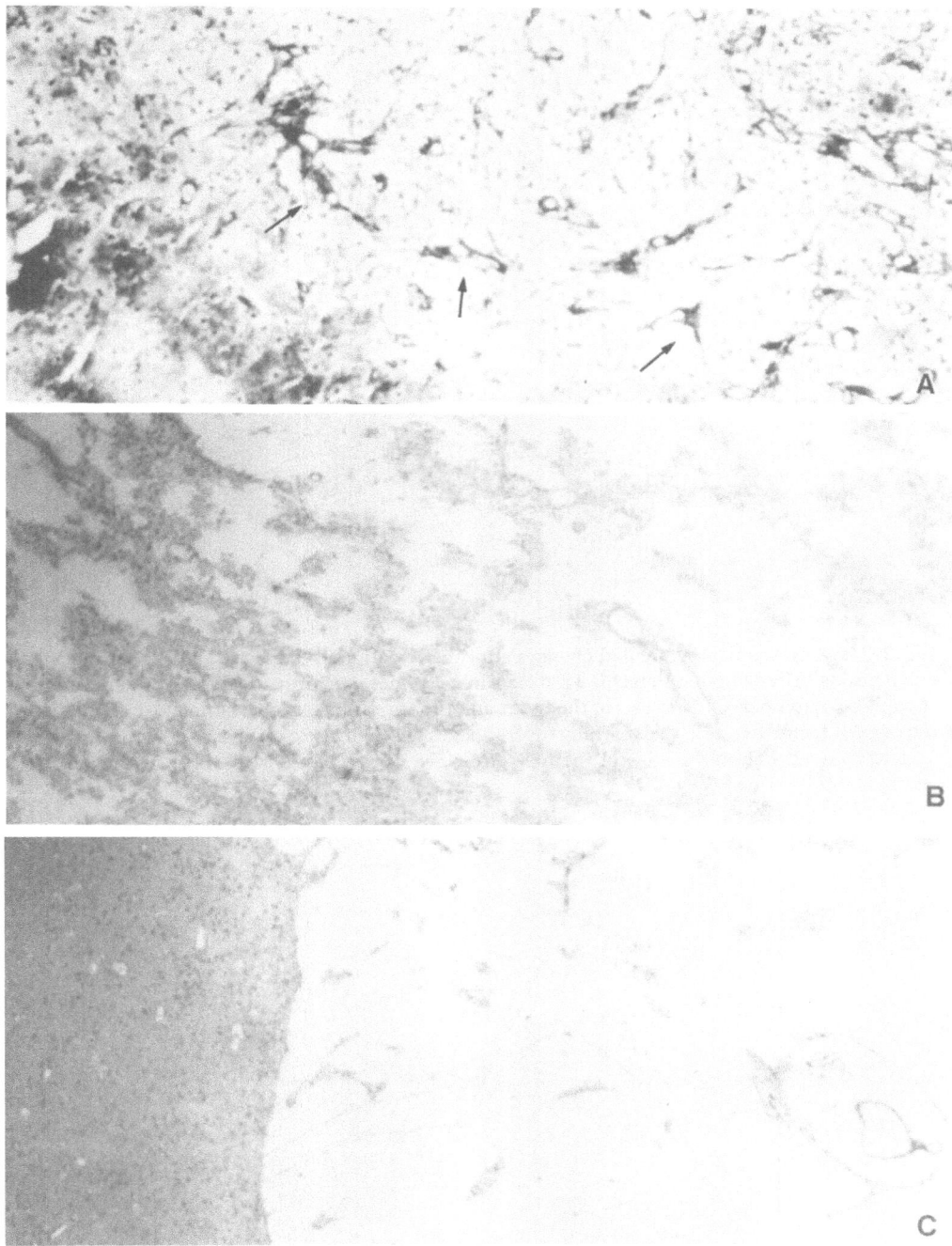

Fig. 2-23. Microscopical findings of ischemic foci (H-E stain × 40). A) 7 days after temporary occlusion. Immature new capillaries are found (arrows). B) 4 weeks after temporary occlusion. Absorption of necrotic tissue and degeneration of new capilaries are found. C) 6 weeks after temporary occlusion. Large cavities are found in the ischemic foci

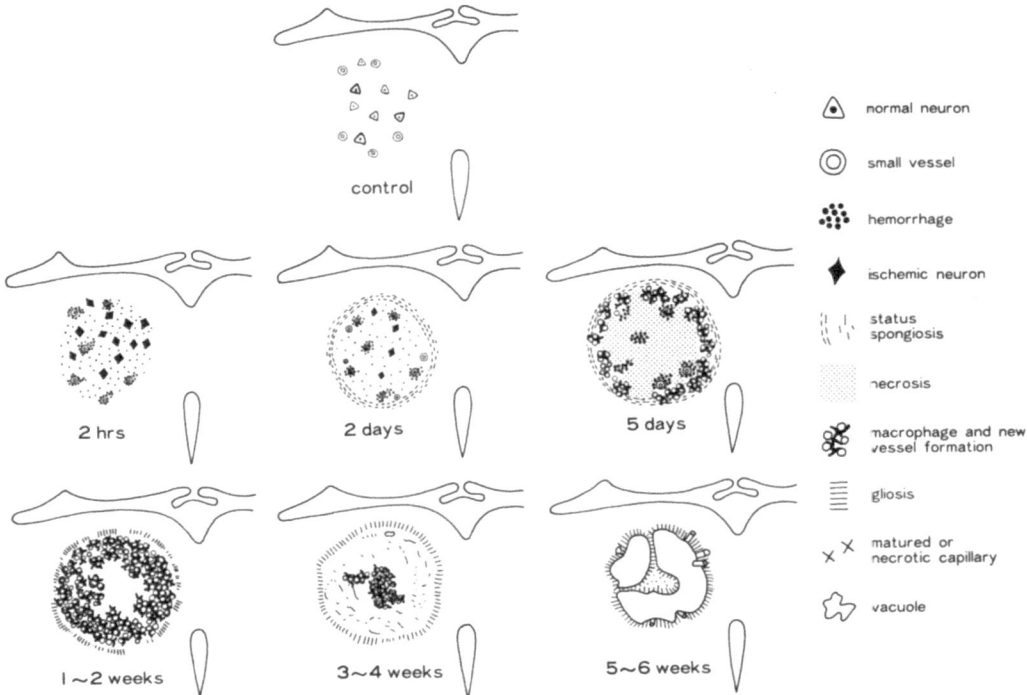

Fig. 2-24. Microscopical sequential changes after recirculation in the foci. Control: Normal vessels and neurons. After 2 hours: Petechiae and ischemic changes in neurons. After 2 days: Petechiae and spongiosis. After 5 days: Necrosis of the tissue and the appearance of a few immature capillaries not the edge of foci. After 1–2 weeks: Many macrophages and immature capillaries. After 3–4 weeks. A few vacuoles not on the edge of the foci. After 5–6 weeks: Cavity wall made of thin gliosis

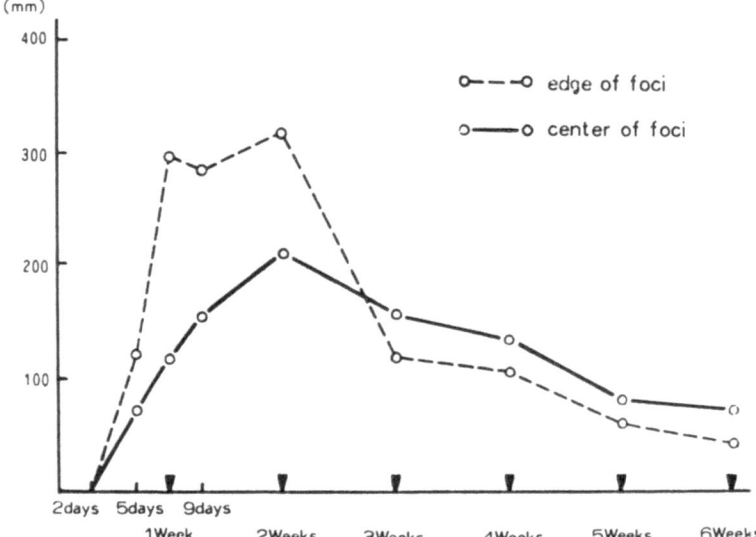

Fig. 2-25. Length of immature capillaries in 1 mm³. Immature capillaries appear after 5 days of recirculation. They increase after 1–2 weeks and begin to decrease after 3 weeks of recirculation. This phenomenon is apparent on the edge of foci

vacuoles first appeared in the peripheral region 5 days after recirculation and were most abundant between 7 and 14 days, after which they rapidly disappeared. In contrast, at the center of the focus neovascularization was abundant from the 7th day, was most abundant after 2–3 weeks and gradually disappeared thereafter (**Fig. 2-25**).

In CT scans, study of the disappearance of contrast enhancement effects showed that enhancement was present 2 hours after recirculation and between 7 and 14 days after recirculation (**Fig. 2-26**). In light of histological findings, the enhancement seen 2 hours after recirculation is thought to have been caused by exudation of the contrast medium from vessels already having undergone ischemic change, whereas that seen between 7 and 14 days is thought to have resulted from exudation of the media from neovasculature.

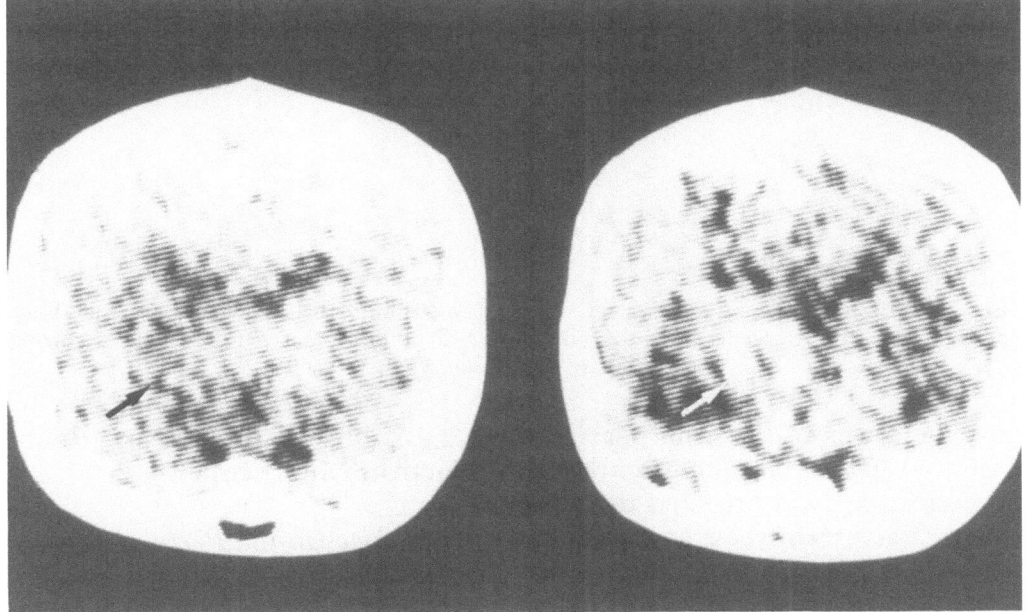

Fig. 2-26. CT findings after 7 days of recirculation. Left: Plain CT shows a low density area in the right thalamus (arrow), Right: Enhanced CT shows a high density area (arrow)

2.2.2 Sequential Changes in Global Ischemia

The three vessel occlusion model in the rat which produces bilateral hemispheric ischemia was used. Using this model, the rCBF of both cerebral hemispheres measured using the ^{14}C-iodo-antipyrine CBF method, was reduced to less than 5% of that of the control animals. Three animals were prepared for each of the groups undergoing 0, 5, 10, 30 and 60 minutes of ischemia. After completion of the experiment perfusion fixation was performed immediately

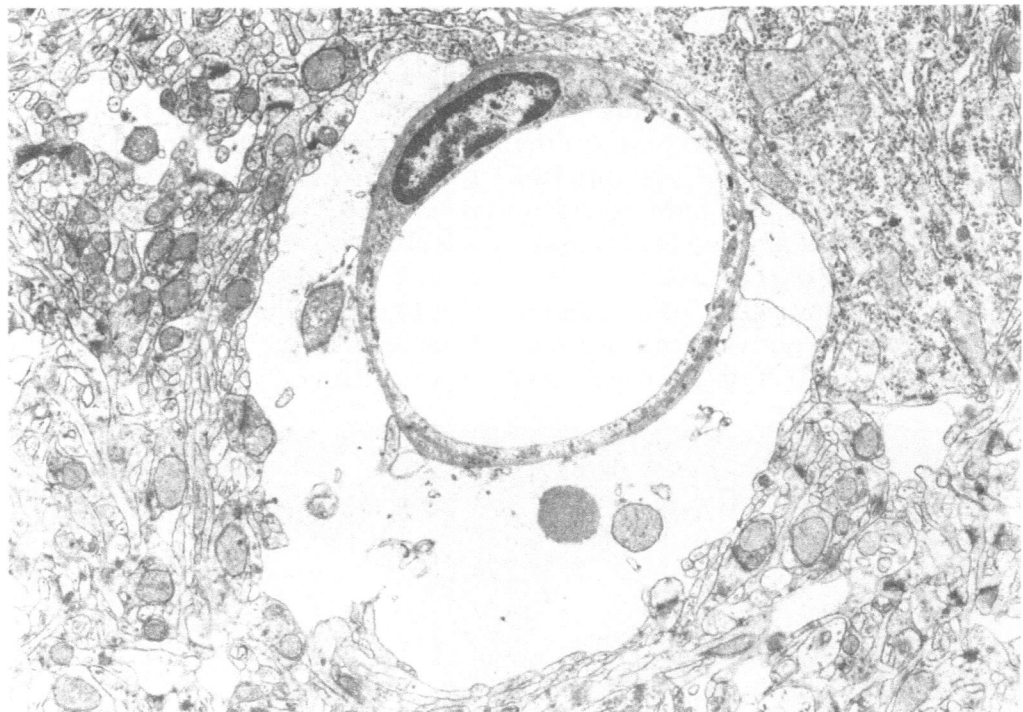

Fig. 2-27. Electron micrograph of a rat subjected to 60 min global ischemia. Severe swelling of perivascular astrocytic feet is noted along the entire wall of the capillary. However, no narrowing of the vascular lumen is detected. × 9,600

and the cortex of the hemisphere was removed. Of two neighboring regions of cortex, one was prepared using the freeze fracture method and ultrathin slices were made from the other. Finally, a coronal section passing through the optic chiasma was stained using the Hematoxylin-Eosin and Klüver-Barrela methods and examined under a light microscope.

2.2.2.1 Observations Using the Light Microscope

Other than the occasional "dark neurons" after 30 or more minutes of occlusion, light microscopy revealed no abnormalities.

2.2.2.2 Observations Using the Electron Microscope

a) Changes in the Perivascular Astrocytic Feet

The perivascular astrocytic foot in the control animals was found to be a thin membrane external of blood vessels, but in the 30 minutes' occlusion group, the cytoplasm had become watery and swelling had begun. After 1 hour of occlusion, severe swelling, such that the astrocytic foot surrounded the blood vessels, was found, but no stenosis of the vascular lumen was apparent (**Fig. 2-27**). In regions where the change was mild, there was enlarge-

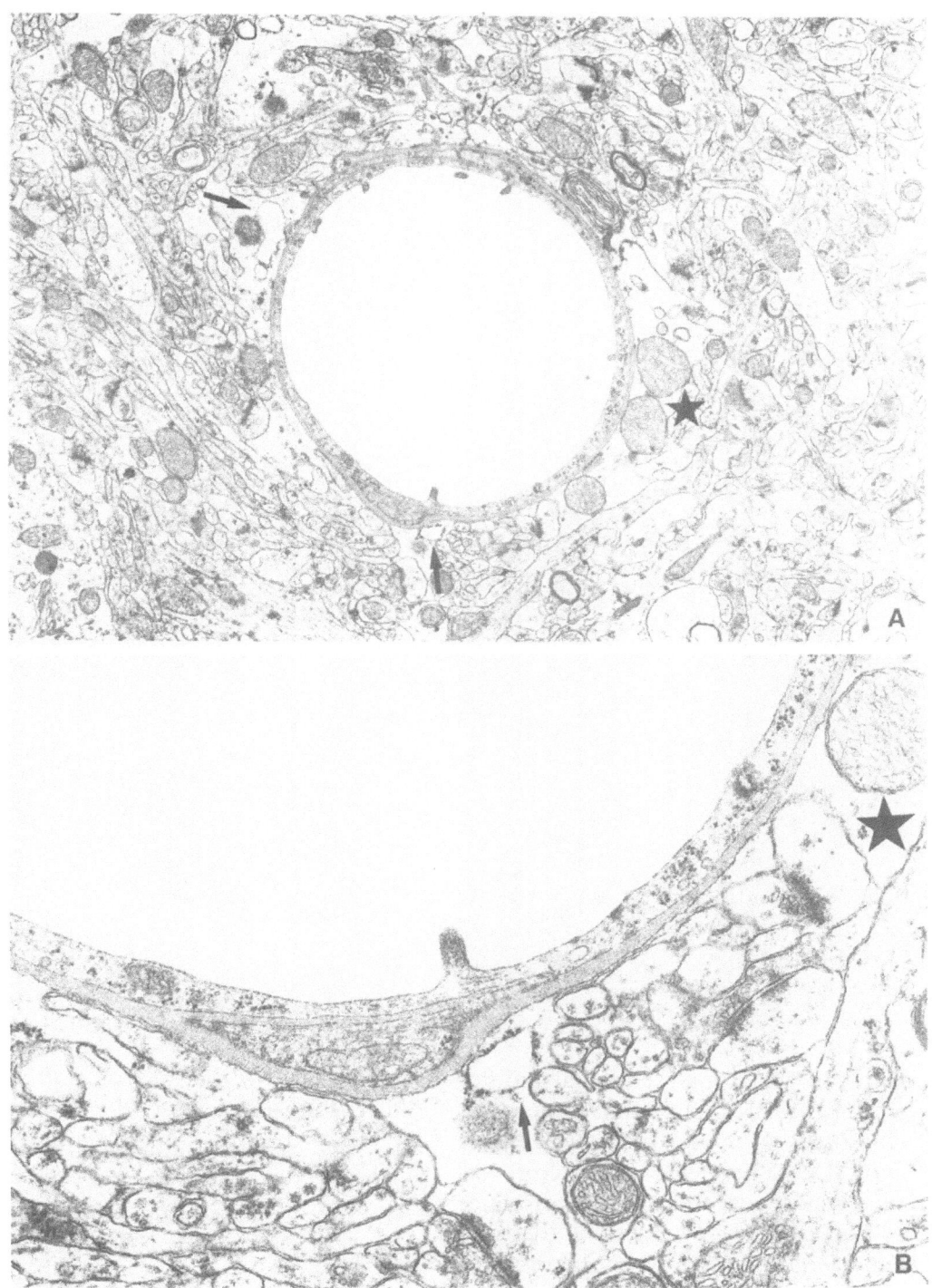

Fig. 2-28. Electron micrographs taken from the same sample as in **Fig. 2-27.** A) The swelling of astrocytic processes surrounding the capillary wall is less prominent but the detachment of ribosomes from the rER is evident (arrows). The mitochondria appear to be intact (asterisk). × 10,000. B) and C) High power view of A). × 36,000

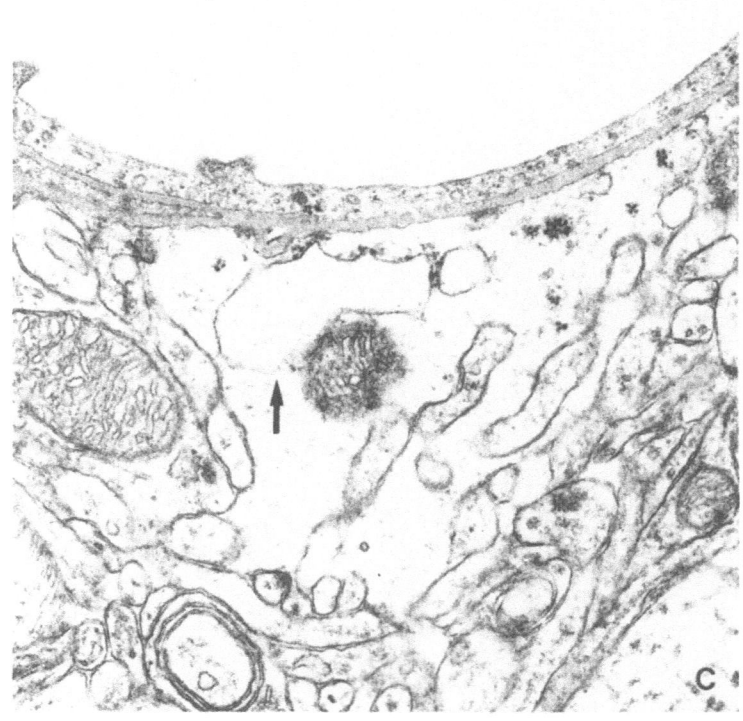

Fig. 2-28—C

ment of the rough endoplasmic reticulum (rER) and detachment of
ribosomes form the rER (**Fig. 2-28A**
and **B**). Where the pathology was
severe, however, there was the presence
of vacuoles of low electron density,
which were thought to be the result of
rER swelling (**Fig. 2-28C**). Changes in
the mitochondria, however, were
slight.

b) Changes in the Cytoplasm and Nucleus of Astrocytes

In the animals undergoing 1 hour
occlusion, swelling of the perikaryon of
astrocytes, a decrease in the electron
density of the nucleus, and disruption
of the nuclear membrane were found.
Although there was swelling of mitochondria and rER and disruption of the

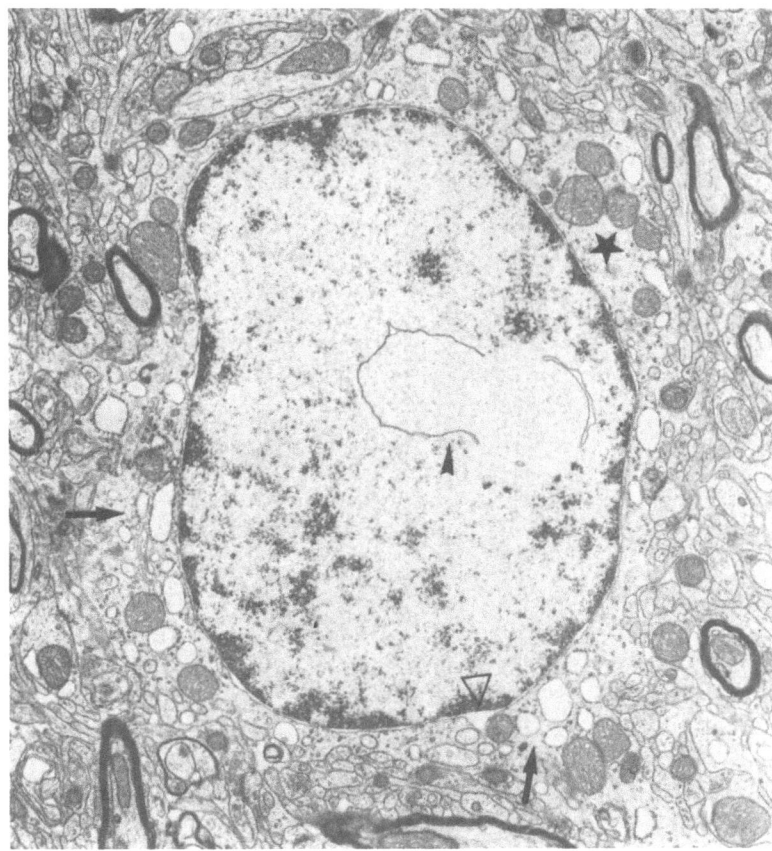

Fig. 2-29. Electron micrograph of an astrocyte in the rat subjected to 60 min global ischemia. Intact mitochondria (asterisk), enlarged rER (arrow), disintegration of the nuclear membrane (arrow head) and enlarged nuclear cistern (open triangle) are evident. × 8,700

unit membrane structure, the changes in the mitochondria were comparatively mild (**Fig. 2-29**).

c) Changes in the Endothelial Cells

In the occlusion groups, widening of the juxta-junction space of the tight junction, swelling of the cytoplasm and a decrease in the electron density and swelling of the cytoplasm were seen. In regions where there was a decrease in electron density, the low density regions were limited to a single endo-thelial cell found within the wall of blood vessels. Significant changes in such changes were not found, however, with increases in occlusion time.

d) Changes in the Neuron

Neuronal changes were first apparent in the animals undergoing 1 hour of vascular occlusion. Slight swelling of mitochondria and rER was seen, there was detachment of ribosomes from the rER and an increase in free ribosomes was noted. These neuronal

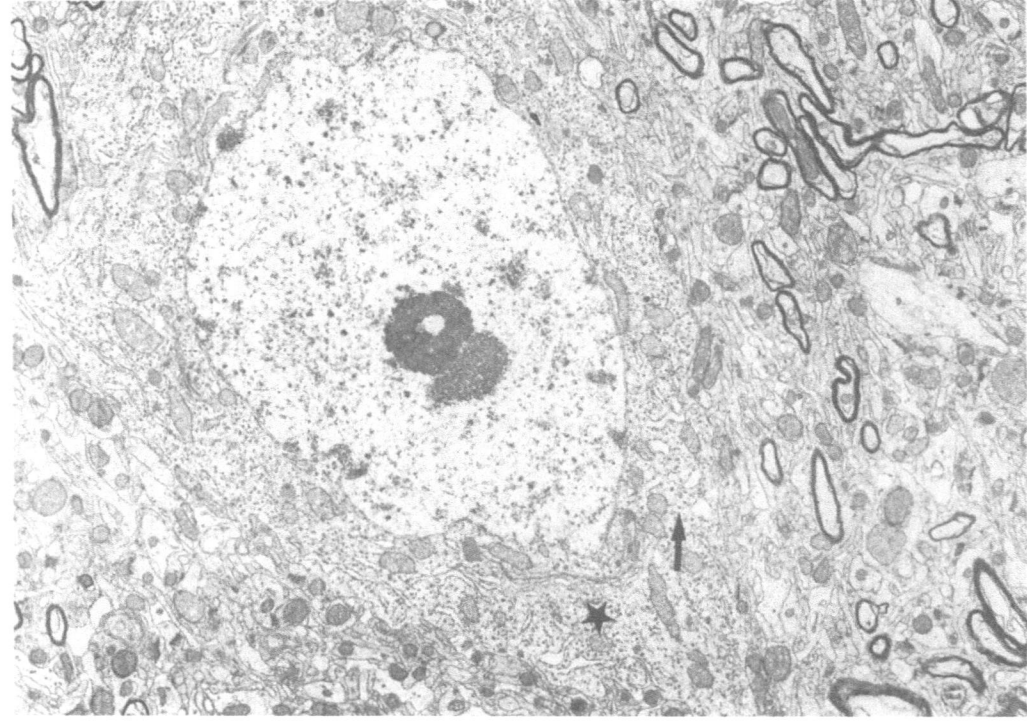

Fig. 2-30. Electron micrograph of a cortical neuron in the rat subjected to 60 min global ischemia. Rough endoplasmic reticuli are slightly swollen (arrow). Detachment of the ribosomes from the rough endoplasmic reticuli is suspected although the rosette formation of the ribosomes is still retained (asterisk). × 17,000

changes were, however, milder than those seen in astrocytes (**Fig. 2-30**).

e) Changes in the Neuropile

In comparison with changes in other cerebral regions, those in the neuropile in the 1 hour occlusion group were characteristically seen around large blood vessels. Although some swelling of the astrocytic foot around such vessels was also seen, most of the changes were in the mitochondria of the neuropile, where the swelling was so severe that the cristae could almost no longer be identified (**Fig. 2-31**).

2.2.2.3 Observation Using the Freeze Fracture Method

It is known that in the perivascular astrocytic foot, there is an orthogonal array (OA) and so-called "pits". The OA is thought to be a multienzyme complex involved in the transport of substances and in the osmosis of such substances (**Fig. 2-32A**). It has been found that the number of these characteristic intramembrane components decreases with increases in the occlusion time. There is also a tendency for the structure of the components to change from an orthogonal array to an

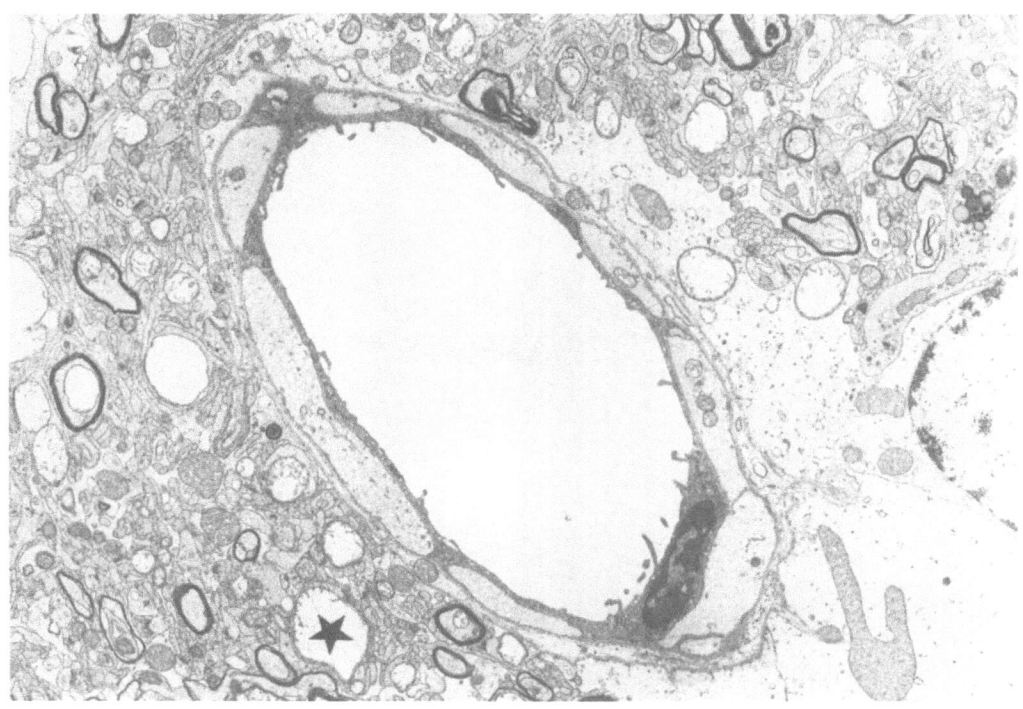

Fig. 2-31. Electron micrograph of a larger vessel in the rat subjected to 60 min global ischemica. The mitochondria in the neurites are markedly swollen (asterisk). × 6300

amorphous array, although even after 60 minutes of occlusion the pits are still present in some regions (**Fig. 2-32B**).

Comparisons were made of the state of 50 blood vessels per rat in control and occlusion groups. In the control group, 32.2 ± 5.4 (mean $\pm$ S.D.) vessels showed OA or pits, whereas after 5 minutes of occlusion this number had fallen to 26.0 ± 2.2. After 10 minutes of occlusion 17.3 ± 3.3 vessels showed OA or pits, after 30 minutes 9.7 ± 4.7 and after 60 minutes 5.7 ± 2.1. Although a significant difference between the control and 5 minutes' occlusion group was not found, a highly significant difference ($p < 0.001$) was seen after 10 or more minutes of occlusion (**Fig. 2-33**). There were no changes in the tight junctions of endothelial cells or the gap junctions of astrocytes, no aggregation of intramembrane particles could be seen in the cellular membrane, and the normal fluidity of the cell membrane was maintained.

From these experiments it was concluded that changes in the perivascular astrocytic foot were the earliest changes in global ischemia, and that reduction in the orthogonal array of membrane proteins was occurred in an extremely early period after only 10 minutes of ischemia; these changes precede the

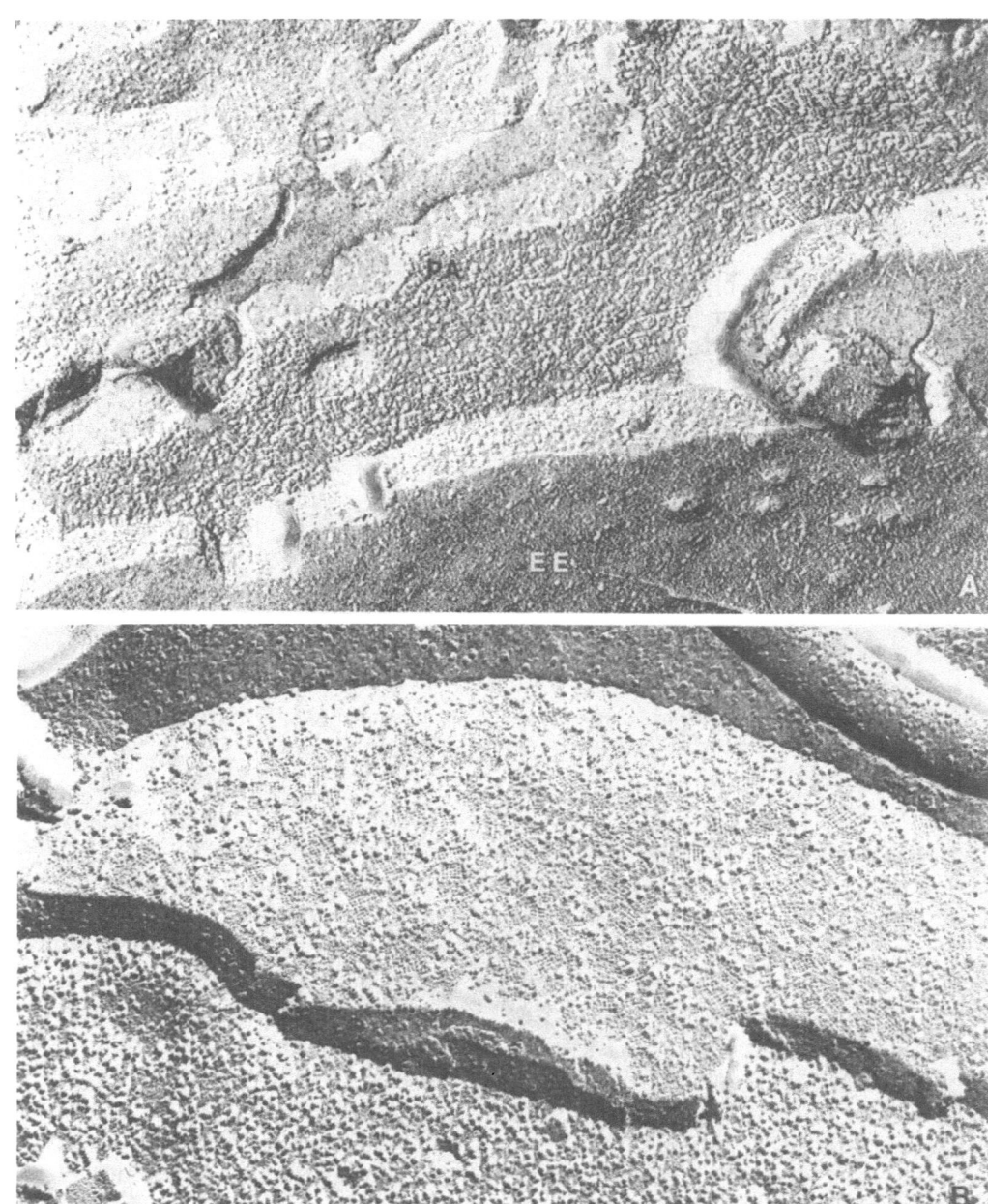

Fig. 2-32. Freeze-fracture profile of perivascular astrocytic process in rat cerebral cortex. A) 5 min ischemia. Note coexistence of orthogonal arrays and large disintegrated particles in the P face of the process (*PA*). *EE*, E face of abluminal membrane of an endothelial cell. × 100,000. B) 60 min ischemia. Pits in astrocytic membrane. × 100,000

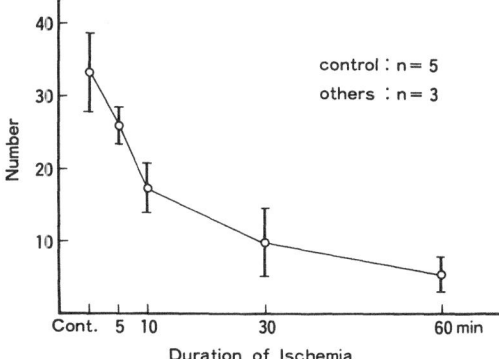

Fig. 2-33. The number of freeze-fracture profiles of perivascular astrocytic processes with orthogonal arrays or pits. Fifty profiles were examined in each animal. The bar represents mean ± S.D.

swelling of the cytoplasm of the astrocytes. At the level of the cellular organelle, the rER showed earlier changes than the mitochondria. Even after a lengthy period of 60 minutes of cerebral ischemia, the neurons showed only slight change.

3. CEREBRAL BLOOD FLOW

3.1 Introduction

Research on the nature of brain infarction has proceeded using various approaches, including those of histopathology, biochemistry and electrophysiology. One of the most fundamental parameters in such research is cerebral blood flow (CBF)—which is an essential barometer for determining the level of brain ischemia.

In recent years, methods for determining the level of regional cerebral blood flow (rCBF) have become established and many significant studies on the pathophysiology of the ischemic brain have been reported. There have, however, been differences of opinion concerning even the results of experimental studies on the ischemic brain under strictly controlled conditions. One reason for the lack of consensus is that, until recently, it has been impossible to produce an animal model in which a lesion can be consistently produced at a known site and, moreover, in which the severity of the pathology can be controlled. Indeed, the models used by various researchers have produced lesions those distribution and severity have differed considerably, depending upon the individual animals used.

More recently, attention has been focused on the pathology of the regions surrounding an ischemic focus. For such research, however, it is essential that the sites at which the measurements are made be accurately known to lie within or without the focus of ischemia.

We have studied various aspects of the pathology of brain ischemia and brain infarction and believe that our so-called "canine thalamic infarction model"[49] is ideal for investigating brain hemodynamics. Since the pathological focus is confined to the anterior half of the unilateral thalamus when using this model, it is a simple matter to monitor the hemodynamics within the ischemic focus and within its immediate surroundings. Moreover, because the focus is small, little or no brain edema occurs during the experiment and it is possible to avoid secondary pathological effects such as changes in intracranial pressure. Brain infarction can thus be produced without manipulating blood pressure and the duration of the vascular occlusion can be freely adjusted. Therefore it is thought that this canine model is highly appropriate for studying the dynamically changing pathology of the ischemic brain.

Our technique for measuring blood flow has been the hydrogen clearance method[6, 7], which is known to be accurate over a radius of some millimeters from an implanted electrode[8]. It is thus an effective means for making sequential and quantitatively precise measurements of the changing pathology of brain.

The present chapter is divided into five additional sections which discuss our experimental work on cerebral hemodynamics with reference to related work in the literature. Section II discusses the correlation between brain electrical activity and cerebral blood flow at an ischemic focus; section III discusses the hemodynamics following recirculation found at the center of the ischemic focus and at its peripheral regions. In section IV, the sequential changes in the vascular reactivity within an ischemic focus are outlined from the perspective of autoregulation and the CO_2 response. In section V the hemodynamics of hemorrhagic infarction are covered.

Since our experimental work on each of these topics was undertaken using the canine thalamic infarction model[49], details of this model will be presented first[27, 35].

The experimental animals were mongrel dogs weighing roughly 10 kg each. Under *i.v.* administration of pentobarbital anesthesia (35 mg/kg), an endotracheal tube was inserted, the dogs were immobilized with an *i.v.* dose of pancronium bromide (0.04 mg/kg/hr), and artificial respiration was instituted. Intermittent sampling of arterial blood was also done and blood pH, $PaCO_2$ and PaO_2 levels, measured using a blood gas analyzer, were maintained within physiological

limits. Blood pressure was measured continuously via a catheter inserted into the abdominal aorta, and the rectal temperature was maintained at normothermia. In order to maintain a constant mild state of anesthesia pentobarbital (2.5 mg/kg/hr) was administered intravenously (**Fig. 3-1**).

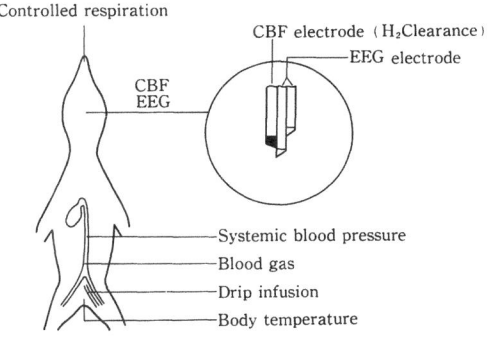

Fig. 3-1. Diagramatic representation of the experiment

The value of rCBF was determined using the initial slope method of the clearance curve following 3 minutes of inhalation of 5–10% hydrogen gas. The rCBF was calculated as follows:

$$rCBF = 0.693 \times \lambda/T_{1/2} \times 100 \, ml/min/100 \, g$$

with $\lambda = 1$ taken as unity, following Aukland *et al.*[3]. The electrode for EEG recording was a bipolar electrode made of two 0.25 mm stainless steel wires. For rCBF measurements, a needletype platinum electrode (0.3 mm diameter) was cemented to the EEG electrode. Finally, a deep recording electrode was placed in the anterior portion of the nucleus ventralis thalami, using the stereotaxic atlas of Lim *et al.*[19].

3.2 Correlation Between Regional EEG and rCBF

When studying the pathology of the ischemic brain, the brain's electrical activity has long been used as an index of its functional impairment during ischemia. Clinically, as well, an important theme of clinical research has been the correlation between cerebral hemodynamics and brain function in temporary occlusion of blood flow in neurosurgical operations and during extracorporeal circulation.

A large number of studies have been undertaken from this perspective—specifically on the relationship between the EEG and rCBF during ischemia. Suzuki et al.[39] investigated the changes in the EEG occurring with decreases in cortical blood flow due to exsanguination and found that a 50% decrease in rCBF resulted in EEG slow waves, whereas an 80% or greater decrease resulted in complete attenuation of electrical activity. By means of vascular clipping of the middle cerebral artery (MCA) in monkeys, Sundt et al.[38] reduced cortical blood flow by 20–50% and found that there was a decrease only in the voltage of the basic pattern of the EEG. With increasing severity of the ischemia, however, a low voltage pattern emerged. In contrast, Heiss et al.[10] used a feline MCA occlusion model and investigated the spontaneous activity of single cortical neurons. They found that brain electrical activity disappears at rCBF values lower than 20 ml/100 g/min.

Using a baboon MCA occlusion model, Branston et al.[4] studied evoked potentials and found that the potentials decrease at rCBF values below 20 ml/100 g/min and completely disappear at values below 15 ml/100 g/min. Using an experimental model in which the bilateral common carotid arteries (CCA) were ligated and venous blood drawn from the rat, Mabe et al.[20] have studied power spectral patterns. They reported a negative correlation between cerebral blood flow and the δ-power percentage in the EEG. When the residual blood flow was 6–8 ml/min/100 g or less, the EEG became completely attenuated.

It must be said, however, that most of the models used by various investigators have been either global ischemia models or MCA occlusion models, and in either case the extent and severity of the ischemia varies considerably among individual animals. For this reason, we have developed a canine thalamic infarction model in which a focus of ischemia can be consistently produced at a defined portion of the brain[49]. After performing vascular occlusion, the relationship between the EEG and rCBF at the center of the ischemic focus can then be studied[33].

The electrical activity measured at the anterior half of the thalamus prior to occlusion shows a typical waxing and waning at 7–10 Hz[31], but following occlusion four different patterns can be distinguished. As illustrated in **Fig. 3-2**, when the electrical activity is the same as that prior to occlusion, it is scored as "no change". When there is some decrease in EEG voltage, but little change in frequencies, it is scored as "slight change". When there is dis-

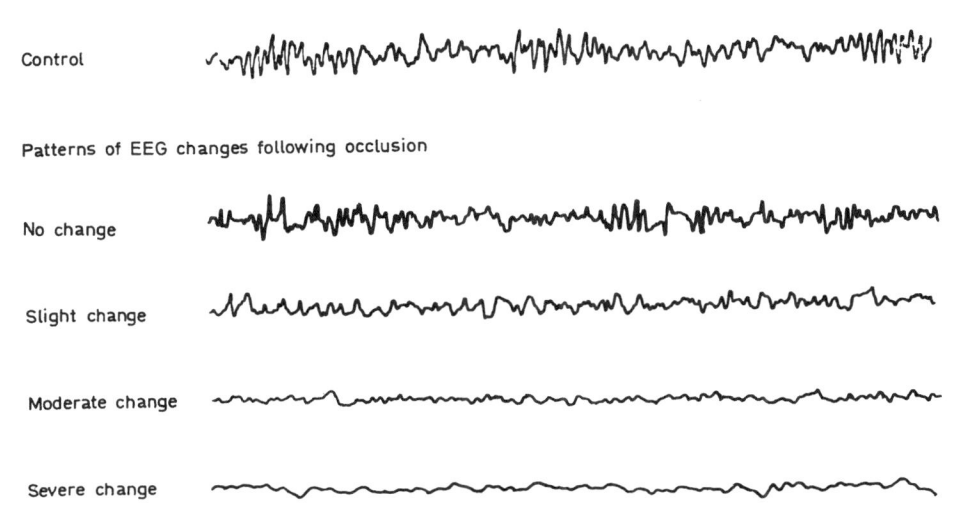

Control

Patterns of EEG changes following occlusion

No change

Slight change

Moderate change

Severe change

Fig. 3-2. Patterns of the thalamic depth EEG changes following occlusion of 4 arteries at the base of the brain. EEG changes are classified into 4 patterns

tinct attenuation of EEG voltage and an increase in the slow wave components, it is scored as "moderate change", and when there is complete attenuation with a disappearance of fast wave components, it is scored as "severe change".

Study of the relationship between these levels of EEG activity and the level of rCBF produced the following results. In the "no change" group, the pre-occlusion rCBF level was 32 ± 6 ml/100 g/min, whereas the post-

occlusion level was 28 ± 7 ml/100 g/min (86%). In the "slight change" group, the rCBF decreased from 40 ± 12 ml to 26 ± 12 ml (63%) and in the "moderate change" group, the rCBF fell from 41 ± 14 ml to 14 ± 7 ml (34%). Finally, in the "severe change" group, rCBF decreased from 33 ± 11 ml to 9 ± 3 ml (29%) (**Fig. 3-3** and **Table 3-1**).

Using this canine model, it is apparent that level of rCBF which brought about slight decreases in the

Table 3-1. Relationship between rCBF and EEG pattern after cerebrovascular occlusion

rCBF / EEG	before occlusion (mℓ/100g /min)	after occlusion (mℓ/100g /min)	after occlusion before occlusion (%)
no change	32 ± 6	28 ± 7	86 ± 18
slight change	40 ± 12	26 ± 12	63 ± 17
moderate change	41 ± 14	14 ± 7	34 ± 14
severe change	33 ± 11	9 ± 3	29 ± 12

amplitude of the electrical activity recorded at deep electrodes was approximately 63% that found prior to vascular occlusion. As the EEG changes became more severe, the decreases in rCBF became greater: when there was virtual disappearance of fast-wave components and marked attenuation of the electrical activity, the rCBF was found to have decreased to approximately 30% of the pre-occlusion level.

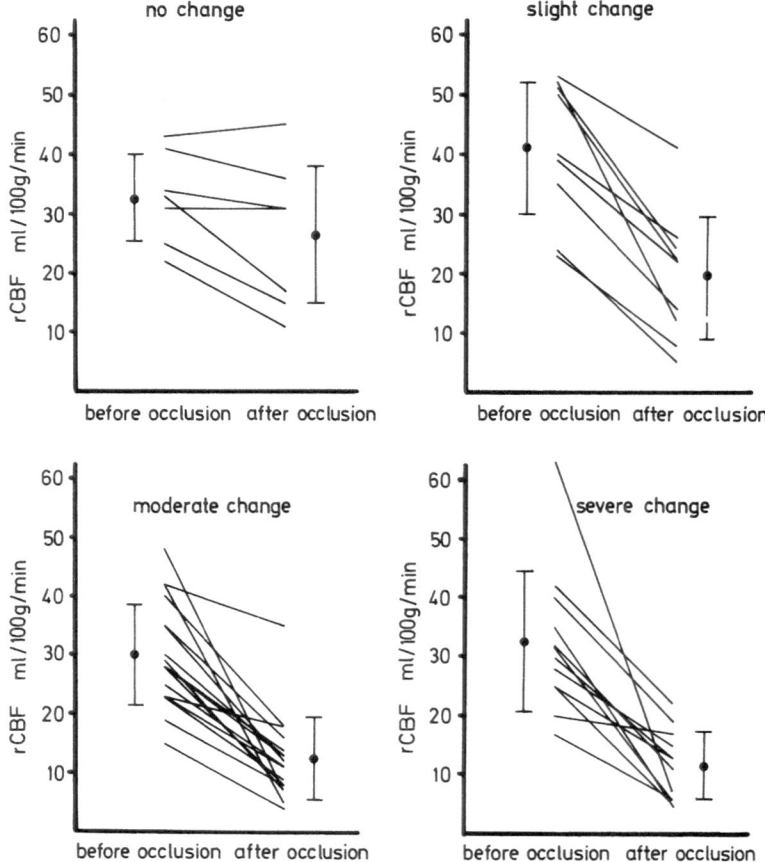

Fig. 3-3. Relationship between patterns of EEG changes and rCBF changes following occlusion of 4 arteries

3.3 Regional Cerebral Hemodynamics Following Recirculation

3.3.1 Hemodynamics at the Center of an Ischemic Focus

Research devoted to elucidation of the pathophysiology of cerebral ischemia using cerebral hemodynamics as an index has progressed considerably at the hands of many workers using various experimental models. A question of central importance to such research is: What are the conditions which make the pathology of cerebral ischemia an irreversible process? That is, what level of ischemia continuing for how long a duration will result in infarction? Furthermore, what kinds of hemodynamics are present following recirculation and what significance do they have? The studies of several research groups have addressed these questions.

Morawetz et al.[25] occluded the MCA for 2–3 hours in nonanesthetized monkeys, allowed recirculation for two weeks, and then studied the autopsied brains. They found a close correlation between the percentage fall in rCBF and the size of the infarctic focus. In addition, they found that the monkeys with high levels of post-recirculation rCBF—showing either a return to pre-occlusion levels or hyperperfusion—also had severe brain infarction. In their experiments, when the rCBF was maintained at 12 ml/100 g/min or lower for 2 hours, infarction was later found.

Using the cat, Tamura et al.[42] performed 2 hour occlusion of the MCA followed by recirculation. They maintain that the maximum amount of rCBF needed to induce brain damage due to recirculation is 12–15 ml/100 g/min. However, the actual level of rCBF following recirculation in the group which suffered severe brain damage varied widely among the animals—from low to high levels. Traupe et al.[43] occluded the MCA of cats for 15–120 minutes and studied the brain hemodynamics following recirculation. Among the animals undergoing 15–30 minutes of occlusion, some showed hyperperfu-

sion during recirculation, but this was transient in all cases and blood flow returned to the pre-occlusion levels. Among the 60 minute occlusion animals, most showed hyperperfusion following recirculation, and eventually hypoperfusion. Although some showed hyperperfusion, most animals undergoing 120 minute ischemia showed hypoperfusion. These workers suggested that the hyperperfusion aggravated the degree of ischemic brain damage.

Using a canine thalamic infarction model, we also have studied the changes in thalamic EEG activity and cerebral hemodynamics at the center of a focus of pathology produced by vascular occlusion and subsequent recirculation. These results were then compared with histological findings. The focus of our work has been the study of the differences between animals which succumb to infarction due to recirculation and those which eventually recover[27, 28].

Only those animals which showed complete attenuation of the thalamic EEG voltage and reduction in fast wave components in the EEG following vascular occlusion were used in the experiments[33]. Study was made of four dogs undergoing 30 minute occlusion, four undergoing 60 minute occlusion and five undergoing 120 minute occlusion. Observations of the thalamic EEG and rCBF were made during occlusion and following recirculation, and histological study of the autopsied brains was made after the completion of the experiment.

Our results were as follows. The thalamic rCBF fell from 22–

44 ml/100 g/min to 8–10 ml, in the four dogs undergoing 30 minute occlusion. Following recirculation, two dogs showed recovery of rCBF to a pre-occlusion level, whereas the other dogs showed transient hyperperfusion. All dogs, however, recovered normal blood flow within one hour from the start of reflow (**Fig. 3-4**).

In the four dogs undergoing 60 minutes of occlusion, rCBF fell from 23–38 ml/100 g/min to 6–10 ml/100 g/min. Following reflow, these values rose to 2–3 times their pre-occlusion levels and remained above normal for 12–24 hours thereafter. The CO_2 response was disturbed after recirculation (**Fig. 3-7**).

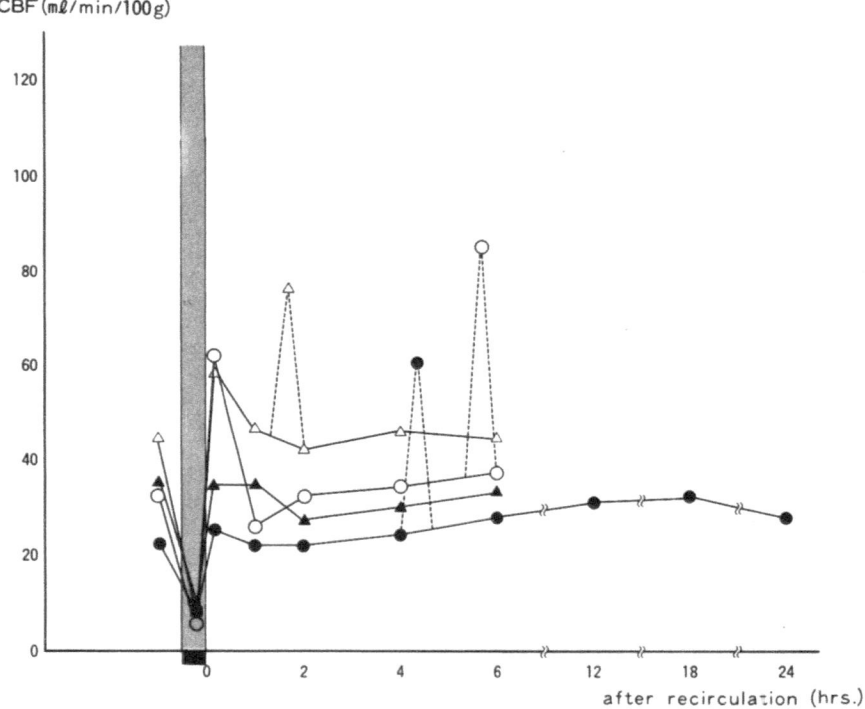

Fig. 3-4. Sequential changes in rCBF following 30 min temporary occlusion of 4 arteries at the base of the brain. (Thin dotted lines indicate CO_2 responses)

In the EEG, there was the appearance of fast wave components and recovery of low voltage activity following recirculation, and after 6–12 hours of reflow, virtually normal EEGs were found (**Fig. 3-5**). Histologically, no dogs showed signs of infarctic foci (**Fig. 3-6**).

No recovery of the EEG was seen following recirculation and there was continued worsening and attenuation of electrical activity (**Fig. 3-8**). Histologically, all dogs had foci of cerebral infarction (**Fig. 3-9**).

Among the five dogs undergoing 120 minutes of occlusion, rCBF de-

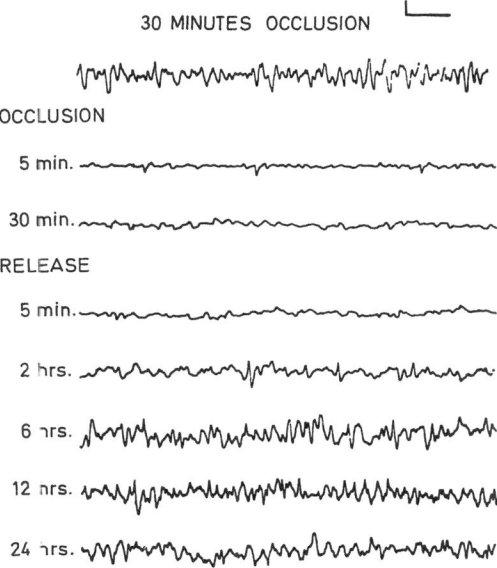

30 MINUTES OCCLUSION

OCCLUSION

5 min.

30 min.

RELEASE

5 min.

2 hrs.

6 hrs.

12 hrs.

24 hrs.

Fig. 3-5. Sequential changes in EEG pattern in the thalamus following 30 min temporary occlusion of 4 arteries at the base of the brain

creased from 23–52 ml/100 g/min to 8–10 ml/100 g/min. Reflow resulted in rCBF values some 2–4 fold greater than the pre-occludion levels. Although there was some gradual decrease thereafter, 6–24 hours later all five digs showed abnormally high rCBF levels. Disturbance in the CO_2 response following reflow was evident in all dogs (**Fig. 3-10**).

Recirculation produced continuous deterioration in the EEG in all dogs, resulting eventually in complete attenuation (**Fig. 3-11**). Histologically, all five dogs showed cerebral infarction (**Fig. 3-12**).

With regard to the degree of cerebral ischemia produced in these experiments, all thirteen dogs showed decreases to less than 1/3 the rCBF values found prior to occlusion, but

after reflow signs of infarction were not seen following 30 minutes of vascular occlusion, but were seen after 60 or 120 minutes of occlusion. These differences in cerebral hemodynamics are thought to be due to the fact that in the 30 minute occlusion group, blood flow returned relatively quickly to its pre-occlusion level, whereas in the animals with 60 or 120 minute occlusion, there were markedly high blood flow values. Although a gradual decrease in blood flow occured thereafter, continued hyperperfusion was evident after 24 hours of reflow. Moreover, while the CO_2 response remained normal in the 30 minute occlusion group, it was disturbed in the other groups.

Tissue acidosis progresses during a state of cerebral ischemia and it is thought that when recirculation is allowed under such conditions, a state of hyperperfusion is likely to follow. In fact, transient hyperperfusion was observed following reflow in two of the animals undergoing 30 minute occlusion. In light of the fact that no disturbance of the CO_2 response was found following recirculation in this group and none of the dogs showed signs of irreversible histological damage, it is thought that the transient hyperperfusion seen in this group was due to a normal vascular response in a state of ischemic acidosis.

On the other hand, it is thought that the hyperperfusion seen in the 60–120 minute occlusion groups have been due to ischemic acidosis, but is more likely to have been caused primarily by disturbance in vascular reactivity (*i.e.*, vasoparalysis) since the hyperperfusion persisted for a lengthy period following

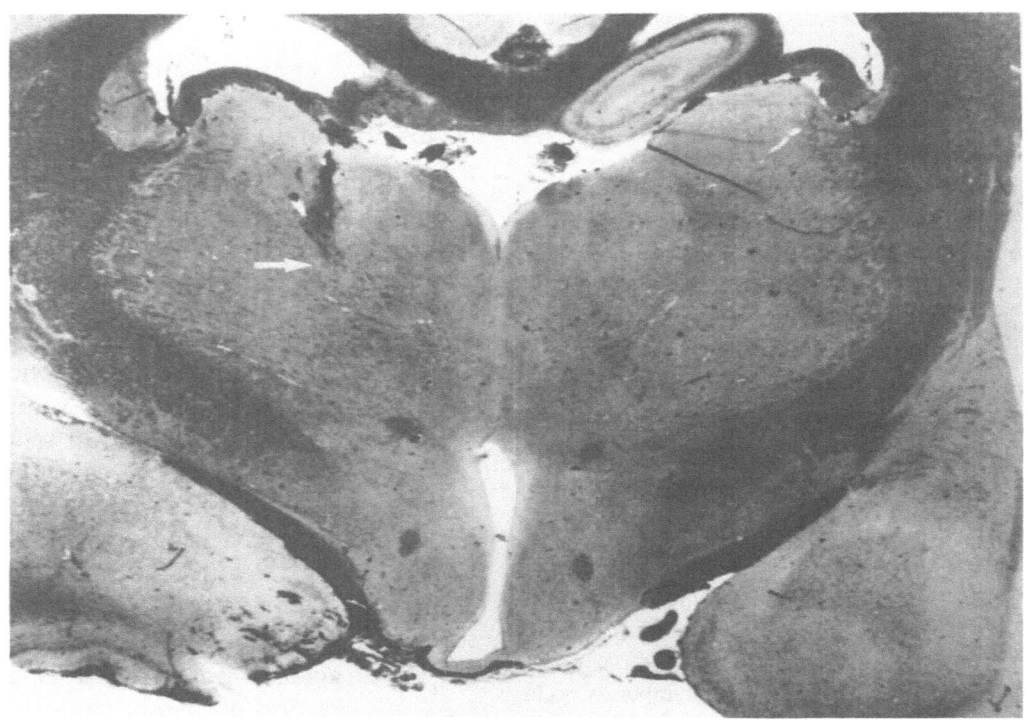

Fig. 3-6. A frontal slice through the anterior part of the thalamus. The case was subjected to temporary occlusion for 30 min. No infarction was observed around the electrode. The arrow marks the location of the tip of the electrode

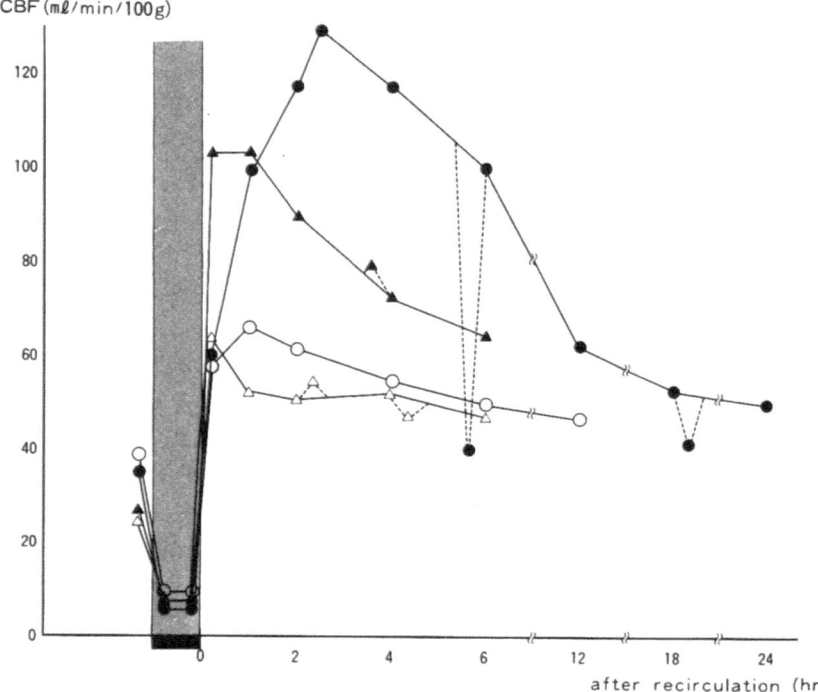

Fig. 3-7. Sequential changes in rCBF following 60 min temporary occlusion of 4 arteries at the base of the brain. (Thin dotted lines indicate CO_2 responses)

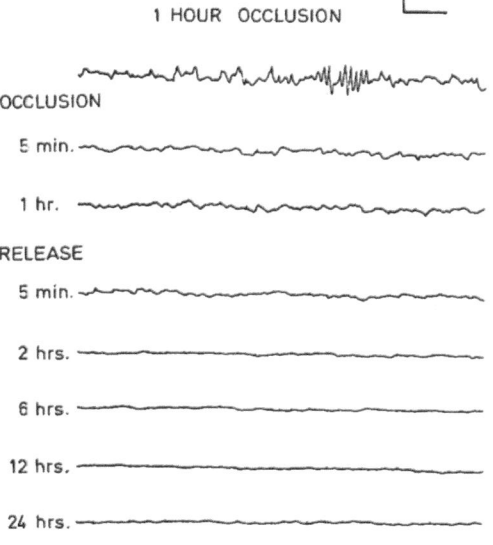

Fig. 3-8. Sequential changes in EEG pattern in the thalamus following 60 min temporary occlusion of 4 arteries at the base of the brain

reflow and there was a disturbance in the CO_2 response. It is also noteworthy that four of the nine dogs showed a peak of blood flow immediately following the release of vascular occlusion. The remaining five animals showed gradual increases in hyperperfusion following reflow with one dog showing increases for 12 hours.

If the vasoparalysis is progressive with no signs of recovery, the rCBF values would be expected to continue to rise, but in the present experiments, rCBF values decreased. In general, increases in tissue pressure are known to produce decreases in rCBF[9, 21]. It is thought that the decrease in hyperperfusion seen in our experiments therefore was related to an increase in tissue

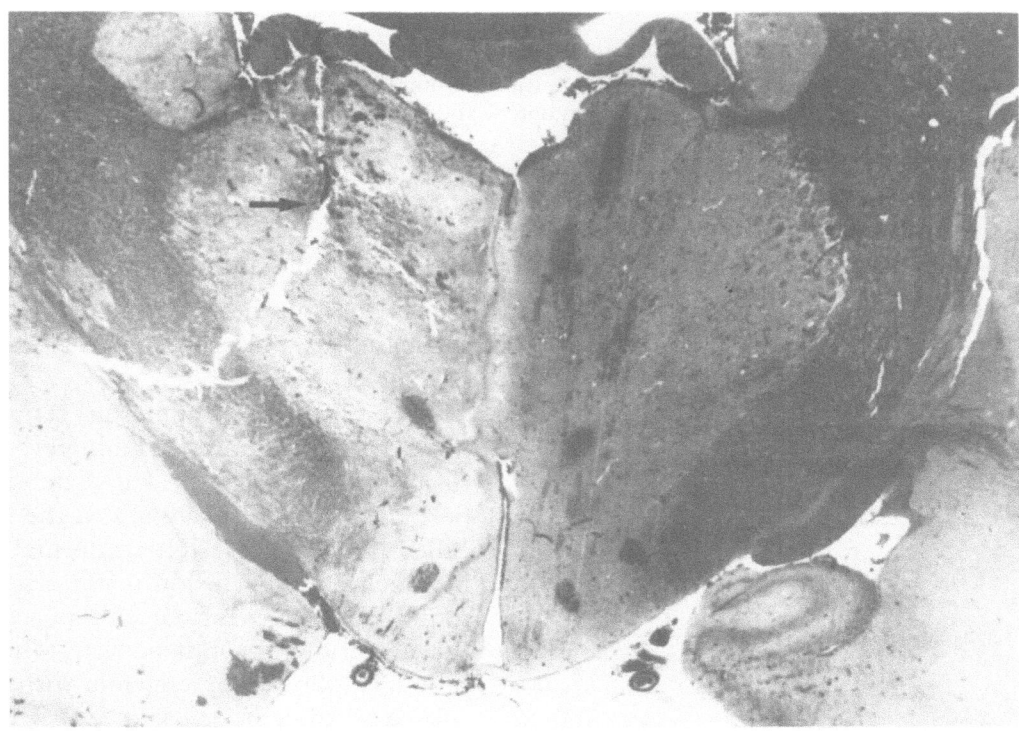

Fig. 3-9. A frontal slice through the anterior part of the thalamus. The case was subjected to temporary occlusion for 60 min. Infarctic focus was observed around the electrode. The arrow marks the location of the tip of the electrode

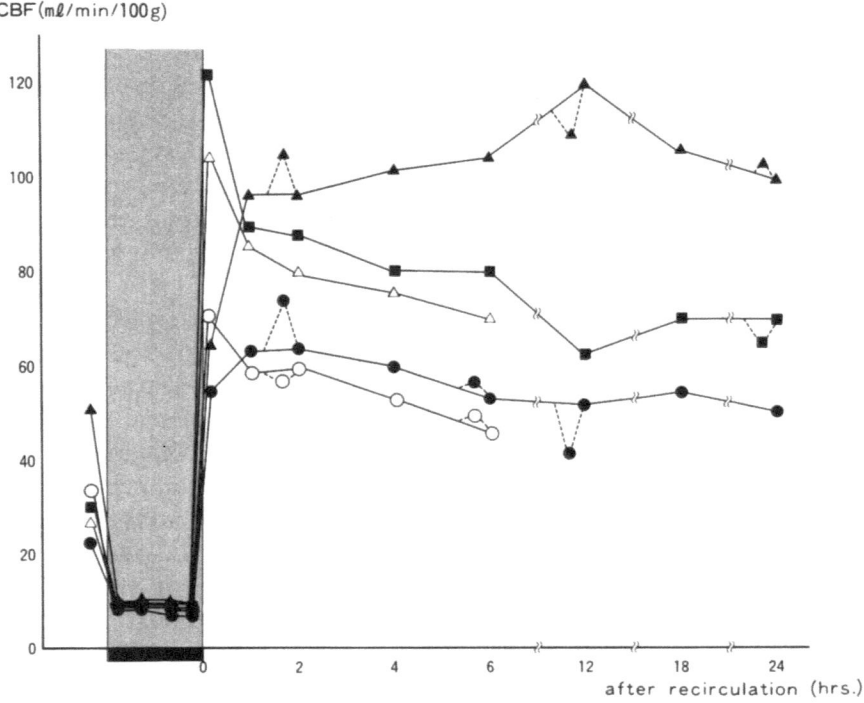

Fig. 3-10. Sequential changes in rCBF following 120 min temporary occlusion of 4 arteries at the base of the brain. (Thin dotted lines indicate CO_2 responses)

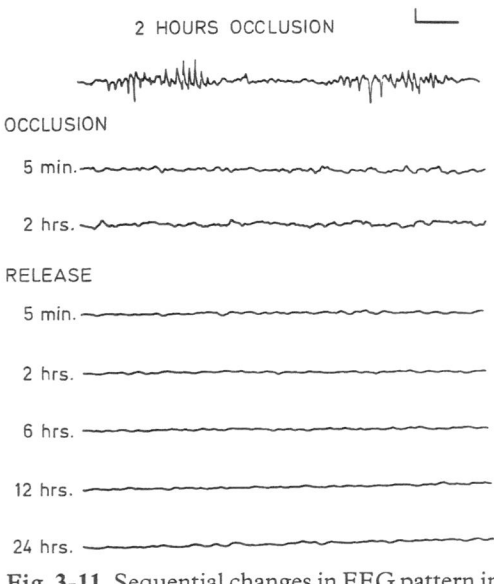

Fig. 3-11. Sequential changes in EEG pattern in the thalamus following 120 min. Temporary occlusion of 4 arteries at the base of the brain

pressure due to the appearance of vasogenic edema. In other words, the time difference between the peaks of hyperperfusion is thought to have been due to the relationship between the degree of vasoparalysis and the degree of tissue pressure.

There have been numerous reports of studies which have found hyperperfusion continuing for an extended period following ischemia, *i.e.*, the so-called "luxury perfusion syndrome" [18] but such hyperperfusion is not always found. There have also been reports of the no-reflow phenomenon [1] in which reflow is followed by oligemia without subsequent hyperperfusion [14]. It is believed, however, that in cases of global ischemia or in cases where the pa

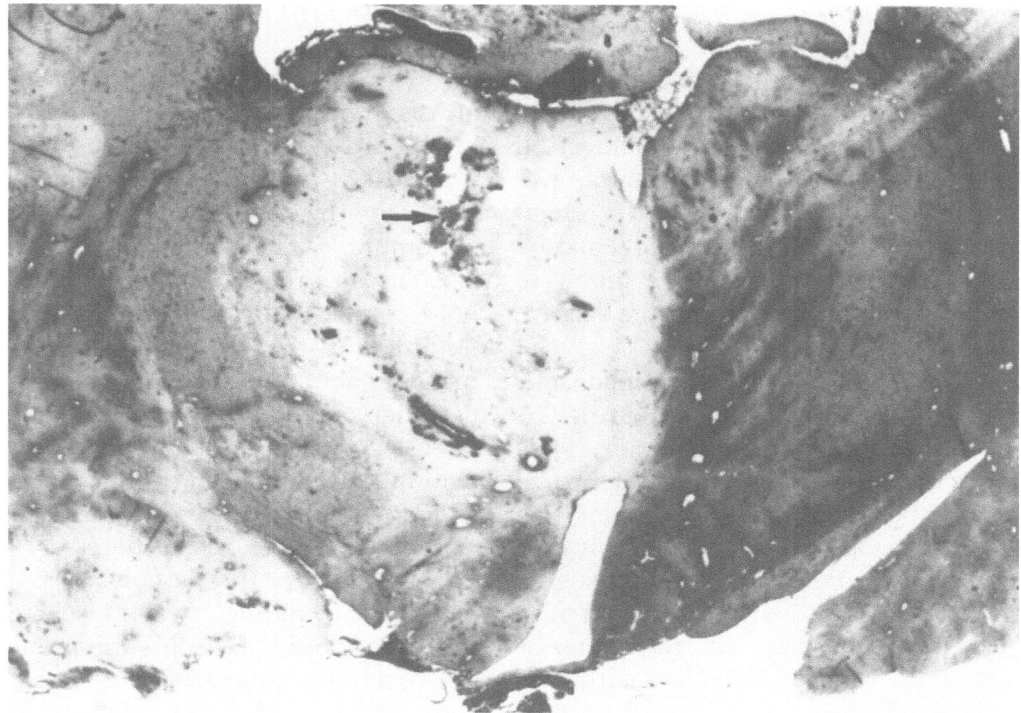

Fig. 3-12. A frontal slice through the anterior part of the thalamus. The case was subjected to temporary occlusion for 120 min. Infarctic focus was observed around the electrode. The arrow marks the location of the tip of the electrode

thological focus in focal ischemia is large, increases of the intracranial pressure may suppress hyperperfusion. With the present experimental model, however, the pathological focus is small and confined to a specific location, thus even if there were local increases in the tissue pressure, no clear signs of brain swelling would be evident and there would be no significant changes in intracranial pressure. It is consequently thought that, using this technique, the cerebral hemodynamics observed in the process of cerebral infarction can be seen in a relatively pure form, without the contamination of secondary effects brought about by increases in intracranial pressure.

We have discussed the two varieties of hyperperfusion which can appear at the center of an ischemic focus due to reflow, but it is also known that hyperperfusion occurs in the periphery of an ischemic focus. For the study of such peripheral hemodynamics, a different experimental approach is required. Discussion of the changes in hemodynamics occurring at the center of a focus of cerebral infarction, in its border region and in the immediate periphery of the focus will be carried out in the following section.

3.3.2 Hemodynamics of the Border and Periphery of an Ischemic Focus

Among the cases of cerebral infarction encountered daily, almost all are examples of focal infarction—regardless of their extent or the size of the focus. As a consequence, the nature of the cerebral pathology is likely to differ depending upon the site relative to the center of the focus.

Symon *et al.*[40] have defined four zones of infarction using an MCA occlusion model in the baboon. They are: (i) the Sylvian opercular region, (ii) a neighboring intermediate zone, (iii) a parieto-occipital region in the watershed between the middle and posterior cerebral arterial distributions, and (iv) a parasagittal region at the anterior cerebral arterial distribution. Although they studied the rCBF and vascular reactivity of each region, hyperperfusion was not found at any site.

Using an MCA occlusion model in the monkey, Jones *et al.*[15] using the hydrogen clearance method, placed five electrodes at 2.5–3.0 mm intervals into the region between the globus pallidum and the subcortex to making rCBF measurements. They examined the relationship between the degree of ischemia and the length of occlusion at each site and the formation of infarctic foci. They reported having been found hyperperfusion at all such sites following release of the occlusion.

Among reports of clinical cases as well, there are several studies in which hyperperfusion was found. Olsen *et al.*[30] studied the rCBF of 41 cases of acute cerebral infarction. Sixteen cases showed local hyperemia and, notably, all cases with cortical infarction exhibited border zone hyperemia.

In 1981, Astrup, Seijo and Symon[2] published their concept of the ischemic penumbra. They emphasized the existence of a region between the threshold of rCBF bringing about energy failure and the threshold of rCBF bringing about electrical failure in the periphery of an ischemic focus. Subsequently, several papers on the ischemic penumbra have been published[11, 37] and, at the same time, the significance of reflow has been reconsidered.

In this way, the pathology of the regions surrounding a focus of ischemia has come to be viewed with importance, but it should be noted that already in 1962, Meyer *et al.*[23] had demonstrated the existence of three distinct pathological conditions from experimental investigation of embolization using the monkey. That is, they observed that (i) there was a notable decrease in cerebral blood flow as well as disturbed energy metabolism at the center of an ischemic focus, (ii) there was a neighboring middle region where a moderate decrease in cerebral blood flow was found, but metabolic activity was maintained, and (iii) there was a peripheral region in which a state of hyperemia was found due to the diffusion of CO_2 gas which had accumulated in the middle region.

We also have performed experiments to elucidate the nature of the pathology in the surrounding region and have made observations of rCBF changes, vascular reactivity, EEG changes and histological findings[27, 29].

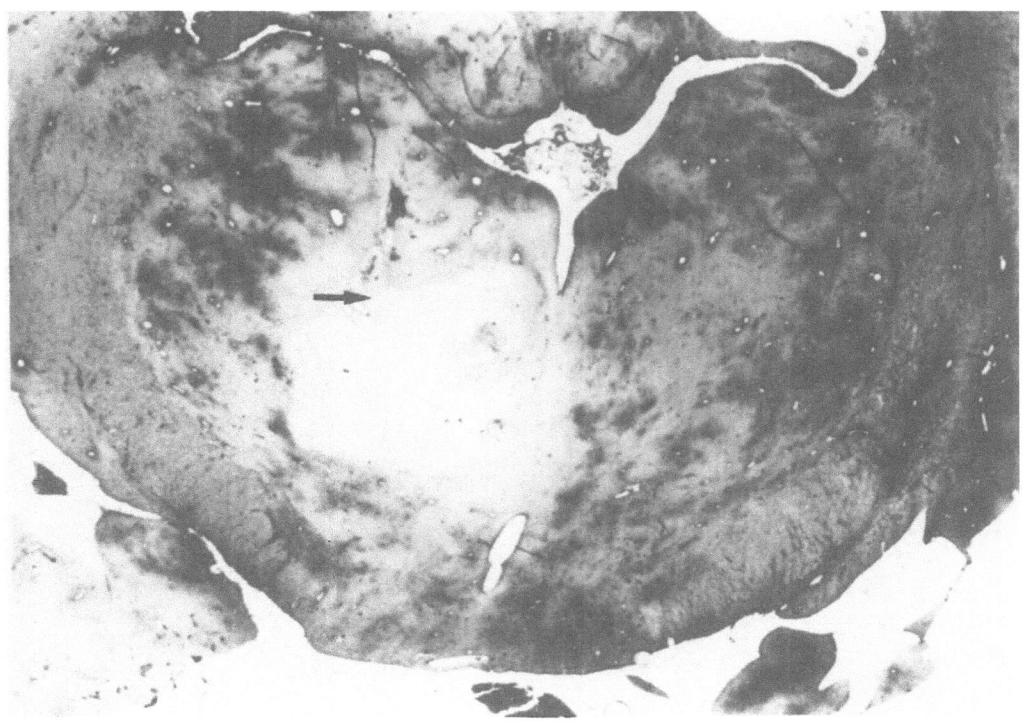

Fig. 3-13. A frontal slice through the anterior part of thalamus. The electrode was inserted into the center of the infarctic focus. The arrow marks the location of the tip of the electrode

In our canine thalamic infarction model, the focus of pathology is a region at the center of the nucleus ventralis anterior of the thalamus. The focus has a diameter of less than 10 mm[49].

Taking advantage of this feature and the hydrogen clearance technique for measuring rCBF, we inserted two electrodes 4 mm apart into the center and into the border region of the ischemic focus. By means of retrospective histological examination, we later confirm that the anterior electrode was indeed within the center of the focus (Fig. 3-12). In two animals, the posterior electrodes were 1–1.5 mm medial of the outer border of the focus (Fig. 3-13), while in two others they were 1–1.5 mm outside of the outer border (Fig. 3–14). Therefore, it was possible to examine the relationship between the infarctic focus and the three sites—*i.e.*, the center of the focus, within its outer border and just outside its outer border. In the animals which showed a decrease in rCBF to about 10 ml/100 g/min at the anterior electrode due to vascular occlusion, recirculation was allowed after two hours of occlusion. Study was then made of the changes in rCBF and EEG activity during occlusion and following reflow at the above-mentioned three regions of the brain.

It was found that the rCBF at the

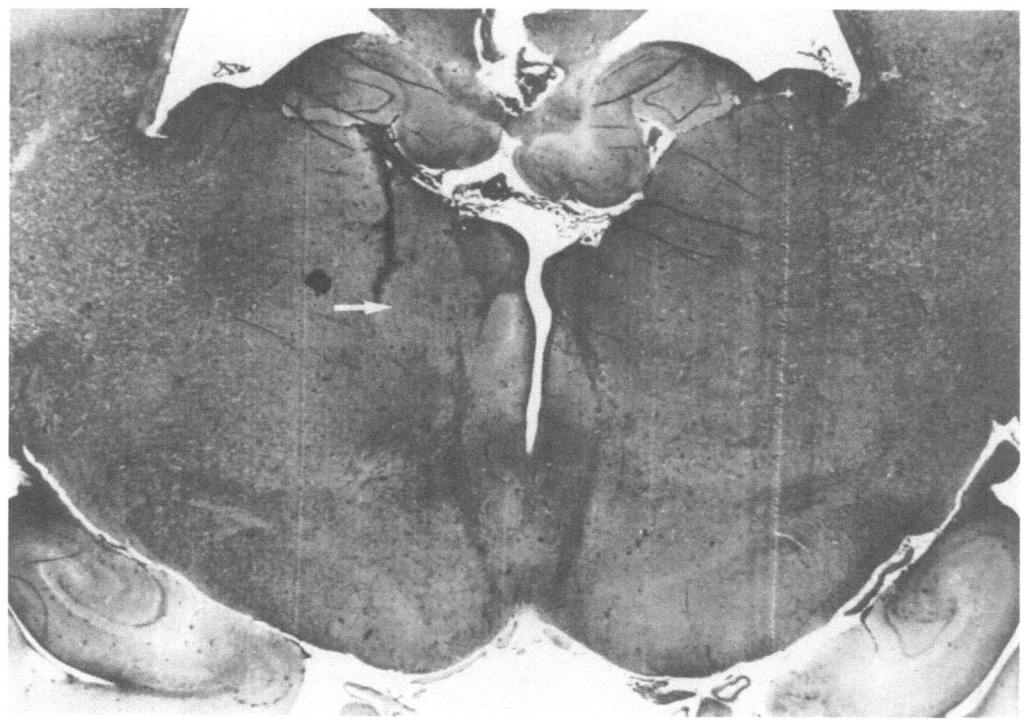

Fig. 3-14. A frontal slice through the posterior part of the thalamus. The electrode was inserted into the peripheral portion surrounding the infarctic focus, where infarctic change was observed. The arrow marks the location of the tip of the electrode

center of the focus in four dogs fell from between 30 and 52 to between 8 and 12 ml/100 g/min due to the vascular occlusion. During the occlusion, there was virtually no change in the rCBF, but following reflow, rCBF rose to high values between 60 and 106 ml. Thereafter, there was an trend toward gradual decline, but high values were still obtained after 6 and 24 hours of recirculation. The CO_2 response following reflow was disturbed in all animals (**Fig. 3-10**). Starting with such changes in cerebral hemodynamics, there then followed EEG changes (**Fig. 3-11**), and histological changes (**Fig. 3-12**) similar to those found at the

center of the focus in the animals subjected to 120 minutes of occlusion.

The changes in hemodynamics within the border region were slight in two animals. The pre-occlusion rCBF values of 31 and 35 ml/100 g/min did not decreases, substantially due to the occlusion, but following recirculation the rCBF increased to 66 and 78 ml, respectively—roughly a 2-fold increase. Although there were animals which showed a gradual decrease after reaching a peak rCBF levels one hour after reflow, abnormally high levels were maintained for 6–24 hours. The CO_2 response following recirculation was already disturbed after two hours

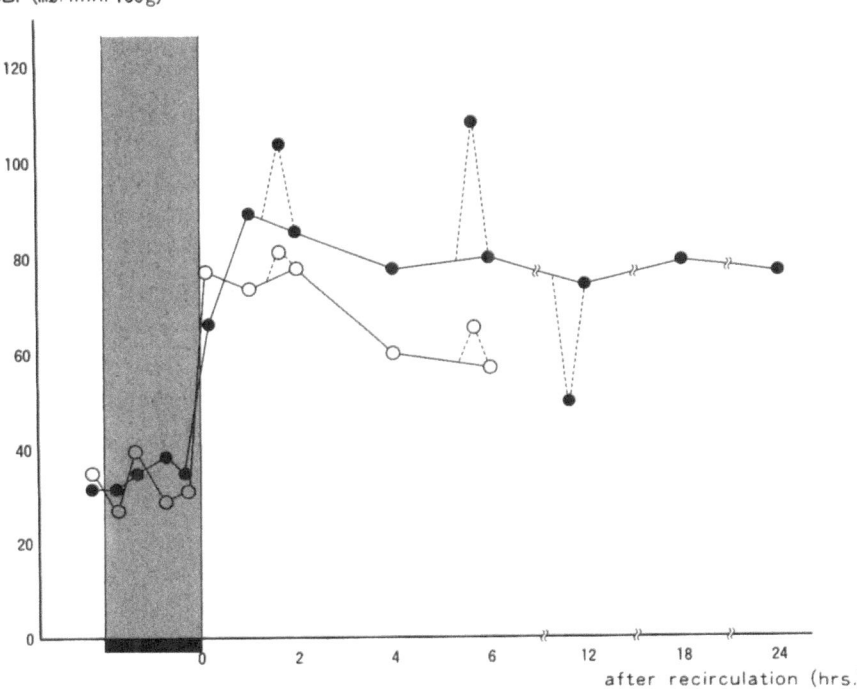

CBF (ml/min/100g)

after recirculation (hrs.)

Fig. 3-15. Sequential changes in rCBF in the peripheral portion of the infarctic focus following 120 min temporary occlusion of the arteries (dotted line indicates CO_2 responses)

in one dog, but although no notable disturbance was apparent in the other dog after 2–6 hours, a paradoxical response was found after 12 hours (**Fig. 3-15**).

In the EEG, fast-wave components were reduced immediately following occlusion and there was gradual appearance of slow-wave activity in both dogs. Although the reappearance of fast-wave components was evident five minutes after the start of reflow, both animals showed greatly attenuated electrical activity after two hours (**Fig. 3-16**).

The rCBF values prior to occlusion in the two dogs with electrodes in the immediate periphery of the focus were 34 and 25 ml/100 g/min. After 30, 60,

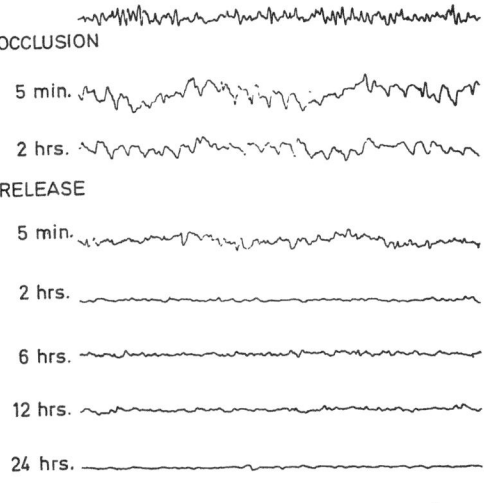

OCCLUSION

5 min.

2 hrs.

RELEASE

5 min.

2 hrs.

6 hrs.

12 hrs.

24 hrs.

Fig. 3-16. Sequential changes in EEG pattern in the peripheral portion of the infarctic focus following 120 min temporary occlusion of the arteries

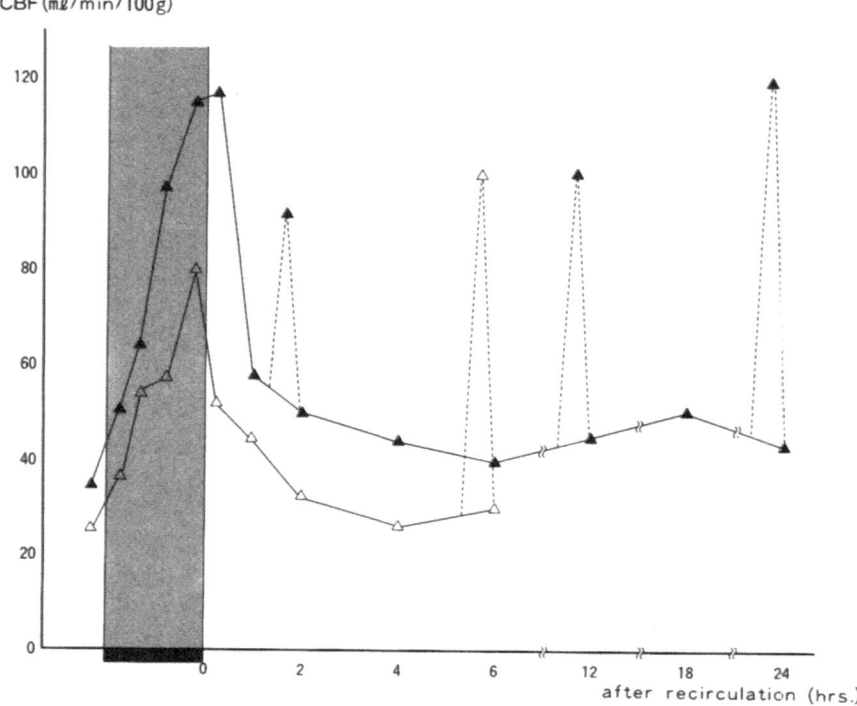

Fig. 3-17. Sequential changes in rCBF in the portion surrounding the infarctic focus following 120 min temporary occlusion of the arteries (dotted line indicate CO_2 responses)

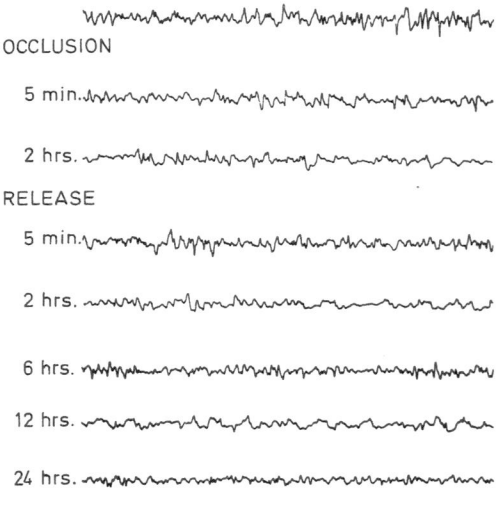

Fig. 3-18. Sequential changes in EEG pattern in the portion surrounding the infarctic focus following 120 min temporary occlusion of the arteries

90, and 110 minutes of occlusion, one dog had rCBF values of 52, 64, 98, and 116 ml/100 g/min, whereas the other dog had values of 36, 54, 56, and 80 ml/100 g/min, respectively. In other words, both showed sequential increases during occlusion. After the start of reflow, however, both animals showed rCBF decreases until two hours of reflow had been allowed—when the rCBF returned to 55 and 33 ml—*i.e.*, nearly to the pre-occlusion levels. Thereafter, these levels were maintained. There was no obvious disturbance of the CO_2 response following recirculation (**Fig. 3-17**). In the EEG, vascular occlusion produced a slight attenuation of voltage and a slight reduction in fast wave components. Following recirculation, the mild attenuation of voltage was maintained, but there was a complete recovery of fast wave components (**Fig. 3-18**).

As discussed earlier, the pathology at the center of the focus was characterized by hyperperfusion appearing after recirculation, which was due to vasoparalysis, which—in turn—is thought to result in luxury perfusion, as suggested by the histological damage[28].

In the border regions, however, marked ischemia similar to that seen during occlusion at the central region was not found. Since there were EEG changes immediately following the vascular occlusion, it is thought that the border region suffers a decrease in tissue O_2 pressure and a disturbance of cerebral metabolism. This eventually leads to the accumulation of CO_2 and acidic metabolites. As a consequence, it is believed that there is severe vasodilatation of vessels which have succumbed to vasoparalysis[23], similar to that of the vessels at the center of the focus.

Investigation of the significance of recirculation at the peripheral region indicates that recirculation causes worsening of the cellular damage at this site. This is suggested by the fact that, although there is a transient reappearance of fastwave components after five minutes of recirculation, there is eventual attenuation of the EEG and sequential worsening of the CO_2 response following recirculation.

In the region just outside the outer region of the ischemic focus, there is a gradually progressing hyperperfusion during vascular occlusion. This is thought to be due to the diffusion to the surrounding brain tissue of CO_2 gas which has accumulated within the ischemic focus[22, 23] and to represent a normal cerebrovascular response to these substances. In other words, this hyperperfusion is believed to be a normal vascular response in light of the fact that it gradually returned to normal together with the continuation of recirculation and there after no disturbance of the CO_2 response was found.

Vascular reactivity is particularly important for elucidating the pathology of these states, and this problem will be discussed more fully in Section IV. In any case, in the present experiments, the CO_2 response was disturbed at sites where, histologically, the formation of an infarctic focus was seen and the CO_2 response was maintained at sites where infarction did not appear. A correlation was also found between the disturbance of the CO_2 response after recirculation and cellular damage.

Finally, with regard to the relationship between rCBF and EEG findings, it can be said that in the border region, there were marked changes in the EEG despite the fact that there were no changes in rCBF due to vascular occlusion. At the region just outside the focus, no changes in the EEG were found, despite the fact that there was marked hyperpefusion. Moreover, at both the center of the focus and at the border region, following recirculation there was complete attenuation of the EEG despite the fact that there was severe hyperperfusion. These findings clearly indicate the uncoupling of rCBF and EEG activity. Already in Section II we have reported that, at the center of an ischemic focus, there is a correlation between the EEG changes immediately following vascular occlusion and the decrease in rCBF. It is important to note, however, that such correlations between rCBF and EEG measurements are found only at the center of an ischemic focus during vascular occlusion.

As seen from the above, it is evident that differing pathological processes are at work of various sites in and around of focus of ischemic pathology, but each of these processes is likely to have influences on the others. Furthermore, various factors, such as the severity of the ischemia at the center of the focus, its duration, and the presence or absence of reflow, are likely to be involved in complex permutations in the formation of focal cerebral infarction.

3.4 Sequential Changes in Vascular Reactivities

It is well known that autoregulation and the CO_2 response are heavily involved in the regulation of cerebral hemodynamics and that in various pathological conditions these regulatory mechanisms become disturbed. Clinically, findings concerning vascular reactivity are used in determining the indication of vascular reconstruction and are applied to inducing hypertension or hypercapnia therapy in cases of ischemic pathology. For these and other reasons, the state of vascular reactivity has come to be viewed with importance.

There have also been a number of experimental reports on findings concerning the vascular reactivity in cerebral ischemia, but the majority of such papers have dealt only with the most general features of such reactivity. Few have been concerned with the change in vascular reactivity during the development of the ischemia or with temporal changes in the vascular state. In addition, there are as yet no convincing hypotheses concerning the meaning of the deficits in such reactivity in brain disorders.

For these reasons, we have studied the autoregulation and CO_2 response through each stage of cerebral ischemia and made quantitative observations of the cerebral hemodynamics using the canine thalamic infarction model. We have also investigated the changes in autoregulation in the region peripheral to the ischemic focus.

Since it is believed that the nature of autoregulation and the nature of the CO_2 response are fundamentally different, they are frequently viewed as two independent phenomena in pathological conditions. The following discussion will consequently deal with autoregulation and the CO_2 response separately.

3.4.1 Autoregulation

In reports of clinical cases of cerebral ischemia, nearly all such patients show disturbance of autoregulation, although most of this disturbance is local. The disturbance is usually detected within several days of the ischemia attack.

Using a canine MCA occlusion

model, Kogure *et al.*[16] found that, although autoregulation was maintained immediately following vascular occlusion, disturbance emerged after 4 hours. Waltz *et al.*[46] reported that, using a feline MCA occlusion model, rCBF values increase together with the mean systemic blood pressure up to 120 mmHg, but, above that level, rCBF decreases.

Using a baboon MCA occlusion model, Symon *et al.*[41] investigated autoregulation at three sites (at the center of a focus of severe ischemia, at a nearby border region of moderate ischemia, and at a peripheral region of mild ischemia) following either an increase in blood pressure due to aramine administration or a decrease due to exsanguination. In the acute stage, autoregulation in response to the hypertension was maintained at the middle and border regions and disturbed at the central focus, but in response to the induced hypotension there was disturbance in autoregulation at all three sites. They also found that, in the chronic stage, disturbances in autoregulation confined to those sites which eventually showed signs of infarction. In addition, they noted a positive correlation between the degree of the ischemia and the severity of the disturbance of autoregulation.

In order to study the changes in autoregulation with occur within a limited focus of cerebral ischemia, we first examined the changes in autoregulation in relation to the degree of ischmia at the center of an ischemic focus. Next, we made comparisons of the changes in autoregulation which occur at the three sites having differing hemodynamics[32] (as defined above, the center, border and peripheral regions of an ischemic focus).

3.4.1.1 Sequential Change in Autoregulation Occurring at the Center of an Ischemic Focus

In this model, an electrode is inserted into the anterior half of the thalamus at the center of an ischemic focus and the experimental animals are divided into three groups depending upon the level of the rCBF following vascular occlusion[33]. Specifically, there were seven dogs in the severe ischemia group (rCBF reduced to less than 40% of the pre-occlusion level), six in the moderate ischemia group (40–70%) and five in the mild ischemia group (> 70%). Autoregulation was investigated in each animal 1, 3, and 5 hours following the occlusion and the autoregulation index ($\triangle$ rCBF/$\triangle$ mBP) was calculated following the administration of angiotensin II (hypertensin ®) to induce an increase in mBP of approximately 30 mmHg.

It was found that, in the severe ischemia group, the reduction in rCBF due to occlusion was from 33.1 ± 9.6 to 9.9 ± 2.1 ml/100 g/min. In the moderate ischemia group, it was from 34.0 ± 10.3 to 16.9 ± 4.6 ml/100 g/min and in the mild ischemia group from 29.6 ± 7.2 to 22.1 ± 4.2 ml/100 g/min (**Fig. 3-19**). The changes in the autoregulation index in each group were as follows. In the severe ischemia group, the indices prior to the occlusion and after 1, 3, and 5 hours of occlusion were 0.01 ± 0.05, 0.05 ± 0.06, 0.04 ± 0.05 and 0.03 ± 0.07 respectively. In the moderate ischemia group, these values

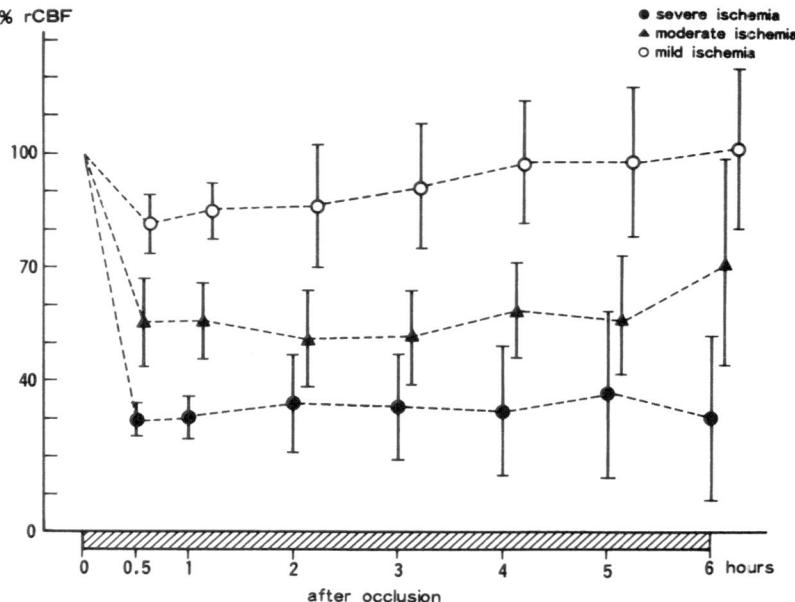

Fig. 3-19. Sequential changes in rCBF in the center of ischemic focus

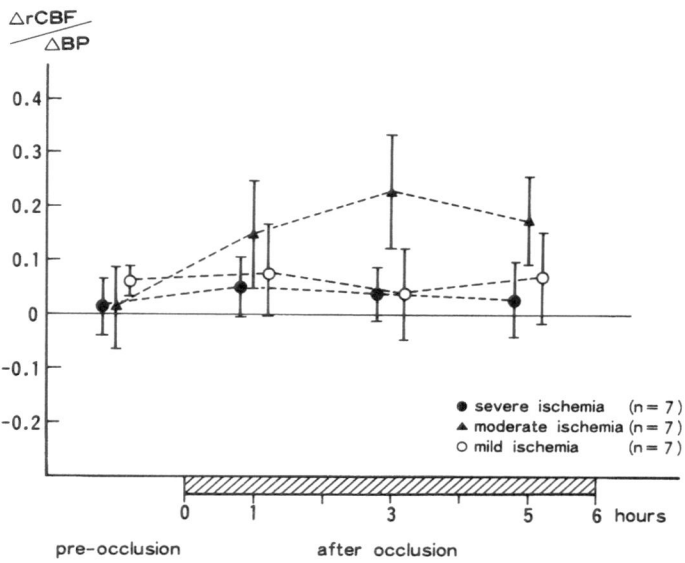

Fig. 3-20. Sequential changes in autoregulation in the center of ischemic focus

were 0.02 ± 0.09, 0.11 ± 0.07, 0.22 ± 0.12 and 0.18 ± 0.09. And in the mild ischemia group, they were 0.05 ± 0.04, 0.04 ± 0.02, 0.06 ± 0.10 and 0.08 ± 0.11 (**Fig. 3-20**).

In the group in which rCBF levels remained at > 70% of their pre-occlusion levels, no disturbance in autoregulation was observed, but in the group with moderate ischmia (40–70% of the pre-occlusion levels) disturbance in autoregulation was seen after one hour of occlusion, suggesting that disturbance will arise when a certain level of severity of ischemia is reached. In severe ischemia (< 40% of the pre-occlusion rCBF levels), however, there were no increases in rCBF following the induction of hypertension, and autoregulation was maintained. These findings are in striking contrast to those of earlier studies by several research groups[41, 46].

It is generally thought that disturbances in autoregulation at foci of cerebral ischemia are due to vasoparalysis caused by the cellular damage of ischemic anoxia. Theoretically, however, this mechanism would imply that the severity of the vasoparalysis would increase as the ischemia progressed from moderate to severe. In fact, though, as will be discussed below, the CO_2 response in severe ischemia using this model was disturbed and, although there was vasoparalysis, Miller et al.[24] labeled the phenomenon of no increase in rCBF in response to induced hypertension "false autoregulation", as distinct from autoregulation in the normal brain. With regard to the mechanisms involved, it has been argued that there is no change in the cerebral perfusion pressure due to an increase in intracranial tension caused by systemic hypertension, and it has also been argued that there is an increase in local tissue pressure due to the induced hypertension. It is currently unclear how the false autoregulation found using this model should be interpreted, but the following three hypotheses are worthy of consideration.

First, it is possible that the collateral circulation in this model occurs not via leptomeningeal anastomosis, but via the intracortical capillary network and that, in the severe ischemia group, these collateral pathways are extremely scarce. Consequently, even when there is an increase in systemic blood pressure, there would be no increase in the perfusion pressure in the brain. According to this mechanism as discussed below, although an increase in rCBF would occur due to the administration of mannitol in the moderate ischemia group, there would be almost no increase in the severe ischemia group[35].

Secondly, it may be considered that, as indicated in the section on histology (Chapter 2), in the severe ischemia group, there may be the appearance of histological damage to the vascular system, such as bleb formation and swelling of the capillaries within the ischemic focus, together with prolongation of the ischemia[47]. As a consequence, following an increase in tissue pressure due to the appearance of vasogenic edema, the perfusion would not increase with the hypertensive load and neither would the rCBF increase.

Thirdly, it is possible that the collateral pathways which enter and leave

76

3. Cerebral Blood Flow

the ischemic focus themselves succumbed to dysautoregulation and, as a consequence, blood flow is stolen by the tissue surrounding the ischemic focus due to the induced hypertension.

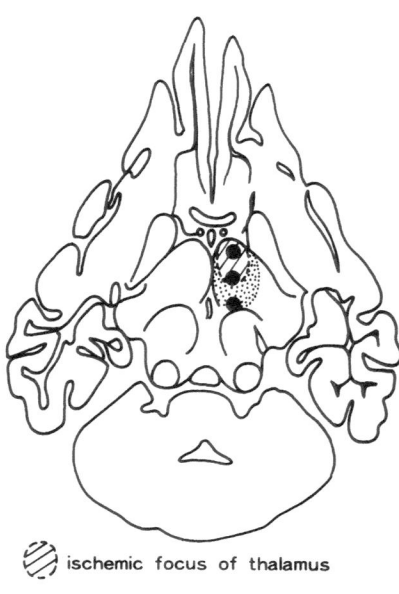

ischemic focus of thalamus

● electrode of rCBF & EEG

Fig. 3-21. Diagram of a horizontal slice of our experimental model

Whichever the case may be, it is apparent that for the elucidation of the mechanisms involved, further research on the vascular reactivity and microcirculation in and around the ischemic focus will be required using both histological and biochemical techniques.

3.4.1.2 Sequential Changes in Autoregulation at the Border and in the Periphery of an Ischemic Focus

In order to observed the pathology surrounding a focus of ischemia, three groups of electrodes were used. The first (anterior) electrode was stereotaxically inserted the nucleus ventralis anterior of the right side of the thalamus, the second (middle) was inserted 3 mm posterior to the first and the third (posterior) was placed another 3 mm posteriorly (**Fig. 3-21**). Using only dogs showing severe ischemia at the anterior electrode, the changes in rCBF and autoregulation were measured at the sites of the middle and posterior electrodes (**Fig. 3-22**).

By means of an autopsy study done 6 hours after vascular occlusion, it was confirmed that all five dogs had infarctic foci and all five animals had the tip of the anterior electrode in the center of the infarctic focus. The site of the middle electrode was determined retrospectively to be located either within the outer border of the focus (border[29] four dogs) or outside the outer border (periphery[29] outside one dog). All of the posterior electrodes were located at peripheral sites. A total of five dogs with posterior or middle electrodes in the peripheral region of the ischemic foci were chosen and the results were taken together as representative of the hemodynamics in the peripheral region.

The results were as follows. The autoregulation indices at the center of the focus prior to occlusion and 1, 3, and 5 hours after occlusion were -0.02 ± 0.05, 0.02 ± 0.05, 0.04 ± 0.04 and 0.02 ± 0.07 respectively. In the border region, these values were -0.04 ± 0.07, 0.09 ± 0.10, 0.17 ± 0.12 and 0.17 ± 0.12. In the periphery, were 0 ± 0.05, -0.01 ± 0.21, -0.02 ± 0.18 and 0.11 ± 0.11 (**Fig. 3-23**).

In this experiment, a state of false

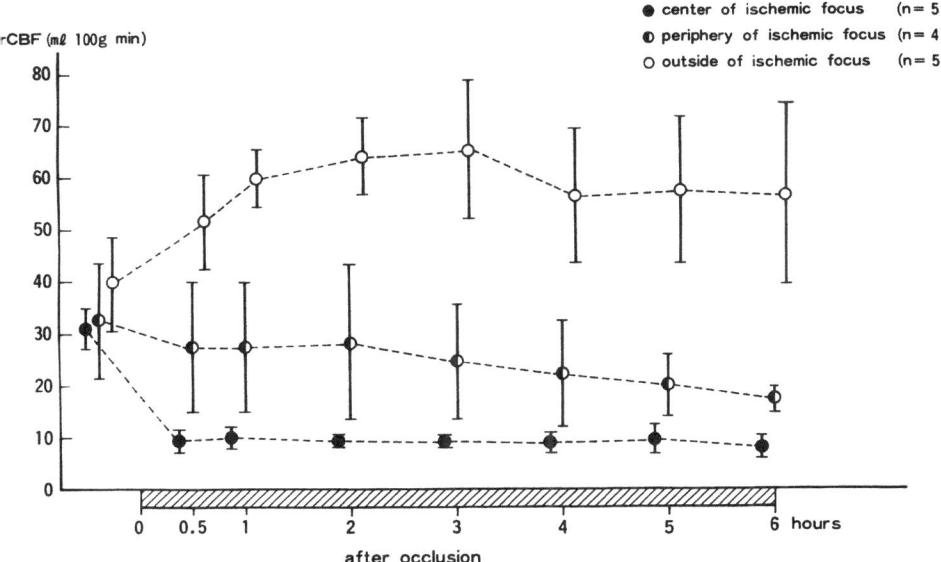

Fig. 3-22. Sequential changes in rCBF in and around the infarctic focus with severe ischemia

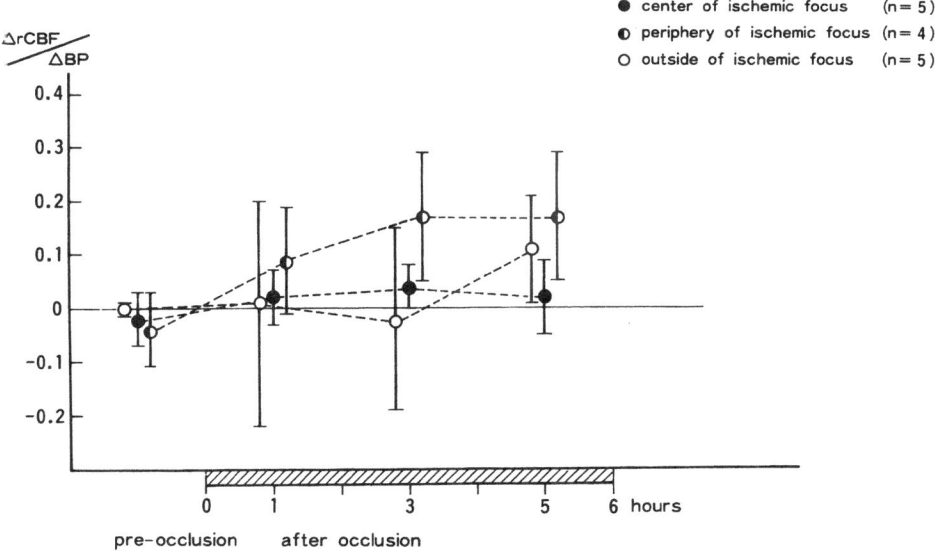

Fig. 3-23. Sequential changes in autoregulation in and around the infarctic focus with severe ischemia

autoregulation was seen at the center of the ischemic focus, as described above, but in the border region after one hour of occlusion disturbance of autoregulation was seen which worsened after 3 to 5 hours. The mechanisms involved are thought to be as follows. Similar to what occurs at the center of the focus, there is accumulation of acidic metabolites due to ischemic anoxia at the border region and this produces a state of vasoparalysis. There is, however, a greater abundance of collateral pathways to the border region than to the center of the focus, which means that the perfusion pressure more easily increases in response to an increase in the systemic blood pressure. The rCBF therefore increases and this is thought to lead to sequential worsening of the vasoparalysis at the border region.

Although autoregulation was maintained in the peripheral region, there were also animals in which rCBF decreased following an increase in blood pressure. This process is believed to be similar to the so-called paradoxical response seen during CO_2 loading. That is, the blood flow in the peripheral region where autoregulation is maintained is stolen by the border region, which is in a state of dysautoregulation due to increases in blood pressure.

As pointed out in the previous section, the border region succumbs to infarction in the model where two

hours of vascular occlusion is followed by reflow. The infarction is thought to indicate that the hyperperfusion due to reflow produces cellular damage. In the present experiment, using a six hour occlusion model without reflow, however, despite the fact that the decrease in rCBF was mild, there was the development of infarction. As discussed in the following section, since it is thought that the rCBF threshold for causing irreversible changes at the center of the focus after six hours of occlusion is approximately 50% of the pre-occlusion rCBF level[36], it is impossible to explain the pathology of the border region solely in terms of the fall in rCBF. The results of the experiments described in this section suggest that the irreversible cellular damage is brought about by the continuation of the vasoparalysis beyond a certain threshold period.

A search of the literature by us has produced only one previous experimental study on the vascular reactivity in the peripheral region surrounding a focus of cerebral ischemia and a sequential study of the hemodynamics was not undertaken (the study of Symon et al.[41], described above). Many unanswered question therefore remain concerning the hemodynamics of the border and peripheral regions of an ischemic focus, and further experimental research is required.

3.4.2 CO₂ Response

Some studies in the past have dealt with the CO_2 response during brain ischemia. Using an experimental model of middle cerebral artery occlusion in

the baboon, Symon et al.[40] found a disturbed CO_2 response in the vicinity of the ischemia focus produced by the occlusion. They observed that in regions of severe ischemia, CO_2 loading produced a paradoxical response, where rCBF actually decreased. Waltz[45] measured the CO_2 response using an experimental model involving ligation of the middle cerebral artery in the cat, and showed the responses in the ischemic cortex were disturbed or paradoxical. Kogure *et al.*[16] ligated the middle cerebral artery of dogs and found an increase in rCBF due to CO_2 loading—*i.e.*, no paradoxical response.

Thus far, most reports on the CO_2 response following recirculation have been concerned with models in which "global ischemia" is produced. Hossmann *et al.*[12] produced global ischemia of 1 hour duration in cats and Nemoto *et al.*[26] produced ischemic foci of 15 min duration in dogs. Both groups found disturbance in the CO_2 response measured over a few hours following recirculation.

These differences are not thought to be due solely to differences of opinion among various research groups, but rather to various differences in methodology, as pointed out by Waltz[45]. In fact, there have been no studies of sequential changes in the CO_2 response after different grades of ischemia. In the present study[34], we have investigated the relationship between impaired CO_2 reactivity and the severity and duration of experimentally induced cerebral ischemia. We have also studied the sequential changes in the CO_2 response, effects on the CO_2 response due to recirculation, and the relationship between the development of infarction and changes in the CO_2 response.

Similar to the experiments in autoregulation, the animals were divided into three groups according to the severity of the ischemia: severe ischemia (where rCBF fell to 40% of the pre-occlusion level), moderate ischemia (where rCBF fell to 40–70%), and mild ischemia (where rCBF fell to 70% or more of the pre-occlusion level). Dogs were prepared from each group with temporary vascular occlusion lasting 30, 60, 120 or 360 minutes and the CO_2 response was determined prior to occlusion, during occlusion and during the 4 hours of recirculation. Brains were removed 6 hours after release of the vascular occlusion and placed in a 10% formalin solution for 2 weeks. The formation of foci of infarction was judged by the shrinkage or the swelling of nerve cells and severe spongiosis of the neuropile.

Values for CO_2 responsiveness were expressed as the changes in rCBF due to increases in $PaCO_2$ (% $\triangle$ rCBF/$\triangle$ $PaCO_2$). Disturbance of the CO_2 response was judged as any value below 1% $\triangle$ rCBF/$PaCO_2$.

It was found that the vascular occlusion resulted in severe ischemia in 11 dogs, moderate ischemia in 8 dogs and mild ischemia in 2 dogs. Among the 11 dogs with severe ischemia, the occlusion was maintained for 30 minutes in 2 dogs, 60 minutes in 4 dogs, 120 minutes in 3 dogs and 360 minutes in 2 dogs. Although the CO_2 response recovered following 30 minutes of occlusion, in 8 of the 9 remaining animals undergoing 1 or more hour of occlusion, there was disturbance in the CO_2 response during occlusion and after recirculation (**Fig. 3-24**). Autopsies showed foci of infarction in all 8 of these animals.

Among the 8 dogs with moderate

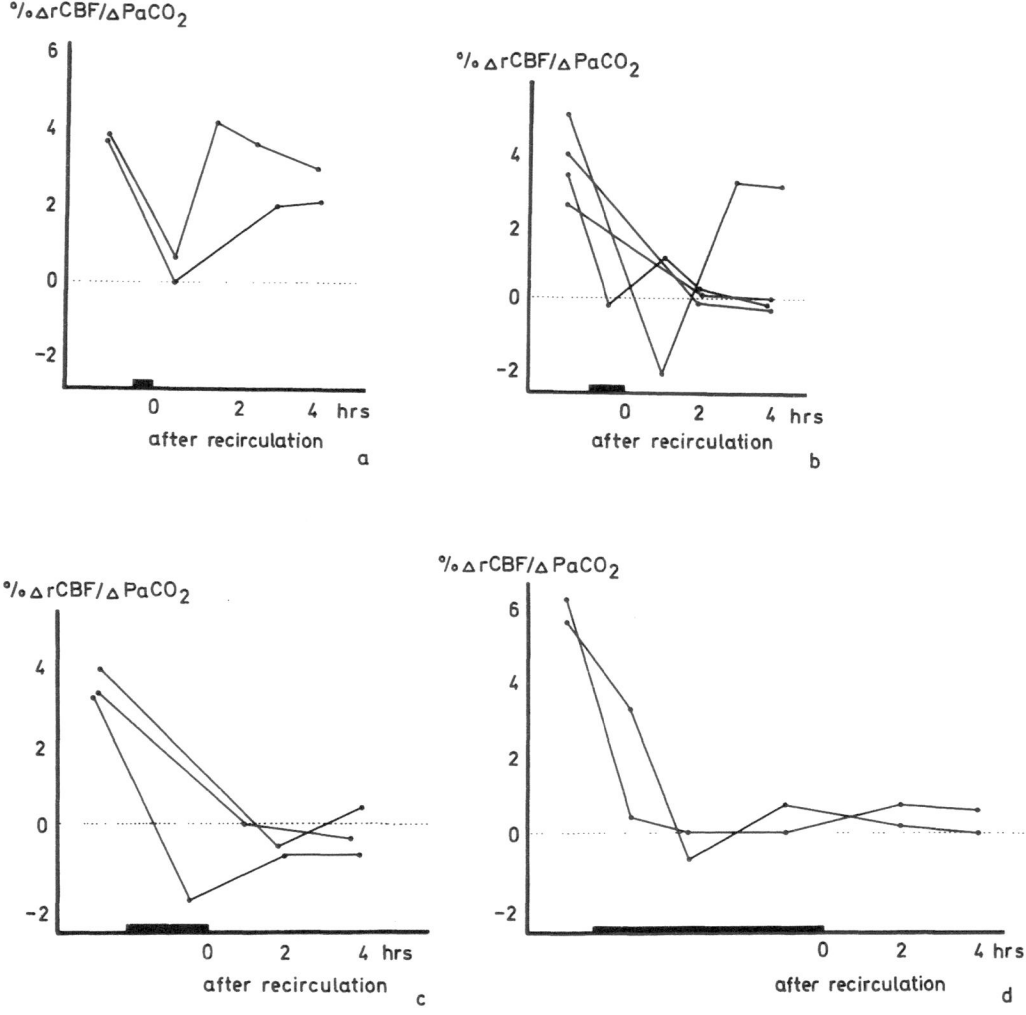

Fig. 3-24. Sequential changes in the CO_2 response in severe ischemic group

ischemia, study was made of 3 dogs undergoing 1 hour of occlusion, 3 undergoing 2 hours of occlusion and 2 undergoing 6 hours of occlusion. Although most of these animals showed disturbance in the CO_2 response during the occlusion, no obvious trend was apparent following reflow—some showing complete recovery of the CO_2 response and other continuing distur-

bance (**Fig. 3-25**). Among the 3 animals which had disturbance of the CO_2 response after 4 hours of reflow, 2 showed histological signs of infarction. In contrast, in 4 of the 5 animals that maintained the CO_2 response, no signs of infarction were found.

Finally, in both of the animals with mild ischemia, the vascular occlusion was continued for 6 hours. The CO_2

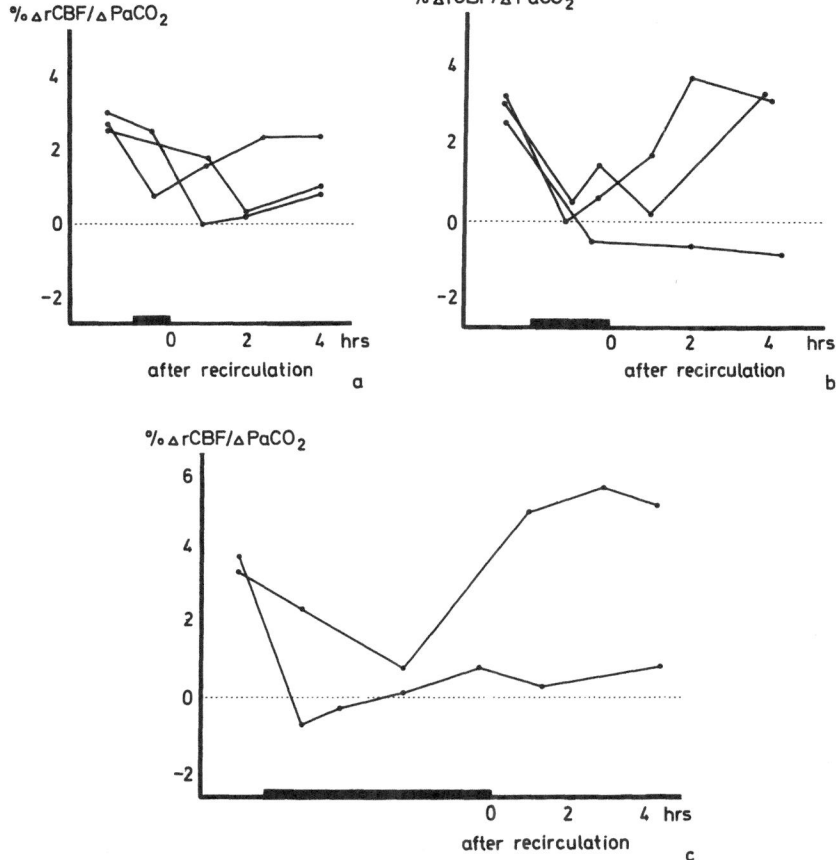

Fig. 3-25. Sequential changes in the CO_2 response in moderate ischemic group

response was well-maintained during both occlusion and reflow (**Fig. 3-26**) and neither dog showed signs of infarction.

Our experimental results in the severe ischemia group during occlusion showed disturbed CO_2 responses at all measurements except one. In contrast, in the mild ischemia group, all measurements except one showed no disturbance of the CO_2 response. In the moderate ischemia group, however, no consistent trend was apparent. These findings are thought to indicate that, to

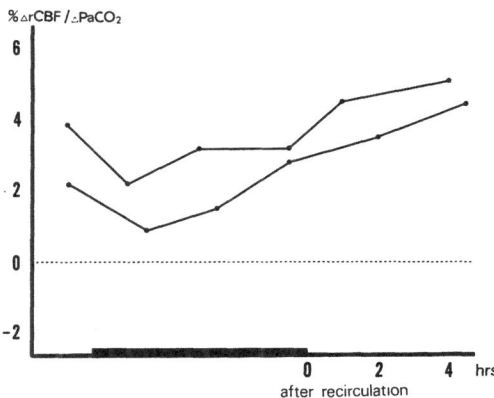

Fig. 3-26. Sequential changes in the CO_2 response in mild ischemic group. 6 hour occlusion

a limited extent, there is a correlation between the severity of ischemia and disturbance of the CO_2 response.

In order to investigate the effects of recirculation on the CO_2 response, the response prior to the release of occlusion and the response following 4 hours of recirculation were compared. In the severe ischemia group, none of the dogs in which CO_2 response measurements were made during occlusion, (1 dog undergoing 1 hour of occlusion, 1 dog undergoing 2 hours of occlusion, and 2 dogs undergoing 6 hours of occlusion), showed recovery of the response after recirculation. In contrast, in the case of even severe ischemia during 30 minutes of occlusion, although measurement of the CO_2 response is not possible because of the brief period of occlusion, transient disturbances in the response were noted following recirculation, but these soon recovered. These findings are also in agreement with our previously published results[27, 28, 48] on EEG findings and histological studies.

As summarized in **Fig. 3-27**, study was also made of the CO_2 response immediately prior to autopsy, *i.e.*, after 4 hours of recirculation. Among the dogs with severe ischemia, no disturbance in the CO_2 response was seen in either of the dogs undergoing 30 minute occlusion, but in 3 of the 4 dogs with 1 hour of occlusion, all 3 of the dogs with 2 hours of occlusion and in both of the dogs with 6 hours of occlusion, disturbance of the CO_2 response was found. Among the dogs with moderate ischemia, a definite trend in the CO_2 response was not seen, whereas in both of the dogs with mild ischemia, no disturbance of the CO_2 response was

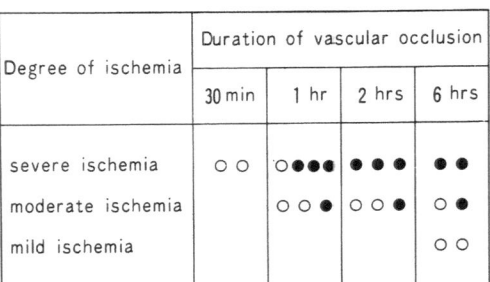

| | Duration of vascular occlusion | | | |
Degree of ischemia	30 min	1 hr	2 hrs	6 hrs
severe ischemia	○ ○	○●●●	● ● ●	● ●
moderate ischemia		○ ○ ●	○ ○ ●	○ ●
mild ischemia				○ ○

● Deficit of CO_2 response
○ No dificit of CO_2 response

Fig. 3-27. Summary of the results: The CO_2 response after 4 hours of recirculation

seen in either dog after 6 hours of vascular occlusion. In other words, the greater the fall in rCBF due to occlusion, the more likely it's that a disturbance in the CO_2 response will be found after 4 hours of reflow. On the other hand, it was found, with regard only to the dogs with severe ischemia, that the longer the duration of the temporary vascular occlusion, the larger was the percentage of animals showing disturbance of the CO_2 response.

The relationship between findings for the CO_2 response immediately prior to autopsy and the histological findings were investigated (**Fig. 3-28**). Infarction was not found in 8 of the 9 animals having in undisturbed CO_2 response, whereas 10 of the 11 with disturbance of this response showed infarction. The average value of the CO_2 response of the animals with infarction was— $0.3 \pm 0.7\%$ ΔrCBF/PaCO$_2$, whereas that of the animals with no infarction was $3.0 \pm 1.4\%$ ΔrCBF/PaCO$_2$. This difference is statistically significant at the < 0.01 level. It is therefore apparent that by measuring the CO_2 response after 4 hours of recirculation, it

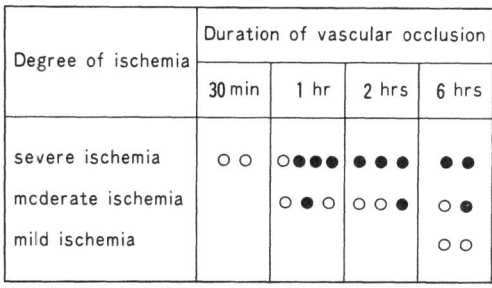

Degree of ischemia	Duration of vascular occlusion			
	30 min	1 hr	2 hrs	6 hrs
severe ischemia	○ ○	○ ● ● ●	● ● ●	● ●
moderate ischemia		○ ● ○	○ ○ ●	○ ●
mild ischemia				○ ○

● with infarction

○ without infarction

Fig. 3-28. Summary of the results: The histopathological findings at autopsy

can be determined whether or not the ischemic focus has resulted in infarction after recirculation.

It must next be asked what factors are involved in the disturbance in the CO_2 response. Waltz *et al.*[44, 45] have suggested that the CO_2 response is affected by 1) the complete dilatation of blood vessels at the ischemic focus due to accumulation of CO_2 and acidic metabolites, 2) pathological changes in the vascular walls, 3) vascular constriction due to ischemia, and 4) changes in the autonomic nervous system. It is also possible that there are influences due to 5) increases in intracranial pressure and 6) cerebral edema. The CO_2 response is thought likely to be influenced by many of these factors in combination. It is worth noting that Kogure *et al.*[16] also found that 7) the size of the ischemic focus, 8) differences in the mount of collateral circulation, 9) the site and timing of vascular stenosis, and 10) chemical substances released from thrombi and emboli are also factors involved in the disturbance of the CO_2 response.

Let us next discuss the possible mechanism by which the CO_2 response becomes disturbed in light of the present experimental results and previously reported research on the pathophysiology of focal cerebral infarction[17, 27, 36, 47, 48]. In all 8 of the dogs with severe or moderate ischemia which not showing signs of cerebral infarction, the CO_2 response was disturbed at least during the occlusion and immediately following the vascular release, and the response recovered during several hours following release of the vascular occlusion. In these experiments, the mechanism of the CO_2 disturbance is believed to be as follows:

Vascular occlusion causes acidosis within the ischemic focus and a normal vascular response to CO_2 loading cannot then occur. Recirculation results in gradual improvement in the tissue acidosis and the CO_2 response then recovers. Such changes in the CO_2 response are in good agreement with the following data. As we have previously reported[27, 28] in cases where foci of infarction do not develop, although there is the transient appearance of hyperperfusion after recirculation, rCBF values similar to those found prior to occlusion are soon obtained. In contrast, in 10 of the 11 dogs in the present study that developed infarction, recovery of the CO_2 response was not found either during occlusion or recirculation. One of the causes of this phenomenon is thought to be tissue acidosis, but more importantly the tissues including the cerebral vessels are believed to be histologically damaged. We have previously reported[47, 48] on electron microscopic observations on the sequential changes in small vessels made following the production of severely ischemic foci. There, after 1 or 2 hours of vascular occlusion, swelling of glial cells and pericytes in the vicinity of blood vessels were found. In contrast, following recirculation, the swelling of the endothelial cells of capillaries and pericytes becomes even more remarkable and stenosis of the vascular lumen can be observed. As time elapses such small vessels begin to show signs of necrosis. When 6 hours of occlusion is performed[17, 47], these changes as well as bleb formation and opening of tight junctions[47] can be seen. Following recirculation, there is the appearance of diapedesis. In this type of case in which recovery of the CO_2 response is not seen in the acute period of recirculation, irreversible histological change are thought to have occurred already.

3.5 Hemorrhagic Infarction

Hemorrhagic infarction is one of the most important factors of aggravation of symptoms in cerebral infarction and is frequently encountered in the clinic[5, 13]. It is for example, sometimes found that in cases of surgical vascular reconstruction in the acute stage or in cases of spontaneous reflow following cerebral embolism, hemorrhagic infarction is brought about, together with the deterioration of consciousness and neurological symptoms. Unfortunately, there have been few experimental studies addressing questions concerning the pathophysiology of hemorrhagic infarction, and particularly few reports on circulatory dynamics.

For this reason we report an investigation of the pathophysiology of hemorrhagic infarction, as studied from the perspective of thalamic EEG and circulatory dynamics, in relation to the presence or absence of hemorrhagic infarction in the autopsied brain[36].

3.5.1 Conditions Conductive to Hemorrhagic Infarction

Using the thalamic infarction model, we have made a series of studies on the effects of occlusion and recirculation on brain tissue. It has been found that in brain samples obtained after 30 min–24 hours of vascular occlusion without subsequent recirculation, no hemorrhagic infarction occurs[47]. Similarly, after 24 hours of recirculation following 2 hours of occlusion, the autopsied brain does not show signs of hemorrhagic infarction[27, 48]. In contrast, however, hemorrhagic infarction is produced[17] when recirculation is allowed after 6–24 hours of vascular occlusion (**Fig. 3-29**). These results are thought to indicate the importance of both the period of vascular occlusion and the period of recirculation in the development of hemorrhagic infarction.

In the present study we investigated

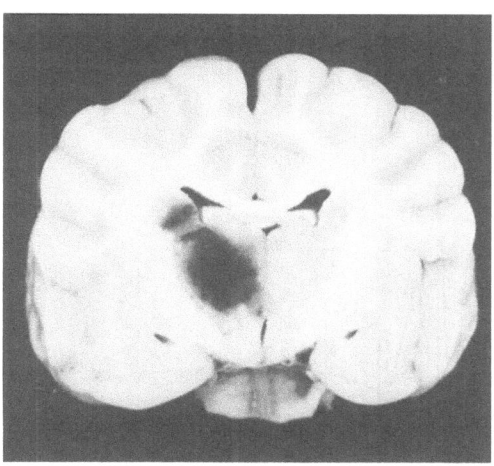

Fig. 3-29. Autpsied section. Hemorrhagic infarction is clearly seen in the right thalamus

We found that the mean rCBF values in 4 animals showing hemorrhagic infarction fell to 40 ± 13% of the pre-occlusion value immediately following vascular occlusion and to 44 ± 12%, 42 ± 12%, and 40 ± 13% after 2, 4, and 6 hours of occlusion, respectively. In contrast, among 3 animals in which hemorrhagic infarction was not found, the rCBF values immediately following occlusion and 2, 4, and 6 hours later were, respectively, 68 ± 13%, 67 ± 16%, 70 ± 14%, and 67 ± 6% (**Fig. 3-30**).

the relationship between the degree of fall in rCBF due to vascular occlusion and the appearance of hemorrhagic infarction, when recirculation was permitted following 6 hour occlusion.

These findings are thought to indicate that, in this experimental model, the emergence of hemorrhagic infarction requires 6 or more hours of cerebral ischemia such that rCBF falls to less than approximately 50% of the pre-occlusion level, which is then followed by recirculation.

3.5.2 Hemodynamics Following Recirculation

In order to determine the nature of the hemodynamics at the focus of hemorrhagic infarction, a comparison was made between the circulatory condition following recirculation in animals with hemorrhagic infarction and those without hemorrhagic infarction. By making the pre-occlusion rCBF value 100%, it was found that immediately following recirculation, rCBF was at 159 ± 63% among the animals ultimately showing hemorrhagic infarction, but was 82 ± 8% in the animals without hemorrhagic infarction. Two hours after the start of recirculation, the two groups showed similar levels of rCBF, but after 4 and 6 hours of recirculation, the

hemorrhagic infarction group had relatively low rCBF levels, whereas the animals without hemorrhagic infarction showed a return of rCBF to the pre-occlusion condition (**Fig. 3-30**). These findings suggest that a characteristic feature of the animals showing hemorrhagic infarction is the appearance of transient hyperperfusion immediately following recirculation. Such hyperperfusion is followed by a fall in rCBF; after 2 hours of recirculation, rCBF levels are below the pre-occlusion level[36].

Details of the CO_2 response have been published elsewhere[34], but it is worth noting that there were distur-

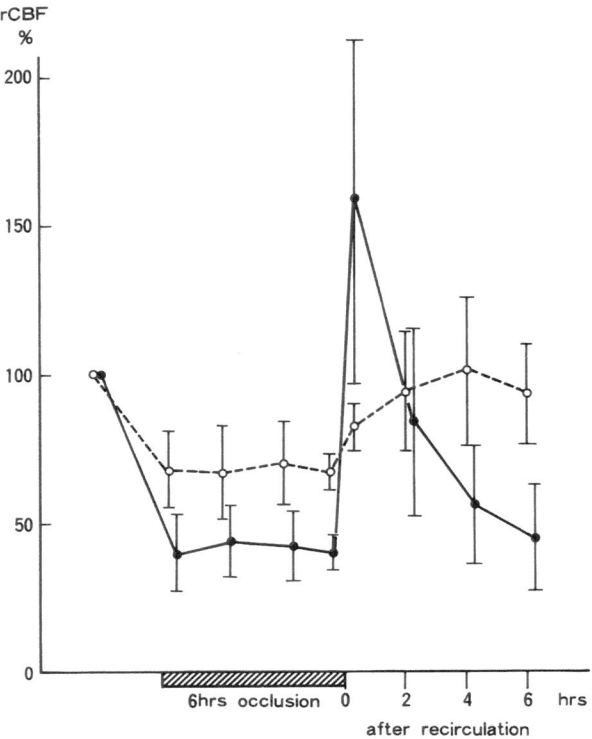

Fig. 3-30. Hemodynamics with hemorrhagic infarction and without hemorrhagic infarction

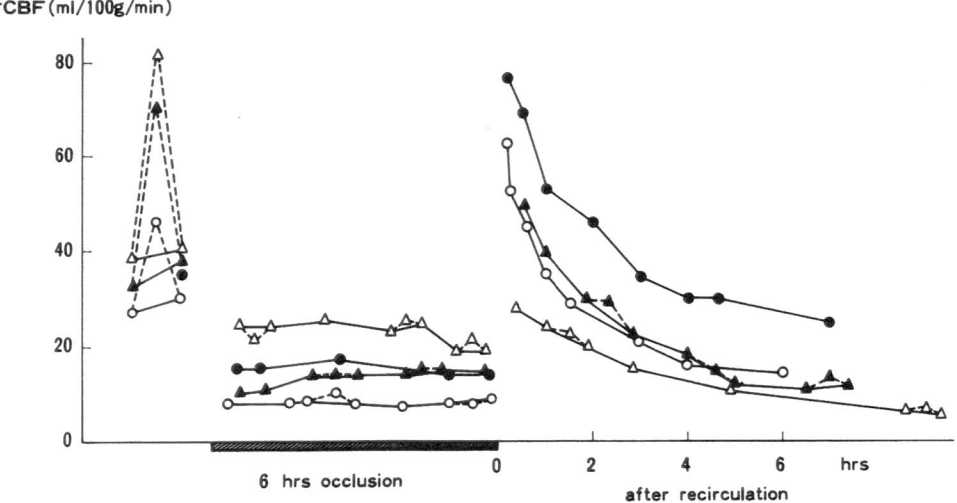

Fig. 3-31. Hemodynamics with hemorrhagic infarction

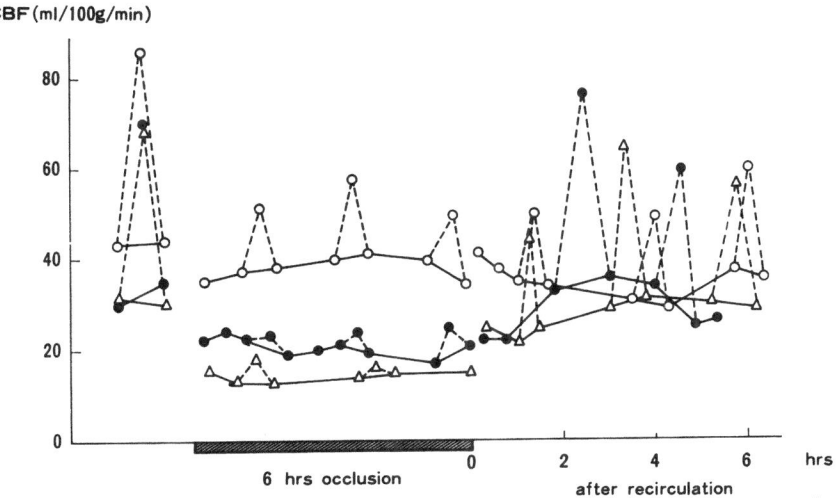

Fig. 3-32. Hemodynamics without hemorrhagic infarction

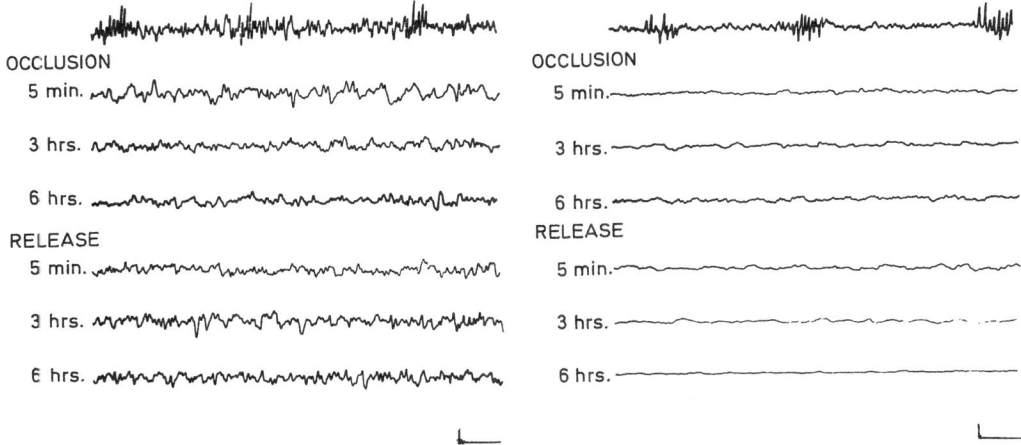

Fig. 3-33. EEG changes in the right thalamus without hemorrhagic infarction

Fig. 3-34. EEG changes in the right thalamus with hemorrhagic infarction

bances in the CO_2 response both during and following vascular occlusion among the animals ultimately showing hemorrhagic infarction (**Figs. 3-31, 3-32**). With regard to thalamic EEG, it was found that the electrical activity following recirculation recovered to almost normal (**Fig. 3-33**) in all of the dogs not ultimately showing foci of hemorrhagic infarction. On the other hand, none of the animals showing hemorrhagic infarction demonstrated recovery of normal EEG activity following recirculation (**Fig. 3-34**).

These experimental findings are thought to indicate that a fall in rCBF due to vascular occlusion to below one half the level found prior to vascular occlusion and continuation of such a hemodynamic state for 6 hours using this model brings about disturbances in the regulation of cerebral circulation, as indicated by the disturbances of the CO_2 response. Electron microscopically, dehiscence of tight junctions in capillaries has been recognized[47]. By means of sudden increase in perfusion pressure due to recirculation, transient hyperperfusion is brought about and blood components leak from the cerebral vessels at all weak points, resulting in hemorrhagic infarction. Consequently, a rise in tissue pressure is invited, there is a sharp fall in rCBF and severe damage to the surrounding brain tissue is brought about by prolonged ischemia.

4. ISCHEMIC BRAIN EDEMA

4.1 Introduction

Among the cerebrovascular occlusive disorders, brain edema following brain ischemia, in which the brain volume increases and the brain suffers herniation, is the most common cause of death. Since, as medicions, we must treat the ischemic brain and try to prevent brain edema, a complete understanding of the pathogenesis of brain edema is extremely important. In this chapter, results which we and others have obtained in animal experiments on the pathophysiology of ischemic brain disease will be introduced.

4.1.1 The Definition of Ischemic Brain Edema

The earliest record of the phrase "oedema of the brain" is to be found in a paper by Buchill and Tuke in 1874[7], but they meant a state of the brain in which water or blood serum had entered the brain tissue. Thereafter, Reichardt (1905)[55] asserted that there are fundamentally two forms of brain edema—one in which there is intracellular retention of fluid components (brain swelling) and one in which there is extracellular retention (brain edema). Spatz[64], Selbach[59] and Zülch[74] reported similar conclusions. In practice, however, it has been found that the pathology of brain edema cannot be neatly divided into two groups and, depending upon the stage of the pathology, both conditions can be found simultaneously[9, 61].

Brain edema has also been simply defined as an increase in brain volume, and it is known that in such cases factors such as increases in the cerebral vascular bed play a role in the retention of water within the brain tissue.

4.1.2 Pathophysiology of Ischemic Brain Edema

4.1.2.1 Causal Factors

In 1967, Klatzo[26] suggested that the pathogenesis of brain edema can be divided into two groups: (**Table 4-1**) cytotoxic edema and vasogenic edema.

He maintained that, in the former case, there is no deficit of the blood-brain-barrier, but, due to metabolic abnormalities of the cell itself, there is rapid accumulation of intracellular Na^+

Table 4-1. Types of brain edema

	Vasogenic	Cytotoxic
Site of primary injury	Cerebral blood vessels	Brain tissue parenchyme
Localization	Predominantly white matter	Gray or white matter depending on type cytotoxic agent
Vascular permeability	Increased permeability, including leakage of proteins and extravasation of conventional BBB indicators	Vascular permeability remains undisrupted; no protein leakage; no extravasation of conentional BBB indicator
Ultrastructural features	Enlargement of extracellular spaces in the white matter, cellular swelling affecting predominantly astrocytes	No enlargement of extracellular spaces; swelling of various cellular tissue components depending on the type of cytotoxic ager.t
Biochemical features of edema fluid	Basically plasma filtrate including serum protein	Basically plasma ultrafiltrate, composition influenced by the type of cytotxic agent

caused directly by abnormalities in the Na$^+$ pump (the ATP-dependent Na-pump). As a result, there is an increase in the intracellular water volume caused by an influx of water in order to maintain the balance of osmolarity. In the case of vasogenic edema, however, there are local disturbances of the blood-brain-barrier and a consequent exudation of serum proteins into the extracellular space within the brain. As a means for thinking about the pathology of brain edema, the concept of these two groups is useful, but in actuality the independent existence of pure vasogenic edema or pure cytotoxic edema is not found. Although one form of the disorder may in fact precede the other, it is generally thought that both kinds of pathology exist together in the human brain[27].

The most important factors involved in the pathogenesis of ischemic brain edema are the metabolic state of the brain and the state of the blood-brain-barrier. For the brain to maintain normal functions, it is well-known that a large volume of energy is required[62, 63] and that the energy be produced by the hydrolysis of ATP (by ATPase) which has been synthesized in the mitochondria[60, 62, 63]. That is, all of the enzymatic reactions involved in energy consumption are closely related to the functions of mitochondrias. Since, in the ischemic brain, there are deficits in the supply of both mitochondrial enzymes and substrates, ATP production is disturbed[60, 62, 63]. Thus, in order to compensate for the reduced ATP supply, abnormal reactions occur in the brain due to anaerobic glycolysis[21, 58, 62]. It is known from numerous animal experiments that immediately following the production of an ischemic brain, there are decreases in phosphocreatinine, ATP, glucose and glycogen, and marked increases in AMP, ADP and lactate. This phenomenon has also been confirmed in the ischemic human brain and is known to lead to deficits in the Na^+ pump, which regulates the intracellular/extracellular balance of various substances[5, 9, 23]. Specifically, it is known that the ratio of intracellular Na^+/K^+ is maintained at a low level—despite the fact that the extracellular medium has a high Na^+/K^+ ration—due to the control of ion distribution by the ATP-dependent Na^+ pump present in the cell membrane. When there is insufficient ATP, however, the Na^+ pump cannot work and the extracellular Na^+ enters the cell—leading, in turn, to the inflow and retention of H_2O[5, 9, 23]. This phenomenon is thought to occur in the early period of brain edema[25, 27].

The ischemic focus where water retention occurs is a form of focal ischemia and when the energy charge returns to normal due to collateral circulation, the Na^+ pump begins to work and the extracellular water content is returned to normal. For this reason, it is known as reversible ischemia. In other words, there is a close correlation between the pathogenesis of brain edema in the ischemic brain and energy metabolism, and the more severe or the more prolonged the ischemia, the greater the increases in water content of the brain due to the above mechanism[12].

With regard to the relationship between the blood-brain-barrier and brain edema, it is thought that the blood-brain-barrier is an interface between the blood and brain tissue which maintains a constant environment for the neural tissues of the central nervous system[3, 72]. In its broadest definition, the blood-brain-barrier is a blood-CSF barrier[3, 48] but in its narrower definition, it is a barrier acting between cerebral capillaries and the substance of the brain (**Fig. 4-1**). According to the horse radish peroxidase (HRP) studies of Reese and Karnovsky[72], the blood-brain-barrier is morphologically due to the presence of tight-junctions between endothelial cells of the cerebral capillaries and to vesicular transport. The tight junctions are comprised of structures containing one or two proteins, located at the point of membrane fusion of two neighboring cells. If HRP or microperoxidase is injected intravenously, HRP molecules can then be observed with arterioles of 15–30 µn diameter, but HRP is not found to exudate to the extracellular space—thus indicating that passage over the tight junctions between endothelial cells at the capillaries is being prevented[3, 72]. In the ischemic brain, however, the permeability of the blood-brain-barrier to HRP passage is increased and, indeed, when there are increases in blood pressure, the passage of HRP is facilitated. In light of the fact that there are increases in the number of endothelial transport vesicles when hypertension is induced, it is thought that increased permeability may produce increases in vesicular transfer[22, 24, 30, 32, 35, 50, 51, 68, 72] (**Figs. 4-2, 4-3**).

Moreover, it is known that carotid arterial injection of a hyperosmolaric solution results in increased permeability of the Evans-blue-albumin complex[17, 54]. There are in fact two distinct hypotheses concerning the mechanisms involved. On the one hand, some believe that there may be

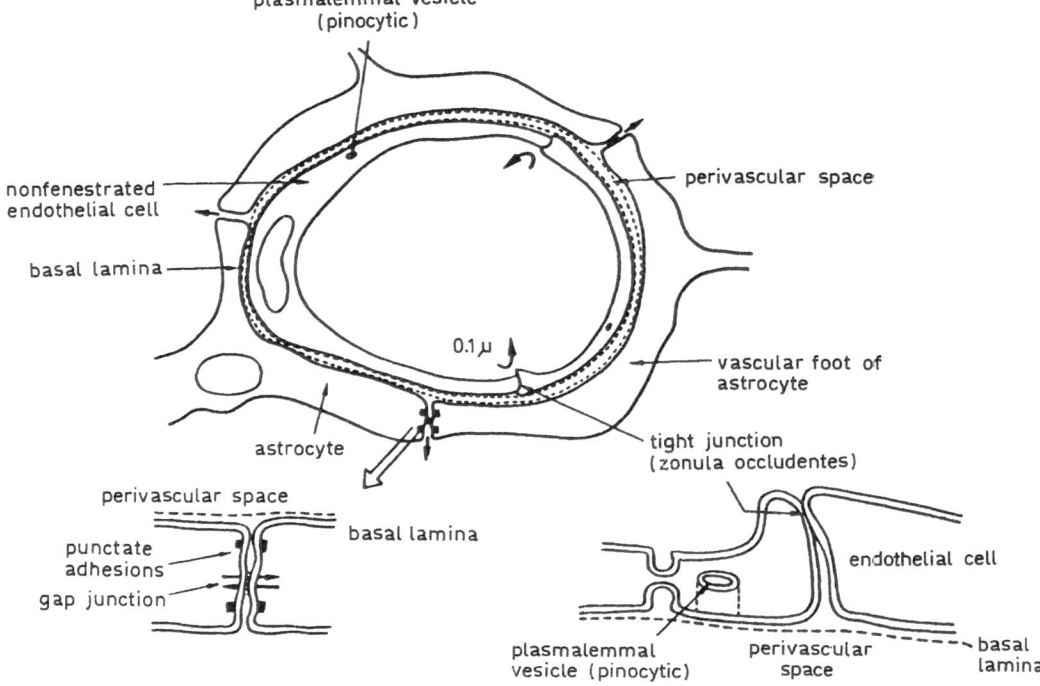

Fig. 4-1. The morphological aspects of the blood-brain barrier (by A Hirano)

osmotic opening of the blood-brain-barrier due to the opening of tight junctions following endothelial cell shrinkage[6, 17, 54]. On the other hand, other researchers maintain that there is no opening of tight junctions, but instead an increase in vesicle transfer[22, 32, 35, 72]. Either way, it is believed that, in both ischemia and other pathological conditions, the blood-brain-barrier allows for increased peremability and the retention of abnormally high levels of water and consequently the swelling of the brain.

These mechanisms are involved in the early stages of the pathogenesis of ischemic brain edema, but in recent years a great deal of research has been undertaken on pathological processes which are secondary to the early stages

of the pathology. Among the secondary processes thought to be important in the pathology of ischemic brain edema are the abnormalities in glutamate produced by cell necrosis[36], serotonin released by platelets[66] and the Kallicrein-kininogen-kinin(KKK) system[17], as well as the accumulation of free radicals[53] and fatty acids[8] which occurs with increasing severity of tissue damage. With regard to glutamate, it is known to increase along with free fatty acids with increases in the permeability of Na^+ through the cellular membrane and, due to deficits in cellular energy metabolism, it is said to bring about edema[36]. Serotonin and the KKK system are said to increase pinocytosis and cause paralysis of precapillary vascular segments. This in turn strength-

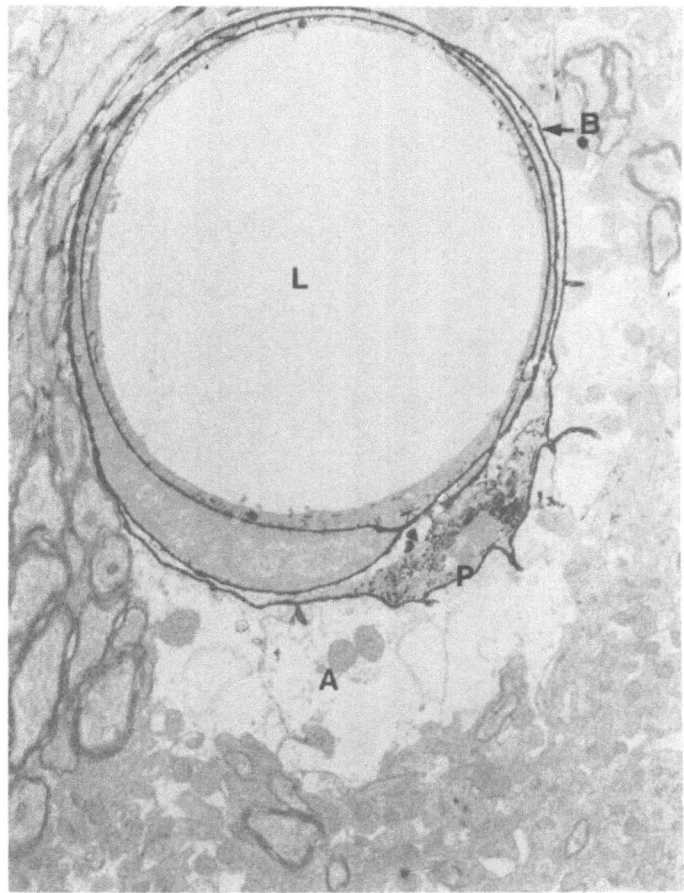

Fig. 4-2. This figure shows a vessel of the thalamus of the rat after yielding to 30 min ischemia-10 min reflow. Horseradish peroxidase, which was added as a macromolecular tracer, shows extravasation around the vessel, the perivascular astrocytic endfeet of which are mildly swollen. *L* Lumen of the vessel, *A* Perivascular astrocyte (endfeet), *B* Basal lamina filled with horseradish peroxidase, *P* Pericyte. (Magnification × 15,000)

ens the hydrostatic filtration of the already permeable blood-brain-barrier. The accumulation of fatty acids and free radicals is known to be closely linked with a decrease in energy charge and it is thought that this produces edema by destroying or increasing the peremability of cell membranes. It should be noted, however, that many uncertainties concerning the mode of action remain. That is, it is not known which factors are the most important in the genesis or aggravation of brain edema and, without detailed examination of the various experimental models used by various research groups, it is impossible to draw firm and unified conclusions about the mechanisms involved.

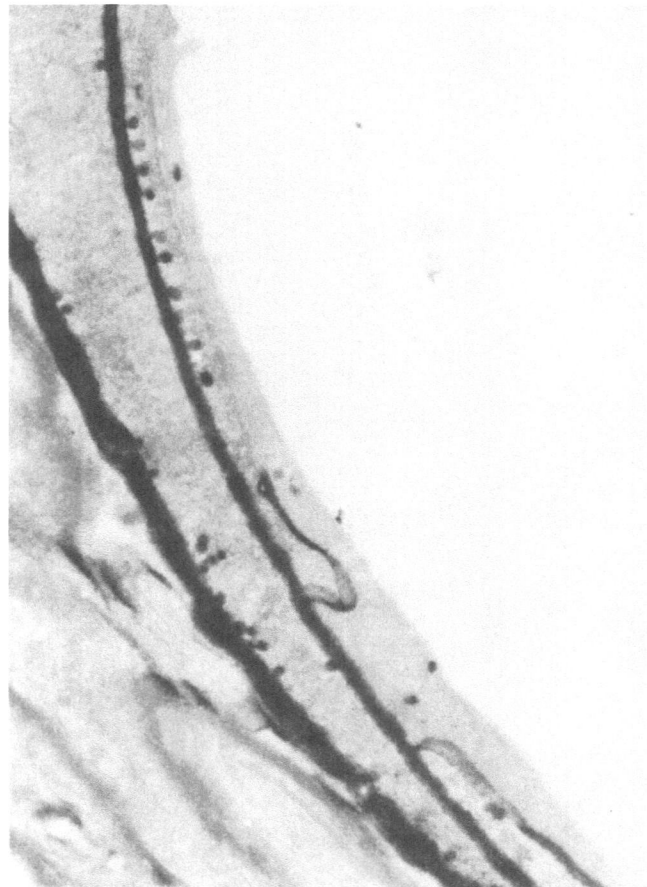

Fig. 4-3. Greater magnification shows retrograde filling of the interendothelial space with horseradish peroxidase to the zonula occludens at the antiluminal sike. An increasing number of transcytotic vesicles are filled with horseradish peroxidase. × 24,000

4.1.2.2 The Onset and Natural Course of Ischemic Brain Edema

When the brain becomes ischemic, the causal factors discussed above come into play. The subsequent-pathology of the brain, however, depends heavily on various other factors, such as the duration and scope of the ischemia, its location (focal or global) and whether the edema occurs during ischemia or following recirculation. It is conse-quently extremely important to know the dynamics of brain edema due to these factors.

a) Brain Edema in the Completely Ischemic Brain

Hossmann has used the cat for an experimental model said to be a completely ischemic brain[18]. He re-ported that, since the increase in water content of the brain during ischemia is

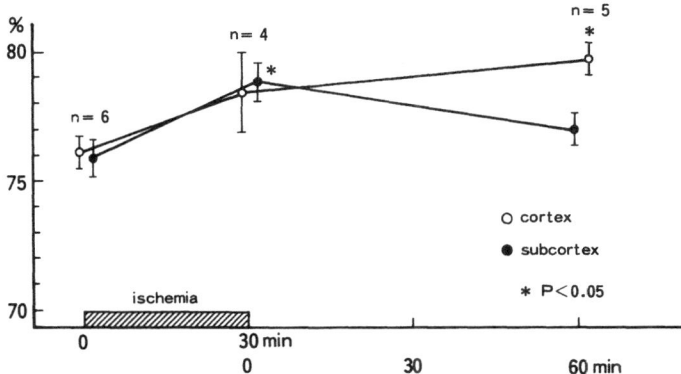

Fig. 4-4. Time sequential change in total water

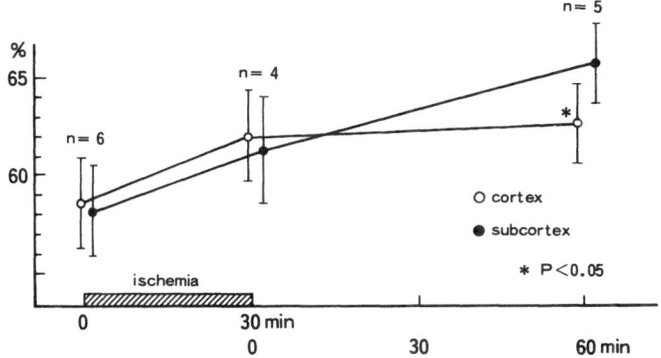

Fig. 4-5. Time sequential change in free water

slight and there is a reduction in the gap between cells, there is consequently almost no brain edema. In contrast, Tomita *et al.*[69] found that there was swelling of the completely ischemic brain in a rat preparation following cardiac arrest. They attributed the swelling to an increase in osmotic pressure within the brain tissue. We have also employed the 3-vessel occlusion model for bilateral hemispheric infarct using the rat and have found increases in the water content of the brain, which we attribute to anaerobic glycolysis[24] (**Figs. 4-4, 4-5**).

b) Brain Edema in the Focally Ischemic Brain (Incomplete Brain Ischemia)

In the focally ischemic brain with a focus of ischemia surrounded by normal tissue, it is known that a hydrostatic pressure gradient is produced between the surrounding open vessels and the center of the ischemic focus[28] and, moreover, that the flow of water pressing through the arterial side of the capillaries, entering the brain tissue and reaching the venous side of the capillaries obeys Starling's law[28]. That is, since most of the edematous fluids are supplied from open vessels, at the focus of the ischemia, water and Na$^+$ flow according to the hydrostatic gradient from the open vessels. Since there is no action of the Na$^+$-pump (which has been disturbed due to the ischemia) at these vessels, the outflow of water and Na$^+$ cannot occur. Thus, a state of edema is produced. In that condition, when the serum components which have flowed in reach the brain parenchyma, it is thought that the above-mentioned serotonin, glutamate, fatty acids and KKK system further increase membrane permeability[36, 44, 53, 70]. In addition, the free radicals released from the condition of enzyme deficiency are thought to cause destruction of cell membranes and further increase the edema by facilitating membrane permeability[8, 53]. In other words, in focal ischemia the hydrostatic pressure gradient is an important factor in the aggravation of brain swelling. When the hydrostatic pressure gradient between opened and closed vessels becomes larger, serum protein and water are exudated within a short period of ischemia, but regions of intense exudation are said to form a border region between themselves and regions of normal tissue[15, 20].

With regard to the border regions between the focus of ischemia and the surrounding normal tissue using this incomplete ischemia model, the longer the duration of the ischemia, the greater is the damage to the cells within the ischemic focus and to the cellular membranes. As discussed above, due to increases in factors which facilitate the edema (serotonin, the KKK system,

glutamate and free radicals), the permeability of the cell membrane increases. In the case of incomplete ischemia, the supply of edematous fluids increases geometrically and the degree of brain edema becomes greater than that in complete ischemia[2, 47].

4.1.2.3 Ischemic Brain Edema Following Recirculation

We have discussed the fact that brain edema in the surrounding tissue commonly occurs during ischemia, but it is well-known that post-ischemic reflow of blood further facilitates the emergence of brain edema[57]. The factors which produce brain edema during ischemia are also at work during reflow, but study of the changes following recirculation suggests that there are in fact additional factors then involved. That is, it is thought that at the time of reflow a colloid pressure gradient is produced which reduces or even reverses the normal gradient between the perfused blood and the extra-vascular osmotic pressure (which increases during ischemia), and this colloid pressure gradient works in addition to the hydrostatic pressure gradient produced between the lumen of the patent vessels and the center of the ischemic focus. In other words, the additional factors which are produced in the edematous brain due to postischemic recirculation are: (i) the generation of a hydrostatic pressure gradient from within the vessels to the extravascular space, and (ii) an increase in the extravascular colloid pressure[20, 37].

Using the canine model for incomplete ischemia of a cerebral hemisphere (**Fig. 4-6**), we have also studied the brain swelling occurring after recirculation[67]. As shown in **Fig. 4-6**, a right temporal craniotomy is performed in this model, followed by the occlusion of certain vessels at the base

ing can then be observed through a bone window during three hours of reflow.

In animals which have not been given any preventive therapy, severe brain swelling is seen after one hour of

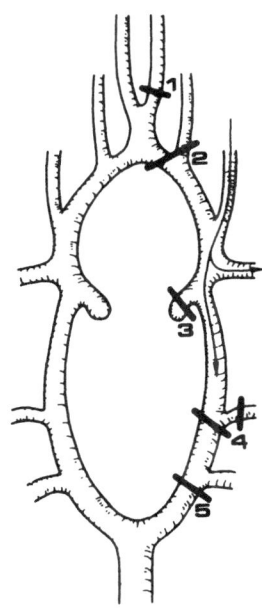

Fig. 4-6. Sites of occlusion for incomplete ischemia of the cerebral hemisphere in the canine model. *1* A2 portion of the anterior cerebral artery. *2* Bifurcation of the ethmoidal artery in the A1 portion. *3* Internal carotid artery. *4* Bifurcation of the posterior communicating artery and the posterior cerebral artery. *5* Anterior cerebellar artery

of the brain. The regions normally perfused by the posterior communicating artery and right middle cerebral artery through the A_1 portion of the anterior cerebral artery then receive only the small volume of blood which enters via the external carotid artery from the ophthalmic artery. Using this model, six hours of occlusion is followed by recirculation, and brain swell-

reflow and the swelling is so severe that the substance of the brain is damaged after three hours. It may be thought that, in this experiment, since the preparatory step of the brain edema is prolonged (*i.e.*, six hours of incomplete ischemia), causal factors related to both Klatzo's cytotoxic and vasogenic edema have already been produced (*e.g.*, free radicals, serotonin, glutamate, the

KKK system, etc.). Consequently, re-circulation results in an increase in the hydrostatic pressure gradient and therefore a severe degree of swelling. As seen from the above, allowing reflow after ischemia—particularly after severe ischemia—results in progression and aggravation of brain edema[20, 37].

4.1.2.4 The Disappearance of Edema

Little is known concerning the mechanisms by which edematous fluids are absorbed from the extracellular space, but the following possibilities have been suggested: (i) absorption by the CSF, (ii) reabsorption by the capillaries, (iii) absorption by cells, particularly glial cells, and (iv) disappearance due to the metabolism of the components of the edematous fluids[37, 56]. These mechanisms have been suggested to work alone or in combination to result in the absorption of the edema.

It is possible to demonstrate the uptake of fluids into the CSF by measuring the disappearance of the edema due to changes in the intraventricular pressure, i.e., the pressure gradient. Indeed, it has been shown[4] that if the pressure gradient is lowered, the recovery of the edema is poor, whereas if it is raised the recovery is improved.

Experiments concerned with the reabsorption of edematous fluids by capillaries have involved comparisons of the clearance of ^{14}C-dextran and ^{14}C-sucrose. A capillary effect has been surmised from the fact that the sucrose, with its low molecular weight, disappears rapidly[65].

Uptake of fluids by glial cells has been inferred from the fact that HRP is transported retrogradely from axon terminals toward the cell body[17].

Clearly, all of these mechanisms remain hypothetical and are in need of further experimental verification.

4.1.3 The Treatment of Ischemic Brain Edema

Therapy for ischemic brain edema generally consists of two main objectives: treatment of the ischemic brain and treatment of the ensuing brain swelling. In general, it is known that the larger the brain damage due to ischemia, the more severe will be the brain edema. Consequently, therapy for brain ischemia is undertaken primarily to suppress the factors resulting in the brain edema, which itself

cannot be removed, and to suppress secondary pathology. Since these issues have been discussed in the previous chapter, here we will discuss therapy undertaken to combat directly the brain swelling.

Therapeutic measures which should be undertaken in the treatment of brain swelling include the following: (i) increasing the pressure gradient, (ii) recovery of normal functions of the blood-brain-barrier, (iii) raising the clearance ratio of edematous fluids, and (iv) improving the cellular metabolism of tissues which have been injured due to the ischemia. Currently, no single therapeutic method will suffice to accomplish all of these objectives, and—whether drug therapy or surgical therapy is to be employed—several therapeutic measures must be used simultaneously. Moreover, among the appropriate drugs, it is unusual for a specific agent to have a single mechanism of action, but rather multiple effects have been frequently reported.

4.1.3.1 General Therapeutic Steps

It is often found that, once the brain succumbs to swelling, other abnormalities such as hypoxia and the inability to control hypertension ensue. Hypoxia causes deficits in the oxidative phosphorylation in the mitochondria (where ATP is produced) together with the release of substances which disturb energy metabolism and activation of lysosomal hydrolyzing enzymes. As a consequence, changes—particularly rapid increases in the blood pressure—which are major factors in the production of edema can bring about the destruction of the blood-brain-barrier and the hydrostatic pressure gradient between the blood and the brain tissue will increase—thus bringing about the leakage of and increases in the edematous fluids. Therefore, it can be said that the first step in the treatment of brain edema is to take measures to restore the normal physiological state of the biological organism—especially,

the normal functioning of the blood-brain-barrier.

4.1.3.2 Drug Therapy

a) Hyperosmolar Therapy

As its name implies, osmotherapy using hyperosmolar solutions makes use of the osmotic pressure which is produced by hyperosmolar solutions. By means of a bolus or continuous venous injection, it is possible to raise the blood's osmotic pressure and due to an increase in the osmotic pressure gradient between the brain and the blood, brain tissue water molecules can be made to enter the blood, thus leading to an improvement in the brain swelling[4].

Historically, Weed *et al.* in 1919[71] were the first to report the effectiveness of such therapy, and since then various hypertonic solutions, such as glucose, urea, mannitol and glycerol, have been used. It must be said, however, that since the contents of these hypertonic solutions enter the tissue no longer protected by the blood-brain-barrier, the pressure gradient is not increased, and it has been reported that water is thus removed only from normal, not damaged, brain tissue[49]. Moreover, when the osmotic pressure of blood has been lowered following an increase, the distribution of the osmotic pressure at the site of the damaged blood-brain-barrier reverses and it is known that there is actually transferral of water from the blood to the damaged tissue (the so-called rebound phenomenon)[31, 40]. As a consequence, drugs such as mannitol and glycerol, which provoke only a small rebound phenomenon, are currently the primary agents used for drug therapy.

In 1962, Weiss[73] introduced mannitol as a cerebral hypotensive agent, and since then many clinical and experimental results have been published. It is now thought that mannitol, which is an "alcohol sugar" with a molecular structure similar to the six-unit sugar of glucose and with a molecular weight of 182, is not hydrolized in the body due to its non-reactiveness and does not enter the cell, but remains in the blood stream. As a consequence, it can be used as a hyperosmotic agent.

Since dehydration is due to the osmotic gradient of the functional blood-brain-barrier[19, 52] and effectively works to counter the edema, it is thought to work more efficiently in regions of normal tissue than in damaged regions where the blood-brain-barrier is also disrupted. It is said that the cerebral hypotensive effect of mannitol begins 20–45 minutes after administration when given in a dose of 1.0–2.0 g/kg of body weight[14]. Clinically, the effects are found to coincide with the effects on the ICP. In addition, mannitol is found to maintain its effectiveness for 3 and 1/2 hours following administration. With regard to the rebound phenomenon which follows a decrease in intracranial pressure, it is thought to be stronger with mannitol than with glycerol[46], but it has also been reported that, depending upon the means of administration, the rebound phenomenon can actually be weaker than that with glycerol.

Care must be taken when using mannitol because electrolyte abnor-malities or hypovolemia can arrise with dehydration. Nonetheless, by means of appropriate administration, favorable effects—including those related to the rebound phenomenon—can be obtained with mannitol[38, 39]. Its mechanism of action as a cerebral hypotensive agent is reported not to be solely the removal of water from cerebral tissue due to the osmotic gradient, but to be also due to the suppression of CSF production[16]. In addition, it has also been found that peripheral circulation is also improved due to mannitol administration and that that effect alone would have an anti-edematous result. In other words, it is thought that mannitol decreases blood viscosity and/or activates the autoregulation functions of the cerebral blood vessels[45].

In a series of animal experiments, we also have studied the protective effects of mannitol on the brain. We have found that it acts not only raises the osmotic gradient, but also suppresses the free radicals which, as described above, are secondary factors produced by the ischemia. We previously believed that mannitol remained in the blood stream without entering brain tissue in the normal brain and thereby raised the osmotic gradient, but recently we have administered H^1-labeled mannitol to the ischemic brain and have found that a small volume of mannitol does indeed enter the damaged brain tissue.

Using a brain tumor model, Lantos et al.[10] also found that there is uptake of C^{14}-labeled mannitol by the tumor tissue, but the uptake is only 1/100th that of normal brain tissue. In other words, it is apparent that there is uptake of mannitol into tissue when the blood-brain-barrier has been destroyed—such as in the ischemic brain or in brain tumor tissue. Mannitol, then, enters damaged brain tissue—suggesting the possibility that it suppresses the free radicals generated there due to ischemia.

Like mannitol, glycerol raises the osmotic gradient and has been used as a cerebral hypotensive agent. Unlike mannitol, however, it is taken up by cells and is metabolized during the synthesis of glucose[42, 43]. In comparison with mannitol, glycerol has a weaker dehydrating effect when given in similar doses, and it is said to produce a weaker rebound phenomenon. It should be noted, however, that in recent reports of Node *et al.*, mannitol has been found to produce a weaker rebound phenomenon than that of glycerol[43].

b) Steroid Therapy

Galicich and French[13] were the first to use steroids in the treatment of brain swelling, and they reported that glucocorticosteroids were effective in treating the swelling of brain tissue around a brain tumor. Since then, a variety of experimental studies have been reported and it is currently thought that the steroids have their effect on brain swelling due to four likely mechanisms: (i) They maintain the blood-brain-barrier due to stabilization of cell membranes and lysosomal membranes. (ii) They reduce the edema due to the effects of a shift in the distribution of electrolytes and due to improved cerebral metabolism. (iii) They reduce the intracranial pressure due to suppression CSF production. And (iv) they bring about improvements in regional cerebral blood flow[8, 29, 33, 34, 41]. All of these mechanisms remain hypothetical and, indeed, the subject of the effectiveness of glucocorticoids is still controversial. Specifically, although it may be the case that the glucocorticoids are effective in reducing the brain swelling around brain tumors, the response to various other forms of brain swelling appears to differ depending upon the nature of the pathology[13]. Treating doubts therefore remain concerning its effectiveness in the cellular damage at the center of an ischemic focus and in the case of so-called cytotoxic edema[11, 21].

5. CEREBRAL METABOLISM AND FREE RADICAL PATHOLOGY

5.1 Introduction

The most fundamental factors involved in the pathology of cerebral infarction are the decreased supply of oxygen and glucose to the brain tissue due to disturbances in cerebral blood flow. The term "cerebral ischemia" has long been used to describe a state of disrupted cerebral blood flow and metabolism. Biochemical research on the pathology of cerebral ischemia is less than 50 years old, which causes us some surprise. Indeed, one of the first usages of the term "neurochemistry" was at the First International Neurochemistry Symposium in 1954[71]. Prior to that date, of course, brain proteins and lipids had been discovered by A. F. Fourcroy (1755–1809)[85], and brain lipids had been discovered by L. N. Vauguelin (1763—1829), but not until the time of J. L. W. Thudichum (1828—1901)[229] was systematic research on the biochemistry of the nervous system undertaken. He made analyses of some 130 substances found in the brain and undertook lipid analysis of various phospholipids, including sphingomyelin and phosphatide. Biochemical study of the ischemic brain, however, lagged far behind investigations in other areas and was not the focus of much research until the 1960s, when Lowry's (1964)[143] analysis technique based upon the absorption spectrometer was reported. Making use of enzyme reactions, that method allowed for quantitative measurement of carbohydrates and energy metabolism in samples of brain tissue. Thereafter, together with the development of various photometric analytic techniques—including gas chromatography, high performance liquid chromatography (HPLC), nuclear magnetic resonance (NMR), electron spin resonance (ESR), computer, X-ray and electron diffraction, and the sensitive photomultiplier.—applications using biological materials have proliferated and progress in the biochemistry of the ischemic brain has been remarkable. Today, a massive amount of data and various hypotheses concerning the pathology of cerebral ischemia are to be found in the literature.

From the time of Lowry until very recently, research on the biochemistry of cerebral ischemia focused primarily on energy metabolism. Among several outstanding studies, the works of Siesjö[209, 210], Kogure[133, 134] and their colleagues are worth mentioning. In fact, measurements of cerebral energy metabolism underlie all considerations of the metabolic state of the ischemic brain and, in light of new findings concerned with the role as neurotransmitter of ATP (Krishtal OA, 1983[137]); first reported by F. A. Holton and P. Holton in 1954[108], there has been renewed interest in this old, but still current topic of cerebral energy metabolism.

With regard to the glucose metabolism of the brain, Sokoloff et al.[212] developed a quantitative autoradiography technique using ^{14}C-2-deoxyglucose. Using that method it has become possible to measure the rate of glucose consumption at various sites within the brain, and many studies have now been done on the glucose metabolism of the hypoxic brain and/or ischemic brain.

In light of the known phospholipid bilayer structure of biomembranes and the mechanisms of the electron transport system of the inner membrane of mitochondria (based upon the chemiosmotic hypothesis of Mitchell et al.[155, 156]), Demopoulos et al.[62–66, 182, 183] suggested that it would be possible to occur the breakdown of phospholipids due to free radical reaction initiated by coenzyme Q radicals (i.e., autoxidation) in the ischemic brain. Subsequently, a considerable number of studies have been done to elucidate the pathology of ischemic brain edema from the perspective of the damage done to biomembranes due to lipid peroxidation.

Bazán[17–20], Rehncrona[192–195] and Yoshida[247–249] experimentally demonstrated that during ischemia there is an increase in polyunsaturated fatty acids, which are produced by the hydrolysis of the acyl bond of phospholipids. Moreover, it has been shown that in the ischemic brain lipid peroxidation can occur as the results of the enzymatic peroxidation of polyunsaturated fatty acids which are liberated from their polar regions due to the action of phospholipase. These findings have also suggested that the products of this metabolic process (such as prostaglandin, HETE, leukotriene, thromboxane and the hydroxy radical) can influence the pathology of the ischemic brain.

Together with research on the depletion of energy and the breakdown of phospholipids which constitute biomembranes in the ischemic brain, considerable work has been done on the neurochemistry of ischemic brain edema. This work has included studies of neurotransmitters[73, 121], the proteins (enzymes) which constitute another important component of the cell membrane, membrane transport mechanisms[45, 72, 93, 141, 146] and the movements and functions of water and various ions through the cell membrane[29, 161]. Recently, in order to obtain more fundamental and more conclusive answers to questions concerning the pathology of cerebral ischemia, the activation or suppression of enzymes (such as phospholipase) and ring-nucleotides as 2nd messengers (such as cyclic AMP and cyclic GMP)[19, 125, 132, 249], the involvement of calmodulin, which is a Ca^{2+}-binding protein[49, 130, 131], the Ca^{2+} effects on phospholipase[149, 210], and the involvement of inositol triphosphate[22, 23, 94, 122] (which is a metabolite of membrane phospholipids involved in the intracellular mobilization of Ca^{2+}) have been intensively investigated.

Within this wide range of studies using increasingly specialized techniques, our own contribution is small and the extent of our current understanding is yet incomplete. There are still many unresolved questions concerning the biochemistry of the ischemic brain. Even with regard to the narrow topic of the peroxidation of membrane phospholipids, the following questions remain to be answered. Do enzymatic peroxidation (arachiodonate cascade) and autoxidation actually occur *in vivo* in the ischemic brain? Do these reactions occur during or after the ischemia and is the site of the reactions at the mitochondria, the microsomes or elsewhere? What, furthermore, is the involvement of cations (Ca^{2+}, Mg^{2+}, etc.) in the activation of the various enzymes participating in de-acylation? What is the role of the metabolism of polyphosphoinositides, which is one of the phospholipids involved in intracellular Ca^{2+} mobilization? Finally, what indeed is the essence of the cellular damage which is incurred in the ischemic brain? Considerable work remains to be done before these and other problems are resolved.

For nearly two decades, researchers in our department have been involved in the study of various aspects of brain infarction, with the primary objective of utilizing the results of experimental research for the benefit of clinical patients. In the present chapter, the results of our work concerning neurochemistry will be introduced and an objective overview of current ideas concerning the biochemistry of the ischemic brain will be presented.

5.2 Phospholipids as Biomembrane Constituents

The lipid components constitute some 45–65% of the brain dry weight and, according to O'Brien JS *et al.*[171, 172], 37.6% of the gray matter dry weight and 66.3% of the white matter dry weight. Moreover, they report that approximately 75% of myelin dry weight is made of lipid. Of the total lipid content of the brain, phospholipids constitute 45%, 60–70%, and 30–50% of the whole brain, gray matter and white matter, respectively[171, 172]. These figures clearly indicate the high volume of lipid components in the brain, or, in other words, the importance of the phospholipids as biomembrane constituents.

The phospholipid content of individual cell types is as follows. In nerve cells, some 70–80% of the lipids are phospholipids, in oligodendroglia about 45–62% of them are and in astrocytes they constitute about 70–73%[170, 184]. The phospholipid components in nerve cells are about 40% phosphatidyl choline (PC), 20–40% phosphatidyl ethanolamine (PE), 4–9% phosphatidyl serine (PS) and 4–5% phosphatidyl inositol (PI), whereas in glial cells they are about 30% PC, 17–27% PE, 4.7–10% PS and 1–4% PI. Noteworthy is the high PC content in nerve cells in comparison with the glia[60, 170, 184]. These differences in phospholipid content have significant implications for the physiological functions and metabolism of phospholipids during ischemia. That is, the activity of phospholipase, which acts as a trigger for cellular destruction during ischemia, is several fold greater in neurons than in glial cells, and it is known that phospholipase A_1 has its strongest effect on the PC components—resulting in the liberation of fatty acids[240].

Currently, the most fundamental concept concerning the structure of biomembranes is the fluid mosaic model, proposed by Singer and Nicolson in 1972[211] (**Fig. 5-1**). The

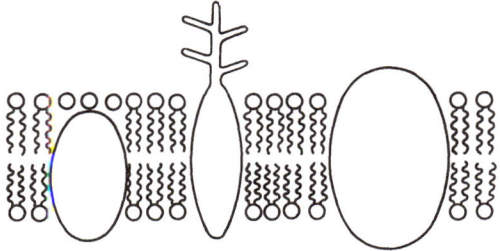

Fig. 5-1. The sectional plan of the biomembrane model. Intrinsic proteins illustrated by ellipses are floating like "islands" in the "sea" of the phospholipid bilayer

hydrophilic head (polar) regions of the phospholipid molecules form the outer and inner layers of the membrane bilayer, whereas the hydrophobic tail regions become aligned with the one another within the membrane itself. This fundamental bilayer structure is said to have a thickness of between 45 and 100 Å[211]. In contrast, membrane proteins can take up positions everywhere within the phospholipid bilayer, penetrate the membrane and extend to both surfaces or exist on

either the inner or outer surface of the membrane. The membrane binding proteins, present on the membrane surface, are bonded with ionic or hydrogen bonds, whereas the so-called intrinsic proteins within the membrane assume an α-helix structure—where the polar parts of the protein are exposed to the internal or external liquid environments and the non-polar parts are buried within the fatty acid carbon chain[86, 95]. Moreover, it has recently been shown that there are differences in the protein and phospholipid content of the external and internal surfaces of the cell membrane, indicating that the bilayer itself is asymmetrical[44, 95]. It should also be noted that it is not the case that the entire membrane is always fluid. The degree of fluidity changes with temperature and it is known that the lipid molecules immediately surrounding a membrane protein molecule (the so-called boundary lipids) are not mobile[123]. As a consequence, in membranes with a high content of enzyme (protein) molecules, such as the mitochondrial membrane, roughly 30% of the lipids are immobile[43, 44, 197].

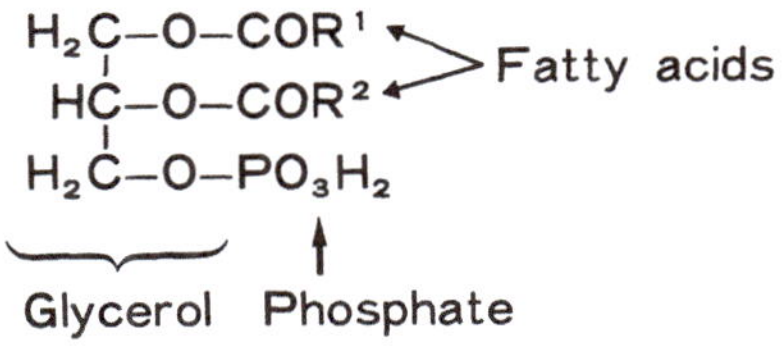

$$H_2C-O-COR^1$$
$$HC-O-COR^2$$
$$H_2C-O-PO_3H_2$$

Fatty acids

Glycerol Phosphate

Fig. 5-2. The structure of phosphatidic acid

The phospholipids which make up the biomembrane are comprised fundamentally of phosphatidic acid which is glycerol with three ester-bonds of two fatty acids (C_1 and C_2) and a phosphate (C_3) (**Fig. 5-2**). The phosphatidic acids

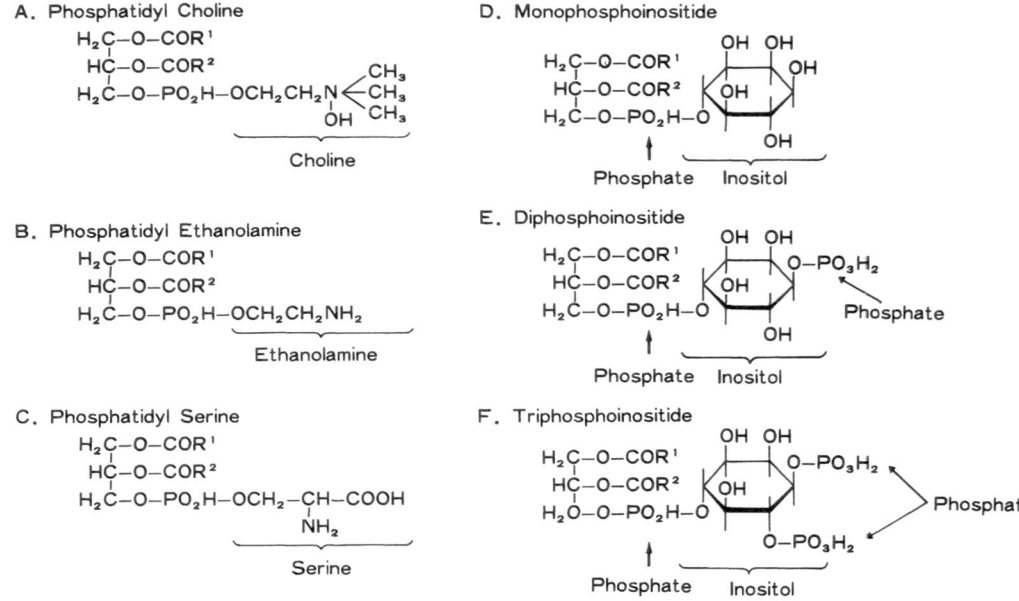

Fig. 5-3. The structural formula of phospholipids

connected with choline, ethanolamine or serine are called phosphatidyl choline, phosphatidyl ethanolamine, phosphatidyl serine. The phosphatidic acid which bind the ringed alcohol, inositol, is called phosphatidyl inositol and, depending upon the number of phosphates present, they are labeled mono-PI, di-PI or tri-PI (**Fig. 5-3**). The glycerol and phosphate structure corresponds to the hydrophilic head (polar) regions of the membrane structure and the hydrocarbon chains of fatty acids correspond to the hydrophobic tail regions (**Fig. 5-4**). Depending upon the presence of double bonds, the fatty acids are subdivided into saturated and unsaturated types. Typical of the unsaturated types are oleic acid ($C_{18:1}$), arachidonic acid ($C_{20:4}$) which contains four double bonds, and docosa-

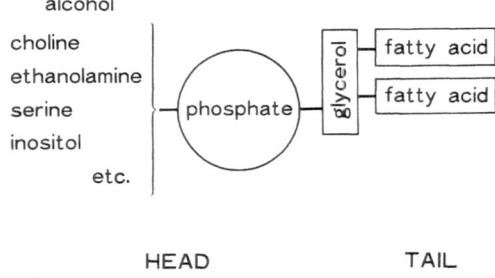

Fig. 5-4. The schema illustrating the structural model of phospholipid

hexaenoic acid ($C_{22:6}$), and typical of the saturated types are palmitic acid ($C_{16:0}$). They become the substrate for the autoxidation resulting in the formation of the alkyl radical, the alkoxy radical the peroxy radical or hydroperoxide (ROOH), while cyclooxygenase or lipoxygenase can convert them into various physiologically active substance, such as prostaglandin, thromboxane and leukotriene. These substances are known to play active roles in various pathological states and are consequently the focus of much current research.

5.3 Ca^{2+} Homeostasis

In the normal condition for mammalian cells, the extracellular concentration of Ca^{2+} is about 10^{-3}M, whereas the intracellular Ca^{2+} concentration is maintained at a level some 10,000-fold lower $(10^{-7}-10^{-8}$M)[9]. In the normal cell, the influx of Ca^{2+} occurs over three routes: via the trans-membrane voltage-dependent Ca^{2+} channels[9, 10, 14], via the Na^{+}–Ca^{2+} exchange transport system[162, 191] and via passive diffusion[50]. The efflux of Ca^{2+} occurs by way of ATP-dependent Ca^{2+}–ATPase and the Na^{+}–Ca^{2+} exchange transport system[13, 22, 162, 191]. For each Ca^{2+} transported out of the cell three Na^{+} enter. Another possible way to increase intracellular free Ca^{2+} is to depend on receptor-linked polyphosphoinositides hydrolysis. Inositol phospholipid metabolism is coupled to several neurotransmitter receptors in mammalian CNS including cholinergic, adrenergic, and histaminergic receptors. Stimulation by some agonist-receptor coupling causes activation of phospholipase C, results in the breakdown of triphosphoinositide and leads to the release of diacylglycerol (DG) and IP$_3$ (inositol 1, 4, 5-triphosphate). DG can then activate protein kinase C, and IP$_3$ has been reported as an intracellular calcium-mobilizing agent (**Fig. 5-5**).

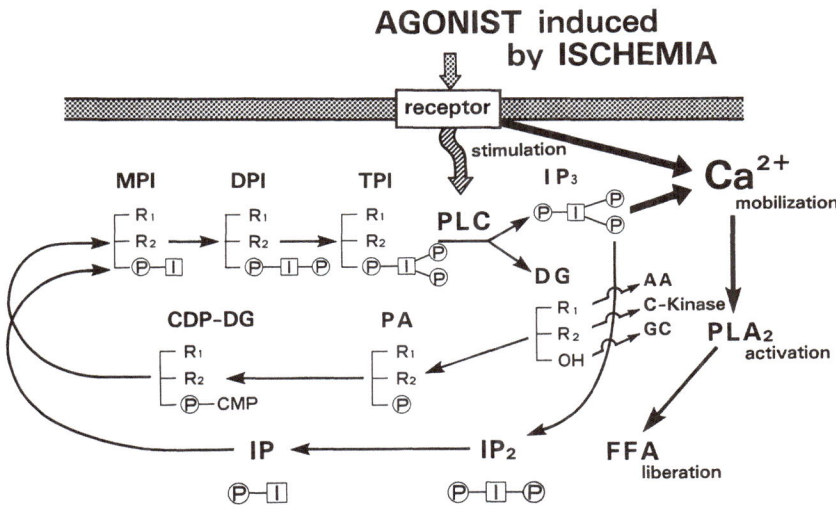

Fig. 5-5. A scheme illustrating the possible role of polyphosphoinositides due to the agonist-receptor coupling mechanism, which mobilizes intracellular Ca^{2+} in neurons. *MPI* phophatidyl inositol, *DPI* diphosphoinositide, *TPI* triphosphoinositide, *IP* inositol phosphate, *IP$_2$* inositol diphosphate, *IP$_3$* inositol triphosphate, *DG* diglyceride, *PA* phosphatidic acid, *CDP-DG* cytidine diphosphate-diglyceride, *CMP* cytidine monophosphate, *FFA* free fatty acid, *PLA$_2$* phospholipase A$_2$, *PLC* phospholipase C

Calmodulin, one of the calcium-binding proteins, is known to have a concentration of about 10 uM in the brain and a Kd value of 2×10^{-6}. Each molecule of calmodulin binds four Ca^{2+} and as a Ca^{2+}-receptor, calmodulin is known to participate in the activation or suppression of enzymes which phosphorylate various proteins[130, 131, 136, 154]. In order to main-tain Ca^{2+} homeostasis at various intracellular sites, a normal supply of ATP and O_2 is required. When there is energy failure due to ischemia, the Ca^{2+} balance is destroyed and it is known that either through Ca^{2+} itself or through calcium-binding proteins such as calmodulin, the above-mentioned enzymes become activated.

5.4 Lipid Metabolism in Cerebral Ischemia

As a result of increased lipid catabolism due to an ischemic load, a rapid increase in free fatty acids occurs[17, 89, 148, 186, 195, 216, 248]. This phenomenon has been demonstrated experimentally in our department. acid $(C_{20:4})$ and docosahexaenoic acid $(C_{22:6})$], whereas 30 minutes following the start of ischemia the free fatty acid content rises to 2,146.2 nmol/g wet weight. Following recirculation, however, free fatty acid levels rapidly

Table 5-1. Free fatty acid contents of rat ischemic brain cortex (mean ± S.E.M., nmol/gm wet weight, n = 5 or 6)

	Total free fatty acids	Palmitic acid	Stearic acid	Oleic acid	Arachidonic acid	Docosahexaenoic acid
Pre-Ischemia	182.0±15.0	92.1± 8.5	50.2± 8.9	24.0± 4.5	23.0± 2.0	10.1± 4.0
Ischemia 5 min	1035.0±61.8**	172.7± 8.9*	320.1±18.7**	113.6±11.6**	364.0±21.2**	33.8± 2.6'
Ischemia 30 min	2146.2±130.9**	530.0±44.4**	588.0±41.6**	343.0± 7.2***	481.0± 8.8**	161.0±35.4'
Recirculation 10 min	1043.2±76.2**	232.9±20.8**	468.0±37.9**	154.9±43.4**	131.2± 8.0**	68.5±19.8'
Recirculation 30 min	769.0±63.0**	160.8± 9.0**	414.9±34.6**	37.1± 3.9***	29.5± 7.8***	13.0± 2.3

Different from pre-ischemia.   * $P<0.001$,   ** $P<0.01$,   *** $P<0.05$ mean±S. E. M

Using a global cerebral ischemia model produced by 3-vessel occlusion, we have found a rapid 7–5 fold increase in free fatty acids, as compared to the normal state. That is, prior to vascular occlusion, there is a total free fatty acid content of 182.0 nmol/g wet weight [the sum of palmitic acid $(C_{16:0})$, stearic acid $(C_{18:0})$, oleic acid $(C_{18:1})$, arachidonic decrease: after 10 minutes of reflow, to 1,043.2 nmol/g and after 30 minutes, to 769.0 nmol/g. The volume of each type of fatty acid is shown in **Table 5-1** and **Fig. 5-6**. The ischemic load produced the largest rate of increases in oleic acid and arachidonic acid—both of which were found to have increased 10 fold after 30 minutes. Those free fatty acids which showed only small rates of increases were the saturated fatty acids, palmitic acid and stearic acid, although they are the highest in absolute values after 30 minutes of ischemia. Following

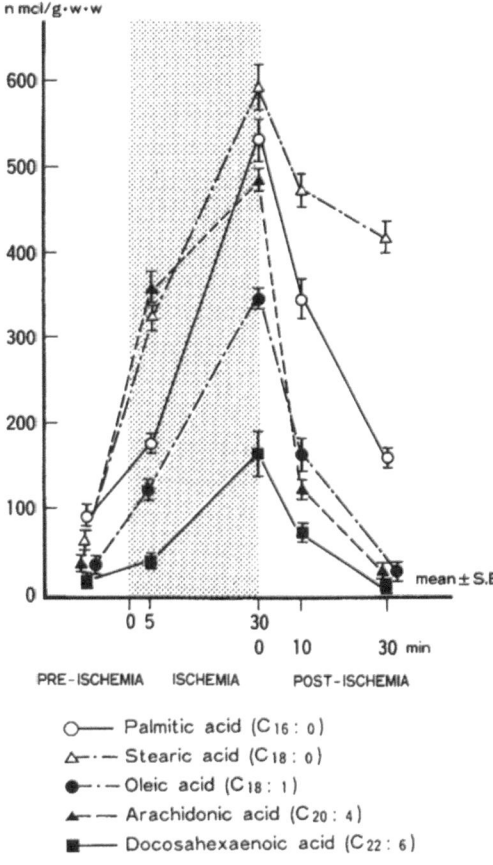

○—— Palmitic acid (C$_{16:0}$)
△—·— Stearic acid (C$_{18:0}$)
●—·— Oleic acid (C$_{18:1}$)
▲—— Arachidonic acid (C$_{20:4}$)
■—— Docosahexaenoic acid (C$_{22:6}$)

Fig. 5-6. The free fatty acid contents of rat forebrain. The free fatty acids accumulated by ischemic insult were immediately reduced following the start of recirculation

reflow, there were rapid decreases in all of the fatty acids, but after both 10 and 30 minutes of recirculation, the greatest decreases were found for arachidonic acid. The tendency for decreases in saturated fatty acid, palmitic acid and

stearic acid, was weak (**Table 5-1** and **Fig. 5-6**).

In the present section, the mechanism and significance of the appearance of free fatty acids will be discussed.

At the start of ischemia, there is a decrease in O_2 and ATP. Initially, Ca^{2+} homeostasis is maintained due to the transient increases in activity of Na$^+$, K$^+$-ATPase[102, 239] and Na$^+$, Ca^{2+}-exchange transport system, but ultimately homeostasis is lost and there are increases in the intracellular Ca^{2+} content[103, 105]. In addition, the hypoxia produces deficits in the mitochondrial redox system, the (H^+)–(OH^-) gradient is no longer maintained, and Ca^{2+} is released into the cellular cytoplasm. Increases in intracellular Ca^{2+} result in the facilitation of deacylation, which in turn causes the release of free fatty acids[23].

As discussed earlier, the hydrolysis of triphosphoinositide begins with the activation of phospholipase C by coupling receptor occupation, resulting in the breakdown of triphosphoinositide into DG and IP$_3$. DG is then broken down into glycerol and free fatty acids by the action of DG lipase and MG lipase. IP$_3$, which is another possible product of receptor-linked inositide acts as an intracellular calcium-mobilizing agent (**Figs. 5-5, 5-7**). Polyunsaturated fatty acids with more than two double bonds are comprised mainly of PE. PI containing 33.8% arachidonic acid (C$_{20:4}$) in particular plays a major role as a substrate of lipid peroxidation and in the arachidonate cascade[59, 109, 148]. In addition, there is deacylation due to phospholipase A$_1$ and A$_2$. As a result of the increases in intracellular free Ca^{2+} due to progressive ischemia and the breakdown of polyphosphoinositides, the activity of phospholipase A_1 and A_2 is facilitated and the fatty acids at positions C_1 and C_2

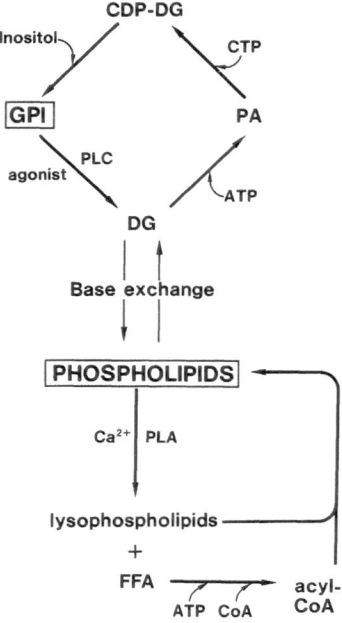

Fig. 5-7. The interaction between the inositol cycle and the metabolism of other phospholipids. It was thought that phosphatidyl inositol (*GPI*) is hydrolized into diglyceride (*DG*) and inositol phosphate moiety as the "phospholipid effect" (Michell 1975). But recently it has been realized that triphosphoinositide (*TPI*) is the main donator of diglyceride in the agonist receptor coupling mechanism rather than phosphatidyl inositol (Berridge 1984). GPI = MPI; See **Fig. 5-5**. This schema was reprinted from **Fig. 1** of Path Biol 30, p 271 (1981, Ref. No. 239) with the permission of T Wieloch and BK Siesjö

of the phospholipid become hydrolyzed—Producing lysophospholipid and free fatty acids[19, 20, 216, 241]. During ischemia, the ATP dependent reacylation of acyl-Co A is suppressed due to a deficiency of ATP[18], and

lysophospholipid is broken down by lysophospholipase into free fatty acids and water soluble substances (**Fig. 5-7**).

Due to the activation of the above two major reaction systems (the inositol cycle and deacylation), there is a rapid increase in free fatty acids during ischemia, as first reported by Bazán *et al.*, in 1971[17]. They found that, using an ischemia model in which the rat was decapitated and the head was kept at room temperature, free fatty acids increased rapidly over the first 30–240 seconds of ischemia and reached a plateau after 5–12 minutes. In 1981, the same decapitation model was used with preservation of the brain at 37 °C, and it was found that the increase in free fatty acids was linear for the first 60 minutes of ischemia (Nemoto)[165, 166, 206, 207]. In addition to results using the decapitation model, Yoshida reported that marked free fatty acids accumulated following 30 minute occlusion of the bilateral common carotid artery in the gerbil (1980)[248], and it was shown that both complete and severe incomplete cerebral ischemia cause a similar increase in the tissue content of free fatty acid (Rehncrona 1982)[195].

Several studies have subsequently been carried out on the decrease in phospholipids during ischemia, since it is known that the increase in free fatty acids during ischemia is brought about by the decomposition of phospholipids by phospholipase and the suppression of reacylation due to an ATP deficiency.

Following 30 minutes of occlusion of the bilateral common carotid artery (CCA) in the gerbil, Yoshida *et al.*,

(1980) found a 4% decrease in PC and a 16% decrease in PE[248]. Marion et al. injected (^{3}H)-arachidonic acid into rat cerebral ventricles and then decapitated the animals[148]. They found that most of the free arachidonic acid was derived from PC and PI. Using an incomplete cerebral ischemia model in the Mongolian gerbil, Majewska et al. found the largest decrease in PE, followed by PC, in the mitochondria of the guinea pig brain approximately 5 minutes after the start of ischemia induced by decapitation[147]. In contrast, Rehncrona et al. reported that regardless of the complete or incomplete nature of the ischemia—there were no changes in phospholipid content due to ischemia, other than a very small decrease in PI and PS[195].

As mentioned above, the liberation of free fatty acids from phospholipids during ischemia implies destruction of the membrane structure, but the free fatty acids have several other effects, some of which are deleterious. First, they produce deficits in mitochondrial respiratory activity. Due to the ischemia, the volume of mitochondrial free fatty acids increases and, acting as an uncoupler, they reduce the ADP/O ratio as a result of a deficit in the oxidative phosphorylation[140, 173—175]. The second effect of the free fatty acids concerns their involvement in brain edema. Chan et al. placed cerebral cortical slices into various solutions containing free fatty acids, and reported the occurrence of edema in the presence of polyunsaturated fatty acids[46]. In addition, in in vivo experiments where extrinsic arachidonic acid has been injected directly into the cerebrum, the occurrence of edema has been observed[188]. Further details of the mechanisms involved are not known, but it is thought that polyunsaturated fatty acids, especially an arachidonic acid, play a key role in the appearance of edema.

The third effect of free fatty acids concerns their involvement in the aggravation of ischemic brain damage. The increased volume of free fatty acids becomes substrate for autoxidation due to free radicals and is converted into alkyl radicals, alkoxy radicals, peroxy radicals and hydroperoxide at the start of reflow, or they become the substrate for enzymatic lipid peroxidation and are transformed into physiologically-active substances, which aggravate the ischemic brain damage.

Free fatty acids, which increase during ischemia, begin to decrease rapidly together with the start of recirculation. This effect is thought to be due to: (i) the activation of reacylation due to a recovery of ATP, (ii) the wash out of the venous system by reflow, (iii) increasing free radical reactions which use free fatty acids as substrates, and (iv) the formation of eicosanoids, such as leukotriene, prostaglandin, TXA_2 and prostacyclin, due to enzymatic lipid peroxidation of arachidonic acid.

In the case of the enzymatic peroxidation of arachidonic acid (cyclooxygenase), arachidonic acid is converted into PGG_2 (which is a PG endoperoxide) and PGH_2, and then into $PGF_{2\alpha}$, PGD_2 and PGE_2[61, 90, 208]. PGI_2 (prostacyclin) is formed from PG endoperoxide and TXA_2 is also synthesized[1, 242]. Due to the action of lipoxygenase, arachidonic acids are also converted into hydroperoxyeicosatet-raenoic acid (HPETE), HETE[213, 228] and leukotriene[56, 129, 160, 163]. In this manner, recirculation produces a variety of physiologically-active substances, which are thought to have a variety of effects on the permeability of blood vessels, the dilatation and constriction of vessels, platelet aggregation and leucotaxis[36, 158]. Details concerning the mechanisms of action remain to be determined.

Finally, it should be mentioned that the autoxidation due to free radicals, which is another form of lipid peroxidation, will be discussed in a later section (Section 5-5).

5.5 The Free Radical Reaction in Cerebral Ischemia

The neurosurgeon who frequently encounters clinical cases of cerebral infarction is often forced to consider why nerve cells are so susceptible to damage following ischemia. Although it is well-known that brain cells from which the energy reservoir has been depleted due to several minutes of ischemia will die, it is not the case that the vulnerability of nerve cells to ischemia can be explained entirely in terms of the depleted energy supply. An example which illustrates this fact is a paradox obtained from energy metabolism research, which was so actively pursued from the 1960s[210]. It was found that, following recirculation, the recovery of energy metabolism was less impaired after complete ischemia than after incomplete ischemia, where a certain level of blood flow was maintained[167–169, 192].

Research on the superoxide anion (O_2^-) reached a turning point during the 1960s when Fridovich et al. discovered the superoxide dismutase[151]. This enzyme allowed for the measurement of the superoxide anion, which previously could not be detected due to its short-lived nature, and subsequently a wealth of data and hypotheses have been published on this topic.

In the early 1970s, Demopoulos et al. suggested that autoxidative lipid peroxidation due to free radicals is an important problem in the ischemic brain[64]. Then many researchers published several influential studies on the measurement of α-tocopherol, ascorbate and other intrinsic radical scavengers and conjugated dienes, fluorescent substances and TBA-RS, which can be used as parameters of lipid peroxidation. These experimental studies strongly suggested the occurrence of lipid peroxidation due to free radicals in ischemic or postischemic brain damage. Here, the destruction of biomembranes due to lipid peroxidation may constitute a significant clue in resolving the above-mentioned paradox and in identifying the important pathological factors which must be dealt with in the clinical treatment of brain ischemia.

In general, lipid peroxidation can be subdivided into two reaction systems: (i) autoxidative lipid peroxidation due to the free radical reaction, and (ii) enzymatic lipid peroxidation (the arachidonate cascade). Although both of these reaction systems produce hydroperoxides, they are independent and little is known concerning their mutual influences. In the present chapter, we will discuss our own findings from experimental research on lipid peroxidation in relation to previously reported data.

5.5.1 Fundamental Knowledge Concerning Free Radicals

The earliest researchers on the problem of lipid peroxidation due to the free radical reaction were Gee, Bolland and Bateman and colleagues at the British Rubber Producers Research Association[15, 28]. Throughout the 1940s two paired electrons, and since a free radical contains an unpaired electron, it has a strong affinity for other electrons and, being unstable, is highly reactive. For example, the oxygen atom or molecule in itself is a free radical. In the

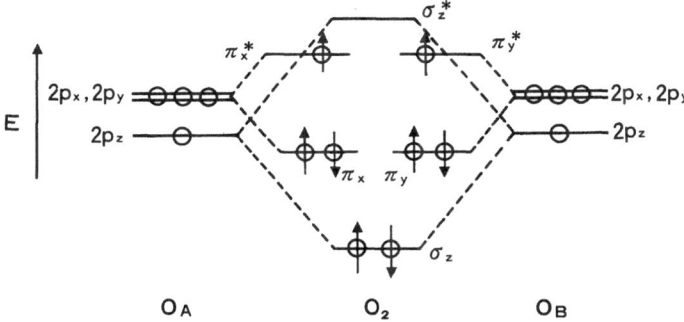

Fig. 5-8. The electron configuration of the oxygen molecule

and 1950s they did excellent research in this field and already clarified several of the fundamental mechanisms involved prior to the 1960s. Although their work was devoted primarily to the problem of the deterioration of rubber due to lipid peroxidation, already at that time, many studies had been carried out on free radical polymerization within the areas of concern to the chemical industry (the development of synthetic rubbers, plastic production, etc.)

The term "free radical" is used to mean an atom or molecule with an unpaired electron in its valence electron orbital. Most electron orbitals contain manner of the atomic orbital or molecular orbital method, of the 8 electrons of the oxygen atom, 2 are found in each of 1s, 2s, and $2p_x$ oribtals, and the remaining 2 electrons are unpaired in, respectively, the $2p_y$ and $2p_z$ orbitals. In the oxygen molecule, 4 electrons enter the K-shell and, the 12 electrons of the L-shell are distributed as follows: 4 in the 2s orbital, 2 in the 2p (σ_z) orbital, 4 in the π_x and π_y bonding orbitals, and 2 in the π_x^* and π_y^* antibonding orbitals (**Figs. 5-8**). These two electrons in the π_x^* and π_y^* are the basis for the general oxygen molecule ($^3O_2$) being biradical. (1, 2: indicate

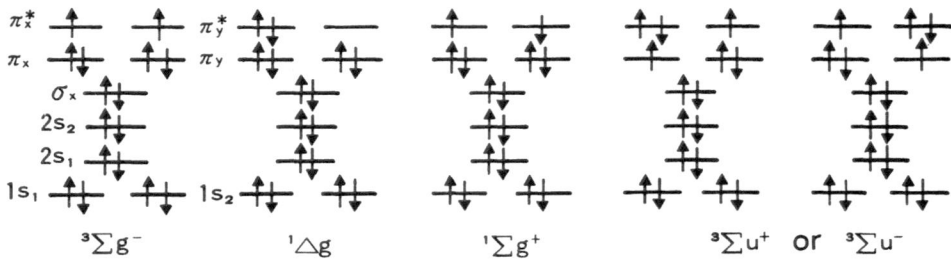

Fig. 5-9. The electron distribution of the oxygen molecule in the ground state and excited state (see Fig. 5-17)

principle quantum number. Smaller the number, inner the orbital is indicated. s, p_x, p_y, p_z indicate the shapes of the orbital. K, L, M-shell indicate the groups of the same quantum number. The K-shell is the most inner orbital group and L-shell is the next one.) In this manner, the oxygen molecule is a biradical contained 2 unpaired electrons. The spins of the unpaired electrons are parallel—implying that the resting state of the molecule is a triplet state (**Figs. 5-9, 5-17**). Although it can be said that triplet oxygen ($^3O_2$) is itself a free radical, within biological systems, some such oxygens are so-called "active oxygens" with still greater reactivity being generated. That is, there are the superoxide anion (O_2^{-}), hydrogen peroxide (H_2O_2), the hydroxy radical (OH·), and singlet oxygen ($^1O_2$)—Most of which can be produced due to the incomplete reduction of oxygen at the termination of the electron transport system. Oxygen-metal complexes, such as the perferryl-oxygen complex ($Fe^{2+}-O_2^{-}$), are also referred to as active oxygens. Among these molecules, H_2O_2, $^1O_2$ and oxygen-metal complexes are not in fact free radicals (**Fig. 5-9**), but they are highly reactive and are consequently involved heavily in lipid peroxidation.

5.5.2 Lipid Peroxidation

The series of reaction mechanisms reported by Bateman *et al.* still form the basis for current ideas on the autoxidation of lipids[15]. That is, the reaction system is comprised of three stages: (i) initiation of the reaction, with the removal of hydrogen from α-methylene carbon, (ii) propagation, with a large increase in radical production, and (iii) termination, with the synthesis of non-radical products and a decrease in radical production. The H· of fatty acids is removed and the synthesized radical is called the alkyl radical (R·). Oxygen is combined with the alkyl radical, producing an peroxyradical (ROO·) and, furthermore, the alkoxy radical (RO·) and hydroperoxide (ROOH) are produced. Here the R symbolizes a lipid (alkyl group) which is a hydrated carbon chain.

As discussed earlier, the lipid autoxidative reaction is initiated by the removal of H· from an α-methylene carbon near a double bond of a polyunsaturated fatty acid. This phenomenon occurs because the dissociation energy of the allyl bond is 85 kcal/mol—*i.e.*, somewhat lower than

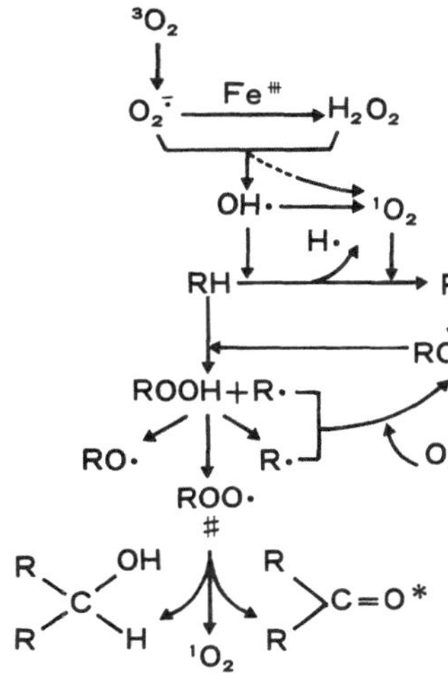

$O_2^{\bar{\cdot}}$	superoxide anion
H_2O_2	hydrogen peroxide
$OH\cdot$	hydroxy radical
$^1O_2$	singlet oxygen
#	Russel's mechanism

Fig. 5-10. Free radical reaction in lipid peroxidation. $R\cdot$ alkyl radical, $ROO\cdot$ peroxy radical, RH unsaturated fatty acid, $RO\cdot$ alkoxy radical, $ROOH$ lipid hydroperoxide. $OH\cdot$ is produced focally by the Haber-Weiss reaction under catalysis of the chelated metal ion in the phospholipid fatty acid part, and initiates peroxidation of membranous lipid. $^1O_2$ is probably branched off by Russel's mechanism (#)

the 94 kcal/mol of the C–H bond[8]. The alkyl radical ($R\cdot$) produced by this reaction forms a conjugated diene due to the transfer of an electron from a carbon

atom[55, 107]. Next, O_2 is added, thus forming a peroxyradical ($ROO\cdot$) **(Fig. 5-10)**. In these processes, the fatty acid is changed from its bent *cis* configuration into a straight *trans* configuration[202]. While the peroxyradical ($ROO\cdot$) becomes stabilized as hydroperoxide ($ROOH$) by the removal of $H\cdot$ from a nearby unsaturated fatty acid, alkyl radicals ($R\cdot$) produced by this reaction repeatedly continue the same reaction in the presence of oxygen. The above steps are thought to be the essence of the lipid autoxidation reaction, but O_2 added to the alkyl radical ($R\cdot$) not only produces an oxygen-centered radical, it is also used for generation of endoperoxide due to electron transfer. Due to Fe^{2+} catalysis, the hydroperoxide ($ROOH$) may be "redox dissociated" into the alkoxy radical ($RO\cdot$) and the hydroxy radical ($OH\cdot$)[8, 12]. In this way, it can be understood that once the reaction has been initiated, the propagation is repeated and the unsaturated fatty acids are consumed while acting as the substrate for the radical reaction **(Fig. 5-10)**. For termination of the free radical reaction, one of the most important actions is the termination of the peroxy radical reaction, as first noted by Russel in 1957[198] **(Fig. 5-11)**:

$$ROO\cdot + ROO\cdot \rightarrow RC = O^* + ROH + {}^1O_2 \ (\text{*indicates the exited state})$$

That is, the peroxyradicals of two molecules form an excited carbonyl compound, alcohol and singlet oxygen. Furthermore, carbon atoms of alkoxy radical ($RO\cdot$) repeatedly receive the attack of the free radical and are converted into aldehydes and alkyl radicals, and long fatty acid chains are fragmented to produce final products such as penthane and ethane[66]. These processes can be expressed in chemical notation as follows:

Initiation:

$$RH \rightarrow R\cdot + H\cdot$$

Propagation:

$$R\cdot + O_2 \rightarrow ROO\cdot$$
$$ROO\cdot + RH \rightarrow ROOH + R\cdot$$

Termination:

$$2\,ROO\cdot \rightarrow$$
$$R\cdot + ROO\cdot \rightarrow \left.\begin{array}{l} \\ \\ \\ \end{array}\right\} \text{Non-radical products}$$
$$2R\cdot \rightarrow$$

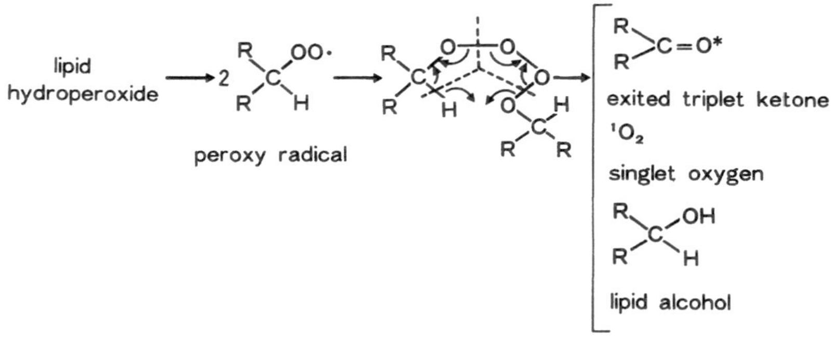

Fig. 5-11. One of the terminations of free radical lipid peroxidation (Russel's mechanism)

5.5.3 Detection of Free Radicals

It is virtually impossible to measure directly the radicals involved in autoxidation due to their short lives and instability. For this reason, various metabolites of lipid autoxidation have been the objects of chemical quantification—including TBA-RS[27, 134, 135, 193, 237, 245], fluorescent substances[193, 225, 226], conjugated dienes[92, 134, 193, 236], and ethane and penthane in the expired gases[97, 196, 226]. Moreover, from the consumption by intrinsic radical scavengers such as α-tocopherol, it has been reported that the state of autoxidation can be indirectly determined[2, 51, 78, 134, 238, 247]. We have elsewhere reported our own experimental research on ischemic brain tissue using chemiluminescence, which has not previously been used in this field, and electron spin resonance techniques. Our studies have included the free radical reaction in both the ischemic brain and the hypoxic brain[112, 113, 114, 231, 232]. Here we will discuss the chemiluminescence and ESR techniques, and refer the reader to other textbooks for details concerning parameters such as TBA-RS, fluorescent substances, etc.

5.5.3.1 Chemiluminescence

The chemiluminescence technique is a method by which the ultraweak luminescence (10^{-14}W) emitted during lipid peroxidation is measured[32, 112, 113, 114, 164, 214, 233]. Unlike the so-called bioluminescence technique, which makes use of the lucipherine-lucipherase system[201], the fundamental principle of chemiluminescence is the detection of photons as electrical pulses, having passed through a photomultiplier with an analyzer (**Fig. 5-12**)[79, 116, 157, 159, 204].

It is thought that Robert Boyle was the first to have reported luminescence during peroxidation by oxygen[31], but the first to have published experimental work on the chemiluminescence during lipid (alkyl groups) peroxidation was Vassil'ev and Vichutinskii of the Soviet Union (1962)[233]. And the first to use the chemiluminescence technique in the measurement of biological materials was also a Russian, Tarusov[227]. Since that time, the quality of chemiluminescence equipment has improved and this technique is now widely used to measure lipid peroxidation in various biological systems[116].

One of the principal advantages of

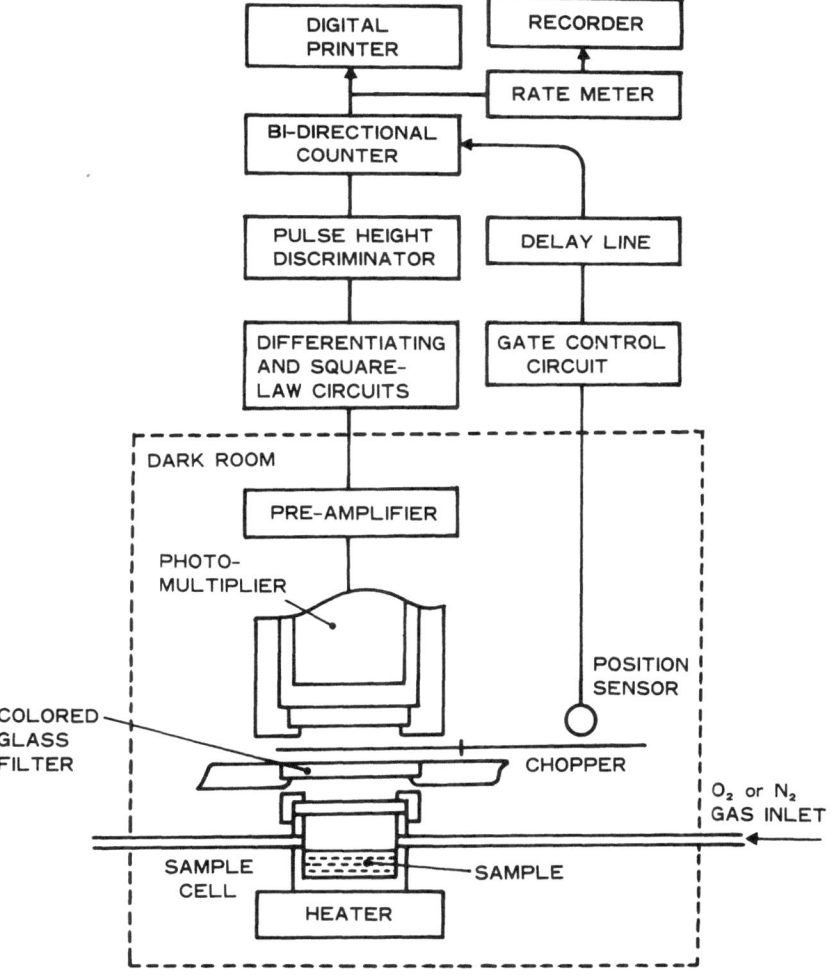

Fig. 5-12. The photon counting apparatus for the measurment of chemiluminescence. This schema was reprinted from **Fig. 1** of Opt Lasers Engineering, 3, p 128 (1982, Ref. No. 116) with the permission of one of the author, H Inaba

the chemiluminescence technique is that it is non-invasive and can be used to measure *in situ* reactions on the surface of internal organs[31]. Several outstanding studies of this kind have been reported by Boveries, Cadenas and Chance in research on liver[30, 31, 32, 37-42]. The chemiluminescence detected by Boveries *et al.*, however, was not a phenomenon due to pathological liver ischemia; that is, it entailed a peroxidation state which was artificially induced by the circulation of hydroperoxides, such as t-BuOOH, through an infusion pump connected directly to the frog liver[30, 32]. In this regard, it is questionable that meaningful chemiluminescence measurements could be made *in situ* from the surface of the ischemic brain, given the level of precision of current analytic equipment. *In vivo* measurement of ischemic brain was impossible in our study. The essence of chemiluminescence is generally thought to be the energy production due to the transitions of singlet oxygens ($^1O_2$) or excited carbonyl compounds, which occur according to Russel's mechanism (one of the termination reactions of lipid peroxidation)[198].

$$ROO \cdot + ROO \cdot \rightarrow ROH + RC = O + {}^1O_2$$
$$ROO \cdot + ROO \cdot \rightarrow ROH + RC = O^* + O_2$$
$$RC = O^* + O_2 \rightarrow RC = O + {}^1O_2$$
$RC = O^*$: excited carbonyl compound
$^1O_2$: singlet oxygen

In addition, some of the $^1O_2$ molecules react with the double bond of polyunsaturated fatty acids and, after the production of an unstable dioxe-

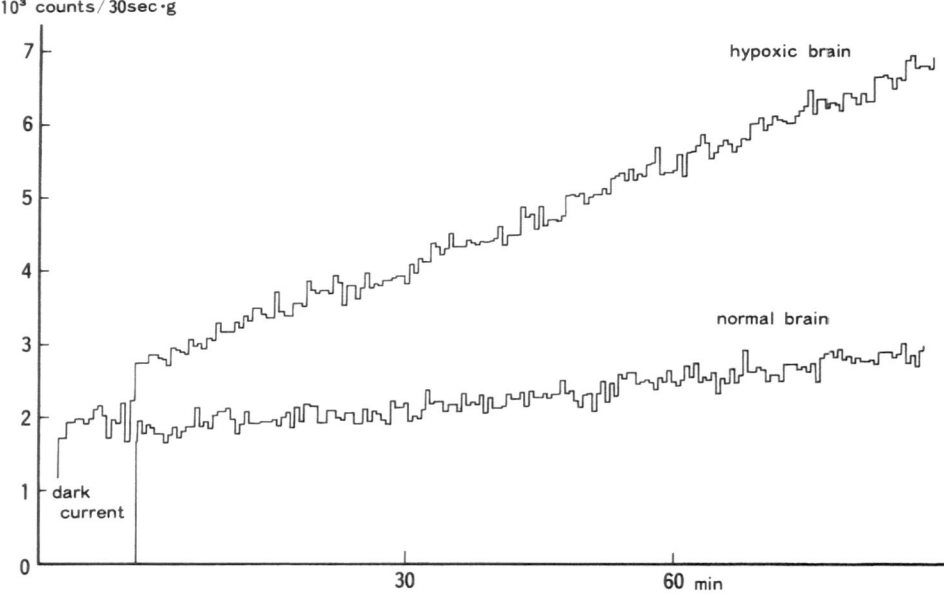

Fig. 5-13. Sequential changes in the chemiluminescence of normal (pre-hypoxic) brain and 5-min post-hypoxic brain (1.0 g) exposed to room air in a heated cell (35–36 °C). There were no increases for 40–60 min in normal brain

tane, produce an excited carbonyl compound[82].

$$-C=C-+\,^1O_2 \rightarrow C=O* + C=O$$

The luminescence phenomenon which accompanies the transitions of the $^1O_2$ and excited carbonyl compounds can be expressed as follows:

$$^1O_2 + \,^1O_2 \rightarrow 2 \quad ^3O_2 + h\nu$$
$$RC=O* \rightarrow RC=O + h\nu$$

where $h\nu$ is the photon energy. The light originating from $^1O_2$ ($[^1\Sigma g^+][^1\triangle g]$, $[^1\triangle g][^1\triangle g]$) is found at 480, 520–530, 580, 635 and 680–700 nm in the spectral field and that from the excited carbonyl compounds at 420–450 and 530 nm[32, 5, 57, 115, 124].

When chemiluminescence measurements were made during incubation in air at 35–36 °C of a rat brain homogenate which had been obtained after 5 minutes of reoxygenation following 5 minutes of hypoxia ($PaO_2 = 17$–22 mmHg), photon counts increased time dependently, but there were no increases in normal (no-loading) brain for 40–60 min (**Fig. 5-13**). This difference in photon counts make the hypoxic brain to be analysed spectrum. That is, if spectral analysis is done at the time when the photon counts of chemiluminescence exceeds 3,000 counts/30 sec · g in hypoxic brain, peaks will be at 480, 520–530, 570, 620–640 and 680–700 nm (**Fig. 5-14**). If a similar analytic technique is used with an ischemic rat brain homogenate obtained after 30 minutes of reflow following 5 minutes of ischemia (due to 3-vessel occlusion)[127], peaks identical to those found in the hypoxic brain are obtained. These findings suggest that

the chemiluminescence obtained from hypoxic and ischemic brain homogenates is due to the $^1O_2$ derived from lipid peroxidation.

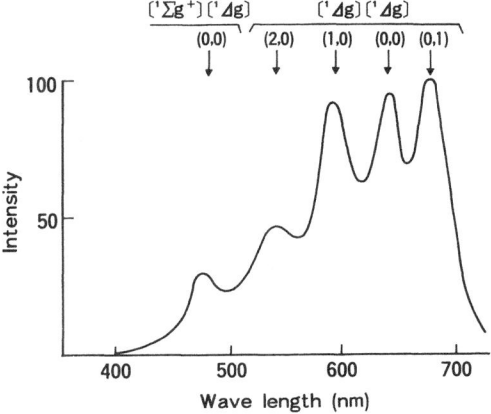

Fig. 5-14. Spectral analysis of chemiluminescence in 5-min post-hypoxic brain *in vitro*. Note that spectral peaks were found at 480, 520–530, 570, 620–640 and 680–700 nm, which means that chemiluminescence was caused by transition of singlet oxygen to the ground state

5.5.3.2 Electron Spin Resonance (ESR)

Due to the fact that ESR characteristically detects only the paramagnetic species in complex reaction systems, it has been widely used in a variety of biological reactions which produce radical species. The fundamental principles of ESR are the same as those of nuclear magnetic resonance. When an atom or molecule with an unpaired electron is placed in a magnetic field of constant strength, the electron orbital is split into specific energy levels. Electron spin resonance (ESR) makes use of this phenomenon by measuring the absorption or radiation of microwaves associated with the transition of electrons between these specific energy levels. Among the techniques employed in the ESR method are detection at room temperature or at ultralow temperatures, by spin-trapping[119] and spin-labeling[21, 200]. As yet, only two studies have been carried out on the ESR measurement of radicals using ischemic brain materials. Using a feline middle cerebral artery occlusion model, Flamm *et al.* found a continuous decrease in the ascorbate radical when measured at room temperature[78] and Fujita *et al.* have reported a spintrapping study (using DMPO) of decapitated rat brain tissue[87]. It is known that the limiting concentration for ESR measurements of materials in water soluble systems is 10^{-6}–10^{-8} M[77]. It is consequently impossible to make direct measurements of short-lived radicals, such as the oxygen radical involved in lipid peroxidation. Several techniques have therefore been developed to circumvent this problem. Since the spin-trapping method allows for measurements of radicals to be made at room temperature, it is thought to be a particularly effective technique for the measurement of the short-lived radicals found in biological materials[77]. The fundamental principle involved is the trapping of short-lived radicals by nitrone- or nitroso-compounds and then their detection as relatively long-lived radicals. By means of analysis of the obtained ESR spectra, the trapped radicals can be identified and their abundance quantified.

Following developments in five research groups, in 1968–1969, various biological applications of ESR have been reported—including radical generation in the ischemic heart, drug metabolism in the liver[187], the detection of the actions of anti-cancer drugs[126, 215] and the phagocytosis of leucocytes[150]. The most frequently used trapping agents are PBN (phenyl-t-butyl nitrone) and DMPO (5,5'-dimethyl-pyrroline-N-oxide). Although PBN is the more stable compound and traps the radical for a relatively longer period, the trapping reaction of PBN is slower and identification of the trapped radical is more difficult[118, 120].

In our studies, we have used PBN as the spin-trapping agent in both the decapitated brain and 3-vessel occlusion model[231, 232]. Since radicals cannot be obtained when only PBN and the ischemic brain homogenate are used, the lipid peroxidation stimulators, NADPH and Fe-EDTA, are added to the reaction mixture and the samples are studied after incubation in nitrogen or air at 37 °C for a defined period of time (**Fig. 5-15**). The signals obtained using these experimental systems are a triplet of doublets (6 peaks)—which can be determined as the hyperfine coupling constants, $A_N = 16.2$–16.5 G and $A_\beta^H = 3.6$–3.8 G. The intensities of the PBN trapped radicals have been found to be oxygen-dependent—with the signals obtained from samples incubated in air being typically greater than those incubated in nitrogen (**Fig. 5-16**).

Since there have not previously been any reports of PBN spin-trapping using ischemic brain homogenates, it is necessary to investigate in detail the nature of the PBN trapped radicals.

Harbour *et al.* have experimentally shown that the hydroxy radical (OH·)

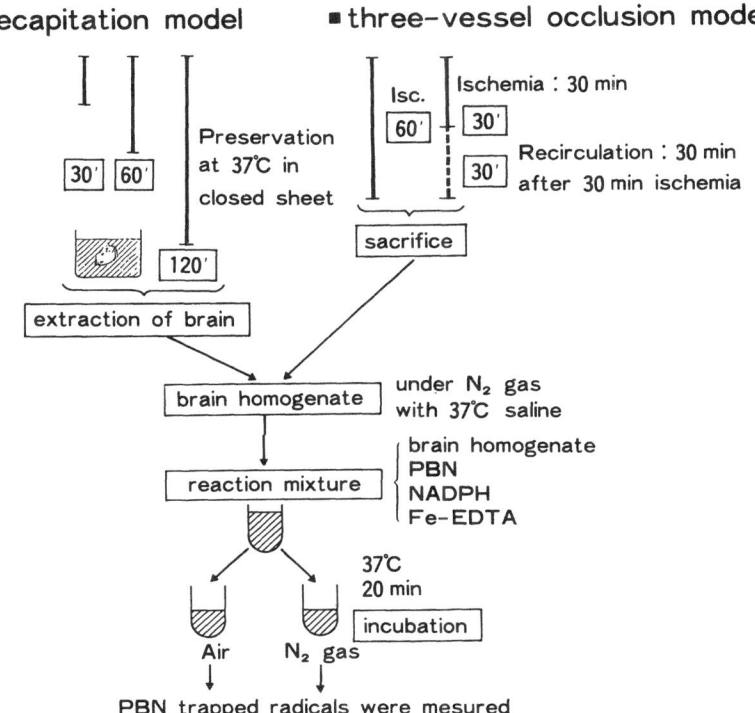

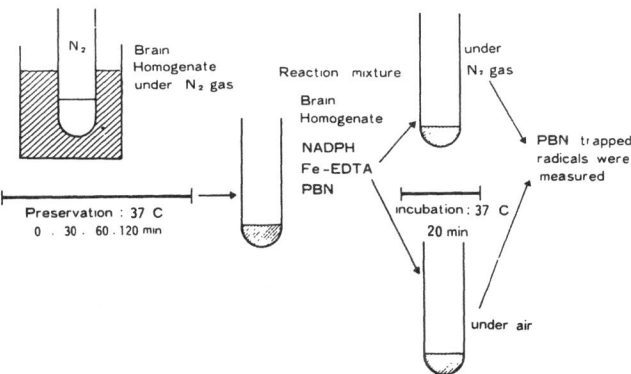

Fig. 5-15. The scheme illustrating the experimental procedures of our ESR study using rat decapitation model, the three-vessel occlusion model and the homogenate-preservation system. The spin trapping technique was applied to the detection of free radicals derived from NADPH-dependent lipid peroxidation in rat ischemic brain homogenate

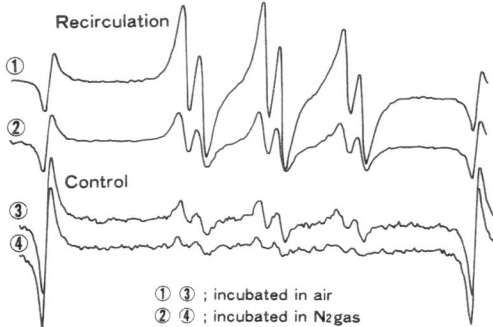

Fig. 5-16. ESR spectra derived from the spin adducts of PBN formed during incubation of reaction mixtures containing of PBN, NADPH, Fe-EDTA and brain homogenate for 20 min at 37 °C. The increase in the signal intensity in the recirculated group (30 min-recirculation following 30 min-ischemia) indicates the susceptibility of reperfused brain to lipid peroxidation

and the hydroperoxy radical (OOH·) are generated by the photolysis of hydrogen peroxide solution and that these radicals are trapped by PBN[104]. In such a case, the hyperfine coupling constants for OH· are $A_N = 15.3$ G and $A_\beta^H = 2.75$ G, and for OOH· they are $A_N = 14.8$ G and $A_\beta^H = 2.75$ G. These values differ from those obtained by us in our study (*i.e.*, $A_N = 16.2$–16.5 G and $A_\beta^H = 3.6$–3.8 G). In other words, the PBN-trapped radicals detected by us are not the active oxygens themselves. Of interest in this regard is a study by Saprin *et al.*[138, 139, 199], who detected PBN adducts of linoleic acid using a lipid peroxidation system, consisting of liver microsomes with NADPH and Fe-EDTA added and incubated at 37 °C. The hyperfine coupling constants obtained by them were analysed as being of two varieties: $A_N = 16.54$, $A_\beta^H = 3.35$ and $A_N = 15.5$, $A_\beta^H = 3.75$, *i.e.*, similar to those obtained by us. They concluded that they had trapped,

respectively, a small alkyl (methyl) group and an alkoxy group. In light of the fact that the production of the PBN-trapped radicals obtained from us also showed an oxygen-dependency in our experimental system, we believe that these radicals were, as indicated by Saprin *et al.*, derived from the peroxidation of free fatty acids.

In order to obtain sufficiently strong signals from the ischemic brain homogenate in the present experiments, it was necessary to add NADPH and Fe-EDTA. With regard to the lipid peroxidation stimulators, NADPH and Fe-EDTA, many experimental findings have been reported. Chan *et al.* found that superoxide anions (O_2^-) were produced by means of adding polyunsaturated fatty acids to brain slices and that the anion production was increased by the addition of NADPH[47]. It has also been reported that the redox reaction due to cytochrome P-450 reductase can be accelerated by the addition of NADPH and Fe-EDTA to liver microsomes—thus resulting in the production of OH radicals and the initiation of lipid peroxidation. Until recent by years, it was thought that cytochrome P-450 is not present in the brain, but by means of the immunohistochemical technique it has been demonstrated that cytochrome P-450 reductase exists in catecholaminergic neurons of rat brain[98]. In our experiments the fact that the intensity of PBN-trapped radicals was dependent upon the concentration of NADPH suggests the possibility that the reductase is mediating the production of this radical in the electron transport system of mitochondria or microsomes.

5.5.4 Initiators of Autoxidative Lipid Peroxidation

As discussed in Section 5-1, it is still unknown whether or not the free radical reaction occurs *in vivo* and when or where it might occur intracellularly. Nevertheless, it has previously been reported that the hydroxy radical (OH·), singlet oxygen ($^1O_2$), the superoxide anion (O_2^-) and oxygen-metal complexes may act *in vitro* as initiators of the removal of H· from polyunsaturated fatty acids. In the liver microsome (which is the organella has been most thoroughly investigated in this field), the participation of an oxygen-metal complex ($Fe^{2+}-O_2^-$) has recently been indicated[8].

While keeping in mind the specificity of the pathology of individual organs (*e.g.*, as a general rule cytochrome P-450 reductase is thought not to be present in the brain), selection and experimental demonstration of the possibility of lipid peroxidation, as reported in other biological systems, must be done in the ischemic brain. It is thought likely that the mechanisms of the appearance of the free radical reaction are closely linked with the problem of the timing (during or after ischemia) of the appearance of lipid peroxidation (to be discussed in Section 5.5).

5.5.4.1 Superoxide Anion (O_2^-)

In contrast to the normal oxygen molecule ($^3O_2$) which is a biradical, the O_2^- molecule is a mono-radical with one electron in an unpaired electron orbital. O_2^- is known to be produced by intact mitochondria as well, and Forman *et al.* have found that, in rat brain mitochondria, O_2^- is produced by dihydroorotate (DHO) dehydrogenase using DHO as a substrate[84]. DHO-dehydrogenase does not alone produce O_2^-, but does so in collaboration with the electron transport system's flavin enzyme found on the inner membrane of mitochondria. Moreover, O_2^- is produced from various substrates by means of key enzymes of the lipid peroxidation system in liver microsomes and by enzymes such as the NADPH-oxidase of leucocytes. On the one hand, the discovery of the superoxide dismutase (SOD) by Fridovich *et al.* has shown that O_2^- is converted into $H_2O_2 + O_2$ by SOD in biological systems, while on the other hand, it has also been shown that $^1O_2$ can be produced from O_2^- via the dismutation reaction[151, 152]. According to Foote[201, 202], the results of quantification of the peroxide products of cholesterol indicate that the percentage of $^1O_2$ produced from O_2^- is only 0.2% of the total volume of oxygen produced. From the above findings it is apparent that O_2^- is an important source of OH radicals, but it is thought that there is a low probability that O_2^- itself acts as an initiator *in vivo*. Chan *et al.*[47], in an experimental study concerning O_2^- with the ischemic rat brain slices, found signs of brain edema when extrinsic polyunsaturated fatty acids were added to brain slices. Using the reduction of nitroblue tetrazolium (NBT) as an index, there was the production of O_2^- was noted.

5.5.4.2 Hydroxy Radical (OH·)

The Haber-Weiss reaction[96], which produces OH·, is well known:

$$O_2^- + H_2O_2 \rightarrow OH· + OH^- + O_2$$

This reaction is based upon the interaction of the ferryl ion with hydrogen peroxide, as discussed by Fenton in 1894[234]:

$$Fe^{2+} + H_2O_2 \rightarrow Fe^{3+} + OH^- + OH·$$
$$Fe^{3+} + H_2O_2 \rightarrow Fe^{2+} + HO_2· + H^+$$

It is, however, doubtful that the Haber-Weiss reaction can take place in a neutral pH environment, such as found in biological systems. It is thought that only in the presence of a chelated metal complex (iron ions) will the reaction proceed and will OH radicals be produced (the so-called iron-promoted Haber-Weiss reaction).

Fong et al. suggested that the initiation of lipid peroxidation in liver microsomes is caused by the OH· produced by the Haber-Weiss reaction by means of the addition of SOD, catalase and OH· scavengers[80]. In addition, using a spin-trapping method with DMPO and PBN, Lai et al. were able to detect OH radicals in a similar liver microsome preparation[138]. When OH· was trapped by the spin-trapping agents or thiourea, MDA production was decreased, so they argued that OH· was an initiator of the free radical reaction.

5.5.4.3 Oxygen-Metal Complexes

Aust et al. reported that in the lipid peroxidation system of liver microsomes and of liposomes prepared from microsomes, ADP-Fe^{2+} is a powerful initiator of the free radical reaction[8]. With regard to the chelating structure of ADP-Fe^{2+}, it has been found that the electron density of iron atom is increased due to ADP chelation and the molecule is stabilized as a perferryl ion with a strong bond between the iron and an oxygen.

$$ADP\text{-}Fe^{2+} + O_2 \underset{2e^-}{\overset{2H^+}{\rightleftharpoons}} ADP\text{-}[FeO]^{2+} + H_2O$$

ADP-Fe^{2+} is reduced when cytochrome P-450 reductase is present—as in the case of liver microsomes—or it is reduced due to O_2^- in the peroxidation system with xanthine-xanthine oxidase, and a complex with oxygen is then formed. Svingen et al. argued that when a chelate is formed with EDTA in place of ADP, the breakdown of hydroperoxide is facilitated and the alkoxy radical produced from the hydroperoxide (ROOH) is involved in the initiation of the free radical reaction[221].

Enzymes which are imbedded in the inner membrane of the mitochondria, such as flavoprotein (Fp), coenzyme Q and the cytochromes (b_1, c_1, c, a and a_3), also contain large amounts of Fe, Cu, and Mn atoms. Particularly cytochrome a_3 (which functions near the termination of the electron transport system)—is known to give an electron directly to O_2, and there is a high probability that an oxygen-metal complex is formed due to contact between a metal and O_2 at that site. In addition, it is also thought that, if other metals in electron transport system were exposed by the destruction of biomembranes in brain ischemia, there is a strong probability of contact with O_2.

5.5.4.4 Singlet Oxygen ($^1O_2$)

It has previously been demonstrated that singlet oxygens appear in the process of lipid peroxidation, but it is still uncertain whether or not $^1O_2$ acts as an initiator of the free radical reaction in biological systems to bring about the removal of H· from polyunsaturated fatty acids. As illustrated in **Fig. 5-9**, when the electron configuration of the oxygen is such that

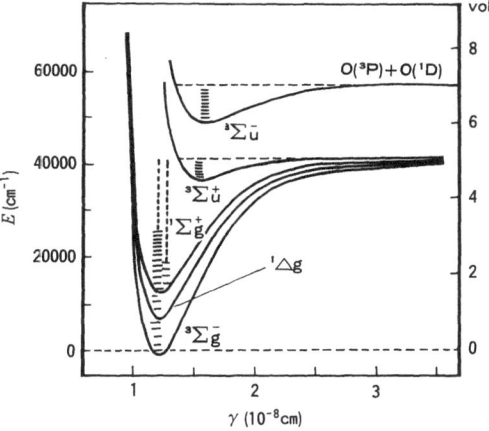

Fig. 5-17. Potential energy of oxygen. A vertical line indicates the energy level and the horizontal line indicates the distance of two atoms. This schema was reprinted from **Fig. 1** of Adv Photochem 7, p 315 (1969) with permission of the author RP Wayne

there are two electrons with antiparallel spin in the π^*_x orbital, it is in a $^1\triangle g$ state, whereas when they have opposite spins within a π^*_x and π^*_y orbitals, it is in a $^1\Sigma g^+$ state. The radical which is thought to be heavily involved in various biological reactions is the long-lived, $^1\triangle g$ state. As illustrated in **Fig. 5-17**, the potential energy of these singlet oxygens, $[^1\triangle g]_2$

and $[^1\Sigma g^+]_2$, is minimized when the internuclear distance between the oxygen atoms is about 1.2 Å. The $^1\Sigma g^+$ state and the $^1\triangle g$ state have higher energies than the resting state of the triplet oxygen ($^3\Sigma g^-$); *i.e.*, 37.5 Kcal and 22.5 Kcal, respectively.

In this regard, it should be noted that the triplet O_2 in an excited state ($^3\Sigma u^-$ and $^3\Sigma u^+$) has the highest potential energy, but the O—O bonds are long (1.534 Å in $^3\Sigma u^+$ and 1.614 Å in $^3\Sigma u^-$). As a consequence, the dissociation to oxygen atoms occurs relatively easily and the reactivity of the triplet state need not be considered when discussing such reactions in biological systems.

The reason why $^1\Sigma g^+$ has a strong reactivity is because it has an empty anti-bonding electron orbital which acts as an electron recipient, as shown in **Fig. 5-9**. It is well known that, due to the strong electron affinity of $^1\triangle g$ (1 eV) and $^1\Sigma g^+$ (2 eV), relative to that of $^3O_2$ (0.15 eV), $^1O_2$ reacts with dienes and olefin (with its abundant electron-donor alkoxy and amino ligands) and brings about ene-type reactions.

Three mechanisms which may be involved in the production of $^1O_2$ in the ischemic brain must be considered. The first is: (i) the generation from Russel's mechanism[198]—a branch of the peroxyradical in the termination of lipid peroxidation. The second is: (ii) dismutation of the superoxide anion[124]. This reaction can be expressed in either of two forms:

$$O_2^- + O_2^- \rightarrow O_2^{2-} + O_2\ (^1O_2)$$

$$O_2^- + O_2^- + 2H^+ \rightarrow {}^1O_2 + H_2O_2$$

Pederson and Aust maintain that, in the induction of the lipid peroxidation reaction due to the xanthine-XOD system, $^1O_2$ produced by the dismutation of O_2^- acts as an initiator[178]. Using 2,5-diphenylisobenzofuran as a trapping agent of $^1O_2$, they maintain that the resultant suppression of lipid peroxidation is due to the above reaction. It is however known that 2,5-diphenylisobenzofuran is not specific for $^1O_2$, and that it also reacts with other radicals. In an experiment using 2,5-diphenylfuran, Svingen et al. found that $^1O_2$ did not act as an initiator of the free radical reaction, but they maintained that, after the generation of this radical in the breakdown of lipid peroxidation, it facilitated the propagation of the free radical reaction[220]. As mentioned earlier (Section 554), Foote et al. found that the volume of $^1O_2$ produced due to the dismutation of O_2^- is no more than 0.2% of the total O_2 produced, and they maintained that many unsolved problems remain concerning the production of $^1O_2$ due to O_2^- dismutation and its activation[83].

The third mechanism of $^1O_2$ production in the ischemic brain is: (iii) the Haber-Weiss reaction. As discussed above, if this reaction takes place within biological systems, it occurs only in the presence of Fe^{2+} ions (i.e., as the chelated metal complex-promoted Haber-Weiss reaction). In experiments on lipid peroxidation using the acetylaldehyde-XOD system, Kellog and Fridovich reported that the oxygen produced by the Haber-Weiss reaction is $^1O_2$ and that it may be involved in the initiation of the free radical reaction[128].

$$O_2^- + H_2O_2 \rightarrow OH^- + OH\cdot + {}^1O_2$$

There is, however, also the view that, since the speed of the Haber-Weiss reaction is slower than that of the O_2^- dismutation reaction, it is inappropriate as a candidate initiator of lipid peroxidation[100].

As seen from the above discussion, there are yet many conflicting views concerning the generation of $^1O_2$ in biological systems and concerning its involvement in the initiation of the free radical reaction. In our own laboratory, we have demonstrated the existence of $^1O_2$ during lipid peroxidation in hypoxic and ischemic brains by means of chemiluminescence measurements and spectral analysis. We believe that the functions of $^1O_2$ may be as suggested by Svingen et al.—i.e., primarily it is involved in a repetitive chain reaction, although a role of $^1O_2$ as an initiator of the free radical reaction has not been proven.

5.5.5 Energy Metabolism and Free Radical Generation

There is a close realationship between changes in energy metabolism due to ischemia and the generation of free radicals. This relationship is due to the fact that the site of free radical generation is limited to the mitochondria and microsomes, and a redox state of the electron transport system is the basis for the generation of free radicals.

We have studied brain metabolism in the Wistar rat using a hypoxic brain model and a global cerebral ischemia model. As an index of lipid peroxidation, we have also simultaneously made chemiluminescence and electron spin resonance (ESR) measurements to determine the relationship between energy conditions and the free radical reaction[112, 113, 114].

ATP levels were found (**Fig. 5-18, Table 5-3**), but when a severe state of global ischemia was produced by means of a nearly complete lack of glucose and O_2 supply (due to 3-vessel occlusion of the basilar and bilateral common carotid arteries), marked decreases in ATP levels were found (**Fig. 5-19, Table 5-4**). In

Table 5-2. The physiological parameters of the rat hypoxic model employed in the chemiluminescence study

		Pre-hypoxia	Hypoxia 5 min	Post-hypoxia 3 min	Post-hypoxia 5 min
MABP	mmHg	143±1	132±3	140±2	138±3
PaO$_2$	mmHg	126±2	19.4±0.4	124±2	127±3
PaCO$_2$	mmHg	38.4±0.9	32.2±0.7	38.2±0.7	36.1±0.9
PH		7.432±0.005	7.321±0.016	7.310±0.006	7.310±0.017
n		34	34	28	26

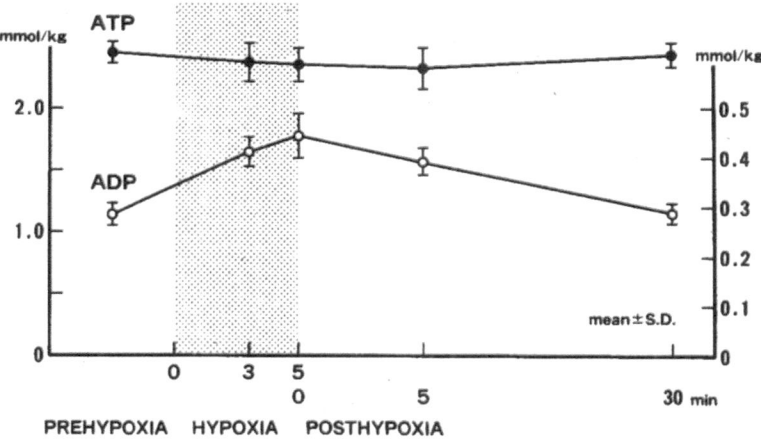

Fig. 5-18. The contents of energy metabolites during and following hypoxia (PaO$_2$ = 17–22 mmHg, mean ± S.D., nmol/gm wet weight, n = 5)

In a hypoxic condition in which the mean arterial pressure was maintained at 100 mmHg or higher, and an arterial hypoxemia load of PaO$_2$ = 17–22 mmHg was induced for 5 minutes (**Table 5-2**), no significant decreases in

addition to ATP, other parameters of hypoxic brain changed as follows: there were increases in lactate, pyruvate and ADP and decreases in glucose. Five minutes after having restored PaO$_2$ to 110 mmHg or more, glucose reached its

Table 5-3. Cerebral contents of energy metabolites and carbohydrates during and following hypoxi (mean ± S.E.M., nmol/gm wet weight, n = 5)

	ATP	ADP	Glucose	Lactate	Pyruvate
Pre-hypoxia	2.44±0.33	0.280±0.007	3.39±0.07	1.69±0.06	0.084±0.003
Hypoxia 3 min	2.37±0.06	0.420±0.013*	2.02±0.06*	5.84±0.44*	0.175±0.051**
Hypoxia 5 min	2.36±0.05	0.445±0.020*	0.84±0.05*	12.07±0.40*	0.200±0.005*
Post-hypoxia 5 min	2.31±0.07	0.392±0.008*	4.71±0.08*	5.87±0.22*	0.136±0.004**
Post-hypoxia 30 min	2.41±0.04	0.287±0.007	3.43±0.10	1.85±0.18	0.087±0.004

Different from pre-hypoxia, *P<0.001, **P<0.01   mean±SE : m mol/kg

Table 5-4. Cerebral contents of energy metabolites and carbohydrates during and followin ischemia (mean ± S.E.M., nmol/gm wet weight, n = 5 or 6)

	ATP	ADP	AMP	E. C.	GLUCOSE	LACTAT
PRE-ISCHEMIA	2.74±0.03	0.243±0.019	0.031±0.004	0.949±0.003	3.51±0.30	1.65±0
ISCHEMIA 3 MIN	0.62±0.06*	0.340±0.038***	1.352±0.097*	0.343±0.012*	0.94±0.14*	13.14±1
ISCHEMIA 5 MIN	0.38±0.05*	0.458±0.014**	1.308±0.124*	0.286±0.012*	0.51±0.03*	16.73±1
RECIRCULATION 5 MIN	1.50±0.07*	0.310±0.028***	0.368±0.035*	0.761±0.015**	1.42±0.15*	9.78±2
RECIRCULATION 30 MIN	1.63±0.11*	0.279±0.004	0.045±0.004***	0.904±0.006***	2.32±0.22**	6.52±0
ISCHEMIA 30 MIN	0.05±0.01*	0.415±0.065**	1.252±0.081*	0.147±0.017*	0.06±0.00*	16.83±0
RECIRCULATION 30 MIN	1.52±0.02*	0.375±0.031***	0.308±0.049*	0.778±0.022**	4.05±0.39**	7.94±0
RECIRCULATION 120 MIN	1.76±0.05*	0.486±0.028*	0.357±0.042*	0.772±0.011**	3.58±0.04	3.78±0
ISCHEMIA 60 MIN	0.018±0.005*	0.253±0.025	0.858±0.069*	0.135±0.012*	0.02±0.01*	19.81±0
RECIRCULATION 10 MIN	0.753±0.060*	0.268±0.024	0.332±0.013*	0.640±0.023*	1.47±0.39*	9.53±0

Different from pre-ischemia. *P<0.001. **P<0.01. ***P<0.05
MEAN±S.E. : M MOL/KG

highest level, while lactate, pyruvate and ADP also showed high levels. After 30 minutes, however, all of the metabolites had returned to their pre-hypoxia levels (Figs. 5-18, 5-20, Table 5-3). In other words, this hypoxic brain preparation is a model in which ATP can be maintained at normal levels while cytoplasmic substrates, lactate and pyruvate, increase.

A characteristic feature of the model is that oxygen supply is maintained at approximately one fifth of its normal value (17–22 mmHg). Under these conditions, the changes in energy metab-olites are reversible and the deficits in the mitochondria respiratory chain are limited to increases in ADP. That is, is a model in which only mild disrup tion of oxidative phosphorylation is brought about. In contrast, when 5 minute ischemia is produced due to 3 vessel occlusion, the energy charge of ATP level which has been drasticall reduced during the ischemia does no return to the levels found prior to th ischemia, even when 30 minutes of recirculation have been allowe (Fig. 5-19, Table 5-4). Similarly there are decreases in brain glucose an

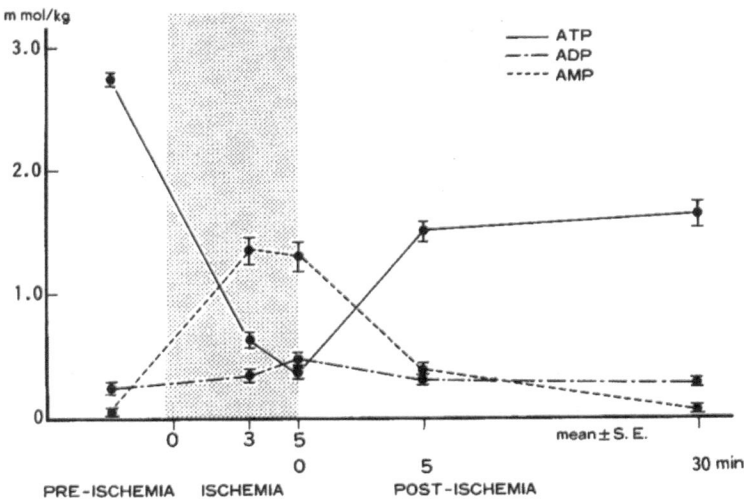

Fig. 5-19. The contents of energy metabolites during and following 5 min-ischemia (mean ± S.E.M., nmol/gm wet weight, n = 5 or 6)

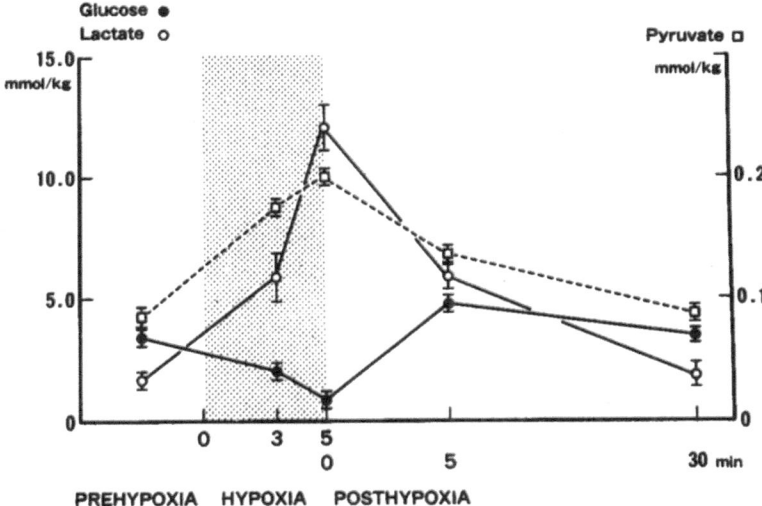

Fig. 5-20. The contents of carbohydrates during and following hypoxia (PaO$_2$ = 17–22 mmHg, mean ± S.D., nmol/gm wet weight, n = 5)

marked increases in lactate, but after 30 minutes of reflow these metabolites do not return to their normal levels (**Fig. 5-21, Table 5-4**).

luminescence increased from the period of ischemia or hypoxia and was maintained at high levels for a certain period following reoxygenation or

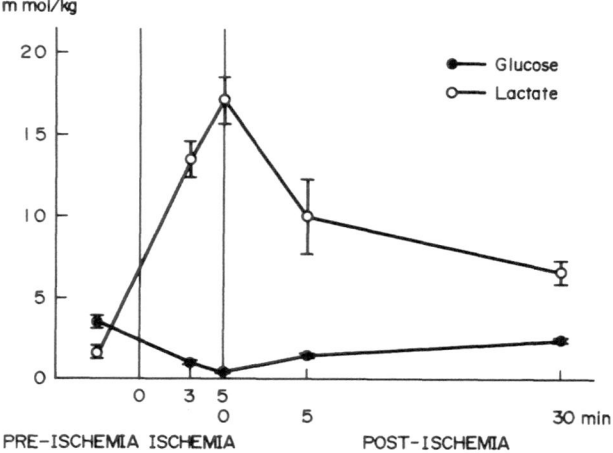

Fig. 5-21. The contents of carbohydrates during and following 5 min-ischemia (mean ± S.E.M., nmol/gm wet weight, n = 5 or 6)

When animals are subjected to 30 minutes of ischemia due to 3-vessel occlusion, the changes in energy metabolities are very similar to those seen in 5-minute ischemic brain, but after 30 minutes of reflow the energy charge is 0.778, suggesting that a lengthy period would be required for recovery (**Table 5-4**). We consequently believe that, in the case of 30 minutes of ischemia, only limited recovery of the deficits in energy metabolism is possible. After 60 minutes of ischemia and 10 minutes of reflow, some indication of recovery of the glucose, lactate, ATP, ADP, and AMP was found (**Table 5-4**), but thereafter there were severe changes in blood pressure, making further measurements impossible.

In the chemiluminescence measurements, it was found that the chemi-

reflow. Together with a return to normal of ADP and carbohydrates, however, there was a decrease in the level of photon counts in hypoxic brain (**Fig. 5-22, Table 5-5**). In other words, in the brain subjected to 5 minutes of hypoxia, there was a positive correlation with the changes in the chemiluminescence value and energy metabolism, and in the 5 and 30 minute ischemic brain, there was a similar correlation with changes in the chemiluminescence value and energy metabolism. Based upon these findings concerning the relationship between chemiluminescence and energy metabolism, it is thought that when there are disturbances of the mitochondrial respiratory chain due to a hypoxic load, such that there are increases in ADP without simultaneous decreases in

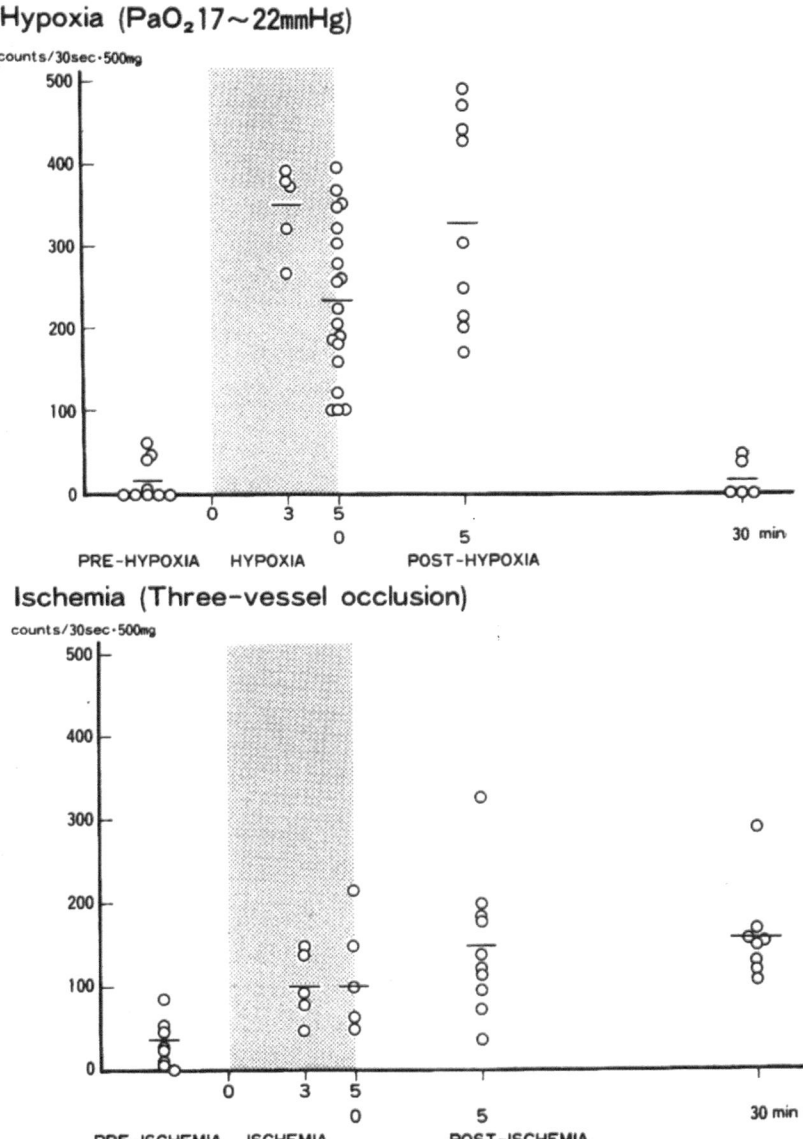

Fig. 5-22. Chemiluminescence values in hypoxic or ischemic brain (photon counts/30 sec. 0.5 gm wet weight). One circle indicates the value of one animal. Note the difference in chemiluminescence values between 30 min-reoxygenated and 30 min-recirculated brain samples

ATP, there is the appearance of the free radical reaction. When these deficits in energy metabolism are mild enough to be reversible, there is the beginning of the propagation in lipid peroxidation.

The high levels of chemiluminescence found following the start of reflow in the ischemic brain mean that the propagation reaction started together with the resupply of oxygen

Table 5-5. Chemiluminescence values in rat hypoxic or ischemic brain (photon counts/30 sec 0.5 gm wet weight)

	PRE-HYPOXIA	HYPOXIA 3 MIN	5 MIN	POST-HYPOXIA 5 MIN	30 MIN
CHEMILUMINESCENCE VALUE	17 ± 8 (N=9)	$347 \pm 23^*$ (N=5)	$231 \pm 4^*$ (N=19)	$326 \pm 40^*$ (N=9)	15 ± 9 (N=5)

MEAN ± S.E. COUNTS/30 SEC·500 MG

	PRE-ISCHEMIA	ISCHEMIA 3 MIN	5 MIN	RECIRCULATION 5 MIN	30 MIN
CHEMILUMINESCENCE VALUE	30 ± 4 (N=8)	$97 \pm 16^{***}$ (N=5)	$112 \pm 27^{***}$ (N=5)	$145 \pm 24^{**}$ (N=10)	$157 \pm 18^*$ (N=8)

* $P < 0.001$, ** $P < 0.01$, *** $P < 0.05$

MEAN ± S.E. COUNTS/30 SEC·500 MG

through the reflowed vessels. Moreover, it is thought that during the measurement process itself there is continual peroxidation due to contact with air, but this condition occurs equally control and ischemic or hypoxic brain (**Fig. 5-13**).

The high chemiluminescence values were obtained in the hypoxic brain, rather than in the ischemic brain with a severe degree of energy failure, but the significance of this difference cannot be easily deduced. What can be concluded is that the modes of generation of the free radical reactions occurring in the hypoxic brain and in the ischemic brain are different (**Fig. 5-22**).

Next let us consider the deficits of cytoplasmic glycolysis and deficits in the mitochondrial electron transport system from the perspective of the generation of active oxygens in the initiation of the free radical reaction.

The following two conditions concerning the mechanism of initiation of the free radical reaction in the ischemic or hypoxic brain must be considered.

(1) Hyper-Reduction of the Respiratory Chain

Due to the decreased supply of O_2, which is the final electron recipient at the terminal group of enzymes imbedded in the inner membrane of the mitochondria, the electron transport system loses its electron recipients and so-called electron-pooling occurs. In a normal state, intracellular NADH is oxidized due to the mitochondrial electron transport system—becoming NAD^+. When electron-pooling occurs in the respiratory chain, the oxidation of NADH is suppressed and the ratio of $NADH/NAD^+$ increases. During a mild period of such deficits, ADP accumulates due to the blocking of the

$ADP + Pi \rightarrow ATP$ reaction in the respiratory chain. Due to the simultaneous suppression of the TCA cycle caused by disturbances of the respiratory chain, there is an increases in cytoplasmic pyruvate and lactate. Moreover, due to the above-mentioned electron-pooling (a state of hyper-reduction) in the mitochondrial respiratory chain, it is thought that there occurs leakage of electrons and/or a reversal of the electron transport phenomenon. The leaked electrons then reduce oxygens within the carbon chains of phospholipids (It is to be noted that the oxygen density of the carbon chains, *i.e.*, the hydrophobic chains, is some 7-fold greater than that of the hydrophilic heads). The reduction is often incomplete and this is thought to facilitate the production of superoxide anions, singlet oxygens, oxygen-metal complexes, etc. Previous researchers, especially Braughler[33] or Demopoulos[61-66] have indicated that they believe free radical mediated peroxidation occurs in this hyper-reductive state. Demopoulos suggests that O_2^- formed by reduced CoQ and O_2 is dismutated to H_2O_2 by SOD in the phospholipid bilayer. The H_2O_2 formed could combine with O_2^- (Haber-Weiss reaction) to form the $OH \cdot$ (Tien *et al.*, 1982)[230] under catalysis of the chelated metal complex, which then would initiate peroxidation of the lipid membrane.

(2) Hyper-Oxidation of the Respiratory Chain

Next, let us consider three abnormal states of the brain: (i) during incomplete ischemia, (ii) following reflow, and (iii) during hypoxia when there is some—albeit a small— supply of oxygen. Under these conditions, the terminal of the respiratory chain enters a state of hyper-oxidation—that is, a state of relative electron deficiency. Consequently, the substrates which normally would be oxidized (*e.g.*, pyruvate) accumulate within the cytoplasm due to suppression of the TCA cycle—despite the fact that O_2 is present at the terminal portion of the mitochondrial respiratory chain. In the respiratory chain with an electron deficiency, the O_2 at its terminal portion is incompletely reduced, and active oxygens are produced. Metals (Fe, Zn, Cu, etc.) found within the enzymes of the electron transport system can form oxygen-metal complexes when contact is made between the metals and O_2 due to the destruction of the mitochondrial inner membrane.

As seen from the above, two diametrically opposite cases must be considered: (i) when the respiratory chain has a full electron charge, and (ii) when it is in a state of electron deficiency. However, in light of the fact that a completely oxygenless condition or complete ischemia is unlikely to occur *in vivo*, it is thought more likely that sequential deficits occur in oxidative phosphorylation during an ischemic load, during which the number of electrons in the respiratory chain gradually decreases following pooling together with deficits in the $ADP + Pi \rightarrow ATP$ reaction. In other words, there are quantitative changes in the supply of electrons in the respiratory chain depending upon the stage of the ischemia and reflow. The possibility of the initiation in the free radical reaction together with energy failure due to the above mechanisms (i) and (ii) is thus indicated.

In the hypoxic brain model, in which a small but significant supply of oxygen is maintained, the number of electrons at the termination of the electron transport system decreases (a state of hyper-oxidation), and active oxygens and oxygen-metal complexes are produced.

In contrast, during an ischemic load (in the 3-vessel occlusion model for bilateral hemispheric infarction using the rat), due to the presence of oxygen—either resolved oxygen or the small amount of oxygen entering the brain over the remaining intact vessels—it is thought that a considerable period at time is required before the oxygen supply at the termination of the respiratory chain will decrease. As a consequence, during the early stage of ischemia, it is likely that active oxygens are generated by mechanisms similar to those of the electron transport system in a state of hyper-oxidation in the hypoxic brain. However, as the ischemia advances, there are slow and continual decrease in the O_2 at the terminal portion of the mitochondrial respiratory chain. Together with this decreases, there will be electron-pooling (hyper-reduction) in the electron transport system, and it is likely that electron leakage, $Q \cdot$ and O_2^- formation, invasion of free radicals to the lipid carbon chains and the generation of oxygen-metal complexes will then occur.

5.5.6 The Problem of "Intra-Ischemic Peroxidation" and "Post-Ischemic Peroxidation"

The destruction of biomembranes is triggered by two mechanisms: The first is autoxidation due to the free radical reaction, as discussed above. The second is enzymatic peroxidation—representative of which is the arachidonate cascade. A review of previous reports concerning these two systems, however, has shed no light on the question of the mutual interactions between them during the metabolism of the ischemic brain.

In addition, together with the resupply of oxygen due to reflow, the fatty acids which have been liberated and have accumulated during ischemia become the substrates for the propagation of the free radical reaction. Here as well, there is speculation that this represents physiological compensation to prevent cellular damages (due to the consumption of polyunsaturated fatty acids as a form of scavenger). As evident from the above, there are still several unresolved problems regarding the essential nature of the cellular deficits during peroxidation in the ischemic brain, the initiation of the free radical reaction and the period during which the propagation reaction can take place.

In 1966, Demopoulos reported the involvement of free radicals in melanoma[63]. By 1973 he had discussed the relationship between free radicals and ischemic brain edema, and thereafter he reported on a series of experimental studies. The view currently advocated by Demopoulos and his colleagues is as follows[62, 65–66].

When there is a sharp decrease in the concentration of O_2 due to ischemia, there is the accumulation of electrons (*i.e.*, a state of hyper-reduction) in the electron transport system at the inner membrane of mitochondria. The accumulation is directly due to a decrease in the supply of the final electron recipient O_2, in a manner similar to the increased traffic on a highway with closed lanes (**Fig. 5-23**). Due to such

ing about the fragmentation of fatty acid carbon chains. As a consequence, the peroxidation reaction due to autoxidation takes place over a wide area.

Alternatively, due to the fact that the non-polar tails (fatty acid carbon chains) contain some 7-fold more O_2 than the polar heads, the reduced coenzyme Q will react with the O_2, thus producing CoQ· and $O_2^{\bar{}}$.

(reduced) $CoQ + O_2 \rightarrow CoQ\cdot + O_2^{\bar{}}$

The generated $O_2^{\bar{}}$ forms H_2O_2 and singlet oxygen, and OH· is produced by the Haber-Weiss reaction in the pres-

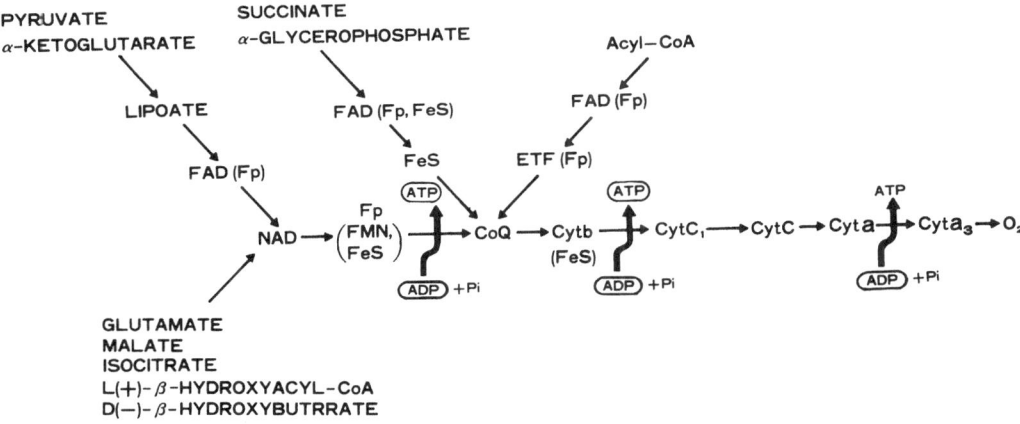

Fig. 5-23. Respiratory chain of mitochondria

accumulation of electrons, coenzyme Q which has had its electron to transport is disturbed—becomes a coenzyme Q radical (CoQH semiquinone) due to reduction by one electron and loses its natural destination. With the removal of H· from an α-methylene carbon site (present about 15—10 Å medial from the membrane surface), the formed radical (R·, ROO·) repeatedly removes H· from nearby fatty acids, thus bring-

ence of metal ions—thus beginning the removal of H· from the carbon chains.

$O_2^{\bar{}} + O_2^{\bar{}} + 2 H^+ \xrightarrow{SOD} H_2O_2 + O_2$

$H_2O_2 + O_2^{\bar{}} \xrightarrow{Fe^{2+}} OH\cdot + OH^- + {}^1O_2$

As evident from the above, Demopoulos argues that at least the initiation of the free radical reaction occurs during ischemia aside from the problem of phospholipids deacylation due to phospholipase. It remains to be determined whether the substrates of lipid autoxidation are liberated free fatty acids or phospholipids (with unsaturated fatty acids containing ester bonds) which are oxidized *in situ*.

It seems that the background of these conceptions is Mitchell's chemiosmotic theory[155, 156]. According to this theory, the electron transport enzymes present in the inner membrane of mitochondria are arranged in such a manner that there is alignment between hydrogen recipients (H) and electron recipients (e). There is then a mechanism which allows for an electron to be removed from the hydrogen (H → $H^+ + e^-$) on the inner membrane of this matrix (the inner side of the inner mitochondrial membrane), and an H^+ to be released externally (into the cytoplasm).

First, a flavoprotein receives an H from $NADH_2$ on the matrix side, becoming $FMNH_2$. Next, FeS proteins remove only electrons from $FMNH_2$ (H → $H^+ + e^-$), thus releasing $2H^+$ into the cytoplasm. This electron then reduces the next electron transport substance CoQ, at which time H^+ is removed from the matrix, giving $CoQ + 2e^- + 2H^+ \rightarrow CoQH_2$.

Next, the $CoQH_2$ releases $2H^+$ into the cytoplasm and sends two electrons to the cytochrome molecules (b, c_1, c, a, and a_3). Unlike $FMNH_2$ and $CoQH_2$, which are transporters of two electrons, the FeS protein and cytochrome are transporters of one electron. As a consequence, free radicals such as FMNH· and CoQH· are generated. It has been reported that, for the same reason,

radicals of ascorbic acid (vitamin C) which reduce cytochrome c generate.

Using the spin-trapping technique with PBN, we have studied the generation of radicals in ischemic rat brain homogenate with NADPH and Fe-EDTA added (**Figs. 5-15, 5-16, 5-24, 5-25**)[231, 232]. As mentioned earlier, we

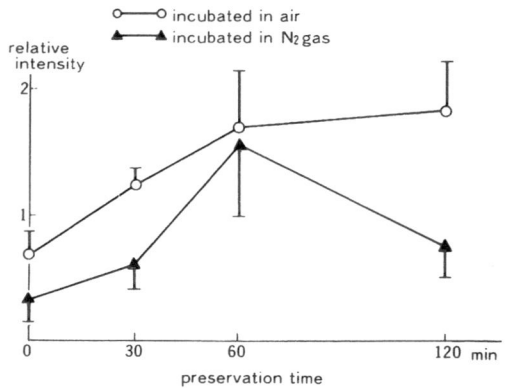

Fig. **5-24**. The changes in ESR signal intensities obtained from the reaction mixtures containing ischemic brain homogenate prepared from the rat decapitation model (mean ± S.E.M., n = 4)

believe that the PBN-trapped radical (A_N = 16.2–16.5 G, A_β^H = 3.6–3.8 G) is derived from the peroxidation of unsaturated fatty acids. In complete cerebral ischemia produced by preservation of the head at 37 °C after decapitation, when the reaction mixture (with PBN, NADPH and Fe-EDTA added) was incubated in air, there was sequential increases in the radical strength when it was left at 37 °C, while in nitrogen gas a peak of radical intensity was found after 60 min of preservation time (**Fig. 5-24**). In the case of 3-vessel occlusion, when the brain homogenate was incubated in nitrogen or air, a pattern of rapid increases followed by rapid decreases was seen, and a peak of radical intensity was found after 30 minutes of ischemia. In addition, a marked increase in radical intensity was observed in reperfused brain homogenate (**Fig. 5-25**).

In either model when the mixture was reacted in nitrogen, the changes in

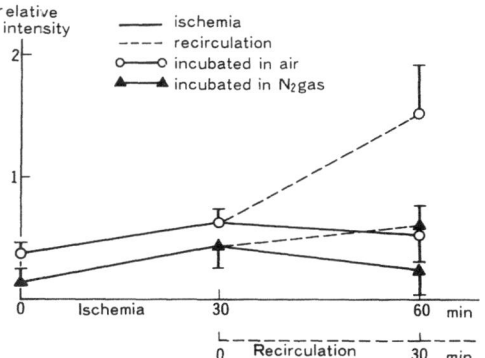

Fig. 5-25. The changes in ESR signal intensities obtained from the reaction mixtures containing ischemic brain homogenate prepared from the rat three-vessel occlusion model. In the recirculated group (30 min-recirculation following 30 min-ischemia) the signal intensities showed a marked increase (mean ± S.E.M., n = 4)

radical intensity suggested that during ischemia with already reduced O_2 levels, it is possible for the free radical reactions to be initiated, although a minute volume of O_2 was present. With regard to the oxygen molecules found during ischemia, it is thought that, in the case of decapitation, there is the involvement of resolved oxygen and, in the case of 3-vessel occlusion, there is the involvement of resolved oxygen plus the small volume of oxygen which enters the brain via intact blood vessels.

In other words, with regard to the changes in intensity in the PBN-

trapped radical in this experimental series, we believe that the intensity increases in parallel with the consumption of oxygen in the brain (resolved oxygen, etc.) and, at the point where all such oxygen has been exhausted, there is a rapid decrease in radical intensity (**Figs. 5-24, 5-25**).

The fact that lipid peroxidation requires oxygen in the process of producing damage to the ischemic brain, as seen from the above findings, has in fact been reported in several other experimental studies. In our measurements of lipid peroxidation using the chemiluminescence method, the deficits in energy metabolism were milder than those seen with global cerebral ischemia due to 3-vessel occlusion, but stronger chemiluminescence was found in the hypoxic brain (17–22 mmHg PaO_2) maintained with a persistent supply of oxygen. These results are thought to reflect the fact that, in addition to differences in the supply of glucose and the degree of disturbance of the electron transport system, oxygen is heavily involved in the intensity of free radical reaction.

With regard to the importance of resolved oxygen in the brain, some implicative suggestions are to be found in the research of Abe et al.[2]. Using a rat decapitation model, they studied the sequential decreases in the intracerebral concentrations of α-tocopherol. They found that there is a 16% decrease at 3 minutes after decapitation and a 20% decrease after 15 minutes, thus demonstrating the possibility that the free radical reaction can be initiated in the process of a rapid decrease in cerebral oxygen.

5.6 Pharmacological Mechanisms of Mannitol, Vitamin E, Glucocorticoids and Phenytoin

5.6.1 Mannitol

Mannitol was first isolated from oleaceae plants in 1806 and has since been found to be abundant in various plant species such as seaweeds. It is currently used as one of the constituents of chewing gum, but its usage in medicinal applications dates from the 1940s when it was first used as a diagnostic chemical in tests on renal function—based upon its diuretic effects.

Subsequently, a large number of pharmacological studies have been carried out, and it is now in general usage in the treatment of acute renal failure, brain edema, intracranial hypertension and glaucoma[142, 177, 185, 217, 235].

The chemical structure of 6-carbon sugar, D-mannitol, is as follows:

$$HOH_2C - \overset{\overset{\displaystyle H}{|}}{\underset{\underset{\displaystyle OH}{|}}{C}} - \overset{\overset{\displaystyle H}{|}}{\underset{\underset{\displaystyle OH}{|}}{C}} - \overset{\overset{\displaystyle OH}{|}}{\underset{\underset{\displaystyle H}{|}}{C}} - \overset{\overset{\displaystyle OH}{|}}{\underset{\underset{\displaystyle H}{|}}{C}} - CH_2OH$$

It is thought to have hydroxyradical (OH·) scavenger effects because of its abundant OH constituents[67, 152, 230]. The consumption of hydroxyradicals is thought to occur by means of the following mechanisms:

$$OH· + OH· \rightarrow H_2O_2 \text{ and}$$
$$OH· \rightleftharpoons H^+ + O^-$$

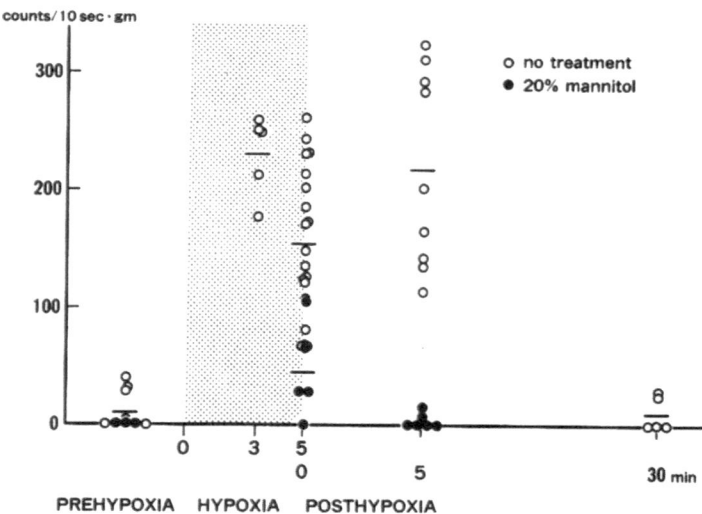

Fig. 5-26. Effect of pre-administered mannitol on chemiluminescence values during and following hypoxia (photon counts/10 sec. 1.0 gm wet weight). One circle indicates the value of one animal

The fact that mannitol, containing many OH constituents, would capture OH· can be deduced from either structural formula.

In the analysis of the intensity of chemiluminescence, the hypoxic brain pretreated with 10 ml/kg of 20% mannitol showed suppression of chemiluminescence (**Fig. 5-26**), whereas in spectral analysis done following the *in vitro* addition of 0.5 g/g brain of 20% mannitol to the brain homogenate, the chemiluminescence was found not to be suppressed significantly at any wavelength (**Fig. 5-27**)[113, 114, 218]. The difference in the suppressive effects in the *in vitro* addition of mannitol to the brain homogenate and the *in vivo* pretreatment is thought to indicate, not that mannitol has its pharmacological effects on the breakdown of lipid peroxidation, but that it possibly suppresses the OH· generated in the mitochondria during hypoxia.

In the measurements of PBN-

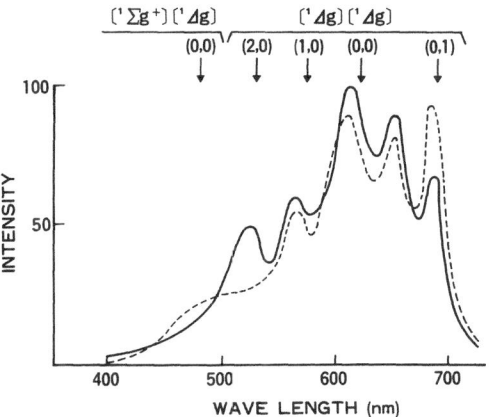

Fig. 5-27. Effect of mannitol on chemiluminescence spectra observed from hypoxic brain homogenate. Little decrease in the spectropeaks was seen after the addition of mannitol

trapped radicals in our ESR experiments (using the brain homogenate preservation system and the 3-vessel occlusion model), there was weakening of the radical intensities when pretreatment with 10 ml/kg of 20% mannitol was done (**Figs. 5-28-A & B**).

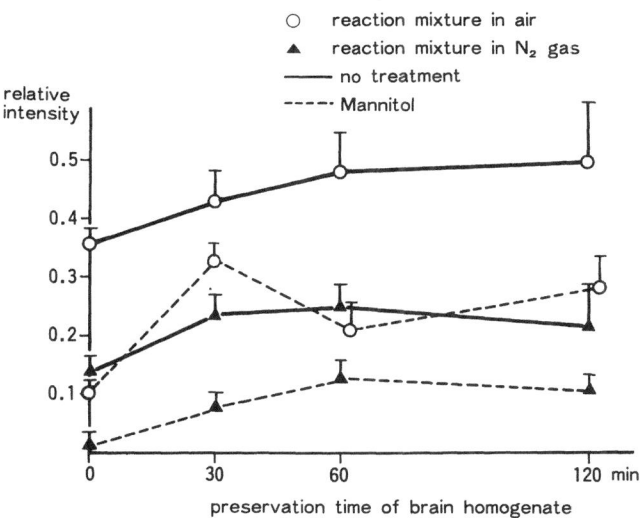

Fig. 5-28-A. Effect of pre-administrated mannitol on ESR signal generation in reaction mixtures containing brain homogenate (mean ± S.E.M., n = 6). The brain homogenate preservation system was used for ischemic load (see Fig. 5-15)

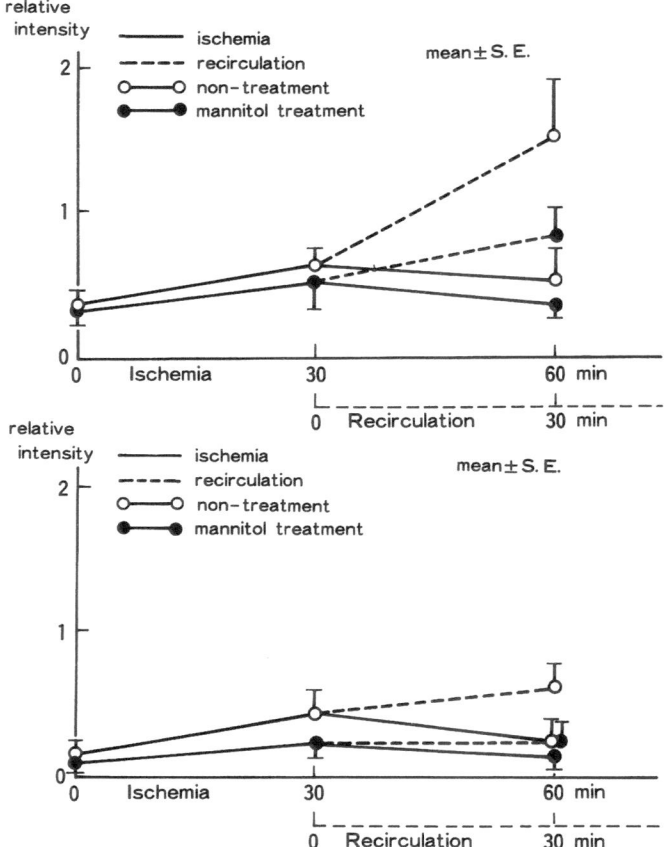

Fig. 5-28-B. Comparison between the untreated group and 20% mannitol preadministrated group in PBN spin adduct intensity during and following 30 min-ischemia produced by three-vessel occlusion. The upper schema indicates incubation of the reaction mixture in room air and lower schema shows the case incubated in nitrogen gas (n = 4)

In light of the above-mentioned binding constants of the PBN-trapped radicals and the fact that there was a dependence of the radical signal intensity on the concentration of resolved oxygen, it was concluded that the radical is derived from the peroxidation of unsaturated fatty acids (*cf.* p 204). Pretreatment with mannitol is, therefore, thought to suppress the generation of these radicals and consequently suppress lipid autoxidation.

With regard to the cortical water content following 30 minutes of ischemia caused by 3-vessel occlusion as well, 10 ml/kg of 20% mannitol was administered and measurements were made 30 minutes after the start of reflow. In contrast to 80.52% in the untreated, there was some decrease to 79.26% in the animals treated with mannitol (**Table 5-13**). However, since the mannitol was a 20% hypertonic solution, it is thought that the mechanism of an effect in this experiment may have been primarily due to a hyperosmolar action.

These findings indicate that mannitol has a free radical scavenging action which captures the hydroxy radical and supresses further reactions at an early stage of the lipid peroxidation reaction.

5.6.2 Vitamin E (α-tocopherol)

Vitamin E was discovered by Evans and Bishop in 1922 as an anti-sterility drug. Many studies have since been made on its pharmacological properties. Tappel (1962) was the first to unambiguously demonstrate the antioxidant action of vitamin E as a free radical scavenger[222-24]. Among the radicals with which vitamin E reacts are the superoxide anion[76], the peroxyradical and $^1O_2$[81, 111, 189, 190]. Due to the reaction with such radicals, vitamin E is thought to act as a free radical scavenger by converting as follows:

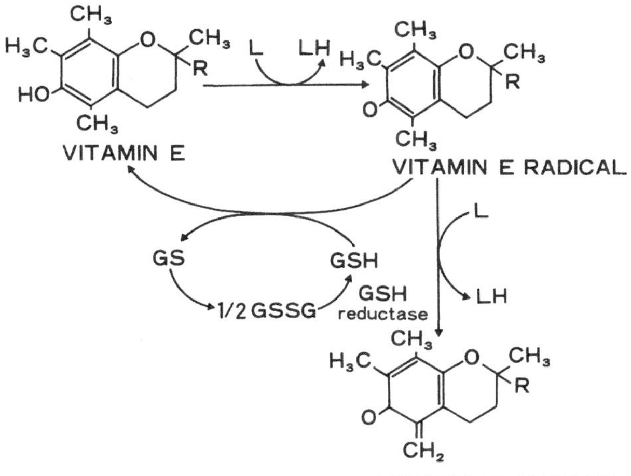

be reduced by means of glutathione, resulting in the regeneration of α-tocopherol and its reutilization.

In addition to its action as a free radical scavenger, the hypothesis developed by Lucy and Diplock states that vitamin E may function as a membrane stabilizer, and as a protector of a membrane-bound antioxidant[68-70, 144, 145]. Formation of the arachidonic acid-vitamin E complex and coexistence with selenium in a reduced form as selenide or with sulphide are the bases of this theory.

CH₃
H₃C, O, CH₃
HO, R
CH₃
VITAMIN E

L, LH

CH₃
H₃C, O, CH₃
O, R
CH₃
VITAMIN E RADICAL

L

GS, GSH
1/2 GSSG, GSH reductase, LH

CH₃
H₃C, O, CH₃
O, R
CH₂
VITAMIN E QUINONE

In other words, vitamin E is first oxidized into the tocopherol radical and then into the tocopherol quinone. The quinone can then be converted into a semi-quinone. In this way, the oxidation of 1 mol of α-tocopherol to produce a quinone results in the consumption of 2 mol of radicals. Tocopherol itself can

In our experimental work, we have found that a 30 mg/kg dose of vitamin E administered at 30 minutes prior to the hypoxic load results in a lower level of chemiluminescence than that found in the untreated group (**Fig. 5-29**). After the addition of 8 mg/g brain of vitamin E to the brain homogenate, spectral

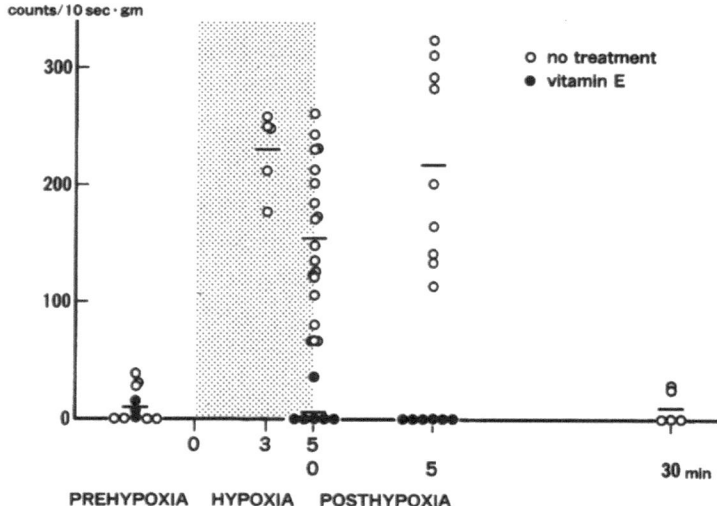

Fig. 5-29. Effect of pre-administered vitamin E on chemiluminescence values during and following hypoxia (photon counts/10 sec. 1.0 gm wet weight). One circle indicates the value of one animal

analysis shows the suppression of chemiluminescence at all wavelengths (**Fig. 5-30**). These findings indicate that vitamin E has a scavenging action on the breakdown of lipid peroxidation process.

Moreover, in our ESR measurements made using the decapitation-

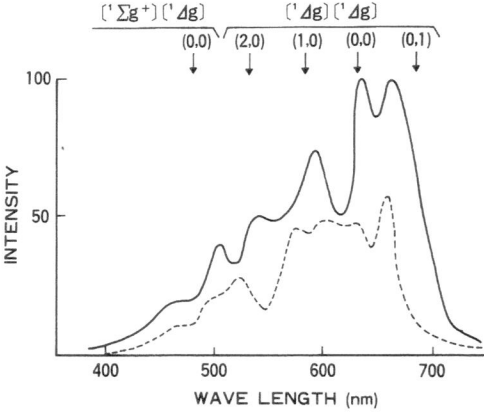

Fig. 5-30. Spectral analysis of chemiluminescence generated from rat hypoxic brain homogenate showed diminishing of spectropeaks after the addition of vitamin E

homogenate system as well, the signal intensities of the PBN-trapped radicals were found to be significantly lower in the treated than in the untreated animals at each time period sampled (**Fig. 5-31**). These findings are in fundamental agreement with those of the chemiluminescence experiments (**Figs. 5-29, 5-30**). Since it is believed that the PBN-trapped radical is a radical converted from free fatty acids, it is likely that, in the lipid autoxidation process, vitamin E acts primarily at the termination reaction. However, with regard to the anti-edemous effects of vitamin E when administered alone, only a very small decrease in the water content of the brain was seen 30 minutes after reflow in the global ischemia model (**Table 5-13**). The above findings indicate that vitamin E has a scavenging action on the radicals ($^1O_2$, O_2^-, ROO·, etc.) produced in the breakdown process of autoxidation in the ischemic or hypoxic brain.

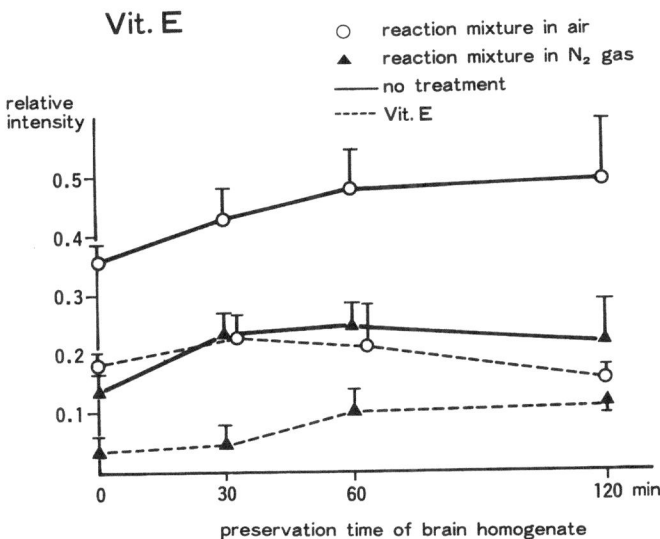

Fig. 5-31. Pre-administration of vitamin E resulted in a decrease in the ESR signal intensities derived from NADPH-dependent lipid peroxidation in rat ischemic brain homogenate (mean ± S.E.M., n = 6). The brain homogenate preservation system was used for ischemic load (see **Fig. 5-15**)

5.6.3 Glucocorticoids

It is well-known that one of the actions of the glucocorticoids, such as betamethasone, is the facilitation of the synthesis of peptides which block phospholipase A$_2$[101, 106, 110]. As a consequence there is the suppression of the liberation of free fatty acids (arachidonic acid liberation) and indirect inhibition of PG biosynthesis. It is, however, thought that a future topics of research must focus on the question of whether or not all of the effects of the glucocorticoids can be explained solely in terms of their effects on the synthesis of peptides which block phospholipase A$_2$, because other studies have demonstrated the stimulatory effect of glucocorticoid on phospholipase A$_2$ activity and cyclooxygenase activity with increase in prostaglandin synthesis[26, 48].

It is known that glucocorticoid produces alterations in the enzyme activity or receptor activity involved in the remodeling of the composition of cell membranes[52, 246]. Some studies have suggested that there may be a direct interaction of this drug with cell membranes wihtout entry into the cytozol and combination with receptors[16, 58, 117]. Namely, it has been reported that high concentrations of glucocorticoids have the effect of stabilizing lysosomes and other membranes.

There have previously been several reports indicating the possibility that glucocorticoids have a free radical inhibiting action in the autoxidation reaction system, without having any effects via enzymes[33, 99, 203].

Seligman and Demopoulos have

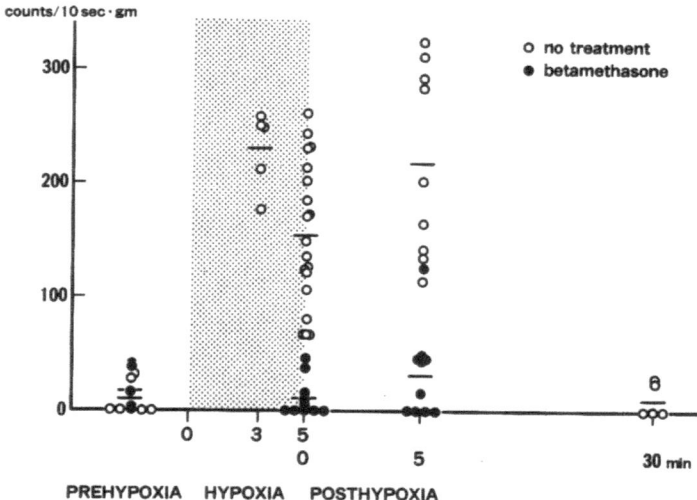

Fig. 5-32. Effect of pre-administered betamethasone on chemiluminescence values during and following hypoxia (photon counts/10 sec. 1.0 gm wet weight). One circle indicates the value of one animal

performed experiments using liposome membranes in which UV irradiation has been performed. By means of sequential measurements of free fatty acids in the liposome suspension following UV irradiation, they found sharp continuous decreases in $C_{20:4}$ (arachidonic acid) and $C_{22:6}$ (docosahexaenoic acid), and a smaller decrease in $C_{16:1}$, $C_{18:1}$, and $C_{18:2}$. In contrast, no changes were seen in the saturated free fatty acids, $C_{16:0}$ and $C_{18:0}$. These findings indicate that the free fatty acids were consumed due to the free radical reaction. MDA was found to increase with time and peak after 120 minutes, after which there was a gradual decrease. In this experimental system, the addition of methylprednisolone immediately prior to the UV irradiation caused a marked suppression of the MDA values. The MDA levels were found to be dose-dependent up to methylprednisolone concentrations of

1.58 mM. These results are particularly noteworthy in so far as they demonstrate the possible effect of glucocorticoids acting as an inhibitor of free radical mediated peroxidation.

In our experiments, pretreatment with 1 mg/kg of betamethasone administered at 30 minutes prior to the hypoxic load resulted in chemiluminescence values which were significantly lower than the untreated values (**Fig. 5-32**). Moreover, addition of 1.0 mg/g brain of betamethasone to the homogenate with increased chemiluminescence produced results in the spectral analysis indicating the suppression of photic intensity at all of the wavelengths for each of the singlet oxygens $[^1\Sigma g^+]$ $[^1\triangle g]$, $2[^1\triangle g]$ (**Fig. 5-33**). These results can be interpreted to imply (i) the suppression of the peroxidative reaction due to blockage of phospholipase A_2 by the treatment with betamethasone, and (ii) in the spectral analysis, the

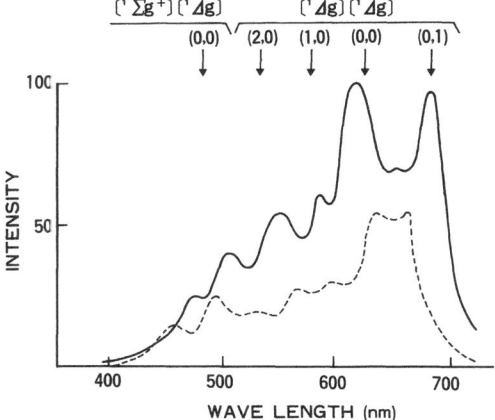

Fig. 5-33. Effect of betamethasone on chemiluminescence spectra observed in rat hypoxic brain homogenate. All the spectropeaks diminished after the addition of the drug

suppression of fatty acid liberation from phospholipids present in the homogenate. Alternatively, it is also possible that, in addition to the blockage of phospholipase A$_2$, the betamethasone may have had scaveng-

ing effects on the peroxidative breakdown of lipids, in light of the fact that there was suppression of spectral strength at all wavelengths.

In the decapitation-homogenate preservation system in which PBN-trapped radicals were studied using the ESR technique, the signal intensity was reduced in the case of administration of 1 mg/kg betamethasone at 30 minutes prior to decapitation, in comparison with strong signals found in the untreated group (**Fig. 5-34**). It should be noted, however, that the degree of reduction was less than that seen following mannitol or vitamin E administration. These findings suggest that the pharmacological activity of the glucocorticoids is not unitary, but, in addition to the deactivation of phospholipase, may also include a inhibiting effect on free radical mediated peroxidation.

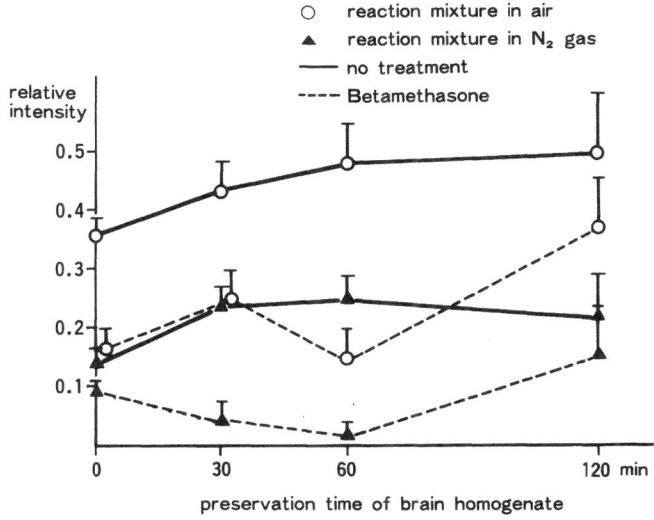

Fig. 5-34. Effect of pre-administered betamethasone on ESR signal generation in rat ischemic brain homogenate with the addition of NADPH and Fe-EDTA (mean ± S.E.M., n = 6). The brain homogenate preservation system was used for ischemic load (see **Fig. 5-15**)

5.6.4 Combined Therapy with Mannitol, Vitamin E and Glucocorticoid

The combined usage of the above three drugs—known as the Sendai Cocktail—is expected to find significant usage in clinical cases of cerebral infarction[219]. Experimentally, we have found that administration of the combination of these drugs to the hypoxic brain results in reduced levels of chemiluminescence at all stages (**Fig. 5-35**).

The ATP, energy charge and glucose levels in the 5-minute ischemic brain to which the Sendai Cocktail is administered, returned to normal soon after the start of reflow. Moreover, the increase in lactate found during ischemia was notably suppressed (**Figs. 5-38, 5-39, Tables 5-8, 5-9**).

The effects of the Sendai Cocktail

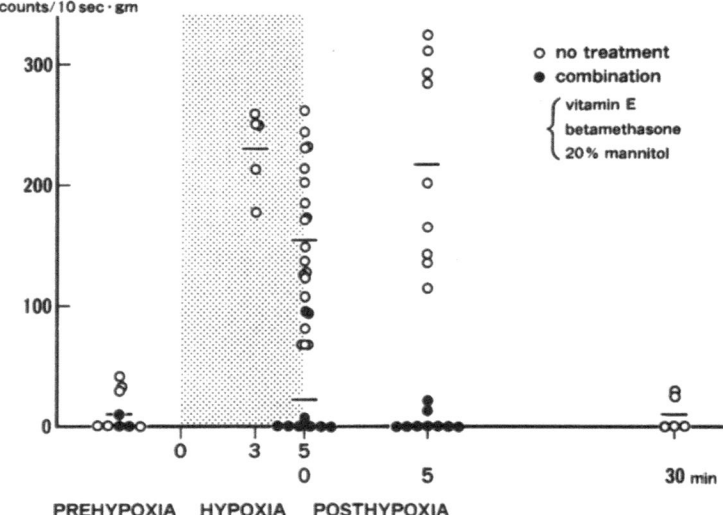

Fig. 5-35. Effect of combined therapy (20% mannitol 10 ml/kg, vitamin E 30 mg/kg, betamethasone 1 mg/kg) on chemiluminescence values during and following hypoxia (photon counts/10 sec. 1.0 gm wet weight). One circle indicates the value of one animal

When these drugs are administered in global cerebral ischemia (due to 3-vessel occlusion), the chemiluminescence is suppressed throughout the period of 5-minute ischemia and 30-minute reflow (**Fig. 5-36, Tables 5-6, 5-7**). In ESR measurements as well, the weakest intensities of PBN-trapped radicals were found following administration of these drugs (**Fig. 5-37**).

on carbohydrates and energy metabolism are thought to be a secondary effect brought about by the scavenging action of free radicals by mannitol, vitamin E and glucocorticoids and/or a protective effect on the mitochondrial respiratory chain due to a suppressive action on lipid peroxidation. In addition, it was found that pretreatment with the Sendai Cocktail in both the 5-minute and the 30-minute ischemic brain produced a marked decrease in the cortical water content (**Fig. 5-40, Tables 5-10, 5-11, 5-12, 5-13**). Although a possible osmolarity effect may be responsible for the actions of mannitol in these cases, the combined usage of these drugs was found to be extremely effective in combating ischemic edema.

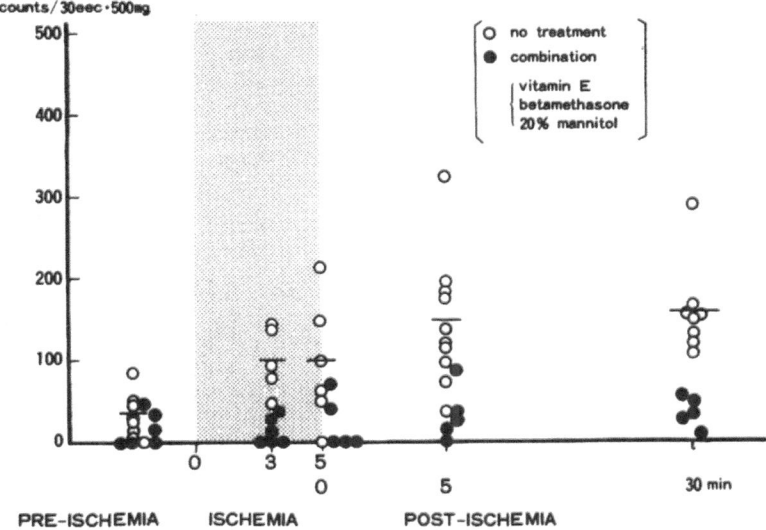

Fig. 5-36. Effect of combined therapy (20% mannitol 10 ml/kg, vitamin E 30 mg/kg, betamethasone 1 mg/kg) on chemiluminescence values during and following 5 min-ischemia (photon counts/30 sec. 0.5 gm wet weight). One circle indicates the value of one animal

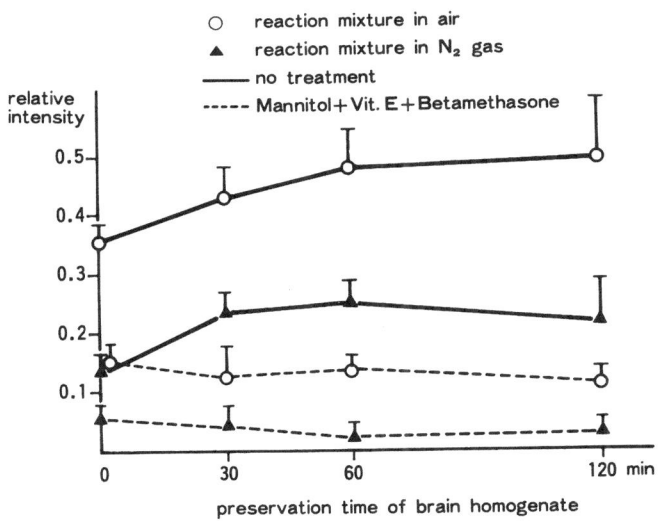

Fig. 5-37. Combined administration of mannitol, vitamin E and betamethasone suppressed the generation of the PBN trapped radical in NADPH-dependent lipid peroxidation system containing ischemic brain homogenate (mean ± S.E.M., n = 6). The brain homogenate preservation system was used for ischemic load (see **Fig. 5-15**)

Table 5-6. Effect of combined therapy (20% mannitol 10 ml/kg, vitamin E 30 mg/kg, betamethasone 1 mg/kg) on chemiluminescence values during 5 min ischemia followed by up to 30 min recirculation (photon counts/30 sec 0.5 gm wet weight)

	PRE-ISCHEMIA	ISCHEMIA		RECIRCULATION	
		3 min	5 min	5 min	30 min
NO TREATMENT	30±4	97±16	112±27	145±24	157±18
	(N=8)	(N=5)	(N=5)	(N=10)	(N=8)
TREATMENT	15±7	16±7**	22±13***	34±13***	31±5*
	(N=5)	(N=6)	(N=5)	(N=5)	(N=5)

Different from no treatment, * P<0.001, ** P<0.01, *** P<0.05
MEAN±S. E. COUNTS/30 SEC·500 MG

Table 5-7. Effect of combined therapy (20% mannitol 10 ml/kg, vitamin E 30 mg/kg, betamethasone 1 mg/kg) on chemiluminescence values during 30 min ischemia followed by up to 120 min recirculation (photon counts/30 sec 0.5 gm wet weight)

	PRE-ISCHEMIA	ISCHEMIA		RECIRCULATION	
		5 MIN	30 MIN	30 MIN	120 MIN
NO TREATMENT	30±4	112±27	173±36	385±61	317±38
	(N=8)	(N=5)	(N=8)	(N=7)	(N=8)
TREATMENT	15±7	22±13**	120±29	73±76*	—
	(N=5)	(N=5)	(N=6)	(N=8)	

Different from no treatment, * P<0.001, ** P<0.05
MEAN±S. E. COUNTS/30 SEC·500 MG

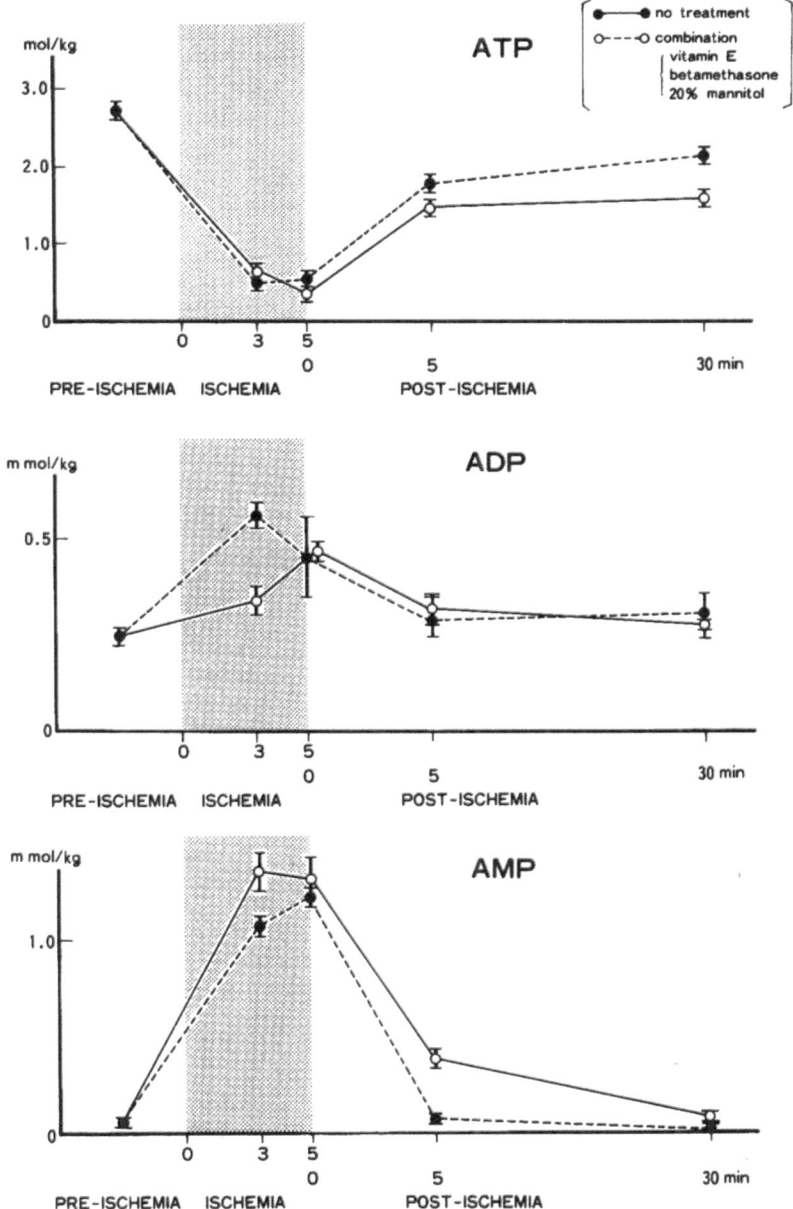

Fig. 5-38. Effect of combined therapy (20% mannitol, vitamin E and betamethasone) on the cerebral contents of energy metabolites during and following 5 min-ischemia (mean ± S.E.M., nmol/gm wet weight, n = 5 or 6)

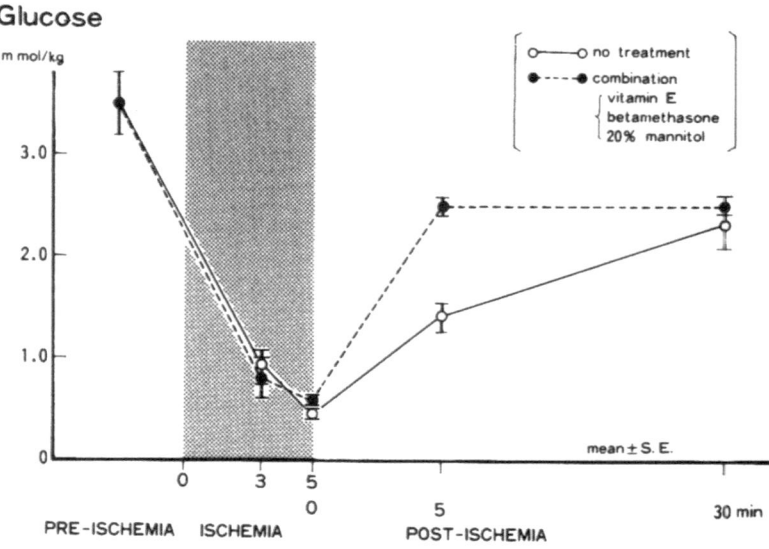

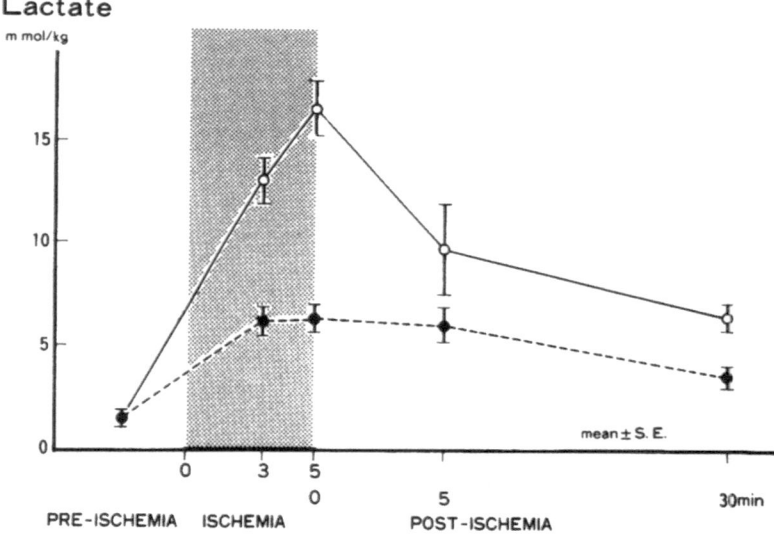

Fig. 5-39. Effect of combined therapy (20% mannitol, vitamin E and betamethasone) on cerebral contents of carbohydrates during and following 5 min-ischemia (mean ± S.E.M., nmol/gm wet weight, n = 5 or 6)

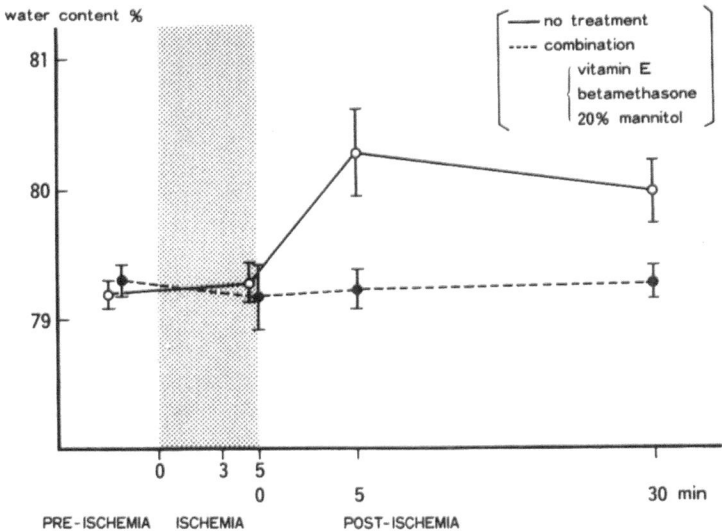

Fig. 5-40. Effect of combined therapy (20% mannitol, vitamin E and betamethasone) on the water content of cerebral cortex during and following 5 min-ischemia (mean ± S.E.M., %, n = 5 ~ 10)

Table 5-8. Effect of combined administration of mannitol, vitamin E and betamethasone on the cerebral contents of energy metabolites during 5 min-ischemia followed by up to 30 min recirculation (mean ± S.E.M., nmol/gm wet weight, n = 5 or 6)

	ATP	ADP	AMP	E. C.	Glucose	Lactate
Pre-Ischemia	2.74±0.03	0.243±0.019	0.031±0.004	0.949±0.003	3.51±0.30	1.65±0.32
Ischemia 3min						
No Treatment	0.62±0.06	0.340±0.038	1.352±0.097	0.343±0.012	0.94±0.14	13.14±1.09
Treatment	0.53±0.09	0.564±0.034**	1.077±0.047	0.357±0.016	0.82±0.20	6.32±0.68*
Ischemia 5 min						
No Treatment	0.38±0.05	0.458±0.014	1.308±0.124	0.286±0.012	0.51±0.03	16.73±1.36
Treatment	0.53±0.07	0.452±0.105	1.231±0.039	0.306±0.016	0.58±0.06	6.45±0.67*
Recirculation 5 min						
No Treatment	1.50±0.07	0.310±0.028	0.368±0.035	0.761±0.015	1.42±0.15	9.78±2.20
Treatment	1.80±0.04***	0.286±0.049	0.059±0.015*	0.760±0.032**	2.50±0.09**	6.14±0.85***
Recirculation 30 min						
No Treatment	1.63±0.11	0.279±0.004	0.045±0.004	0.904±0.006	2.32±0.22	6.52±0.70
Treatment	2.16±0.04**	0.298±0.058	0.032±0.002***	0.928±0.007***	2.52±0.08	3.63±0.53***

Different from no treatment, * P<0.001, ** P<0.01, *** P<0.05

MEAN±SE : M MOL/KG EXCEPT E.C.

Table 5-9. Effect of combined administration of mannitol, vitamin E and betamethasone on the contents of energy metabolites during 30 min followed by up to 120 min recirculation (mean ± S.E.M., nmol/gm wet weight, n = 5 or 6)

	ATP	ADP	AMP	E. C.	GLUCOSE	LACTATE
PRE-ISCHEMIA	2.74±0.03	0.243±0.019	0.031±0.004	0.949±0.003	3.51±0.30	1.65±0.32
ISCHEMIA 5 MIN						
NO TREATMENT	0.38±0.05	0.458±0.014	1.308±0.124	0.286±0.012	0.51±0.03	16.73±1.36
TREATMENT	0.53±0.07	0.452±0.105	1.231±0.039	0.306±0.016	0.58±0.06	6.45±0.67*
ISCHEMIA 30 MIN						
NO TREATMENT	0.05±0.01	0.415±0.065	1.252±0.081	0.147±0.017	0.06±0.00	16.83±0.83
TREATMENT	0.07±0.04	0.376±0.620	1.012±0.050**	0.170±0.010	0.18±0.01*	4.81±0.41*
RECIRCULATION 30 MIN						
NO TREATMENT	1.52±0.02	0.375±0.031	0.308±0.049	0.778±0.022	4.05±0.39	7.94±0.60
TREATMENT	2.24±0.10***	0.238±0.043***	0.123±0.048**	0.842±0.047***	4.08±0.04	3.98±0.31**
RECIRCULATION 120 MIN						
NO TERATMENT	1.76±0.05	0.486±0.028	0.357±0.042	0.772±0.011	3.58±0.04	3.78±0.31
TREATMENT	—	—	—	—	—	—

Different from no-treatment, * P<0.001, ** P<0.01, *** P<0.05

MEAN±SE : M MOL/KG   EXCEPT E.C.

Table 5-10. Effect of combined administration of mannitol, vitamin E and betamethasone on the water content of cerebral cortex during 5 min-ischemia followed by up to 30 min recirculation (mean ± S.E.M., %)

	PRE- ISCHEMIA	ISCHEMIA 5 MIN	RECIRCULATION 5 MIN	RECIRCULATION 30 MIN
NO TREATMENT	79.21±0.13 (N=10)	79.26±0.18 (N=7)	80.27±0.34 (N=6)	79.97±0.25 (N=6)
TREATMENT	79.30±0.12 (N=5)	79.17±0.25 (N=5)	79.22±0.15 (N=5)	79.29±0.13 (N=5)

* P<0.01

Table 5-11. Effect of combined administration of mannitol, vitamin E and betamethasone on the water content of cerebral cortex during 30 min ischemia followed by up to 120 min recirculation (mean ± S.E.M., %)

	PRE- ISCHEMIA	ISCHEMIA 5 MIN	ISCHEMIA 30 MIN	RECIRCULATION 30 MIN	RECIRCULATION 120 MIN
NO TREATMENT	79.21±0.13 (N=10)	79.26±0.18 (N=7)	78.96±0.24 (N=10)	80.52±0.27 (N=10)	80.3i±0.18 (N=17)
TREATMENT	79.30±0.12 (N=5)	79.17±0.25 (N=5)	79.42±0.42 (N=5)	79.50±0.19 (N=5)	—

* P<0.01, ** P<0.05

Table 5-12. Electrolyte contents during 30 min ischemia followed by up to 120 min recirculation (mean ± S.E.M., mEq/gm wet weight)

	PRE-ISCHEMIA	ISCHEMIA 30 MIN	RECIRCULATION 30 MIN	RECIRCULATION 120 MIN
Na⁺	246.4±4.2 (N=7)	248.9±4.0 (N=8)	290.7±4.2** (N=8)	319.7±6.9* (N=6)
K⁺	569.1±8.9 (N=6)	549.7±4.0 (N=8)	472.9±4.2** (N=8)	426.4±5.4* (N=6)

Different from pre-ischemia, * P<0.001, ** P<0.01

Table 5-13. Changes in cortical water contents by preadministration of various drugs at recirculation of 30 min following 5 min or 30 min of ischemia

	ISCHEMIA 5 MIN —RECIRCULATION 30 MIN	ISCHEMIA 30 MIN —RECIRCULATION 30 MIN
NO TREATMENT	79.97±0.25 (N=6)	80.52±0.27 (N=10)
COMBINATION	79.29±0.13* (N=5)	79.50±0.19* (N=5)
VITAMIN E	79.65±0.26 (N=5)	79.99±0.18 (N=5)
20% MANNITOL	79.32±0.11 (N=5)	79.26±0.75* (N=6)
BETAMETHASONE	79.70±0.21 (N=5)	80.00±0.24 (N=5)
ISOTONIC MANNITOL	79.38±0.12 (N=5)	79.64±0.16* (N=6)

Different from no treatment, * P<0.05

5.6.5 Phenytoin (Aleviatin)

The first clinical usage of phenytoin was reported by Merritt and Putnam in 1938[153], and it is currently in use as an anticonvulsant. Many excellent studies have been carried out on its anticonvulsant effect and its suppressive effect on membrane excitability. The majority of those studies, however, have focussed on the pharmacological properties of phenytoin in maintaining the intra- and extracellular balance of cations during neuronal excitation[7, 53, 75, 76, 179–181].

Recently, researches have also been done on the effects of phenytoin in protecting brain against ischemia and hypoxia[3, 4, 6, 7, 54]. Our own experimental work was initiated under the hypothesis that phenytoin maintains the balance of cations and stabilizes biomembranes, results in the suppression of free fatty acids liberation in the ischemic brain. Using the 3-vessel occlusion model and pretreatment with 10 mg/kg phenytoin, we have measured changes in contents of oleic acid, stearic acid, arachidonic acid, palmitic acid and docosahexaenoic acid.

In addition to having established the clinical effectiveness of phenytoin in the control of major motor tonic seizures, previous studies have elucidated six other effects or findings. These are: (i) stabilization on the threshold for neuronal membrane hyperexcitability and a reduction in the post-tetanic potentiation (PTP) of synapses[91], (ii) suppression of digitalis-dependent ventricular arrhythmia[76], (iii) no effects on the interseizure EEG[76], (iv) stabilization of rCBF, cortical water content, the cortical Na : K ratio and the extracellular K^+ concentration of CSF in the ischemic and hypoxic brain[6, 7, 11, 54, 88], (v) augmentation of the brain energy reserve and suppression of brain metabolism[35, 243], and finally (vi) phenytoin has no effects on the water or cation content of the putamen or white matter in the ischemic brain[34]. It is thought that the intra- and extracellular movement of cations, such as Na^+, K^+ and Ca^{2+}, are involved in the above effects brought about by phenytoin.

There is an active transport system which consumes energy together with the hydrolysis of ATP into ADP, and which actively works against the intra- and extracellular charge gradient and concentration gradient; it is therefore called a "pump" and is dependent upon the activity of ATPase. ATPase requires Mg^{2+} as a co-enzyme and its main effect is the efflux of Ca^{2+} and Na^+ to the extracellular medium and the influx of K^+ to the cytoplasm.

In contrast, there is also passive transport or "leakage" which is a simple diffusion process related to the permeability ratio.

ATP is hydrolyzed on the inner surface of the cellular membrane, and it supplies the energy needed for the functioning of the Na^+–K^+ and Ca^{2+} pumps. In brief, the movement of these ions under resting conditions can be summarized as follows. Due to the action of these "pumps", K^+ is transported into the cell, and Na^+ and Ca^{2+} are transported out. Due to "leakage" (passive transport), Na^+ and Ca^{2+} enter the cell and K^+ flows out. During membrane excitation, however, there is a local depolarization and (i) increase in intracellular Na^+ and Ca^{2+} concentration, (ii) the depletion of intracellular K^+.

In other words, the changes in the environment of these ions are due to effects caused by the suppression of active transport ("pumps") or facilitation of passive transport ("leakage"). Consequently, if one has a drug which can modify membrane permeability to these ions, it is possible to reduce membrane excitability.

It is thought that phenytoin has the capability to modify the ion imbalance occurring with membrane excitation. Festoff and Appel have argued that, by activating Na^+, K^+-ATPase, phenytoin decreases intracellular Na^+ and thereby quiets the neuronal membrane[74–76]. On the other hand, Woodbury has found that there

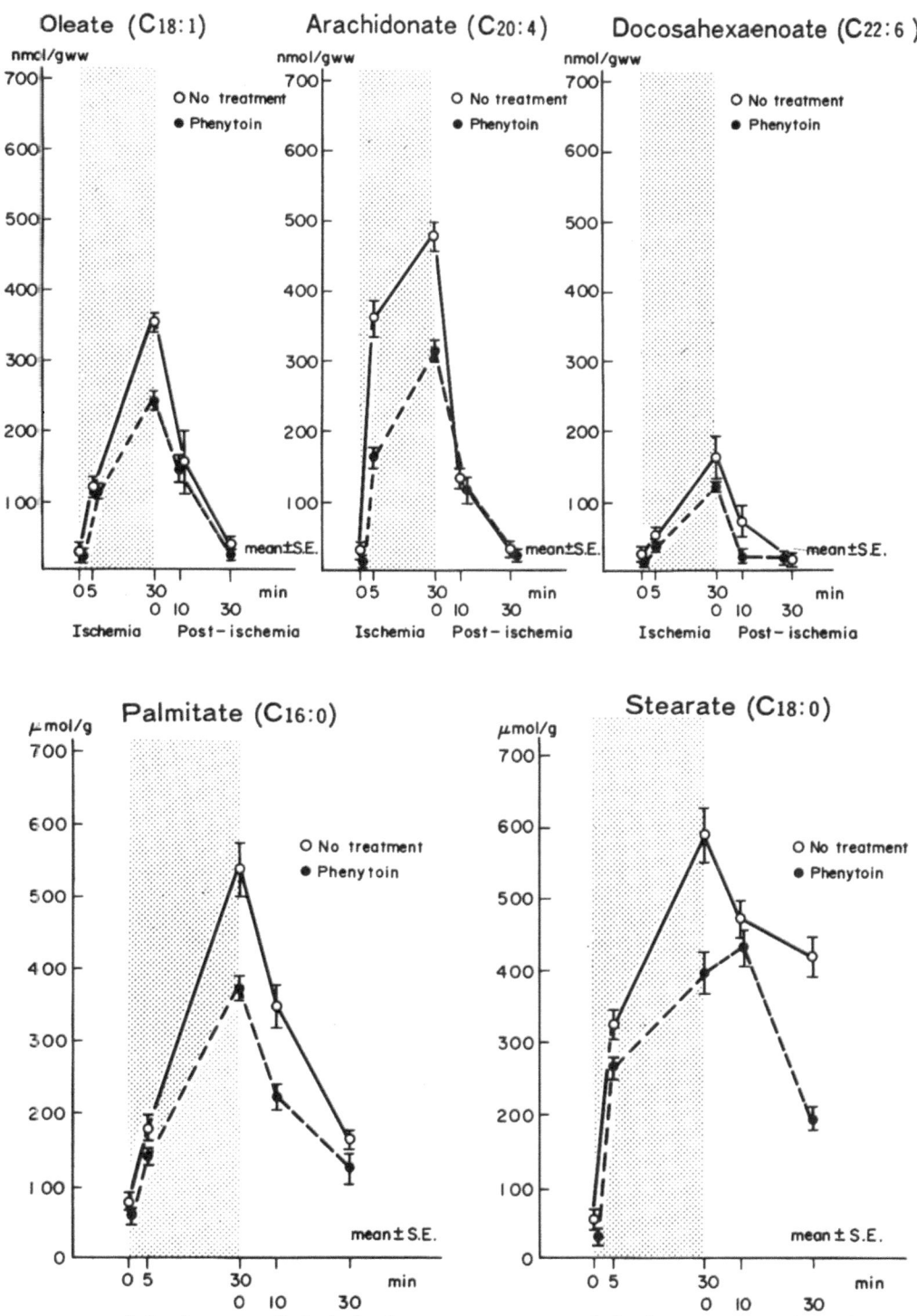

Fig. 5-41. Changes in individual free fatty acids in rat brain cortex during 30 min of ischemia followed by up to 30 min of recirculation. Phenytoin inhibited the free fatty acid liberation (mean ± S.E.M., nmol/gm wet weight, n = 5 or 6)

Table 5-14. Effect of phenytoin on the free fatty acid levels in rat brain cortex during 30 min ischemia followed by up to 30 min recirculation (mean ± S.E.M., nmol/gm wet weight, n = 5 or 6)

	Total free fatty acids	Palmitic acid	Stearic acid	Oleic acid	Arachidonic acid	Docosahexaeno acid
Pre-Ischemia						
no treatment	182.0±15.0	92.1± 8.5	50.2± 8.9	24.0± 4.5	23.0± 2.0	10.1± 4.0
phenytoin	164.2±13.0	70.5± 3.0	40.5± 2.3	23.5± 1.2	18.5± 3.0	11.7±0.9
Ischemia 5 min						
no treatment	1035.0±61.8	172.7± 8.9	320.1±18.7	113.6±11.6	364.0±21.2	33.8± 2.6
phenytoin	745.8±35.9**	145.0± 6.0	274.0± 6.5**	115.2± 3.5	159.7± 8.1*	29.1± 0.5
Ischemia 30 min						
no treatment	2146.2±130.9	530.0±44.4	588.0±41.6	343.0± 7.2	481.0± 8.8	161.0±35.4
phenytoin	1422.1±44.7*	362.6±13.4**	399.6±27.7*	244.2±10.4**	317.3±10.0*	122.9± 6.7
Recirculation 10 min						
no treatment	1043.2±76.2	232.9±20.8	468.0±37.9	154.9±43.4	131.2± 8.0	68.5±19.8
phenytoin	967.5±64.7	220.3±25.1	435.9±31.8	146.0±10.3	119.9±15.8	13.4± 5.2
Recirculation 30 min						
no treatment	769.0±63.0	160.8± 9.0	414.9±34.6	37.1± 3.9	29.5± 7.8	13.0± 2.
phenytoin	356.3±77.8*	126.0±14.6	191.7±17.6*	26.2± 6.2	21.5± 3.2	11.7± 4.0

Different from no treatment   * P<0.01,   ** P<0.05   mean±S.E.M

are not changes in the intracellular K^+ concentration during seizures[243], despite the administration of phenytoin, and has consequently argued that phenytoin is not involved in K^+ movements. It is noteworthy, however, that his conclusion pointed out that care must be taken when interpreting the results of measurements of Na^+ and K^+ in whole brain samples containing both neurons and glia[76]. In experiments on lobster axons, Pincus has also expressed reservations about the activity of Na^+, K^+-ATPase, and maintains that the primary pharmacological activity of phenytoin is the reduction of the intracellular Na^+ concentration. He has emphasized that, as a consequence of that reduction, phenytoin has effects only on the exitable membranes of hypoxic nerves, whereas it has no effect on oxygenated nerves with functions normally maintained[179-181]. In addition, he believes that phenytoin's action in reducing intracellular Na^+ is limited to the influx of sodium (i.e., inhibition of leakage or passive transport)[180].

In our own measurements of free fatty acids in the ischemic brain, we have found potent suppression of free fatty acid liberation in animals which have been given i.v. doses of 10 mg/kg phenytoin. Those free fatty acids which were particularly strongly suppressed were stearic acid and arachidonic acid—the levels of which were 2/3 of those in the untreated groups after 30 minutes of ischemia (**Fig. 5-41, Table 5-14**).

The liberation of free fatty acids has

been found to be dependent upon phospholipase A_2, A_1, and in the early stage, phospholipase C. These enzymes are known to be promoted by Ca^{2+} or the agonist-receptor coupling mechanism. So, the suppressive effects of phenytoin on free fatty acid liberation in the ischemic brain may be due to a decrease in intracellular Ca^{2+} through the Na^+–Ca^{2+} exchange transport system together with a decrease in intracellular Na^- due to ATPase activation, or by the suppression of Ca^{2+} channel. Alternatively, phenytoin may act more directly on the suppression of phospholipases A_2 and A_1 activity or by other mechanisms to bring about greater rigidity of the phospholipid acyl bond. In either case, it is now known that phenytoin suppresses the structural breakdown of phospholipids in the ischemic brain, and is thought likely to be effective in the clinical treatment of cerebral infarction.

6. THE DEVELOPMENT OF NEW BRAIN PROTECTIVE AGENTS

6.1 Introduction

The brain is known to be particularly susceptible to the effects of ischemia and, once brain tissue has fallen into a state of necrosis due to ischemia, no form of therapy has yet been found to be effective. In recent years, various surgical techniques, such as by-pass surgery and embolectomy, have been developed to treat the ischemic brain, but regardless of how quickly such revascularization methods are employed, in the vast majority of cases a considerable period of time will have elapsed from onset until the completion of revascularization. Not only is it common for the ischemic focus to become infarctic during that period, but when recirculation is achieved in the acute period it is not infrequently the case that brain swelling or hemorrhagic infarction ensues and the patient's condition actually becomes aggravated[18, 32, 117]. In light of these facts, we have experimentally investigated—from the clinical neurosurgeon's perspective—the possibility of developing pharamcological means for preventing the irreversible changes in brain tissue in the brief period between onset and surgical treatment.

Our experimental work was initiated and originally motivated by a hint obtained from an accident occurring during aneurysm surgery in 1969 which led to the realization that 20% mannitol, previously used as a hyperosmolar agent, has protective effects on the ischemic brain[103]. We have subsequently developed various experimental models[45, 105, 121] over a period of more than 10 years and investigated the effectiveness of mannitol in protecting the brain. As reported in several studies, we have found that mannitol administration results in (i) an increase in blood flow to an ischemic focus[83], (ii) improvement in the EEG change brought about by the ischemia[122], (iii) suppression of the progressing pathology found in brain tissue following ischemia[112], and (iv) suppression of the development of brain edema following recirculation after ischemia[104].

Moreover, using a biochemical approach, we have also found that mannitol acts as a scavenger of hydroxy radicals, which are known to have the highest chemical reactivity of the active oxygens[108]. We consequently believe that it is precisely mannitol's scavenge capacity which is responsible for its protective effects on the brain.

More recently, we have investigated

the protective properties of the perfluorochemicals (PFC) which have been developed as artificial blood substitutes [20% Fluosol-DA[64]], because they have the capacity to dissolve some 20-fold more oxygen than water and about the same amount as blood. In light of the fact that this PFC has a particle size of 0.1 μm—that is, 1/80 that of the red blood cells, we surmised that it may be applicable as a therapeutic agent in cerebral ischemia. That is, even when the vascular lumen of capillaries is reduced due to the effects of brain ischemia, PFC may be able to pass easily through the vessel and thus supply oxygen to the ischemic focus even under poor circulatory conditions in which the collateral pathways are blocked. The results of a series of experimental studies have indeed shown that the combined administration of 20% mannitol and 20% Fluosol-DA facilitates the recovery of brain functions following ischemia[62], suppresses cerebral edema[106] and prevents the development of hemorrhagic infarction[42] more than the administration of either drug alone.

We have also demonstrated the likelihood that the free radical reaction is a major factor involved in the generation of cerebral tissue damage following ischemia[22, 26]. Moreover, knowing—as mentioned above—that mannitol acts as a free radical scavenger, we have investigated the protective effects of other known radical scavengers, such as vitamin E. The findings from those studies indicated that clinical effectiveness could be expected with the combined administration of mannitol, vitamin E and glucocorticoids. Subsequent study of the effects of those drugs and PFC has indeed shown that their combined administration results in a marked improvement in their effects in protecting the brain from the deleterious effects of ischemia[107].

Based upon this foundation of experimental results, we currently use a combination of 500 ml of 20% mannitol and 300 mg of Vitamin E and 50 mg of dexamethasone or betamethasone during the acute stage in clinical cases of ischemic brain disorder. (More recently, phenytoin has been used instead of steroids.) Thus far, these drugs (known as the "Sendai cocktail") have produced favorable clinical results.

The remainder of this chapter will be concerned with two further issues. The first is the effectiveness of and problems with drug therapies (using various brain protective agents) in brain ischemia, as reported by several research groups. The second topic concerns our own experimental work on the abovementioned combined drug therapy technique. This discussion will deal particularly with the remarkably good effects which we have obtained using the anti-convulsant, phenytoin and calcium antagonists.

6.2 Brain Protective Agents—Short Review

6.2.1 Barbiturates

The barbiturates are the class of cerebral protective agents which have received the most attention in recent years. During the 1960s, a group including Arnfred, Secher, and Wilhjelm[4, 82] reported that the survival times of mice subjected to hypoxia were significantly increased when barbiturates were administered. With regard to the mechanism involved during the hypoxia, they emphasized the role of these drugs in the suppression of cerebral metabolism—i.e., their ability to reduce cerebral oxygen consumption ($CMRO_2$).

Thereafter, many influential experimental studies on the protective effects of the barbiturates in cerebral ischemia, based upon a similar mechanism have been reported. Such research is best reviewed by examining separately the work on focal cerebral ischemia[11, 20, 21, 36, 58, 59, 63, 71, 84, 85, 93] and that on global cerebral ischemia[2, 7, 12, 31, 50, 57, 66, 68, 69, 74, 99, 119] (**Tables 6-1** and **6-2**). Although nearly all research has indicated that the barbiturates are effective in focal ischemia, the results in studies on global ischemia, particularly complete ischemia, have varied and there is currently no consensus about their effects. When considering the source of these differences in experimental results, it is necessary to examine closely the differing mechanisms of action of the barbiturates.

Michenfelder et al.[57] found that administration of thiopental (23 mg/kg/hr) produces a reduction in $CMRO_2$ similar to that brought about by moderate hypothermia[93]. It might therefore be assumed that thiopental deep anesthesia has a mechanism of action similar to 30 °C hypothermia y but such an equation is based naively upon the premise that the protective effects of the barbiturates are identica to the metabolic suppression brough about by hypothermia. In a further study done by Michenfelder et al.[5] using the decapitated dog brain, comparison was again made between the suppressive effect on cerebral metabolism of hypothermia and that of the barbiturates. It was found that although both conditions resulted in similar decreases in $CMRO_2$, hypothermia suppressed the rate of ATF depletion, whereas the barbiturates had no such effect. They believed that the different mechanisms might be due to the barbiturates having effects on cerebral metabolism only when cerebral functions are maintained. In other words, the barbiturates affec $CMRO_2$ by influencing the energy requirements for the maintenance on function, but they have no effect on energy required for maintaining cellular integrity. In contrast, hypothermia suppresses the metabolism of both neuronal function and neuronal cellular integrity. As a consequence, when neuronal functions are abolished and the EEG becomes rapidly attenuated as seen in complete ischemia, hypothermia augments the brain's resistanc to ischemia, whereas the barbiturate have no such effects. In light of these empirical facts, it might be supposed that only in conditions where there is not complete loss of neuronal function, such as in incomplete ischemia or focal ischemia, will the barbiturates have a certain degree of protective effects on the ischemic brain.

In fact, however, when severe cerebral ischemia was induced using a global, complete ischemia model in the dog, Goldstein *et al.* found that the barbiturates to indeed have protective effects[31]. These results suggest that suppression of cerebral metabolism will not suffice to explain the protective effects of the barbiturates and indicate the presence of other mechanisms.

A phenomenon known as the "inverse steal" has been suggested[15, 26, 85] to account for the protective effects of these drugs in focal cerebral ischemia. That is, similar to the therapeutic effectiveness of hyperventilation in brain infarction, the barbiturates are thought to produce constriction of cerebral vessels at non-ischemic regions thereby diverting blood to the ischemic focus and thus reducing the infarction. Nevertheless, currently there is growing evidence denying the effectiveness of hyperventilation in the treatment of focal ischemia and further study is undoubtedly required before proper evaluation can be made of the "inverse steal" due to barbiturates.

Siesjö *et al.*[90] have advocated the view that in cerebral ischemic foci there is an increase in free radicals, particularly active oxygens, which peroxidate lipids and lead to destruction of cellular membranes. The barbiturates may therefore have a scavenging effect on free radicals and consequently protect the brain from the effects of ischemia. Although there have been reports of free radical scavenging effects by barbiturates using indirect experimental methods[54, 95], there have also been reports of experimental results which are entirely negative with regard to the scavenger hypothesis[87, 96].

As possible mechanisms for the protective effects of the barbiturates, suppression of cerebral metabolism, improvement of the distribution of cerebral blood flow, the radical scavenger effect, have been considered, but no single hypothesis has been shown to be able to explain all of the facts concerning the barbiturates in protecting the ischemic brain. Further experimental study of this problem is yet required.

With regard to clinical applications, it must be said that the clinical use of barbiturate therapy is not easily carried out. This is because the patient's state of consciousness cannot be evaluated during barbiturate therapy and since respiratory suppression and hypotension can easily be incurred, strict regulation of the patient's systemic condition is difficult. It is noteworthy that there have been recent reports of the effectiveness of etomidate[111], gamma hydroxybutylate[49], midazolam[70] and nizofenone[72]—which are drugs capable of suppressing cerebral metabolism, but which do not incur the same problems as those encountered with the barbiturates.

Table 6-1. Experimental studies on the effe

Authors	Species	Model
Smith[94] 1974	dog	permanent ICA & MCA occlusion
Hoff[36] 1975	baboon	permanent MCA occlusion
Moseley[63] 1975	rhesus monkey	segmental MCA occlusion (silicon cylinders)
Michenfelder[58] 1975	squirrel monkey	2 hr MCA occlusion
	dog	permanent MCA occlusion
	cat	3 hr MCA occlusion
	gerbil	2 hr CCA occlusion
Michenfelder[59] 1976	Java monkey	permanent MCA occlusion
Corkill[20] 1976	dog	permanent ICA & MCA occlusion
Corkill[21] 1978	dog	permanent ICA & MCA occlusion
Black[10] 1978	cat	permanent MCA occlusion
Simeone[93] 1979	rhesus monkey	permanent MCA occlusion
Selman[85] 1981	baboon	permanent MCA occlusion 6 hr MCA occlusion
Selman[86] 1982	baboon	6 hr MCA occlusion
Ochiai[71] 1982	cat	permanent MCA occlusion

barbiturates in focal cerebral ischemia

arbiturate	Timing of treatment	Dose	Results
entobarbital	preocclusion	56 mg/kg iv	+ (improved function and decreased infarct size)
iopental	preocclusion	20 mg/kg iv plus 20 mg/kg infusion	
iopental	15 min post-occlusion	20 mg/kg iv plus 20 mg/kg infusion	
entobarbital	preocclusion	60, 90, 120 mg/kg iv	+ (decreased infarct size)
entobarbital	30 min post-occlusion	4 mg/kg/hr infusion × 12 hr	+ (improved function and decreased infarct size)
		40 mg/kg ip	+ (less metabolic alterations, improved function and less infarct area)
entobarbital	preocclusion	40 mg/kg iv 70 mg/kg iv 40 mg/kg ip	− (no effect on metabolism, function, or pathology)
entobarbital	30 min post-occlusion	14 mg/kg iv plus 7 mg/kg iv every 2 hr for 48 hr	+ (decreased mortality, decreased infarct size)
entobarbital	1 hr postocclusion	40 mg/kg im	+ (decreased mortality, size)
	3 hr postocclusion 6 hr postocclusion		− (no effect on pathology)
entobarbital	1 hr postocclusion	10–40 mg/kg im	+ (dose dependent reduction in infarct size)
entobarbital	preocclusion 2 hr postocclusion	50 mg/kg iv	+ (decreased infarct size)
entobarbital	30 min post-occlusion	14 mg/kg iv plus 7 mg/kg iv every 2 hr thereafter	+ (supression of brain edema)
entobarbital	30 min post-occlusion	30 mg/kg iv plus infusion to maintain flat EEG	− (detrimental effect on function and pathology) + (complete protection from ischemic damage)
entobarbital	30 min post-occlusion 2 hr postocclusion 4 hr postocclusion	30 mg/kg iv plus infusion to maintain flat EEG	+ (complete protection from ischemic damage) + (improved function and decreased infarct size) − (deleterious effect on function, pathology, and ICP)
entobarbital	30 min postocclusion	25 mg/kg/day ip & sc for 3 days	+ (decreased infarct size, increased CBF)

Table 6-2. Experimental studies on the effe

Authors	Species	Model
Goldstein[31] 1966	dog	8–15 min aorta occlusion
Nilson[65] 1971	rat	5 min hypotension (MABP 25–35 mmHg)
Yatsu[119] 1972	rabbit	hypotension (30–35 mmHg) plus hypoxia (4% O_2)
Michenfelder[57] 1973	dog	9 min hypotension (MABP 25–30 mmHg) 9 min asphyxia
Bleyaert[12] 1978	monkey (Macaca mulatta)	16 min neck tourniquet
Bandaranayake[7] 1978	rat	decapitation
Levy[50] 1978	gerbil	1 hr CCA occlusion
Nordström[68] 1978	rat	permanent both CCA occlusion plus hypotension (50 mmHg)
Nordström[69] 1978	rat	15–30 min both CCA occlusion plus hypotension (50 mmHg)
Pulsinelli[74] 1979	rat	30 min 4-vessel occlusion
Aldrete[2] 1979	rabbit	20 min neck tourniquet
Steen[100] 1979	dog	8–10 min aorta occlusion

6.2.2 Naloxone

In 1981, Baskin and Hosobuchi[8] reported dramatic improvements in the level of consciousness and the disappearance of hemiparesis following naloxone administration in two cases presenting symptoms of cerebral ischemia. (One patient was a case of internal carotid artery aneurysm in which stenosis of the middle cerebral artery developed postoperatively and the other patient was a case of massive brain edema following surgery for an

f barbiturates in global cerebral ischemia

arbiturate	Timing of treatment	Dose	Results
entobarbital	preocclusion	30 mg/kg iv	+ (less neurologic damage)
henobarbital	preocclusion	150 mg/kg ip	+ (less metabolic alterations)
1ethohexital	after onset of isoelectric EEG	5 mg/kg iv	+ (no observable neurological damage)
1iopental	preocclusion	15 mg/kg iv	+ (less metabolic alterations) − (no effect on metabolism)
1iopental	5–60 min post-ischemia	60 or 120 mg/kg iv	+ (improved function)
entobarbital	pretreatment	60 mg/kg ip	+ (suppression of increase in brain osmolality)
entobarbital	1 hr after recirculation	70 mg/kg ip plus 50 mg/kg ip 2 hr later	+ (less ischemic cell change)
entobarbital	preocclusion	150 mg/kg ip	± (less deranged energy state in some animals)
entobarbital	preocclusion	150 mg/kg ip	+ (better metabolic and functional recovery)
entobarbital	just after recirculation 30 min after recirculation 1 hr after recirculation	30 mg/kg ip	− (exacerbation of brain damage) − (no effect on pathology)
hiopental	preocclusion	10 mg/kg iv plus 10 mg/kg im	− (no effect on function and pathology)
entobarbital	preocclusion	30–45 mg/kg iv	− (no effect on function, CBF, and metabolism)

AVM.) The effect of naloxone lasted for only 20 minutes and then quickly disappeared—with reappearance of the symptoms. A repetition of the naloxone infusion produced identical results—indicating that the phenomenon was reversible.

Subsequently, using a unilateral common carotid artery occlusion model in the gerbil, they studied experimentally the effects of naloxone on the ischemic brain. Similar to their experiences with their two clinical patients, they found that repeated administration of naloxone consistently reversed all neurological deficits (which had been apparent for some 30 minutes)[38].

As a result of this research, a series

Table 6-3. Experimental studies on the effect

Timing of treatment	Dose	Results
2–4 hr post-occlusion	1 mg/kg ip or 10 mg pellets sc	+ (improved function)
4 hr–8 days postocclusion	4–7 mg/kg iv	+ (improved function)
4 hr postocclusion	2 mg/kg iv	+ (improved function with decreased CBF)
1 hr postocclusion	2 mg/kg iv plus 2 mg/kg hr infusion	+ (improved SEP and CBF)
30 min post-occlusion	5 mg/kg iv	+ (improved function and decreased infarct size)
6 hr postocclusion	10 mg/kg iv	+ (improved function with unchanged infarct size)
1) immediately after recirculation 2) immediately after occlusion	10 mg/kg ip plus 10 mg pellets sc	− (no effect on function)
10 min post-occlusion	10 mg/kg iv plus 2 mg/kg/hr infusion	− (no effect on function and pathology)
20 min post-occlusion	2 mg/kg iv plus 2 mg/kg/hr infusion or 10 mg/kg iv plus 10 mg/kg/hr infusion	− (no change in function, CBF, or pathology)

of related studies have been undertaken on the effects of naloxone, and the conclusions reached thus far can be broadly classified into positive findings[9, 23, 38, 51] and negative findings[28, 37, 40]. Because there have been a variety of animal species and a variety of ischemic models used for experimental study, and because there have been differences in the timing of treatment and the dosage, the results of all experimental work are not uniform. Given the diversity of experimental results, it is consequently impossible to draw any firm conclusions (**Table 6-3**).

Originally, naloxone was known as an opiate receptor antagonist and has been used as an agent capable of modifying the effects of the endogenous opioids. In light of the fact that naloxone ameliorates the symptoms of cerebral ischemia, Hosobuchi *et al.* have suggested that the endogenous opioids increase in brain tissue during cerebral ischemia and may be one of the factors contributing to the pathology of this condition. It must be said, however, that this mechanism remains hypothetical with, as yet, no experimental proof. In any event, naloxone is known not only to act as an opioid

Authors	Species	Model
Hosobuchi[38] 1982	gerbil	permanent MCA occlusion
Baskin[9] 1982	baboon	permanent ICA, MCA & PcomA occlusion
Levy[51] 1982	cat	permanent MCA occlusion
Faden[23] 1982	dog	CCA air embolization
Zabramski[124] 1984	baboon (Papio anibus)	6 hr MCA occlusion
Baskin[10] 1986	cat	permanent MCA occlusion
Holaday[37] 1982	gerbil	1) 20–30 min blt CCA occlusion 2) permanentCCA occlusion
Hubbard[40] 1983	cat	8 hr MCA occlusion
Gaines[28] 1984	monkey (Macaca irus)	permanent MCA occlusion or 4 hr MCA occlusion

antagonist, but also at high doses it has a number of other actions, including effects on Ca^{++} flux[81] and lipid peroxidation[48].

6.2.3 Prostaglandins and Indomethacin

In order to understand the role of the prostaglandins (PGs) in ischemic cell injury, it is necessary to begin the discussion with a review of the biosynthesis of the PGs, beginning with arachidonic acid (**Fig. 6-1**). Arachidonic acid is converted into a prostaglandin intermediate PGG_2 with the addition of two oxygen molecules due to the catalysis of cyclo-oxygenase. It is noteworthy that, together with and subsequent to the conversion of PGG_2 into PGH_2, there is the liberation of one of the active oxygens, the so-called hydroxyradical ($\cdot OH$). Next, via a variety of metabolic pathways, PGH_2 is converted into thromboxane A_2 (TXA_2) with its powerful platelet aggregating and vasoconstrictive effects and prostacycline (PGI_2) with virtually the opposite effects from those produced by TXA_2. It is thought that these two chemicals,

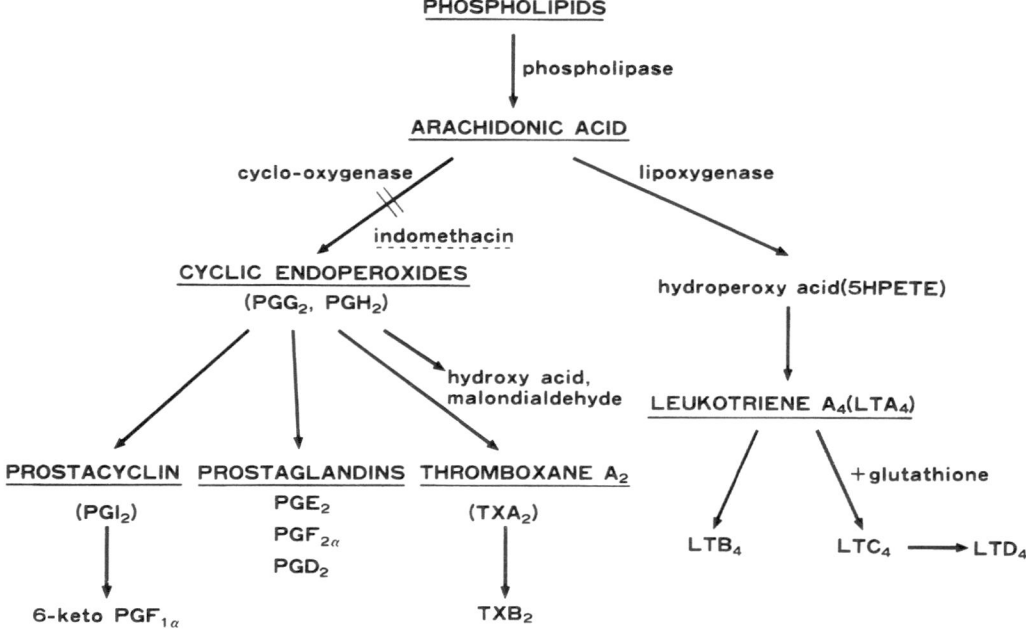

Fig. 6-1. Outline of the pathways of arachidonic acid metabolism

TXA_2 and PGI_2, with their diametrically opposed effects strike a delicate balance and thereby regulate the microcirculation of the brain.

It is known that free fatty acids, particularly arachidonic acid, are liberated from cell membranes during cerebral ischemia. Under conditions where there is no O_2 present, as in the case of complete ischemia, the biosynthesis of the prostaglandins is completely halted at the cyclo-oxygenase step. However, in the case of incomplete ischemia or following recirculation, it is thought that there is activation of PG biosynthesis together with increases in tissue arachidonic acid. As described above, the facilitation of PG synthesis accompanies an increase in the generation of active oxygenes, leading directly not only to tissue damage, but also to the inhibition of PGI_2 synthesis.

As a consequence, the balance of the TXA_2-PGI_2 system is tilted toward TXA_2 and, as a result, it can be presumed that there will occur disturbances of microcirculation due to increased vasoconstriction. Based upon this hypothesis, considerable attention has been drawn to PGI_2 itself and indomethacin—which is known to block specifically cyclo-oxygenase (one of the key enzymes in the PG synthesis pathway)—as therapeutic drugs in cerebral ischemia. Various experimental studies on this topic have been carried out.

In 1979, using a bilateral common carotid artery occlusion model in the gerbil, Gaudet and Levine[29] found that the administration of indomethacin resulted in suppression of the increases in both TXA_2 and PGI_2 in brain tissue which are normally seen following ischemia and, moreover, led to an amelioration of the neurological deficits due to the ischemia. Shohami et al.[89] used a rat incomplete ischemia model in which the bilateral common carotid arteries were occluded and hypotension (MABP 50 mmHg) was then induced by means of exsanguination. They found that indomethacin prevents increases in TXA_2 and improvements in EEG activity. In 1982, using a canine air embolization model, Hallenbeck et al.[33] reported combined treatment with indomethacin, PGI_2 and heparin was effective in improving brain functions.

In contrast, using a middle cerebral artery occlusion model in the baboon, Harris et al. (1982)[34] reported that, administration of indomethacin invites cerebral edema. Although Boulu et al. (1982)[14] found that indomethacin was effective in improving cerebral hemo-dynamics after recirculation, they reported that it was ineffective in improving neurological deficits.

As evident from the above, research on indomethacin and PGI_2 is not yet at a stage where definite conclusions can be drawn with regard to its protective effects on the brain in cerebral ischemia. It should be noted that, since indomethacin blocks the activity of cyclo-oxygenase, it blocks the synthesis of all PG pathway products, including PGI_2. Recently, there have been reports of various drugs which selectively block only the synthesis of TXA_2, and further developments in this direction are eagerly awaited.

Finally, with regard to the role of the other pathway for the metabolism of arachidonic acid, i.e., the leukotrien series synthesized by lipoxygenase, research in this area has just begun[116], although again significant results are expected to appear in this field.

6.3 Development of a New Brain Protective Substance —Our Study

6.3.1 Mannitol

6.3.1.1 The Effect of Mannitol on Cerebral Ischemia—Histological Study

Chronic Experiments—Using the Light Microscope

This investigation was designed to discover whether or not there were significant differences in the appearance of infarctic foci in mannitol-treated and untreated animals, using the canine thalamic infarction model[122]. For evaluation of the infarction, brains were autopsied 7 days after the release of vascular occlusion and studied under the light microscope. It was found 6 of the 10 untreated animals had infarctic foci, but only 1 of the 10 mannitol-treated animals had them—indicating a significant suppressive effect of mannitol on infarction (**Table 6-4**).

Table 6-4. Comparison between control and mannitol groups with 60 min occlusion in dogs

Thalamus infarction	Control	Mannitol
Grade 0	4	9
Grade 1	0	0
Grade 2	2	0
Grade 3	4	1
Total	10	10
	(P<0.03 Fisher)	

Grade 0 : no infarction in thalamus.
Grade 1 : small infarction foci in thalamus.
Grade 2 : infarction affecting about half of thalamus.
Grade 3 : infarction affecting more than two thirds of thalamus.

Acute Experiments—Using the Light and the Electron Microscope

Using the same infarction model, 30, 60, 90, 120 and 180 minutes of vascular occlusion were performed in untreated control animals and in mannitol treated animals. Histological changes were studied using the light and the electron microscopes [112, 113].

After 30 minutes of occlusion, only one of the untreated animal showed no histological changes, but normal brains were found in all of the mannitol-treated groups except the 180 minute occlusion group (**Fig. 6-2A**). In the untreated animals undergoing 60 minute occlusion, all animals showed neuronal shrinkage, whereas none of the mannitol-treated animals showed such pathology. In addition, swelling of glia and spongiosis of the neuropile in the latter group was mild (**Fig. 6-2B**). When vascular occlusion was prolonged for 180 minutes, significant differences between the treated and untreated animals were not found (**Fig. 6-3**).

6.3.1.2 Recirculation in the Acute Period of Cerebral Infarction: Brain Swelling and Its Suppression Using Mannitol

Using the incomplete cerebral hemisphere infarction model in dogs, observations were made on the effect of 20% mannitol on brain swelling in groups of experimental animals following recirculation.

Twenty-seven adult mongrel dogs each weighing approximately 10 kg were used in this study. Under intravenous thiopental anesthesia (1 ml/kg), with Scoville aneurysm clips, we occluded five cerebral arveries simultaneously: the anterior cerebral artery at the A1 portion and at the branching of the ethmoidal artery, the anterior cerebral artery at the A2 portion, the internal carotid artery, the bifurcation of the posterior communicating artery, the posterior cerebral artery, and the anterior cerebellar artery (**Fig. 6-4**).

In ten animals prepared in this manner, the degree of brain swelling

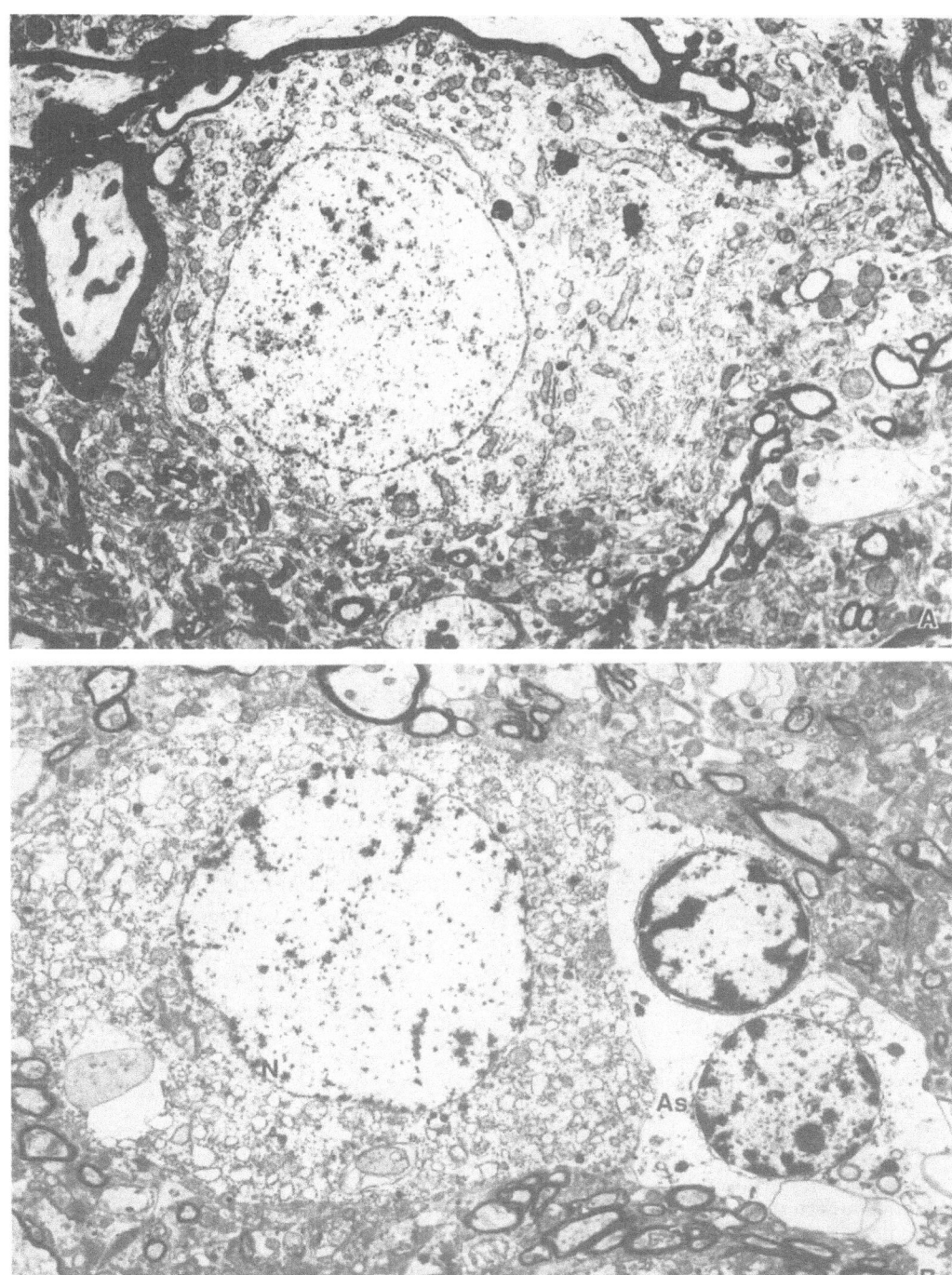

Fig. 6-2. Histological findings in the ischemic foci in the mannitol pretreated group. A) Mannitol pretreated group (2 hours occlusion). A nerve cell and surrounding neuropile had a normal appearance (calibration = 1 μ). B) Mannitol pretreated group (90 minutes occlusion). A nerve cell (N) showed ischemic changes in microvacuolation and a neighbouring astrocyte (As) showed slight swelling (calibration = 1 μ)

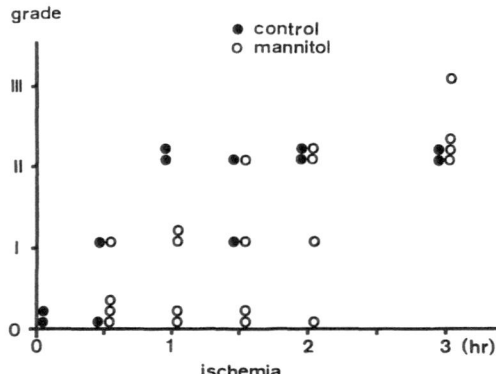

Fig. 6-3. Grading of ischemic changes in controls and animals receiving mannitol

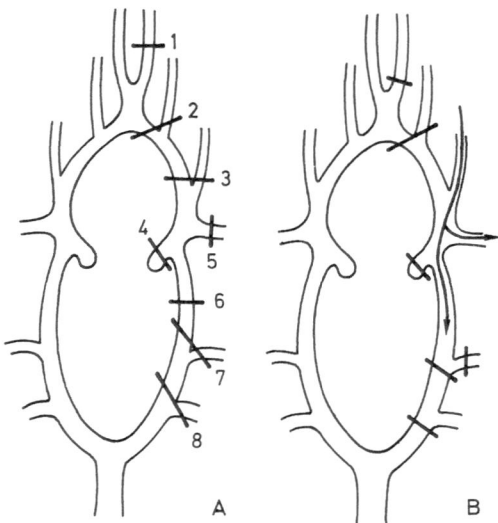

Fig. 6-4. A) Possible sites of occlusion of arteries at the base of the brain. (1) A_2 portion of the anterior cerebral artery (ACA); (2) A_1 portion of ACA at the bifucation of the ethmoidal artery; (3) A_1 portion of ACA at the bifurcation of the ophthalmic artery; (4) internal carotid artery; (5) middle cerebral artery; (6) posterior communicating artery (Pcom); (7) Pcom at the bifurcation of the posterior cerebral artery; (8) anterior cerebellar artery. B) Sites of occlusion used in the present experiment. The ophthalmic artery feeds slight perfusion to the middle cerebral and posterior communicating arteries

after two or six hours of occlusion was observed three hours following recirculation. In order to determine whether or not administration of 20% mannitol can suppress such swelling, the same infarction models were used and mannitol was administered to five dogs which had undergone two-, four-, or six hours of occlusion. Mannitol was administered as follows: In the animals with occlusion of the vessels for two hours, 2 gm/kg of mannitol was given intravenously during a 10-minute period one hour after the vessels had been occluded, 5 minutes before the clips were released, and again one hour after release of the occlusion. In the animals with vascular occlusion for four hours, 2 gm/kg of mannitol was injected four times, 30 minutes, two hours, and three and one-half hours after beginning the occlusion and one hour after clip release. In the animals with six hours of occlusion, 2 gm/kg of mannitol was injected over a 10 minute period four times. 30 minutes, two hours, and four hours after beginning the occlusion and just prior to release of the clips (Fig. 6-5).

For quantitative investigation of the correlation between brain swelling and

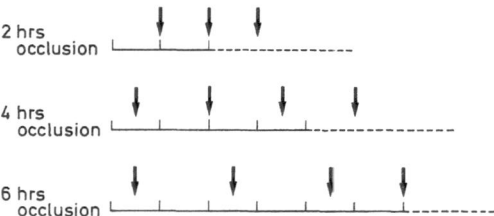

Fig. 6-5. Mannitol administration (arrows) for the canine model of widespread incomplete cerebral hemisphere infarction. (Solid lines = period of occlusion; broken lines = period of recirculation after occlusion)

length of occlusion, the degree of swelling as observed through the window in the bone and dura mater was classified according to the following five grades: Grade 0-space is present between the dura mater and the brain; Grade 1-no space is present between the dura mater and the brain; Grade 2-slight brain swelling is apparent through the bone window but there is little tension on the brain itself; Grade 3-marked brain swelling is seen through the bone window; and Grade 4-rupture of the swollen brain occurs.

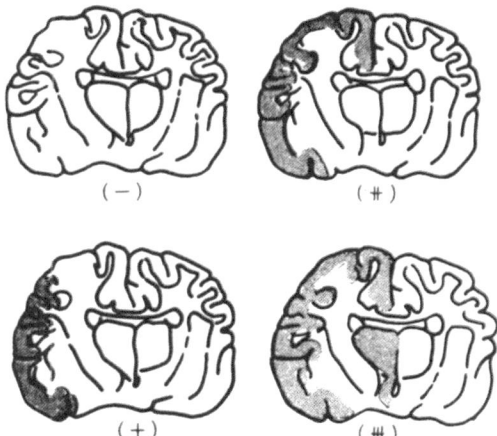

Fig. 6-6. Grading of extravasation of Evans blue. (See the text for details) (—): Grade 0, (+): Grade I, (+ +): Grade II, (+ + +): Grade III

Thirty minutes prior to autopsy, 1 ml/kg of 2% Evans blue was injected intravenously and the brain was removed. The degree of extravasation of the Evans blue was examined following fixation of the brains in formaldehyde and preparation of coronal sections. The following four-grade system was used in evaluation (Fig. 6-6): Grade 0-no extravasation found; Grade I-extravasation found only in the distribution of the anterior cerebral or middle cerebral artery; Grade II-extravasation seen throughout the cortex of the cerebral hemispheres; and Grade III-extravasation present throughout the cerebral cortex, thalamus, hypothalamus, and basal ganglia.

The degree of brain swelling after recirculation following two hours of occlusion is shown in Fig. 6-7A. Among five untreated dogs, all were classed Grade 0 throughout the period of occlusion. However, one hour following recirculation, brain swelling occurred and all animals were Grade 2 after two hours and Grade 3 after three hours. All five dogs treated with mannitol were Grade 0 throughout the period of occlusion. Following release of clips, swelling was minor: only one animal showed slight swelling after one hour. After three hours, four dogs were grade 0 and one was Grade 2.

Brain swelling after recirculation following four fours of occlusion was studied only in a mannitol-treated group consisting of five dogs (Fig. 6-7B). During occlusion, four animals were classed Grade 0 and one as Grade 1. Following recirculation, the dog that showed Grade 1 swelling during occlusion was Grade 3 after 30 minutes and Grade 4 after one hour. However, in the remaining four dogs, no swelling was seen immediately following release of clips, and two of the dogs still showed no swelling after two hours. The final results after three hours were: two Grade 0, one Grade 2, one Grade 3, and one Grade 4.

Brain swelling after recirculation following six hours of occlusion is shown in Fig. 6-7C. During occlusion

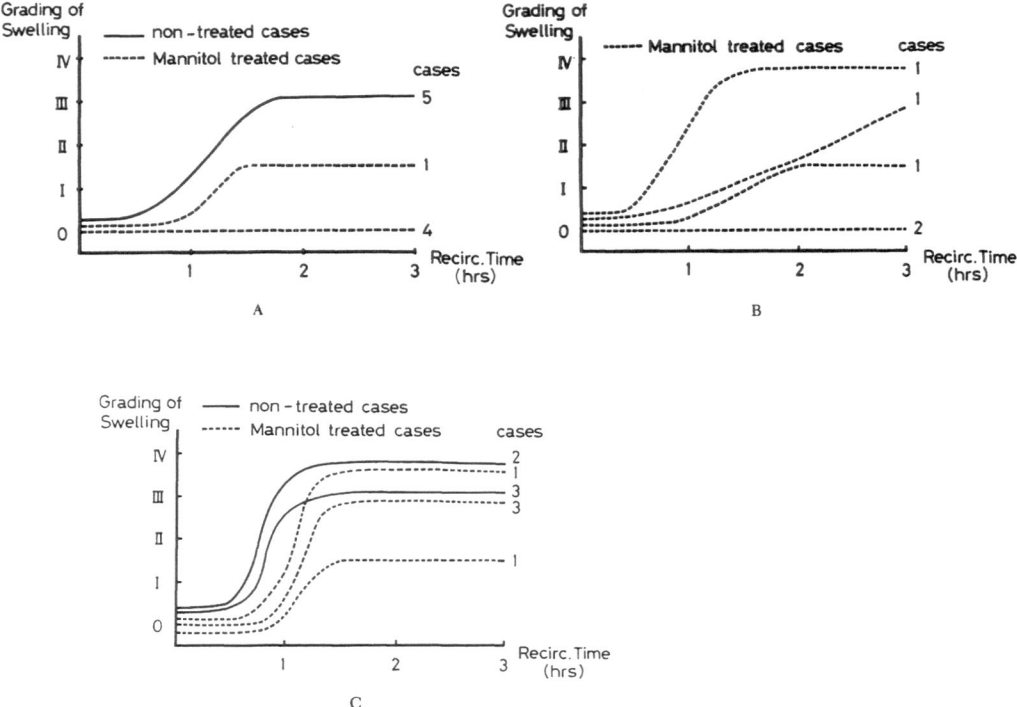

Fig. 6-7. Brain swelling after recirculation following occlusion for (A) two hours, (B) four hours, and (C) six hours

in the untreated group, four dogs were Grade 0 and one was Grade 1. Following the release of the clips, all five animals showed marked brain swelling; three had swelling within one hour of release and all five showed swelling within two hours. The final grades were: three Grade 3 and two Grade 4. Among the mannitol-treated dogs, all five were grade 0 during occlusion and showed brain swelling after release of the clips, but there were substantial differences in the time course. One dog showed swelling after 30 minutes, two after two hours. The final Grades were: one Grade 2, three Grade 3, and one Grade 4.

The degree of destruction of the blood-brain barrier in each group, evaluated by extravasation of Evans blue, is shown in **Table 6-5**. In the five untreated animals undergoing two hours of occlusion, four had Grade I and one had Grade II extravasation; in the five untreated animals undergoing six hours of occlusion, one dog was classed Grade I, one Grade II, and three Grade III. In the mannitol-treated group undergoing two hours of occlusion, however, there were three Grade 0 and two Grade I animals. In the mannitol-treated group undergoing four hours of occlusion, one dog had Grade 0, two had Grade I, and two had

Table 6-5. Degree of extravasation of Evans blue

Occlusion	Grade[a]			
	0	I	II	III
Two hour				
control	0	4	1	0
mannitol	3	2	0	0
Four hour, mannitol	1	2	0	2
Six hour				
control	0	1	1	3
mannitol	0	1	1	3

[a] See text for explanation.

Grade III destruction. In the group undergoing six hours of occlusion there were one Grade I, one Grade II, and three Grade III animals. Therefore, in dogs undergoing two hours of occlusion, mannitol clearly suppressed subsequent swelling; but after six hours of occlusion, mannitol alone had little positive effect.

6.3.1.3 The Effect of Mannitol on Cerebral Ischemia—CBF Study

It is well known that the hemodynamics in the ischemic focus are different from those in the peripheral zone. The canine thalamic infarction model allows for production of an infarcted focus at a definite site; and it is therefore possible to make recordings from the center of the ischemic area, and it is appropriate for making observations on the changes in rCBF at that site following the administration of drugs. We have previously found that 2 hours of vascular occlusion using this model produces irreversible changes and that, at that critical time, rCBF falls to below 30% of the preocclusion level[72, 73]. In the current study, this 30% value was taken as a criterion for obtaining two groups: a severe ischemia group (rCBF less than 30%) and a moderate ischemia group (rCBF between 30 and 60%).

Electrodes were implanted at the center of the ischemic focus at the anterior portion of the nucleus ventralis of the thalamus and 10 ml/kg of 20% mannitol was administered over a 10 minute interval by intravenous drip 30 minutes after occlusion of the 4 cerebral vessels. rCBF measurements were made at 10, 45, 80 and 120 minutes following occlusion[83].

In the four severe ischemia animals, rCBF fell from 38.1 ± 2.5 ml/100 g/min prior to occlusion to 7.6 ± 1.5 ml after occlusion. After completion of mannitol administration, rCBF measured at 45, 80 and 120 minutes following occlusion were 9.3 ± 2.3, 7.5 ± 1.5, 7.3 ± 1.7 ml, respectively. In contrast, among the four moderate ischemia animals (**Fig. 6-8**), the rCBF was 29.1 ± 5.3 ml/100 g/min prior to occlusion and 12.9 ± 2.6 ml after occlusion. The rCBF following administration of mannitol were 21.0 ± 6.7, 15.6 ± 3.0 and 11.3 ± 5.4, respectively at 45, 80 and 120 minutes following occlusion (**Fig. 6-9**).

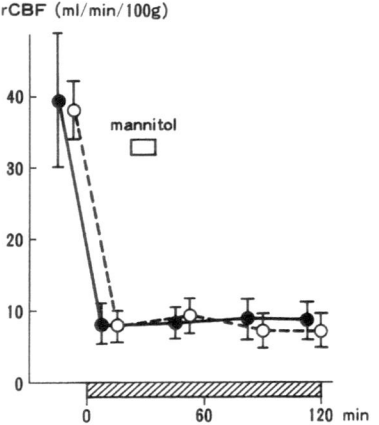

rCBF (ml/min/100g)

Fig. 6-8. Effect of mannitol on rCBF in severe thalamic ischemia

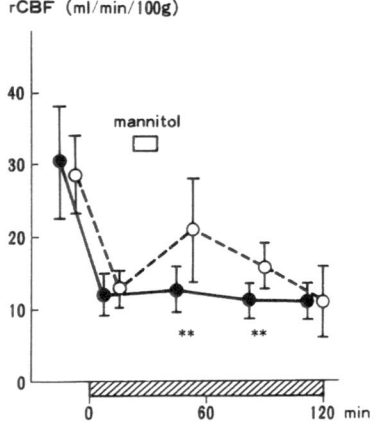

rCBF (ml/min/100g)

Fig. 6-9. Effect of mannitol on rCBF in moderate thalamic ischemia

In other words, mannitol had a considerable effect in increasing the rCBF (8.1 ml on the average) among the animals with moderate ischemia, but had only a minor effect in the severe ischemia animals. In both groups,

mannitol had its effect for 60 minutes. In contrast, it was seen that the serum osmolarity, when 20% mannitol was administered over 10 minutes in a

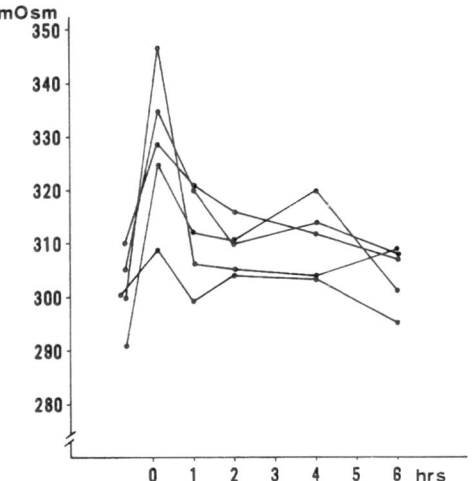

Fig. 6-10. Changes in serum osmolarity following administration of mannitol

similar manner, reached its maximum level immediately following the completion of the venous drip and returned to the preadministration level prior to the elapse of 60 minutes (**Fig. 6-10**).

Various possibilities have been suggested[17, 41, 55] regarding the mechanism of increased rCBF caused by mannitol—including decreases in intracranial pressure, increases in mean arterial pressure, decreases in cerebral vascular resistance, and changes in viscosity of the blood. Recently, Little[52, 53] reported measurements of the size of the lumen of cerebral capillaries using an ischemic model in the cat and showed that mannitol produces improvements in microcirculation of the brain.

In our experiments, an open cranium is required to produce the cerebral ischemia, so it is possible but unlikely that changes in intracranial pressure were responsible or contributed to the effects of mannitol. Since the increase in mean arterial pressure due to mannitol is only 10 mmHg and lasts for about 10 minutes, the possibility of the increase in rCBF being solely due to an increase in perfusion pressure is also thought to be unlikely. The fact that the period of increased rCBF follows closely the increase in osmolarity suggests that this is the principal mechanism whereby tissue pressure decreases as blood osmolarity increases due to dehydration of normal and ischemic brain. The present data from our experiments, however, are not sufficient to be able to draw definite conclusions. Why, for example, the increases in rCBF are not similar in modes of severe and moderate cerebral ischemia remains to be investigated.

6.3.2 Perfluorochemicals (PFC)

6.3.2.1 Introduction

Since the discovery by Clark et al.[19] that one of the PFCs, "FX80", (manufactured by the 3M Company, USA), has a large capacity for dissolving O_2, the PFCs have been studied as potential artificial blood substitutes. It should be noted that the term "perfluorochemicals" is a generic term for fluorinated organic compounds which do not have unsaturated binding within their structures and do not include hydrogen or halogens other than fluorine. Although many varieties of PFC are now known, the discovery of the first PFC without side-effects and the development of a PFC with surfactant properties (necessary to allow for the emulsification of the normally hydrophobic PFC) was carried out in Japan by Mitsuno et al.[61]. Based upon accumulated fundamental research on these chemicals, the Green Cross Corporation (Japan) has recently developed a suitable artifical blood substitute called 20% Fluosol-DA[64]. The main components of Fluosol-DA are perfluorodecaline and perfluorotripropyramine, and the emulsifying agents are pluromic F-68 and yolk phospholipids. The drug also contains an electrolyte solution and a plasma volume expander (Table 6-6). The oxygen dissociation curve of PFC has been found to differ from that of hemoglobin and to follow Henry's law—that is, the dissociation curve changes linearly with changes in the partial pressure of oxygen (PO_2).

At the physiological partial pressure of the lung (100 mmHg), it dissolves "only about 1 volume % of oxygen". Thus, in order for the PFC to supply approximately 5 volume % of oxygen to tissue, as is normally supplied by blood, it is said that the alveolar PO_2 must be increased to 550 mmHg. Consequently, in our experiments, we have raised the arterial partial pressure (PaO_2) to 400 mmHg by means of inhalation of 100% O_2 during PFC administration. In the sections which follow, 20% Fluosol-DA will be referred to simply as PFC.

Table 6-6. Chemical composition of 20% Fluosol-DA

	Fluosol-DA 20%
Perfluorodecalin	14.0 w/v%
Perfluorotripropylamine	6.0 w/v%
Pluronic F-68	2.7 w/v%
Yolk phospholipids	0.4 w/v%
Glycerol	0.8 w/v%
NaCl	0.600 w/v%
KCl	0.034 w/v%
$MgCl_2$	0.020 w/v%
$CaCl_2$	0.028 w/v%
$NaHCO_3$	0.210 w/v%
Glucose	0.180 w/v%
Hydroxyethylstarch	3.0 w/v%

6.3.2.2 PFC and Brain Tissue Partial Oxygen Pressure (PtO_2)

Methods: In order to evaluate the capacity of PFC to deliver O_2 to ischemic brain tissue, the "complete ischemic brain regulated with a perfusion method", as described in Chapter 1, was used. First, the perfusion volume was adjusted to 20 ml/min and continuation of 30 minutes of normal EEG was confirmed. Next, cerebral blood flow was reduced instantaneously to 6 ml/min (30% that of the normal blood flow) and the ischemic state was continued thereafter. The subsequent changes in brain electrical activity and PtO_2 were continuously recorded. Three groups of animals were prepared: group I consisted of 5 untreated dogs; group II was composed of 3 dogs receiving 20 ml/kg of PFC 10 minutes after the start of ischemia; and group III consisted of 3 dogs receiving 10 ml/kg of 20% mannitol and 20 ml/kg of PFC 10 minutes after the start of ischemia.

From the start of these experiments, the animals were made to inhale 100% oxygen to maintain the PaO_2 at 400 mmHg. However, in order to determine whether or not an increase in PaO_2 was required to achieve the effectiveness of PFC, the concentration of the inhaled O_2 was altered, thus bringing about changes in PaO_2. The changes in PtO_2 and EEG activity were then studied. Continous recording of the EEG and power spectral analysis was done using a signal processor (Model 7T07, San'ei Instrument Co.) and PtO_2 was measured using a PO_2 Moniton (Model 636, Roche Co.).

PtO_2 changes: **Fig. 6-11** shows the changes in PtO_2 in representative animals from each group. In all the groups, the PaO_2 was maintained at similar levels, but in group I there was a rapid decrease in PtO_2 from approximately 45 mmHg to 20 mmHg following the start of ischemia. Thereafter the PtO_2 was maintained at a steady low level. In groups II and III, however although there was a decrease in PtO

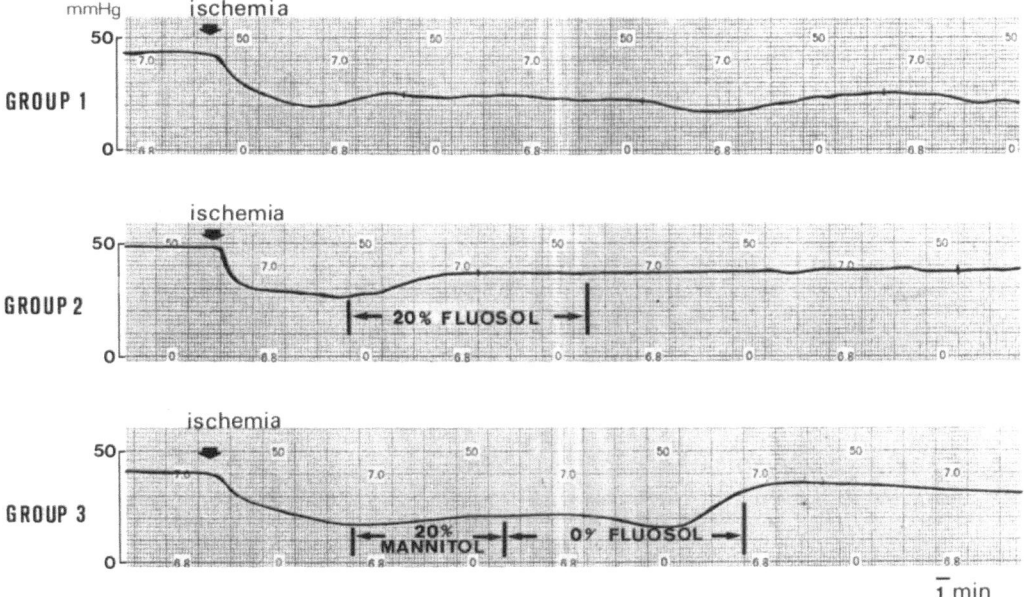

Fig. 6-11. Sequential changes in the partial pressure of O_2 in brain tissue (PtO$_2$). PtO$_2$ decreased rapidly after initiation of ischemia, but rose after PFC infusion

similar to that seen in group I following the start of ischemia, the PtO$_2$ increased by 10–20 mmHg due to the administration of the PFC.

Changes in the EEG: **Fig. 6-12** presents a bird's-eye view of the sequential changes in the power spectrum of the EEG recorded simultaneously with the PtO$_2$. In group I, the EEG showed gradual worsening and became virtually flat after 1 hour. In groups II and III, however, the electrical activity which was in the process of deterioration showed improvements due to administration of the drugs. The recovery of the EEG was particularly good in group III.

Changes in PtO$_2$ and EEG accompanying changes in PaO2: About 2 hours following the start of ischemia in each group, the 100% oxygen which had been delivered was switched off and the animals were made to breath room air, followed by a return to inhalation of the oxygen. It was found that, in group I, the PtO$_2$, which had been reduced to about 20 mmHg, now fell to 10 mmHg due to the inhalation of room air, and when oxygen was again inhaled, the PtO$_2$ returned to 20 mmHg. In groups II and III, however, due to administration of PFC, the PtO$_2$ was maintained at a high level of approximately 40 mmHg even during the ischemia. When, however, the dogs were made to breath room air, the PtO$_2$ fell to the same low level (10 mmHg) as found in group I. With the reintroduction of the oxygen, the PtO$_2$ increased rapidly to the 40 mmHg level (**Fig. 6-13**). In both groups II and III, sharp changes in PtO$_2$ produced dynamic changes in brain electrical activity. **Fig. 6-14** shows a representative power spectrum from group II. There was a strong correlation between the changes in PtO$_2$ and EEG activity, such that a

decrease in PtO$_2$ led to attenuation of the EEG and an increase in PtO$_2$ due to oxygen inhalation resulted in gradual recovery of the EEG.

Since increases in PtO$_2$ due to the

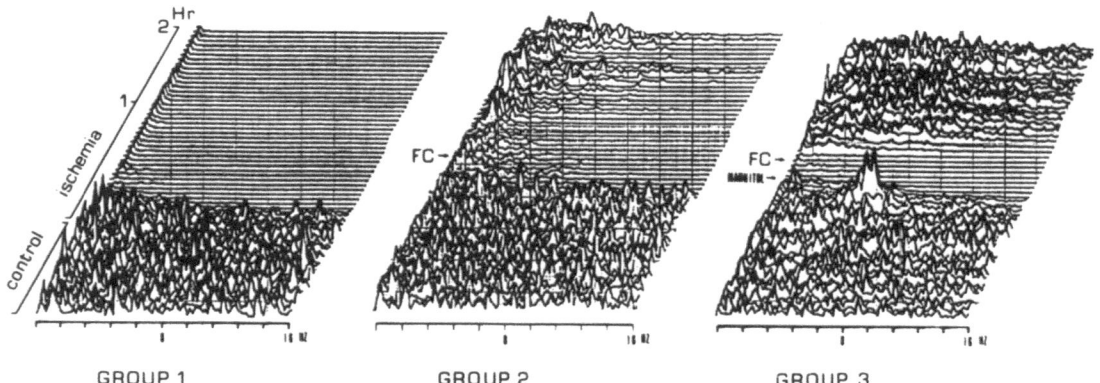

Fig. 6-12. Three dimensional display of EEG power spectrum in each experimental group. In Group 1 animals, EEG records deteriorated and become isoelectric after the onset of ischemia. Group 2 was treated with 20 ml/kg of PFC 10 min after the onset of ischemia. EEG patterns improved after PEC infusion. Group 3 was given 10 ml/kg of 20% mannitol followed by 20 ml/kg of PFC. The EEG improved more in Group 3 than in Group 2 animals

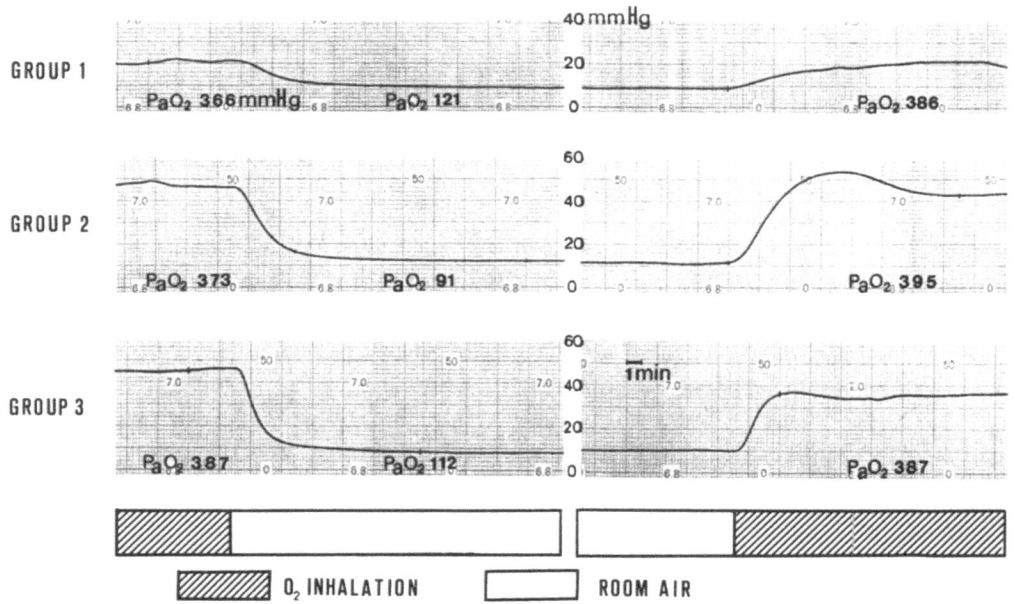

Fig. 6-13. Effect of changes in PaO$_2$ on PtO$_2$. Controlled changes in PaO$_2$ indicated that PaO$_2$ must be raised to about 400 mmHg for PFC to be effective in tissue oxygenation

administration of PFC were confirmed in this experiment, it was concluded that PFC has the capacity to carry oxygen to ischemic brain tissue and, moreover, in order to allow for these effects of PFC, it is necessary to raise the PaO_2 by means of oxygen inhalation.

ing brain ischemia has been increasing of the supply of oxygen to the ischemic focus—no matter how small that increase may be—prior to the ischemic focus succumbing to irreversible tissue damage. In recent years, however, it has often been suggested that this supply of excess oxygen to the ischemic

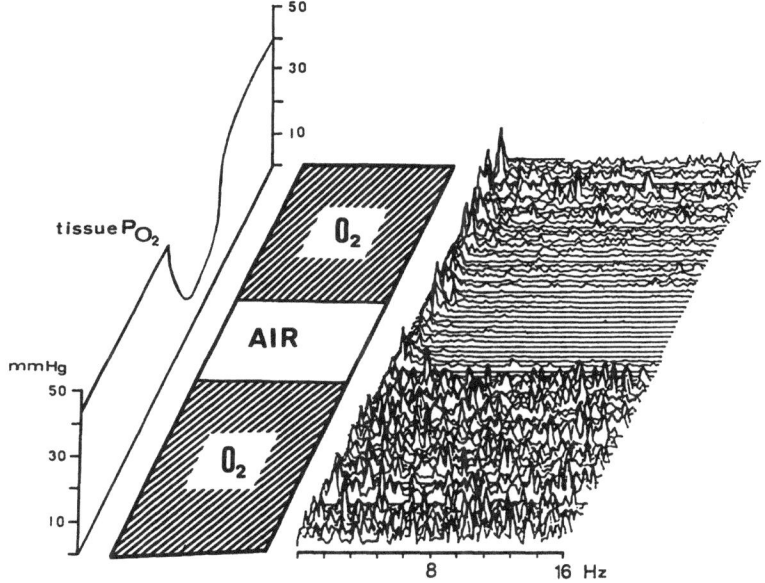

Fig. 6-14. Relationship between PtO_2 and EEG

Although further study of the mechanism of the enhanced effectiveness of combined administration of mannitol and PFC is still needed, we currently believe that it is due primarily to: (i) mannitol's protective effects on cell membranes during ischemia, and (ii) PFC's capacity to deliver O_2 to ischemic tissue.

On therapeutic method that has been considered appropriate for treat-

focus can lead to an increase in the production of active oxygens and the induction of the free radical reaction[29, 30, 120]. As a consequence, treatment of the ischemic brain by means of the administration of PFC alone—with its superior oxygen-carrying capacity—can in fact be dangerous. It is therefore necessary to administer a free radical scavenger, such as mannitol, together with PFC.

6.3.3 Combined Administration of Mannitol and PFC

6.3.3.1 Recovery of Brain Electrical Activity in the Severely Ischemic Brain

Methods: In this experiment "A canine model of completely ischemic brain regulated with the perfusion method"[45] was used (**Figs. 6-15, 6-16**).

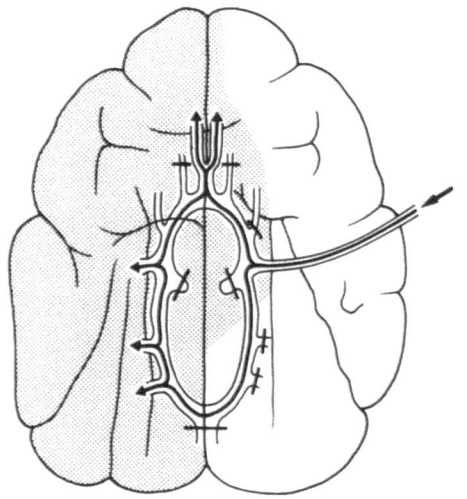

Fig. 6-15. This schema shows the occluded points of the main arteries of the canine brain. The brain (half-shaded area) is perfused by the infusion of arterial blood via a tube cannulated into the right middle cerebral artery

Using adult mongreal dogs weighing approximately 10 kg each, 5 groups of animal were prepared. Group I was an untreated control group (5 dogs); group II consisted of 3 dogs treated with Ringer's solution (30 ml/kg); group III was composed of 5 dogs treated with 10 ml/kg of 20% mannitol; group IV consisted of 5 dogs treated with 20 ml/kg of PFC; and group V composed of 5 dogs treated with 10 ml/kg of 20% mannitol; and 20 ml/kg of PFC.

After preparing the occlusion model animals, the brain was immediately perfused with 20 ml/min of arterial blood using the perfusion pump, and continuation of normal electrical activity of the brain of 30 minutes was confirmed. During this 30 minute period, the various drugs were administered by intravenous drip. Next, the perfused blood volume was instantaneously reduced to 2 ml/min—thus producing a state of severe brain ischemia. After one hour of such

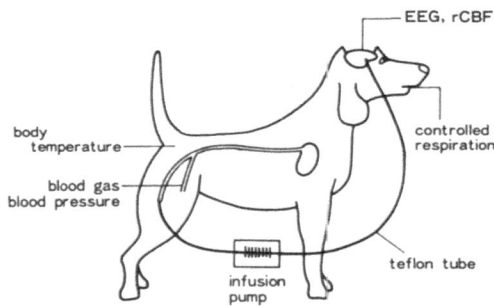

Fig. 6-16. Experimental system of the model

ischemia, blood flow was returned to 20 ml/min and the changes in EEG were recorded for 3 hours thereafter.

It should be noted that normal blood flow in the canine brain has been reported to be 40–60 ml/100 g/min[44, 60]. Since the brain weight of the perfusion area in this model is approximately 40 g, the cerebral blood flow under the infusion of 20 ml/min was calculated to be 50 ml/100 g/min and that during ischemia of 2 ml/min was calculated to be 5 ml/100 g/min.

Results: In both the untreated controls and the animals given Ringer's solution, the electrical acitivty of the brain became completely flat after approximately 4 minutes following the production of ischemia. Observation of the EEG for 3 hours following recirculation revealed absolutely no recovery of normal electrical activity (**Fig. 6-17**). In the animals administered

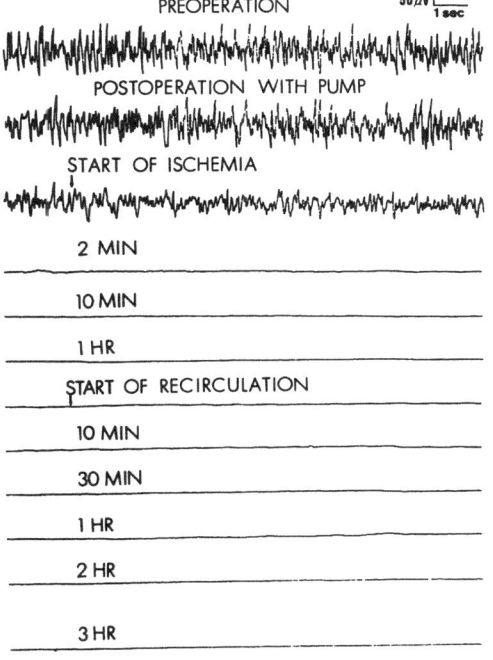

Fig. 6-17. EEG record in Group I (animal No. 1)

ery of brain electrical activity following recirculation commenced within 10 minutes. After 3 hours of observation, although there were some animals in which some slowing of the EEG was apparent, most dogs in the combined treatment group showed electrical activity with voltage approximately the

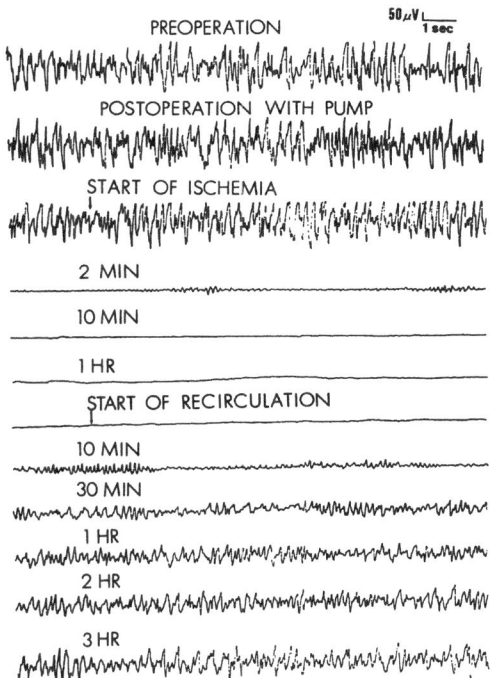

Fig. 6-18. EEG record in Group V (animal No. 22)

mannitol or PFC, the EEG became flat approximately 10 minutes after the start of ischemia, but 20–30 minutes after recirculation low voltage slow waves appeared and thereafter there were some increases in EEG amplitude. In contrast, in the animals given combined treatment with mannitol and PFC, about 18 minutes elapsed before the EEG became flat following the reduction in blood flow and the recov-

same as that seen in the pre-ischemia state (**Fig. 6-18**). The state of the electrical activity of all the dogs in the 5 groups after 3 hours of recirculation is summarized in **Fig. 6-19**.

In light of these experimental findings, it has been concluded that both mannitol and PFC are effective in treating the ischemic brain and that combined administration of these drugs results in markedly greater effective-

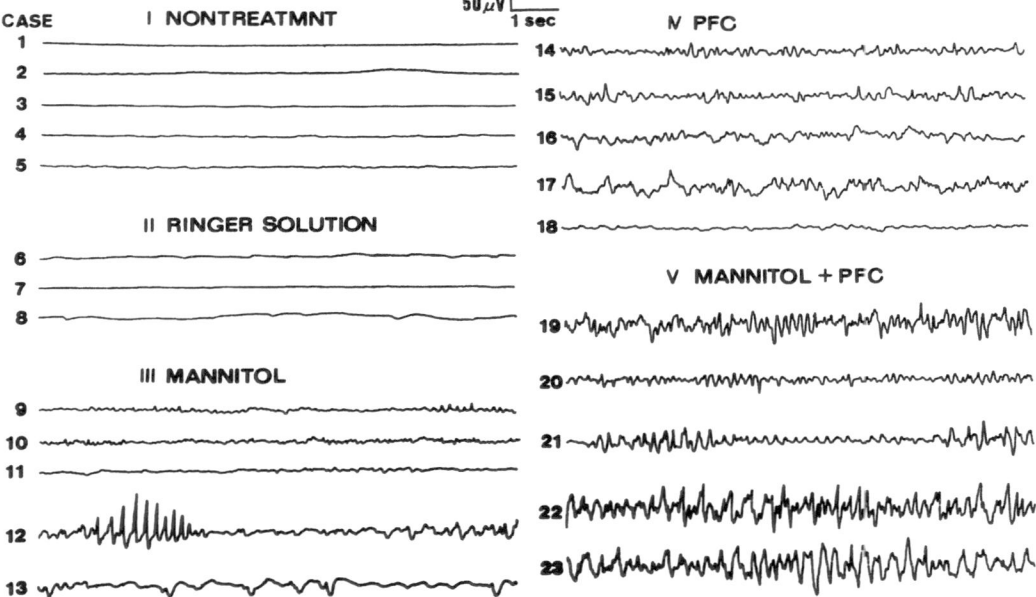

Fig. 6-19. EEG record at 3 hours following recirculation in all experimental groups

ness. Moreover, as mentioned below, we have found that, using various experimental models, treatment with PFC alone has almost no effect, but combined treatment with mannitol and PFC is remarkably effective in suppressing the development of hemorrhagic infarction and brain edema after recirculation in the acute stage of brain ischemia[42, 106].

6.3.3.2 Suppression of Hemorrhagic Infarction

Using the canine thalamic infarction model, the suppressive effects of mannitol alone, PFC alone and mannitol and PFC together were investigated[42]. The animals were prepared in the manner mentioned in Chapter 3 for producing hemorrhagic infarction. That is, after 6 hours of occlusion, 1 hour of recirculation was allowed and the severity of the hemorrhagic infarction was then evaluated as one of 4 grades (0-III, as outlined in **Table 6-7**).

The animal groups included an untreated control group, dogs administered 20% mannitol, those given 20% Fluosol DA, and those given both drugs. Furthermore, 3 subgroups were prepared in the mannitol group—corresponding to different methods of drug administration (**Fig. 6-20**). PFC was administered intravenously over 2 hours beginning 30 minutes after the start of vascular occlusion (**Fig. 6-21**). In addition, in the combined drug group, again 3 subgroups were prepared according to differences in the method of PFC administration (**Fig. 6-22**).

Table 6-7. Degree of hemorrhagic infarction

Treatment groups	Number of animals	Grades of hemorrhagic infarction			
		0	I	II	III
Control	4	. . .	. . .	. . .	4
Mannitol					
30 min after occlusion	4	1	1	1	1
60 min after occlusion	4	2	. . .	2	. . .
120 min after occlusion	4	. . .	. . .	2	2
PFC (40 ml/kg) and O_2 inhalation	3	. . .	. . .	. . .	3
Mannitol and PFC					
PFC (20 ml/kg) and O_2 inhalation	5	. . .	2	1	2
PFC (40 ml/kg) and O_2 inhalation	4	. . .	4	. . .	. . .
PFC (40 ml/kg)	3	2	1	. . .	. . .

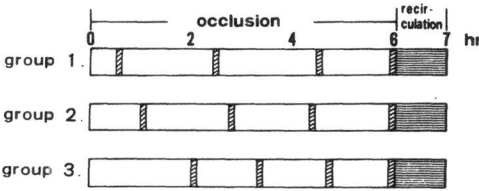

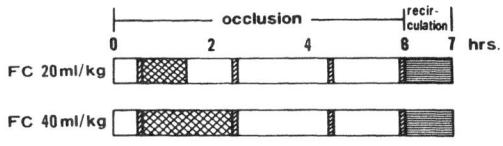

Fig. 6-20. Mannitol-treated. The schedule of mannitol administration. Group 1: the first administration of mannitol is started 30 minutes after the insult of ischemia and administration is repeated 3 times thereafter prior to reflow. Group 2: mannitol is first given 60 minutes after the start of occlusion. Group 3: mannitol iss first given 120 minutes after the start occlusion

Fig. 6-22. Mannitol and fluosol-treated cases. The schedule of mannitol and fluosol administration. Administration of mannitol is begun 30 minutes after the ischemic insult

It was found that PFC alone had no suppressive effect on hemorrhagic infarction, and mannitol alone had suppressive effects only when administered in the early stages of occlusion. Combined therapy (mannitol and 40 ml/kg of PFC) had the strongest suppressive effect (**Table 6-7**).

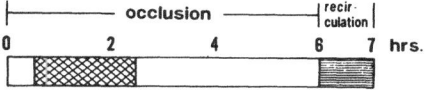

Fig. 6-21. Fluosol-treated cases. The schedule of fluosol administration. Administration of fluosol is begun 30 minutes after the ischemic insult

6.3.4 The Protective Effects of Various Free Radical Scavengers

In light of the importance of the free radical scavenger action of mannitol in protecting the ischemic brain, as mentioned above, a comparative study was undertaken of various other free radical scavengers.

Methods: A canine model of completely ischemic brain regulated with the perfusion method was used. That is, cerebral blood flow was reduced to 10% of that in the normal state and this 10% ischemia condition was maintained for one hour. Thereafter, recirculation was allowed and the effects on the ischemic brain were studied. 10 groups of animals were prepared for this experiment, as follows: group I, the untreated controls; group II, dogs administered 10 mg/kg of vitamin E; group III, dogs administered 30 mg/kg of vitamin E; group IV, dogs administered 2 g/kg of mannitol; group V, dogs administered 3 mg/kg of suloctidil (MY103); group VI, dogs administered 1 mg/kg of dexamethasone; group VII, dogs administered 3 mg/kg of nizofenone (Y9179); and group VIII, dogs administered 1 g/kg of vitamin C. Two other groups received combined administration of mannitol (2 g/kg), vitamin E (10 mg/kg) and dexamethasone (1 mg/kg) (group IX) and, finally, group X received these 3 drugs, followed by 20 ml/kg of PFC. All groups consisted of 5 dogs, and the drugs in all cases were administered intravenously prior to the start of ischemia. The level of recovery of brain electrical activity was evaluated 3 hours after the start of recirculation and classified into 5 grades depending upon the maximum amplitude of the EEG. That is, a completely flattened EEG was evaluated as grade 0, that with voltage of less than 30 μV as grade 1, between 30 and 50 μV as grade 2, between 50 and 100 μV as grade 3 and greater than 100 μV as grade 4 (**Fig. 6-23**).

Results: As shown in **Fig. 6-24** and **Table 6-8**, recovery of EEG activity relative to the untreated control group was found in all of the drug-treated animals. Among the 6 drugs, particularly mannitol, vitamin E, dexa-

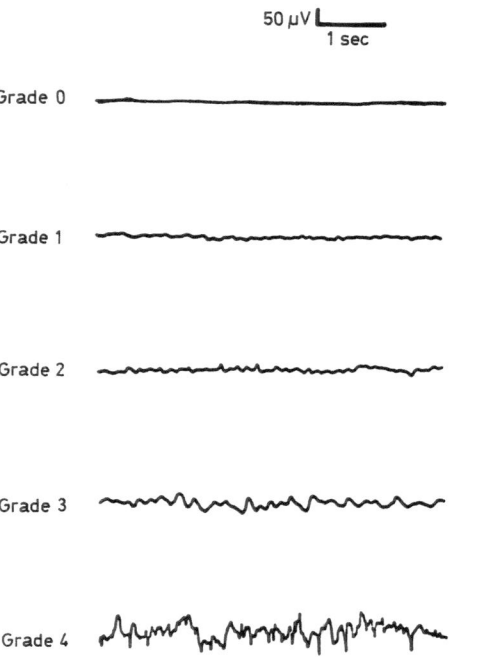

Fig. 6-23. Grading of functional recovery following recirculation in the experimental dogs

methasone and MY103 manifested a capacity to protect the brain against ischemia. Moreover, the combined treatment with mannitol, vitamin E and dexamethasone (group IX) showed recovery of electrical activity following recirculation which was markedly more rapid than that due to the administration of any of these drugs alone. In addition, the level of recovery was to grade 4 in all 5 animals in group IX. In the group X animals, which also received PFC, the electrical activity of the brain did not attenuate during the ischemia and the EEG recovered immediately following recirculation. The difference in the effects in groups IX and X can be clearly seen in a comparison of the power spectra of the continuously recorded EEG (**Fig. 6-25**).

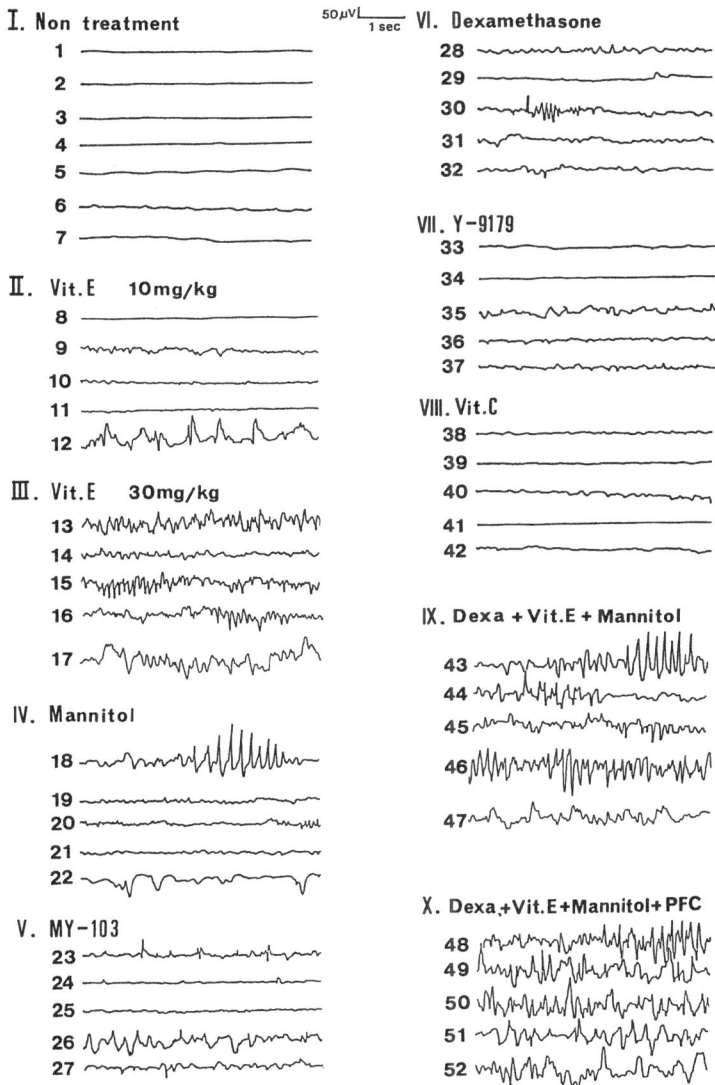

Fig. 6-24. EEG activity at three hours after recirculation in all experimental dogs

Among the pharmacological effects of the drugs used in the present study, the radical scavenger (or anti-oxidant) effects are described in detail in Chapter 5, so only a brief outline will be presented here. It is known that mannitol acts as a scavenger of one of the active oxygen species, the hydroxy radical ($\cdot$OH)[108]. Vitamin E acts as a quencher of singlet oxygens ($^1O_2$)[27] as well as a scavenger of the alkyl radical released by the peroxidation of lipids[110]. It is also known that vitamin C reacts not only with the superoxide anione (O_2-), but also with $\cdot$OH and $^1O_2$, and it has been suggested that it

Table 6-8. Summary of the grading in each group

	Agents	Grade 0	Grade 1	Grade 2	Grade 3	Grade 4
I.	Non treatment	O O O O O O	O			
II.	Vit. E 10mg/kg	O	O O	O		O
III.	Vit. E 30mg/kg				O O	O O O
IV.	Mannitol			O O	O	O O
V.	MY-103		O O		O O	O
VI.	Dexa		O O	O O	O	
VII.	Y-9179	O O	O	O	O	
VIII.	Vit. C	O	O O O	O		
IX.	II.+IV.+VI					O O O O O O O
X.	IX+PFC					O O O O O O O

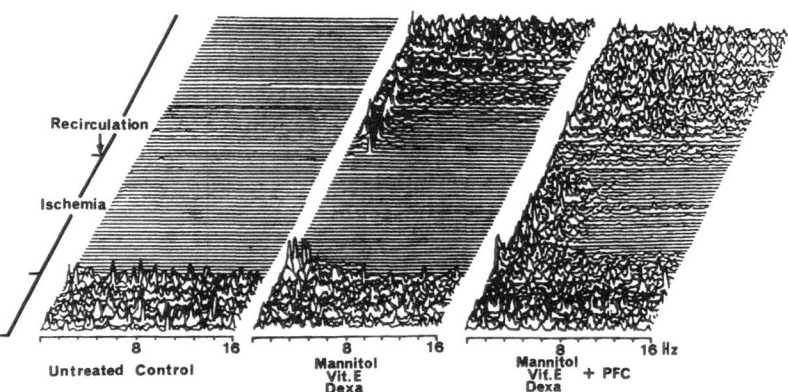

Fig. 6-25. Sequential changes in the EEG power spectrum throughout the course of the experiment in the groups with combination therapies

may act as a quencher of these activated oxygens[13, 67, 76]. However, the fact that vitamin C is not recognized as an antioxidant as powerful as vitamin E is due to the fact that it has a suppressive effect on lipid peroxidation when present in high concentrations. However, when it is found in low concentrations, vitamin C becomes a co-factor in lipid peroxidation[1].

In comparison with the above 3 drugs, research on the antioxidant properties of dexamethasone, Y9179 and MY103 is not as advanced. It is thought that the main effect of dexamethasone is the stabilization of the cellular membrane due to suppression of lysosomal enzyme release[3], but there have also been sporadic reports of anioxidant effects[102]. Y9179 is one of the imidazole derivatives and is known to have a suppressive effect on both energy and oxygen consumption[109]. It is also reported to have antioxidant properties[18]. Although MY103 was developed primarily as a facilitator of capillary circulation[78, 79], in our experiments we have found that this drug also has radical scavenger properties[108].

With regard to the important question why the combined administration of mannitol, vitamin E and dexamethasone has such clearly superior protective effects on the brain, it must be said that many problems remain un-solved. Nonetheless, we now believe that the basic mechanisms may be as outlined in **Fig. 6-26**. That is, these drugs are thought to have suppressive effects at various stages of the reaction process, starting with the generation of various active oxygens and continuing through to the lipid peroxidation.

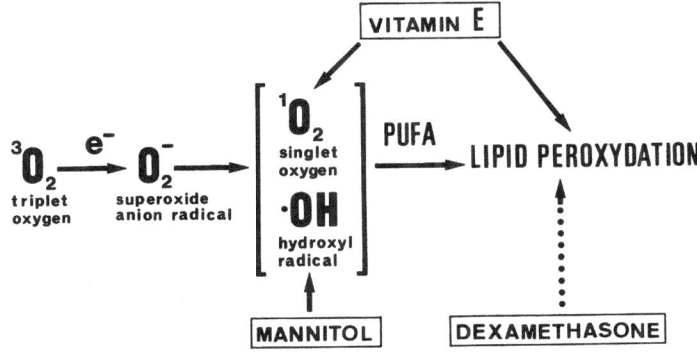

Fig. 6-26. The hypothetical site of action of three component drugs of the "Sendai cocktail". *PUFA:* polyunsaturated fatty acid

6.3.5 Combined Administration of Mannitol, Vitamin E, Dexamethasone and PFC

The experimental results described above clearly demonstrated that combined administration of mannitol, vitamin E, dexamethasone and PFC results in protective effects on the ischemic brain which are markedly superior to those obtained by the administration of any one of these drugs alone. However, it should be emphasized that all of the above experiments involved pretreatment with the drugs prior to producing the ischemia. Unfortunately, in the clinic the opportunity for such preventive administration is exceptional. Moreover, even when treatment is undertaken immediately following onset, there are various problems involved in sending the drugs to the ischemic tissue.

For these reasons, the following experiment was undertaken. First, the natural course of brain functions following the production of various, but constant levels of brain ischemia was observed. Next, investigation was made of the period following ischemia within which protective effects on the brain can be expected when the above 4 drugs are administered subsequent to ischemia.

Methods: Again, the "completely ischemic brain regulated by a pump perfusion method" was used and 4 levels of ischemia were produced. Using EEG as an index of brain function, dogs were observed for 8 hours following the start of the ischemia. The experimental groups included animal given 20% mannitol (10 ml/kg), vitamin E (30 mg/kg), dexamethasone

were readministered at the same doses. The criteria for evaluating the effectiveness of therapy relative to the condition of the untreated controls was based upon the following 4 grades of the continuously recorded EEG power spectrum. Grade 0: similar to the untreated control animals, the EEG showed no recovery from its steady deterioration. Grade 1: Although there

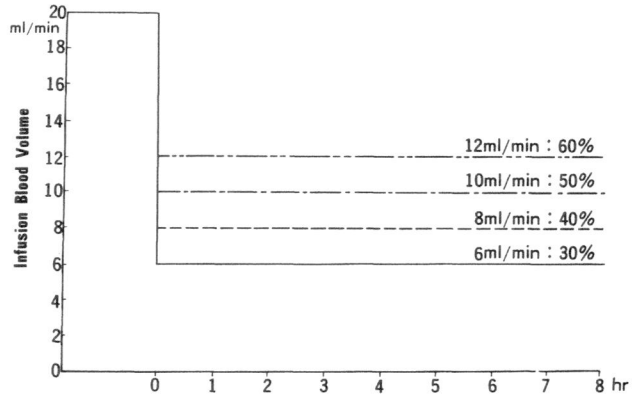

Fig. 6-27. Program of the experiment

(1 mg/kg) and PFC (20 ml/kg) at 1, 2, 3, 4, 5 or 6 hours following the start of the ischemia. The degree of recovery of EEG activity was then studied. Blood flow to the brain using this model was regulated to less than 20 ml/min in all experimental groups. Specifically, the following 4 groups were prepared: groups I, 6 ml/min (30% ischemia); group II, 8 ml/min (40% ischemia); group III, 10 ml/min (50 % ischemia and group IV, 12 ml/min (60 % ischemia) (see **Fig. 6-27**).

From the start of drug administration, the animals were made to inhale 100% oxygen and, at 2 hour intervals , the 3 drugs excluding PFC

was transient recovery of the EEG within 2 hours of therapy, there was ultimately continuing deterioration of the electrical activity (**Fig. 6-28A**). Grade 2: Notable recovery of EEG activity in comparison with the untreated control animals for at least 2 hours following treatment (**Fig. 6-28B**). Grade 3: Recovery of EEG activity to the pre-ischemic state, following therapy (**Fig. 6-28C**).

6.3.5.1   Experimental Results: Untreated Control Group (22 Dogs)

The EEG in the 30% ischemia group become completely attenuated within 1 hour of the start of ischemia

Grade 1

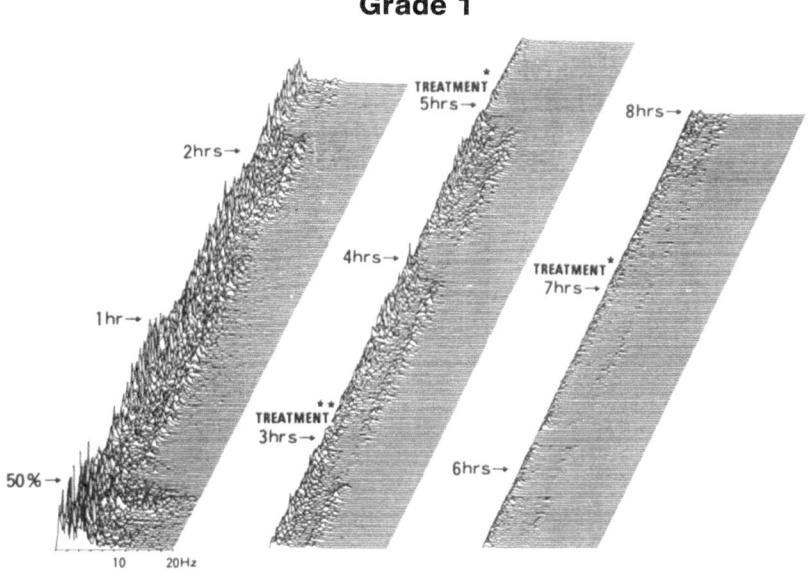

a

Grade 2

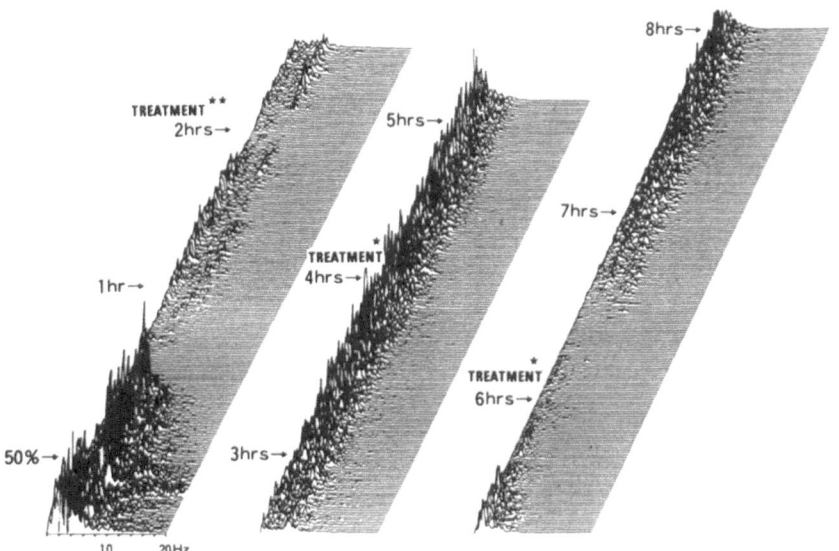

b

Fig. 6-28. Continuous recording of the EEG power spectrum in a) (animal No. 68, the effect of treatment is evaluated as Grade 1), b) (animal No. 65, the effect of the treatment is evaluated as Grade 2), c) (animal No. 63, the effect of treatment is evaluated as Grade 3). Treatment**: mannitol, vitE, dexamethasone and PFC. Treatment*: mannitol, vitE, dexamethasone

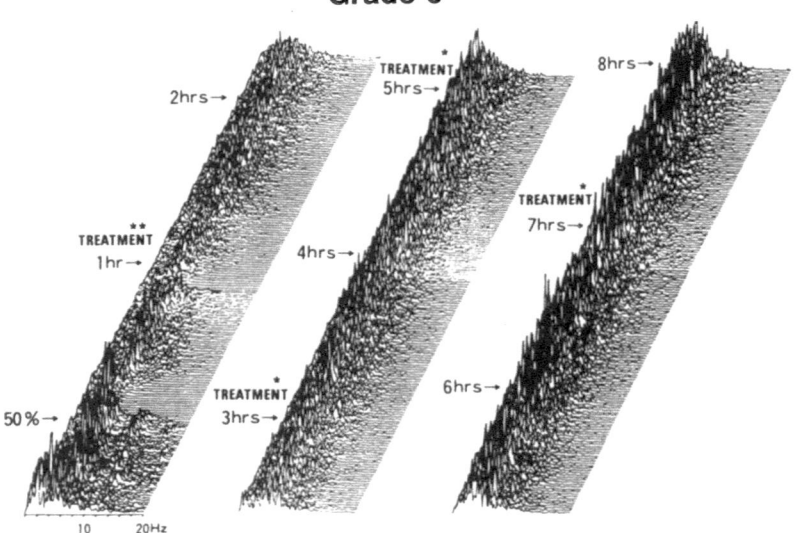

Fig. 6-28-c

(**Fig. 6-29**). In the 40% ischemia group, some animals showed complete attenuation of electrical activity within 1 hour, whereas other animals showed continued low voltage slow waves for 5–7 hours, followed by gradual deterioration and ultimately flat EEGs (**Fig. 6-30**). In the 50% ischemia group, all the animals showed medium voltage slow waves for 8 hours following ischemia (**Fig. 6-31**). Finally, in the 60% ischemia group, no notable deterioration of the EEG was seen over the entire 8 hour period of observation (**Fig. 6-32**).

6.3.5.2 Drug-Treated Animals (61 Dogs)

Among the 30% ischemia animals, all 7 dogs administered the drugs after 1 hour from the start of ischemia showed grade 2 electrical activity, but almost all of the animals treated after 2 or more hours of ischemia showed no favorable

effects of the drugs and became grade 0. Among the 40% ischemia animals, all dogs treated within 4 hours after the onset of ischemia had grade 1 or 2 EEG activity, but some animals after 5 or more hours became grade 0 and no drug effects were discernible. In contrast, all dogs in the 50% ischemia group which had been treated within 6 hours of ischemia showed therapeutic effects of the drugs. Particularly noteworthy were the therapeutic effects in 3 animals which were treated after 1 hour of ischemia. In 2 of these dogs, the EEG returned to a level similar to that seen prior to ischemia, and were evaluated as grade 3. The effects of the drugs in these groups are summarized in **Table 6-9**.

The principal purpose of the present study was to answer the question: "When"—following the onset of brain ischemia—is the latest time at which the combined administration of these drugs can be expected to give beneficial

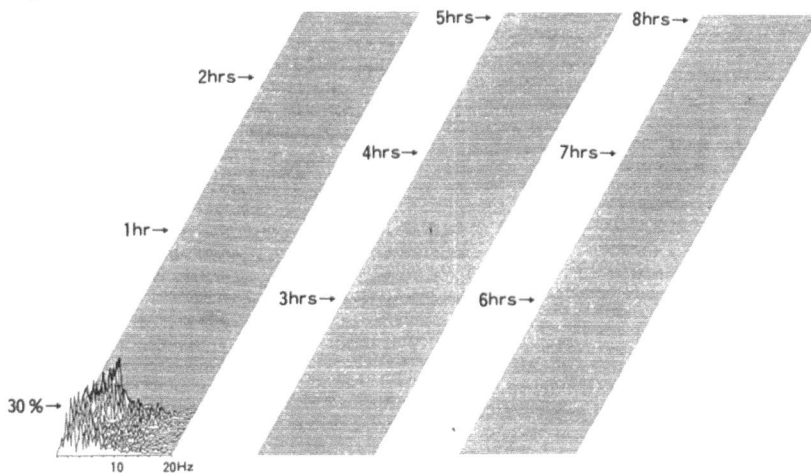

Fig. 6-29. Continuous recording of the EEG power spectrum. Untreated 30% ischemia (animal No. 7)

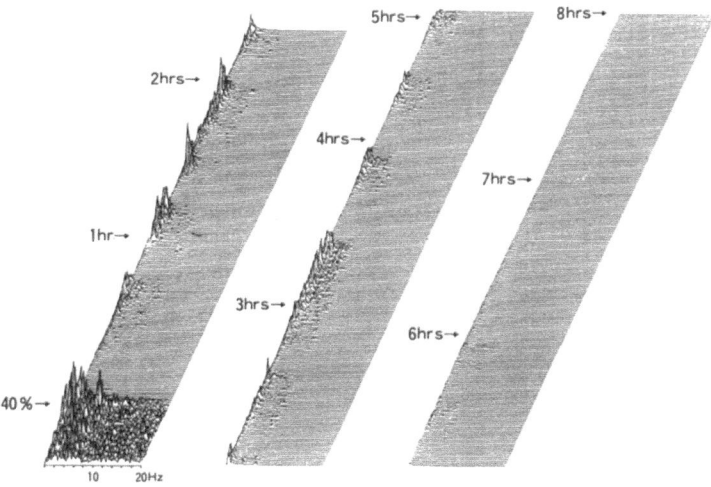

Fig. 6-30. Continuous recording of the EEG power spectrum. Untreated 40% ischemia (animal No. 29)

therapeutic results? In light of the above experimental results, it was concluded that recovery of brain electrical activity can be expected if the drugs are administered within 1 hour of onset of ischemia in the 30% ischemia group, within 4 hours in the 40% ischemia group and within 6 hours in the 50% ischemia group. In those groups treated at an early stage of the ischemia, it is thought likely that recirculation will result in further improvements in brain function.

In light of the above-mentioned findings, we undertook the following experiment. That is, in a group of 30%

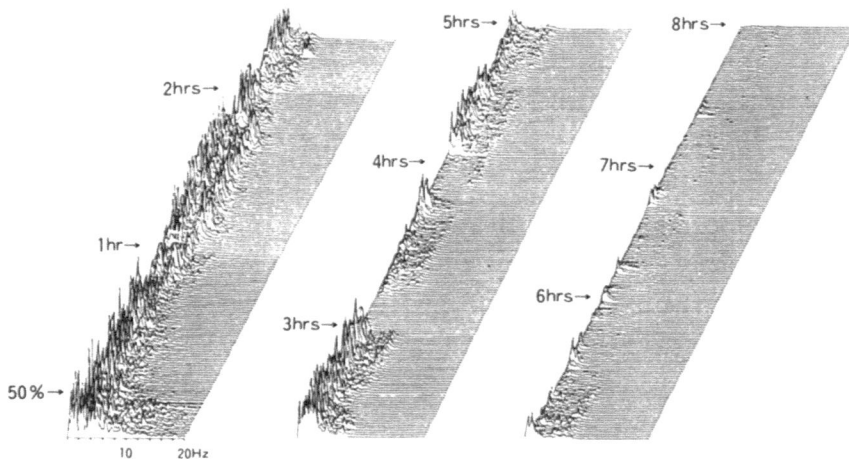

Fig. 6-31. Continuous recording of the EEG power spectrum. Untreated 50% ischemia (animal No. 60)

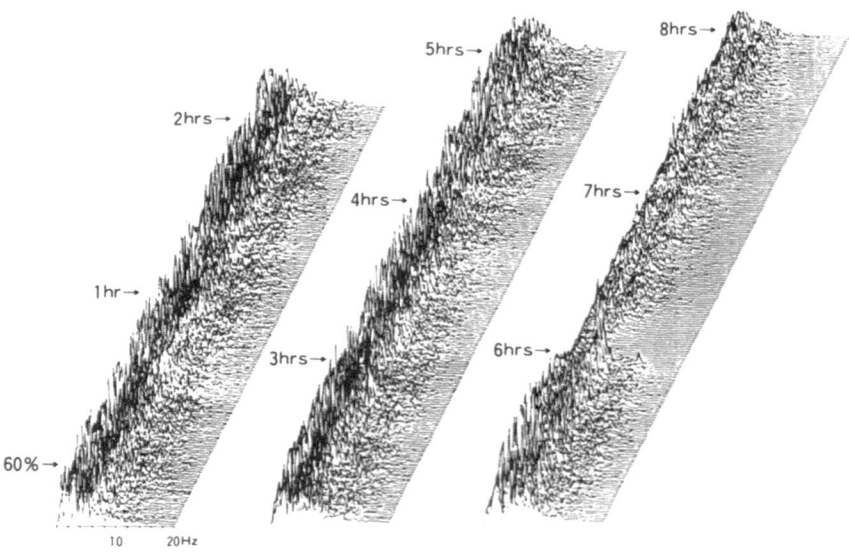

Fig. 6-32. Continuous recording of the EEG power spectrum. Untreated 60% ischemia (animal No. 81)

ischemia animals, the same drugs were administered 30 minutes following the onset of ischemia and 5 hours later cerebral blood flow was restored to its pre-ischemic state. A similar procedure was carried out in untreated dogs which became the control group. Notable differences between the two groups were found. That is, in the control animals the EEG remained completely flat following recirculation (**Fig. 6-33**), whereas in the treated animals there was rapid improvement in the EEG following drug treatment and, follow-

Table 6-9. Summary of the experimental results. The numbers show the number of animals

	30% ischemia				40% ischemia				50% ischemia				60% ischemia	
untreated control	1 2 3	4 5 6	7 8		26 27 28	29 30 31	32		58 59 60	61			81 82 83	
effect (grade)	0	1	2	3	0	1	2	3	0	1	2	3		
start of treatment / 1hr after ischemia			9 13 10 14 11 15 12				33 34				62	63 64		
2hr	16 20 17 18 19	21					35 36 37				65 66 67			
3hr	22					38 39 40	41 42			68	69 70 71			
4hr	23						43 44 45 46 47			72 73	74			
5hr	24				48 49	50	51 52			75	76 77			
6hr	25				53 54		55 56 57			78	79 80			

ing recirculation, the electrical activity returned to a virtually normal level (**Fig. 6-34**). These results clearly indicate that, under administration of the Sendai Cocktail, vascular reconstruction in the acute stage of brain infarction (for which it has previously been thought that surgical treatment is contraindicated) can become an effective therapeutic method. It should be noted, however, that several unsolved problems remain with regard to the timing of the recirculation and the subsequent increases in cerebral blood flow.

6.3.6 Phenytoin (Aleviatin)

The experiments described above were based upon the assumption that the free radical reaction is heavily involved in the tissue damage incurred due to brain ischemia. It was surmised that if mannitol, whose mode of action in ischemia has not previously been clarified, works as a radical scavenger, then other known radical scavengers should also show protective effects on the brain during ischemia. However, such active oxygens and the free radical reaction are not the only factors involved in the pathology of the brain subjected to ischemia. Energy failure has also been indicated, as well as cellular acidosis. In addition, has recently been drawn to disruptions in intracellular ion

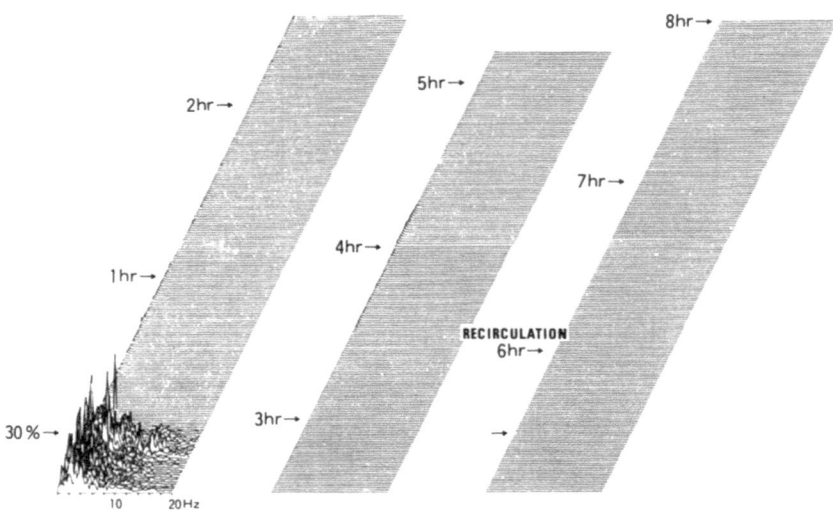

Fig. 6-33. Untreated control with 30% ischemia. Recirculation was performed 5½ hours after the onset of ischemia. The EEG disappeared and remained isoelectric following recirculation

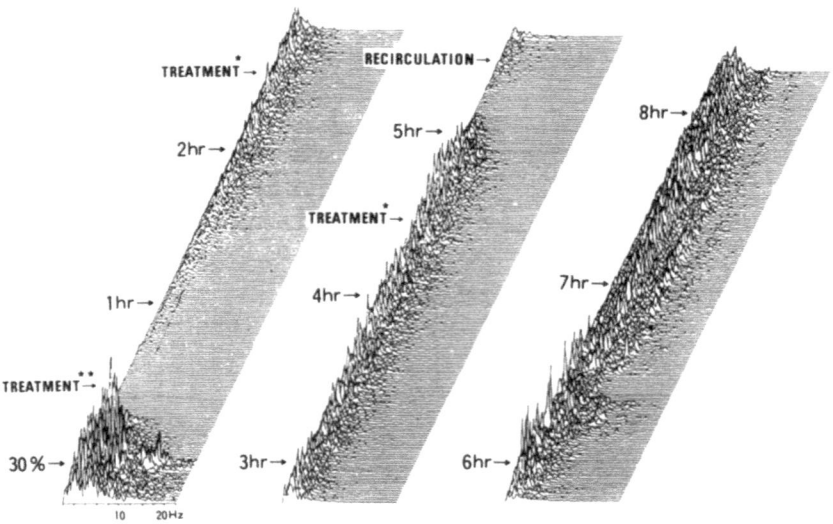

Fig. 6-34. Treatment was started 30 min after the onset of 30% ischemia. The EEG did not disappear during ischemia and recovered immediately following recirculation. Treatment**: mannitol, vitE, dexamethasone and PFC. Treatment*: mannitol, vitE, dexamethasone

homeostasis. That is, it has been argued that the influx of Ca^{++} activates phospholipase, which brings about an increase in free fatty acids, which in turn triggers the progression of the arachidonic acid cascade[75, 91, 92]. Consequently, together with progress in understanding of the pathology of brain ischemia, it can be expected that new therapeutic methods will also be developed.

Phenytoin (Aleviatin), which has found wide clinical use as an anticonvulsant, has recently drawn attention as a brain protective agent in cases of cerebral ischemia. Various suggestions concerning its possible mode of action have been made, but none has yet become established as the most widely accepted. For example, (i) phenytoin is known to suppress intracellular K^+ release, probably due to facilitation of the Na^+—K^+ ATPase system[5, 6]. (ii) It has a suppressive effect on the liberation of free fatty acids from phospholipids in the cell membrane[88]. (iii) It also decrease the rate of oxygen consumption[16, 97]. (iv) Finally, phenytoin increases cerebral blood flow by vasodilatation[47]. Further research in this field is still required in order to determine which of these actions are central to phenytoin's protective effects on the brain.

Recently, we found that, under a hypoxic condition of 96% N_2 and 4% O_2, mice will exhibit convulsions and die within 2–3 minutes, but following pre-administration of the anticonvulsant phenytoin, the mice will survive for an hour or longer. Starting from this observation, we have focused on the protective effects on the brain of phenytoin and performed the following experiment to clarify its effects.

Methods: Using the experimental procedure described above, in which a 10% ischemic state was maintained for 1 hour and followed by recirculation, the effects of pre-administration of phenytoin on the ischemic brain were studied. The following 5 groups of the dogs were prepared: Group I, the untreated control group (5 dogs); group II, 5 dogs given 7 mg/kg *i.v.* of phenytoin; group III, 5 dogs given 10 mg/kg *i.v.* of phenytoin; group IV, 5 dogs given 30 mg/kg *i.v.* of phenytoin; and group V, 5 dogs given 10 mg/kg *i.v.* of phenytoin, 2 g/kg *i.v.* of mannitol and 30 mg/kg *i.v.* of vitamin E.

Results: All of the animals showed completely flat EEGs during the cerebral ischemia. Following recirculation, there was no recovery of brain electrical activity in the untreated control group, but among groups II, III, and IV the EEG recovery was better, the larger the dose of phenytoin. In other words, there was a notable dose dependency for phenytoin in its protective effects on the brain. Using the grading system for evaluating the recovery of the EEG (as described in **Fig. 6-23**), 3 of the 5 dogs in group II and all of the dogs in groups III and IV were evaluated as grade 4, *i.e.*, recovery to the highest grade. In other words, phenytoin was found to have brain protective effects which were superior to those of any of the radical scavengers described earlier (**Fig. 6-35**).

In group V, in which animals were given combined treatment with phenytoin and various radical scavengers, the recovery of the EEG was extremely good—virtually to the level of the normal EEG recorded prior to ischemia (**Fig. 6-36**). It thus can be seen that the combined administration of phenytoin, mannitol and vitamin E has surpris-ingly strong protective effects on the brain. Based upon these latest findings, we believe that the "Sendai cocktail" developed by us is but the first of what may become many clinically useful pro-

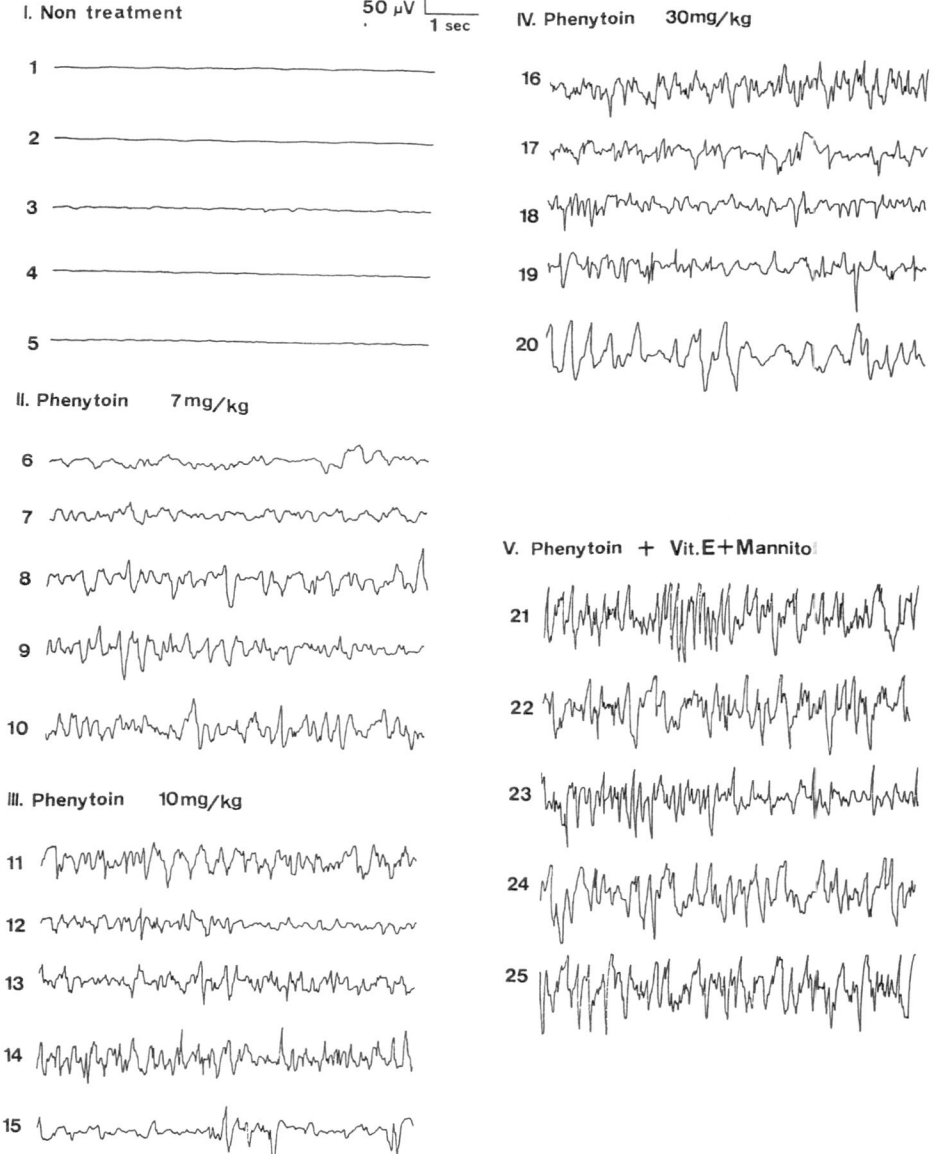

Fig. 6-35. Summary of results. EEG activities recorded at 3 hours after recirculation in all experimental animals treated with various doses of phenytoin

ducts made possible by our experimental models for brain ischemia. The second product, which we call the "new Sendai cocktail", consists of phenytoin, mannitol and vitamin E and is now thought to be at the stage where testing in clinical trials is appropriate.

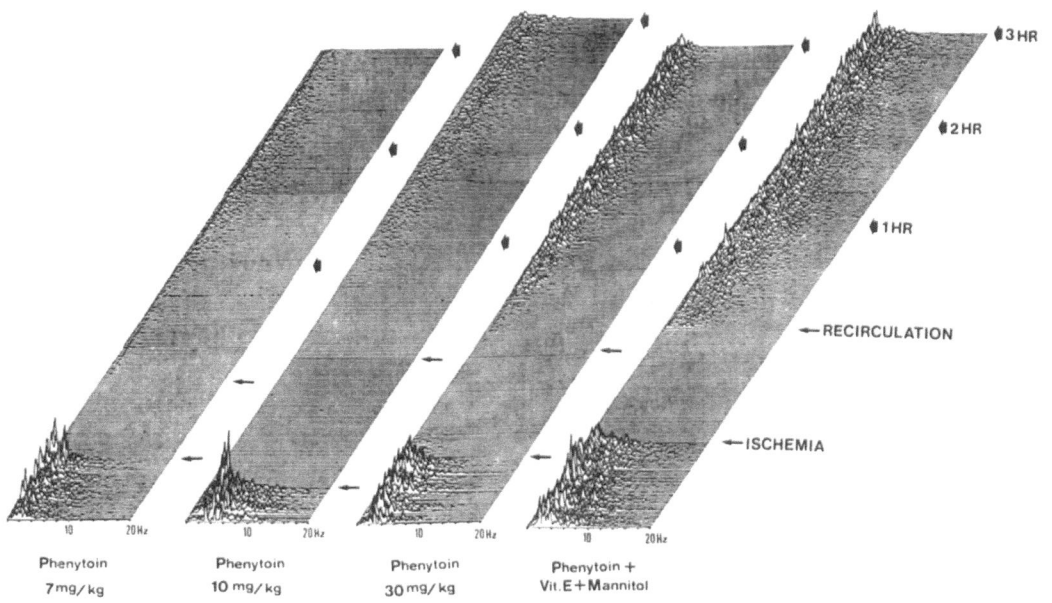

Fig. 6-36. EEG power spectrum analysis of a typical case in each experimental group

6.3.7 Calcium Antagonist (Flunarizine)

In recent years, the hypothesis that the calcium ion (Ca^{++}) might act as a possible trigger for the cellular damage which arises in cerebral ischemia has received considerable attention[75, 91, 92]. Ion flux is known to play a part in the regulation of various cellular functions, and that of Ca^{++} is known to be of central importance. Moreover, only a small imbalance in Ca^{++} homeostasis can lead to severe deficits in cellular functions. At rest when the neuron is receiving no stimulation, the Ca^{++} concentration of the extracellular fluid is known to be 10^{-3} M, whereas the intracellular concentration is approximately 10^{-7} M. In other words, the plasma membrane is the border for a concentration difference in Ca^{++} of more than 10,000 fold. For this reason, there exists an electrochemical gradient for the Ca^{++} entry into the cell. In the normal cell, Ca^{++} shifts across the plasma membrane are controlled by various mechanisms. Specifically, there is the Na^{+}—Ca^{++} antiport system and the ATP-dependent Ca^{++} pump in the plasma membrane, by means of whose action the efflux of cytosolic Ca^{++} to the extracellular fluid is brought about.

Within the cell, there are the Ca^{++}-binding proteins (calmodulin) and mitochondria, which continually take up and release Ca^{++}, thereby controlling its intracellular concentration.

In pathological situations, such as cerebral ischemia, there is a large influx of Ca^{++} into the cells and the cytosolic Ca^{++} concentration rises markedly. In such a state, the above-mentioned phospholipases are activated and there is an increase in free fatty acids due to the hydrolysis of membrane phospholipids. The increase in free fatty acids in turn accelerates the arachidonic acid cascade and the various products (eicosanoids) have untoward biological activity. Simultaneously, various free radicals having toxic cellular effects are released as intermediate metabolites. In this manner, the plasma membrane becomes damaged and there is an increase in membrane permeability. As a result, there is further influx of extracellular calcium and a vicious cycle is established. Due to the excess of mitochondrial Ca^{++}, the energy-producing system also becomes disturbed—eventually leading to cell death.

As is apparent from the above, the reason for expecting Ca^{++} antagonists to have therapeutic effects in cerebral ischemia is that such drugs may suppress these phenomena—initially, the abnormal influx of Ca^{++} and, ultimately, cellular death.

We have studied the protective effects of the Ca^{++} antagonist, flunarizine, on the ischemic brain. A brief outline of that research is presented below.

Methods: Similar to the experimental series described above, cerebral blood flow was reduced to 10% of that found in the normal brain and was maintained at that level for one hour subsequent to the pre-ischemic administration of flunarizine. Recirculation was then allowed for three hours. By means of monitoring the recovery of brain electrical activity and the degree of cerebral edema and by means of an autopsy study of the degree of extravasation of Evans blue, the effectiveness of the drug therapy was evaluated.

Five untreated animals were prepared as a control group and two experimental groups of five dogs each were also prepared (1 mg/kg iv and 3 mg/kg iv of flunarizine, respectively). As described earlier, a five stage grading of the recovery of the EEG subsequent to reflow was used.

Results: Among the control animals, the EEG was judged as grade 0 in four animals and grade 1 in one animal. In contrast, among the animals given 1 mg/kg of flunarizine, one was grade 2, two were grade 3 and two were grade 4. Among those given 3 mg/kg flunarizine, one was grade 3 and four were grade 4. Statistically, the difference between the control and experimental groups was highly significant (Mann-Whitney U-test, **Table 6-10**). With regard to the severity of the brain swelling subsequent to recirculation and the extravasation of Evans blue, there were no significant differences between the control and experimental groups—indicating that the drug therapy had no suppressive effects on the swelling.

Many experimental studies concerning whether or not Ca^{++} antagonists have protective effects on the ischemic brain have recently been reported. Among those which indicate such protective effects are the reports of Karasawa et al.[43], White et al.[114] and Wiernsperger et al.[115]—all three of which found treatment with Ca^{++} antagonists to result in improved neurological recovery after recirculation and in the suppression of post-ischemic hypoperfusion.

Flunarizine is known to have the chemical structure shown in **Fig. 6-37**, and, in relation to the other Ca^{++} antagonists, it has unusual pharmacological properties. That is, other Ca^{++} antagonists have been found to selectively suppress the influx of Ca^{++} through the Ca^{++} channels in the plasma membrane—and for this reason they have been referred to as Ca^{++}

Table 6-10. Summary of EEG grading in each group

	Grade 0	Grade 1	Grade 2	Grade 3	Grade 4
Untreated control	○ ○ ○ ○	○			
Flunarizine 1 mg/kg			○	○ ○	○ ○
Flunarizine 3 mg/kg				○	○ ○ ○ ○

Contradictory results have, however, been reported in several studies in which significant effects of Ca^{++} antagonists were not found (*i.e.*, studies by Hossman et al.[39], Reedy et al.[77], Faden et al.[24], Newberg et al.[65] and Harris et al.[35]). Clearly, among these previous studies there is yet to emerge a consensus on the effectiveness of Ca^{++} antagonists in suppressing the effects of cerebral ischemia, but at least some of the contradictory results may be explained in terms of the differences among the various Ca^{++} antagonists.

entry blockers. In contrast, flunarizine not only blocks the influx of Ca^{++}, it also suppresses the release of Ca^{++} from the inner plasma membrane and from mitochondria and other organelles. In other words, flunarizine prevents the excess accumulation of Ca^{++} in the cytosole—suggesting that this drug should be called a Ca^{++} "overload blocker", rather than a Ca^{++} entry blocker.

Our experimental results clearly indicate that flunarizine has protective effects on the ischemic brain. It is of particular interest that, in this experiment, despite the fact that there was good recovery of brain function (EEG) following recirculation in the treated animals, there were no suppressive effects on ischemic brain edema.

lators can lead to vasodilation only in non-ischemic regions, while blood flow at the ischemic focus itself can actually be decreased (the so-called "steal phenomenon"). It has thus been pointed out that there is a danger that

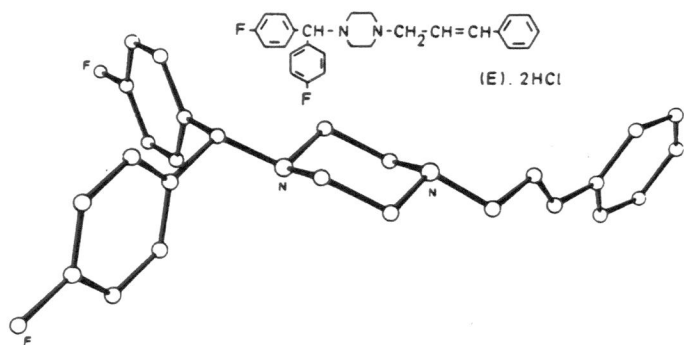

Fig. 6-37. Chemical structure of flunarizine: (E)-1-[bis(4-fluorophenyl)methyl]-4-(3-phenyl-2-propenyl)piperazine dihydrochloride

Both Harris *et al.*[35] and Roy *et al.*[80] found that Ca^{++} antagonists have few suppressive effects on ischemic brain edema, and suggested that these drugs may even have aggravating effects of such edema. In addition to suppressing the release of free fatty acids, as mentioned above, Ca^{++} antagonists are known to be potent vasodilators. In certain pathological states, such as the acute stage of cerebral ischemia where autoregulation and the CO$_2$ response are disturbed, administration of vasodi-

administration of vasodilators may lead to aggravation of the cerebral pathology[80]. In our experiments, the fact that administration of flunarizine did not suppress brain edema may have been caused by an increase in vascular permeability due to the effects of this drug. These possibilities require further experimental study, the outcome of which will heavily influence the ultimate therapeutic value of flunarizine.

Part II
Clinical Study

7. EPIDEMIOLOGY AND SYMPTOMATOLOGY

7.1 Introduction

According to figures published by the World Health Organization (WHO) in 1983[30], diseases of the circulatory system were responsible for some 23% of deaths throughout the world, including both developing and developed nations. In the developed nations alone, however, such diseases accounted for nearly half (48%) of all deaths, among which 21% were due to ischemic heart disease, 13% to cerebrovascular disorders and 14% to other vascular diseases. In other words, cerebrovascular disease (CVD) is second only to ischemic heart disease as a major health problem in the developed nations of the world (**Fig. 7-1**).

Since there have recently been no

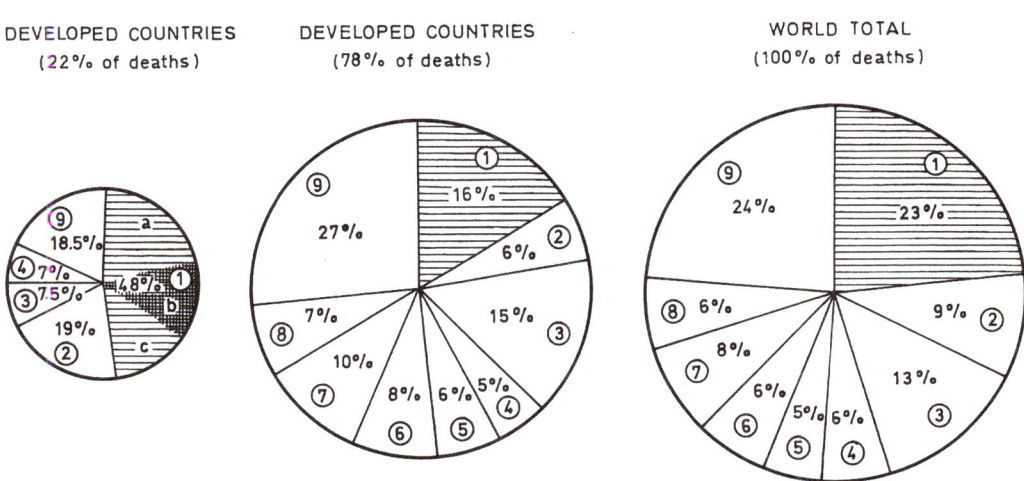

DEVELOPED COUNTRIES (22% of deaths)

DEVELOPED COUNTRIES (78% of deaths)

WORLD TOTAL (100% of deaths)

Fig. 7-1. Estimated distribution of causes of deaths in 1980 (*1*) diseases of circulatory system (*a* ischemic heart disease, *b* cerebrovascular disease, *c* other diseases of circulatory system) (*2*) neoplasms (*3*) diseases of respiratory system (*4*) poisonings and violence (*5*) tuberculosis of the respiratory system (*6*) enteritis and other diarrhoeal disease (*7*) other infective and parasitic diseases (*8*) certain causes of perinatal mortality (*9*) all other and ill-defined causes

reports of the comparative incidence of cerebral infarction among various nations, we will begin our discussion of epidemiology with a brief review[1] of recent national trends based upon WHO statistics concerning death due

for Women in three countries: Portugal, Venezuela and Thailand.

A comparison of Japan with other nations shows that the incidence of CVD was highest for both men and women in 1958, whereas Portugal had

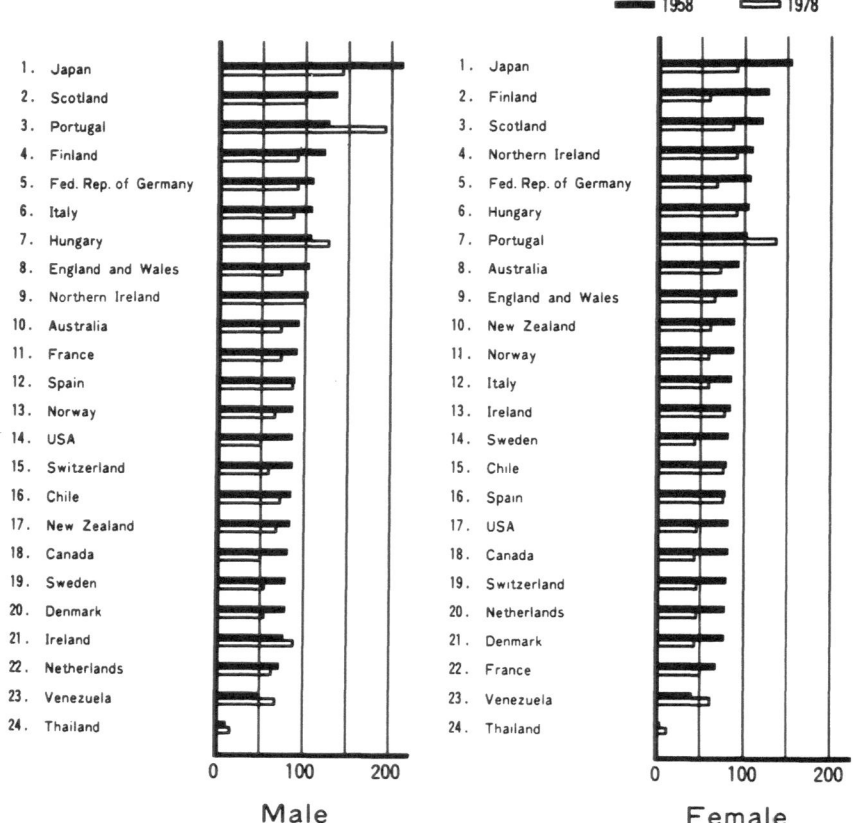

Fig. 7-2. International mortality statistics for cerebrovascular disease

to CVD. When the data from 24 participating nations are considered together during the 20 year period between 1958 and 1978, there were decreases in the age-adjusted death rates from CVD in both men and women throughout the world. Increases were found for men in five countries: Portugal, Hungary, Ireland, Venezuela and Thailand, and

the highest rate by 1978 with Japan next in line. The rate of decrease in incidence over these 20 years, however, was similar in Japan and elsewhere. The decrease was between 30 and 40% for men and women of all ages except those between the ages of 40 and 69 for whom the decrease exceeded 50% (Fig. 7-2).

Yearly trends in the causes of death in Japan show that there continue to be increases in the number of fatalities due to cancer and cardiac disease, whereas there have been decreases in deaths due to CVD. It is of interest that, according to the 1980 statistics[27], the number of deaths due to cancer actually exceeded that due to CVD—which for many years has been the number one killer in Japan. Moreover, judging from the trends in recent years, it is likely that heart disease is now the second greatest cause of death in Japan (**Fig. 7-3**).

With regard to the gross yearly death rate due to cerebral infarction and cerebral hemorrhage, which are principal among CVD cases, there has been a gradual decrease in the number of deaths in Japan due to cerebral hemorrhage since 1960, whereas there has been a rapid increase in the number of deaths due to cerebral infarction up until 1975. In 1975 the number of fatalities due to cerebral infarction actually exceeded that due to cerebral hemorrhage and thereafter the death rate due to cerebral infarction has remained roughly constant. Since deaths due to cerebral hemorrhage have continued to decrease, the percentage of CVD deaths due to cerebral infarction has continued to increase. According to the statistics for 1981[18], among 134.3 deaths (crude death rate) due to CVD, 40.5 were due to cerebral hemorrhage and 63.8 were due to cerebral infarction, resulting in a ratio of 1 : 1.6 (**Fig. 7-4**).

U.S. statistics[2] showed a crude death rate due to CVD between 1968 and 1977 of 20.1%. When this rate was age-adjusted, a 31.5% decrease over

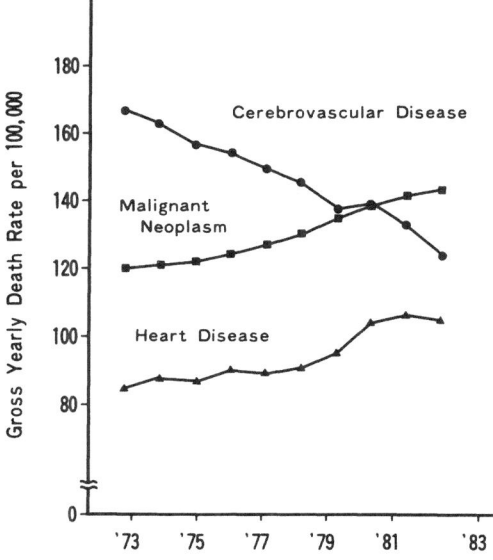

Fig. 7-3. Trend in main causes of death during last 10 years

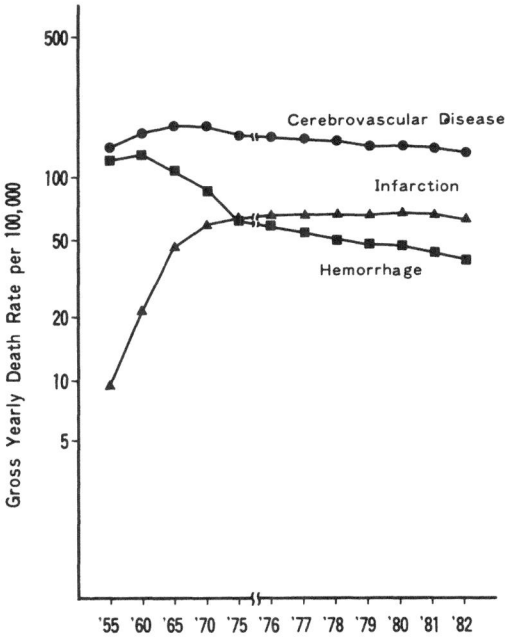

Fig. 7-4. Trend in types of cerebrovascular disease during last 27 years

that period was noted. With regard to differences between cerebral infarction and cerebral hemorrhage, it was found that the age-adjusted death rate due to cerebral hemorrhage fell 53% (from 28.63 to 13.53 per 100,000), while that cerebral infarction fell 45% (from 33.0 to 18.18 per 100,000). Although deaths from cerebral infarction in the U.S. are still more numerous, the more rapid decrease in deaths due to cerebral hemorrhage than that due to cerebral infarction indicates that there is a yearly increase in the percentage of infarction deaths.

As can be seen from the above data, there is an international trend toward fewer deaths due to CVD, and, simultaneously, a trend toward an increasing percentage of CVD deaths due to cerebral infarction. Cerebral infarction must therefore be considered one of the major health hazards today.

7.2 The Natural Course of the Acute Stage of Cerebral Infarction: A Study of 1000 Cases

It is evident from the above statistics that the incidence of ischemic CVD is high and indeed such cases are frequently encountered in routine clinical work. A great deal of fundamental and clinical research has consequently been done and significant progress in the understanding of such diseases has recently been made with the introduction of new diagnostic techniques, such as computed tomography (CT), digital subtraction angiography (DSA),

Table 7-1. Grading of level of consciousness (Japanese Coma Scale)

Grade 0:	fully alert
Grade I:	awake without any stimuli
I-1	orientated but not fully alert
I-2	disorientated to time, place and/or persons
I-3	disorientated to the patient's own name and/or date of birth
Grade II:	be able to awake with verbal and/or painful stimuli
II-1	respond verbally and/or with purposeful movement
II-2	respond consistently with very simple words (ie. yes and no) and/or simple movement
II-3	be able to awake only with strong verbal and painful stimuli; respond inconsistently with very simple words or movements
Grade III:	unable to awake even with strong painful stimuli
III-1	respond with combative or purposeful movements
III-2	respond only with unpurposeful movements or decerebrated posture
III-3	no response except changes in respiratory rythm

magnetic resonance imaging (MRI) and positron emission tomography (PET). Although that greater understanding of ischemic CVD has gradually led to better therapeutic results, it must be said that, due to significant individual differences among ischemic CVD cases, the prognosis of such cases is varied—even when acute stage symptoms are similar. Thus it is difficult to know when functional recovery is due to therapy and when it is due to spontaneous recovery. For this reason, research on the "natural course" of ischemic CVD is absolutely essential in order to demonstrate the effectiveness of acute stage therapeutic methods, particularly medical treatments involving thrombolytic drugs, surgical methods for vascular reconstruction and new combined techniques including the administration of brain protective substances.

Based upon considerations of this kind and in order to clarify the prognosis and clinical picture of acute stage cerebral infarction, a cooperative research project on ischemic CVD was organized at the Eighth Meeting of the Japanese Stroke Society (President: Jiro Suzuki, Sendai, 1983). The results of that cooperative study will be outlined below.

A total of 17 institutes of principally neurology and neurosurgery from throughout Japan participated in the study. Materials were cases of ischemic CVD (excluding Moyamoya disease) which had been brought to these clinics within 24 hours of onset of the disease, for which CT scans and/or cerebral angiograms had been obtained and for which observation of the clinical course (especially deficits in motor function and consciousness) was possible for a period of two months from onset. Data on a total of 1091 cases was collected, but 91 cases were excluded due to the eventual use of barbiturate or surgical therapy—leaving a final number of 1000 cases treated conservatively.

For evaluation of clinical symptoms, the following criteria were used. The state of consciousness was evaluated using the four grades of the Japan Coma Scale (3-3-9 scale) (Table 7-1)[21], motor function was evaluated using Dejong's six-grade classification (Table 7-2)[7], and overall condition was evaluated as one of five

Table 7-2. DeJong's classification of motor function

0	no muscular contraction occurs
1	a flicker or trace of contraction is present without actual movement, or contraction may be palpated in the absence of apparent movement: no motion of joints (10%)
2	the muscle moves the part through a partial arc of movement with gravity eliminated (25%)
3	the muscle completes the whole arc of movement against gravity (50%)
4	the muscle completes the whole arc of movement against gravity together with variable amounts of resistance (75%)
5	the muscle completes the whole arc of movement against gravity with maximum amounts of resistance several times without signs of fatigue; this is normal muscular power

grades two months from onset (Table 7-3). Among the fatalities, a distinction was made between deaths due to complications associated with the CVD and those due directly to cerebral infarction and unrelated to other disorders.

Differential diagnosis of embolism and thrombosis was made following the diagnostic criteria published by the Research Organs of the Ministry of Education (Table 7-4)[22].

Table 7-3. Classification of prognosis at two months after onset

Excellent	the patient recovered completely and returned to work
Good	slight neurological deficits were present but the patient were able to work
Fair	the patient has following handicaps: gait disturbance, psychological disturbance and/or aphasia
Poor	the patient is unable to work even with assistance
Dead	died within two months from the onset

Table 7-4. Differential diagnosis of embolism and thrombosis (according to the criteria of the Ministry of Education of Japan)

A) Cerebral thrombosis

1. Precursor symptoms include repeated attacks of cerebral ischemia, often with complete recovery or improvement in symptoms between attacks

2. The clinical course is progressive, with symptoms appearing gradually over a period of several minutes, hours or longer, or stepwise

3. Demonstration of arteriosclerosis of other organs (particularly of the coronary artery or aorta and peripheral arteries)

4. Presence of other lesions frequently occuring with arteriosclerosis

B) Cerebral embolism

1. Sudden onset

2. In most cases, no precursor symptoms

3. The cause of embolism due to a common heart disease (such as arrythmia or myocardial infarction)

4. Demonstration of embolism is thought to be recent origin
   a) embolism of other organs (pancreas, kidneys, extremities, intestines lungs)
   b) embolism at other cerebral vessels

7.2.1 Analysis of 1000 Cases: Prognosis

7.2.1.1 Description of the 1000 Cases

There were more patients in their 60s than in any other decade of life, followed by those in their 70s and 50s. The mean age was 63.7 years.

Retrospectively, the 1000 cases were diagnosed to be suffering from one of the following three conditions: transient ischemic attack (TIA, 78 cases, 7.8%), reversible ischemic neurological deficit (RIND, 148 cases, 14.8%) and completed stroke (774 cases, 77.4%). Although detailed analyses of the incidence of TIA and RIND have not previously been published, reports of hospital statistics have generally indicated an incidence of 3–6% for TIA[20] and of about 4% for RIND[23]. In contrast, the statistics of the Rochester epidemiological survey[29] indicate an incidence of about 20% for TIA and about 12% for RIND. Since the rate of hospitalization for TIA is naturally low, it is thought that the incidence found in hospital statistics is somewhat lower than its acutal occurrence.

The distribution of lesions in the 890 cases undergoing cerebral angiography was as follows: 183 cases of internal carotid artery (ICA) lesions, 229 cases of middle cerebral artery (MCA) lesions, 9 cases of anterior cerebral artery (ACA) lesions, 6 cases of posterior cerebral artery (PCA) lesions, 58 cases of vertebro-basilar artery (VBA) lesions and 13 cases of multiple lesions. In a total of 284 cases there was no detectable vascular lesion in cerebral angiograms.

7.2.1.2 Lesion Site and Outcome

The outcome in the 1000 cases was as follows: 226 excellent (22.6%), 209 good (20.9%), 183 fair (18.3%), 236 poor (23.6%) and 146 dead (14.6%). Patients capable of returning to normal social lives (excellent and good outcomes) comprised about 40% of the total series.

Relatively few previous studies on the prognosis of acute stage cases have been reported, but the results of the present study are similar to those reported by Jones and Millikan et al.[11]. That is, they found that, among 179 cases of acute cerebral infarction of the carotid arterial system, the prognosis one week from onset was: 11% normal, 37% improved but neurological deficits remaining, 41% with serious neurological deficits remaining and 11% dead.

With regard to the outcome in relation to the site of the vascular lesion, it was found that the prognosis was relatively good for those cases in which a distinct vascular lesion of stenosis, occlusion, etc. was not found. Specifically, among 284 cases, approximately two thirds returned to normal social lives and only six deaths were recorded (2%). In contrast, the prognosis was significantly poorer for those cases with identified vascular occlusion or stenosis, particularly those with occlusion. The prognosis was especially poor for cases with multiple occlusive lesions. None returned to social life and about one half died. In previously reported long term prognostic studies of multiple arterial lesions[26], the survival rate has been found to be low. The death rate among ICA occlusion cases was also about 50% and only 10% returned to normal social lives. The next highest rate of unfavorable outcomes was in occlusive MCA cases. There were rel-

atively few fatalities (15%), but only 20% returned to normal social lives.

The prognosis for ACA and PCA occlusion cases was relatively good, with about two thirds returning to normal social lives. Although some 20% of VBA occlusion cases were fa-

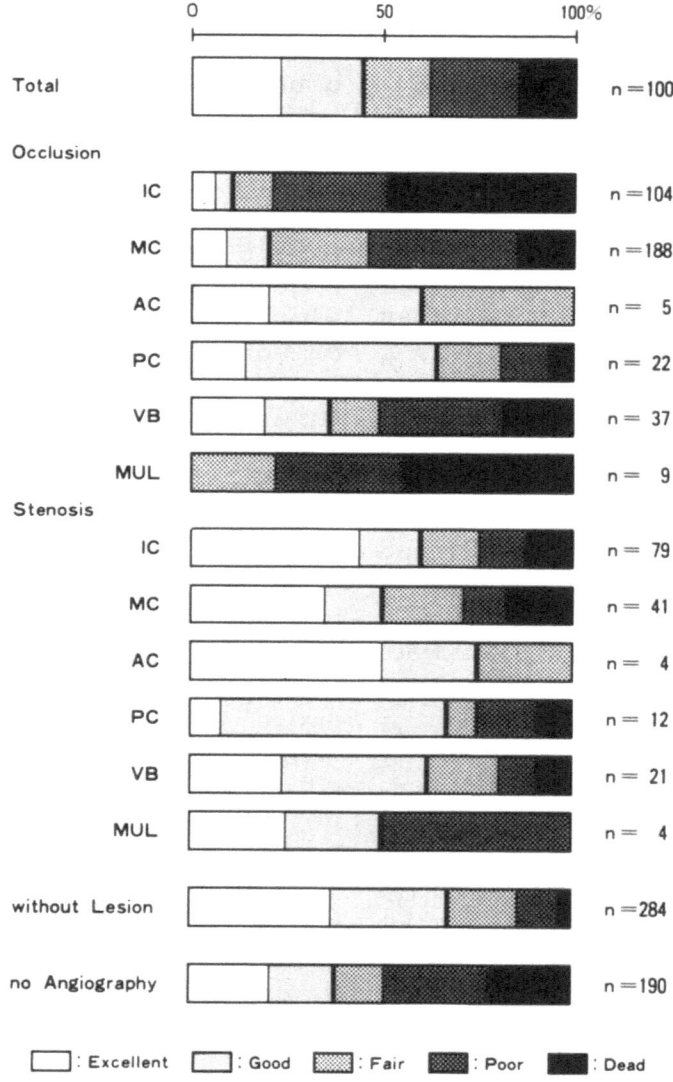

Fig. 7-5. Site of vascular lesions and prognosis at two months after the onset—from an analysis of 1,000 ischemic CVD cases

talities, about one third recovered—
with the prognosis tending toward
either one extreme or the other
(**Fig. 7-5**).

7.2.1.3 Age and Prognosis

It is well-known that age is an
important factor unfluecing the prognosis in cases of cerebral infarction. In
the present study, a close correlation

patients over 80 years of age, whereas a
return to social life was more frequent
the younger the patient. More than half
of the patients less than 50 years old
returned to normal social lives (**Fig. 7-
6**).

7.2.1.4 Study of Fatalities

In order to clarify what factors were
involved in unfavorable outcomes in

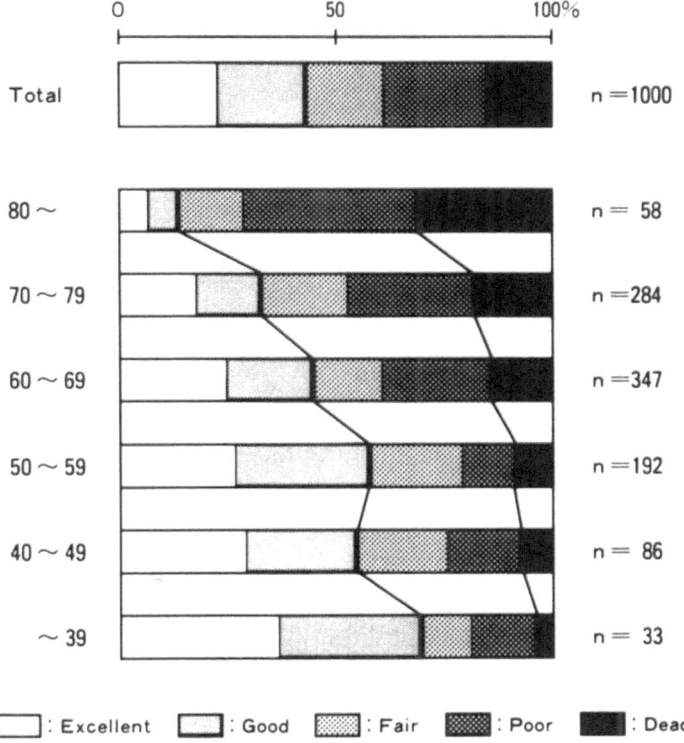

Fig. 7-6. Age distribution and prognosis at two months after the onset—from an analysis of 1,000
ischemic CVD cases

between age and prognosis was
evident—with fatalities increasing with
age, as expected. About one third of the
patients over the age of 70 died within
two months of onset. There were virtually no cases of full recovery among the

cerebral infarction cases, a separate study
was made of the fatalities. The death
rate due to acute cerebral infarction has
long been known to be lower than that
for other cerebrovascular diseases—
usually some 10—20%[12, 16, 19] of in-

farction cases have fatal outcomes within the first month. In the present series as well, a death rate of 14.6% (146 of the 1000 cases) was found over the first two months from onset.

Deaths were due directly to the cerebral infarction (rather than to associated disorders) in 98 cases and due to extracranial complications in 48 cases. In only two of 284 cases (0.7%) in which no vascular lesion could be identified angiographically was the cause of death attributed directly to the cerebral infarction. In contrast, death was attributed to the site of indentified occlusion in 45% of the cases of multiple lesions, 40% of the ICA cases, 14% of the VBA cases, 6% of the MCA cases an 5% of the PCA cases. Deaths due to extracranial complications were most common among cases with lesions of the ICA, MCA and VBA.

Although the mean age of the 1000 cases was 63.7 years, that of the fatalities due to infarction was 67.5 years and that of the fatalities due to complications was 71.8. In other words, fatalities tended to be more frequent among elderly patients, especially those with associated disorders.

a) Death Directly Due to Cerebral Infarction

Of the 98 cases of death attributable directly to the cerebral infarction, 76 were cases of non-hemorrhagic infarction and 22 were hemorrhagic infarction.

A significant correlation was found between the state of consciousness at the time of admission and the incidence of death due to the infarction. Among

the 474 cases of normal consciousness on admission (Grade 0), there were only 6 direct deaths (1.3%), whereas among 330 Grade I cases, 137 Grade II cases and 59 Grade III cases, there were, respectively, 20 (6.1%), 37 (27%) and 35 (59.3%) deaths—the death rate increasing with increasing severity of disturbances of consciousness (**Fig. 7-7**).

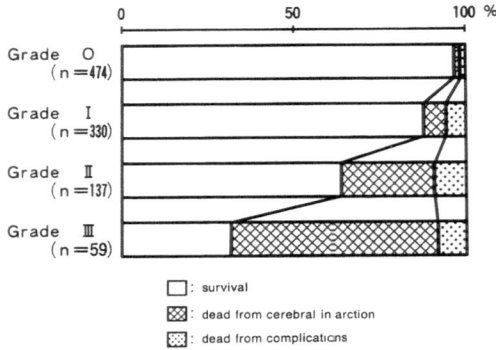

Fig. 7-7. Level of consciousness on admission and mortality rate—from analysis of 1,000 ischemic CVD cases

In a study carried out by Marquardsen et al.[15], it was reported that 65% of the patients who were deeply comatose on admission died within 24 hours, 87% of those who were semicomatose ultimately died at the end of three weeks, but no more than 24% of those who were alert died within the same period. They argued that the most important factor involved in the short-term prognosis of cerebral infarction patients was indeed the state of consciousness in the acute period following onset, and the results form the present cooperative study tend to support that view.

With regard to the timing of death in these cases, it was found that there was a peak incidence on the fourth day from onset in both cases of hemorrhagic and non-hemorrhagic infarction. In the case of non-hemorrhagic infarction, however approximately 70% of the fatalities occured within one week, whereas in the case of hemorrhagic infarction, there was a tendency for death to be somewhat delayed—with only 58% occurring in the first week. Other studies[3, 25] have found a peak in the death rate on the 4th or 5th day from onset and this period is known to coincide with the peak of maximal cerebral edema in cerebral infarction cases. As has been reported in autopsy studies[9, 14], the cause of death in these cases of acute stage cerebral infarction is thought to be cerebral herniation due to edema.

With regard to the vascular lesion in cases of death directly due to hemorrhagic or non-hemorrhagic infarction, one fourth of the 98 deaths were hemorrhagic, whereas, particularly among the MCA lesions, two thirds were hemorrhagic. In so far as MCA lesions are not normally accompanied by hemorrhagic infarction, direct death due to the infarction is therefore thought to be exceptional.

Diffuse LDAs covering regions fed by several of the trunk arteries of the brain were seen supratentorially in CT scans in about two thirds of the deaths due directly to the cerebral infarction. In the remaining one third of such deaths, there were diffuse LDAs extending over the entire region fed by the MCA. There were also, however, cases of VBA occlusion in which death occurred soon after onset, but in which LDAs were not found.

Hemorrhagic infarction was confirmed by CT in 125 of the 1000 cases. Study of the relationship between the period of onset and death showed that there was a tendency for the hemorrhagic infarction to occur within one week of onset among the fatalities, and there were few deaths directly due to the infarction after the start of the second week.

Previous CT studies[28] of the timing of hemorrhagic infarction have shown two periods of infarction—the first within the first week of onset, for which the prognosis is poor and the second around the third week. The results of the present study support such findings.

b) Deaths Due to Complications

It has been pointed out that deaths due to complications account for a large percentage of the fatalities in cerebral infarction and, although not dealing solely with acute cases, it has been reported[4] that some 60% of cerebral infarction deaths are due to complications. In the present series of patients, 4.8% died of complications—that is, 33% of the deaths were of this kind. The site of the responsible lesion in those 48 deaths was predominantly on the ICA and MCA and the lesions were normally accompanied by severe neurological deficits, including disturbances of consciousness.

As has been reported in previous studies[5], death was most commonly due to respiratory complications (pneumonia, respiratory failure, atelectasis)—such deaths accounting for one third of the cases. Heart complications, such as myocardial infarction, were found in nine cases (19%) and gastrointestinal complications, such as GI bleeding, in six cases (13%). Deaths occurred at various times following onset: six in the first week (13%), 11 in the second week (23%), four in the third week (8%), six in the fourth week (13%) and 21 during the fifth through the eighth weeks (43%).

Complications were more commonly encountered among patients with moderate disturbances of consciousness. Just prior to the development of complications, eight patients were in a state of normal consciousness (Grade 0), 14 were in Grade I, 25 were in Grade II and one was in Grade III.

7.2.2 Occlusion of the Internal Carotid Artery[13]

Among cases of cerebral infarction, the higher incidence was found for infarction due to ICA occlusion and, as mentioned above, the prognosis was poor. Among our 1000 cases there were 104 cases of infarction due to ICA occlusion. Sixty-three were male and 41 female; ages ranged from 44 to 85, with a mean of 66.1 years. Seven patients were in their 80s, 34 in their 70s, 37 in their 60s, 18 in their 50s and eight were less than 50 years old.

7.2.2.1 ICA Occlusion and Prognosis

The outcome two months from onset in these 104 cases was excellent in five patients (4.8%), good in six (5.8%), fair in 11 (10.6%), poor in 31 (29.8%) and dead in 51 (49%). As can be seen from these figures, about one half died, only 10.6% returned to useful social lives, and the prognosis was worse than that of any other group with lesions on a single vessel. The prognosis in these cases was markedly poorer than that for similar cases reported[6] in the West— one of several aspects of cerebral infarction where significant differences have been found that among Japanese

cases of cerebral infarction, particularly thrombosis cases, there are relatively few due to extracranial vascular lesions and, as a result, the percentage of embolism cases and of ICA infarction cases is high. It is therefore thought that the relatively poor prognosis in Japan is due to the higher number of cases of infarction due to ICA occlusion caused by embolism.

Among the 78 patients 60 years of age or older, only 11% returned to useful social lives and 56.4% died. In contrast, among the 26 patients less than 60, 23% returned to normal social lives and only 26.9% died. In comparison with the prognosis of the 1000 cases and in comparison with that of lesions on other vessels, the prognosis of the more elderly ICA patients was significantly worse (p < 0.01).

The site of the occlusion of the ICA was at the cervical portion in 68 cases, at the siphon in 19 and at the terminal portion in 17 (**Fig. 7-8**). In comparison with the distribution in cases reported in the West[24], the number of intracranial lesions was large. Although it has previously been reported[10] that the prognosis of occlusive ICA cases is worse among those with intracranial lesions, in the present series there were no significant differences in prognosis among the cases with occlusion at the cervical, siphon or terminal portions.

showed that the outcome was significantly poorer among the embolism cases (p < 0.01) **(Fig. 7-9)**. It is therefore evident that the particularly poor prognosis of occlusive ICA cases in

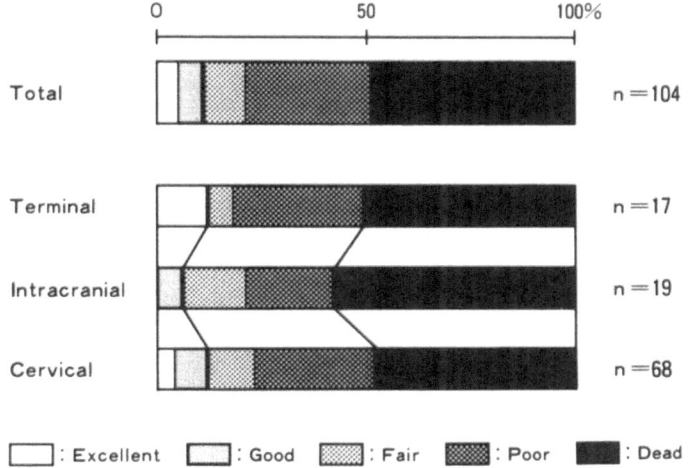

Fig. 7-8. Site of vascular lesions and prognosis at two months after the onset in cases of ICA occlusion

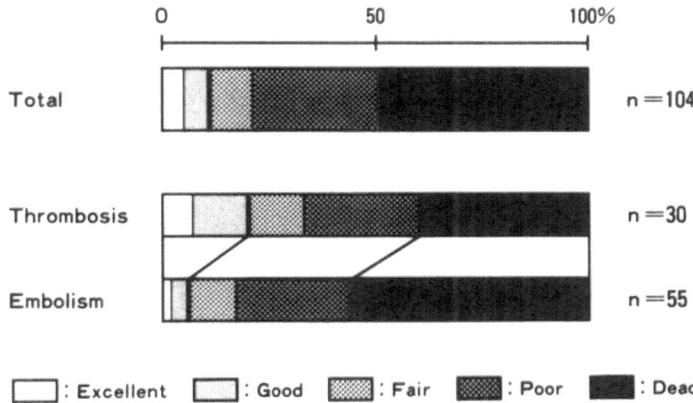

Fig. 7-9. Type of occlusion and prognosis at two months after the onset in cases of ICA occlusion

The ICA occlusion cases included 55 of embolism, 30 of thrombosis and 19 in which the nature of the lesion was uncertain. Study of the outcome of the 85 cases with definitive diagnosis

Japan is due to a combination of factors: first, as has been shown by others, the prognosis of embolism cases is poorer than that of thrombosis cases and, second, the incidence of embolism is much higher than that among occlusive ICA diseases in the West.

7.2.2.2 Acute Stage Disturbances of Consciousness and Prognosis

Mild to moderate disturbances of consciousness were frequently found among the 104 cases of infarction due to

within two months. In contrast, nearly half of the patients in normal states of consciousness (Grade 0) on admission had favorable outcomes. Very few patients who were admitted with impaired consciousness, however, eventually returned to normal social lives (**Fig. 7-10**).

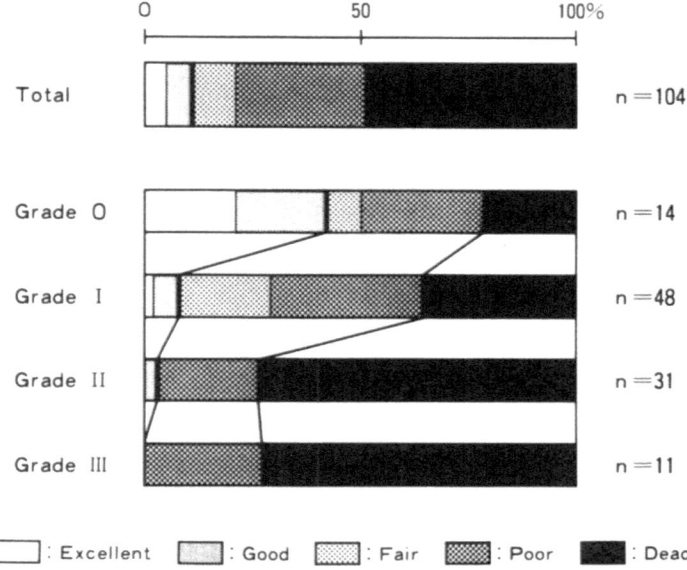

Fig. 7-10. Level of consciousness on admission and prognosis at two months after the onset in cases of ICA occlusion

ICA occlusion at the time of admission to hospital (*i.e.*, within 24 hours of onset). Specifically, there were 14 Grade 0, 48 Grade I, 31 Grade II and 11 Grade III cases. As was true for the entire series of 1000 cases, the greater the disturbances of consciousness on admission, the poorer was the outcome after two months. Patients with severe disturbances had significantly poorer prognoses. Fully 75% of those with moderate or severe disturbances died

7.2.2.3 Acute Stage Motor Deficits and Prognosis

The 104 ICA cases were classified according to the severity of motor paresis on admission as follows: three Grade 5, 17 Grade 4, 12 Grade 3, 16 Grade 2, 34 Grade 1 and 22 Grade 0. The majority of patients had severe motor deficits (Grades 2, 1 and 0). Again, the severer the paresis on admission, the poorer was the prognosis.

Among the Grade 2 or worse cases (in which motor resistance to gravity was impossible), virtually none returned to normal lives. This fact indicates a strong correlation between the degree of motor disturbance in the acute stage of cerebral infarction and the functional prognosis after two months (**Fig. 7-11**).

7.2.2.4 CT Findings in ICA Occlusion

The CT findings in these 104 cases were classified into four groups depending upon the size and location of LDAs: Group A, no LDA (4 cases); Group B, LDAs confined to a portion

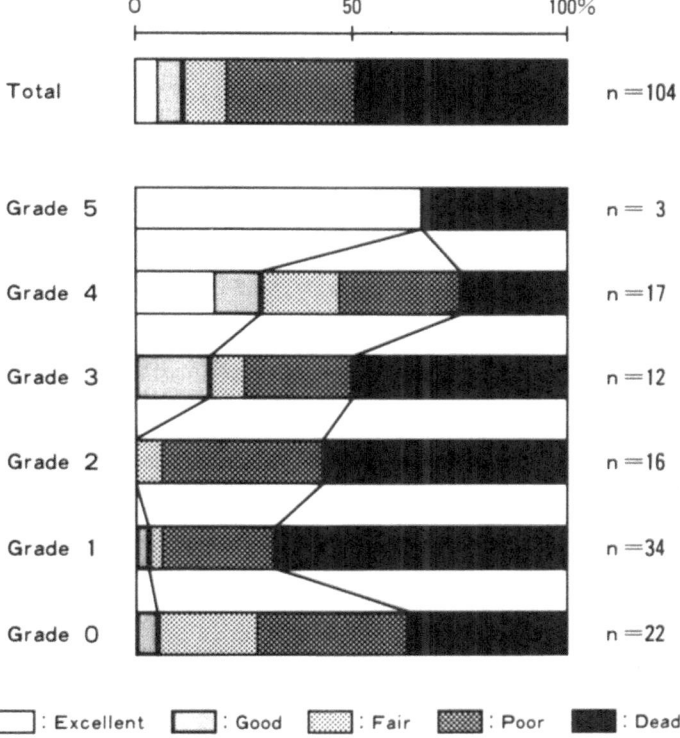

Fig. 7-11. Motor disturbance of upper limb on admission and prognosis at two months after the onset in cases of ICA occlusion

Jones and Millikan[11] previously found a similar correlation between the severity of paresis at the time of admission and the severity of neurological symptoms after one week. They also found a strong correlation between motor symptoms and the death rate.

of the brain fed by the MCA or a portion of the basal ganglia (35 cases); Group C, LDAs covering the entire region of the MCA (26 cases); and Group D, LDAs covering an entire cerebral hemisphere (39 cases). In other words, about two thirds of these 104 cases showed LDAs extending over wide regions of the brain. The outcomes were as follows. Among the four Group A cases, there was one death (25%) and three patients in excellent condition after two months (75%). Among the Group B patients, 22% returned to social lives and 14.3% died.

7.2.2.5 Fatalities Among the ICA Cases

The survival rate of the ICA cases fell sharply from around the fourth day from onset and about one third of the ICA patients had died by the end of the second week. Although the survival

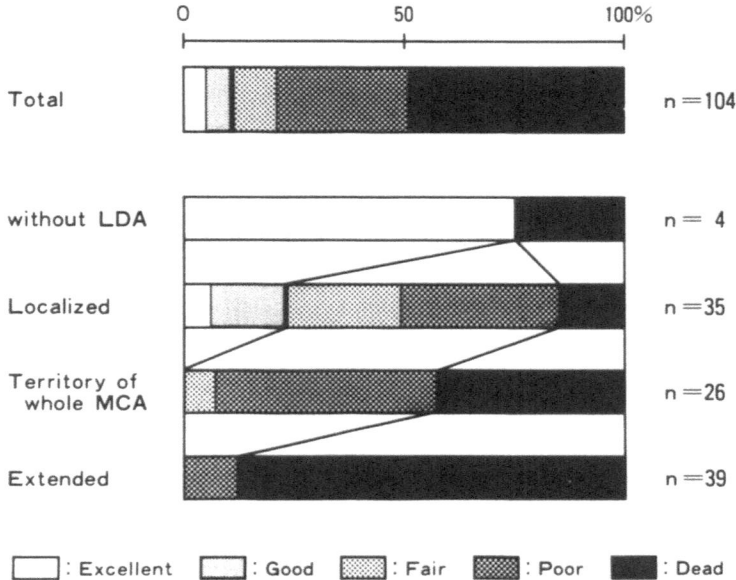

Fig. 7-12. Range of low density area on CT and prognosis at two months after the onset in cases of ICA occlusion

None of the 26 Group C patients or the 39 Group D patients returned to normal social lives and 10 from Group C (38.5%) and 34 from Group D (87.2%) died. In general, the broader the region of LDA, the worse was the prognosis (**Fig. 7-12**).

Excluding the four cases without any LDAs, there were 19 cases of hemorrhagic infarction out of the 100 remaining cases of ICA occlusion. The prognoses of those 19 cases was not significantly different from those of the 81 cases of non-hemorrhagic infarction.

rate fell slowly thereafter, at two months from onset approximately one half of these patients had died. There were no significant differences in the death rates over the two month period for embolism and thrombosis cases, but the survival rate for the embolism cases was in fact lower (**Fig. 7-13**).

Forty of the 51 deaths were due directly to the cerebral infarction (78.4%). There was a peak in the death rate directly due to cerebral infarction on the fourth day from onset and about three fourths of the deaths occurred within one week. In contrast, among the deaths due to other causes, there was no obvious peak incidence: five patients died during the first month and six during the second month. Most of those deaths were due to respiratory complications (five cases), two were due to heart disease, two to sepsis, one due to GI bleeding and one to suffocation. Patients dying of indirect causes were more elderly than those dying of direct causes, and were particularly numerous among patients with prolonged disturbances of consciousness.

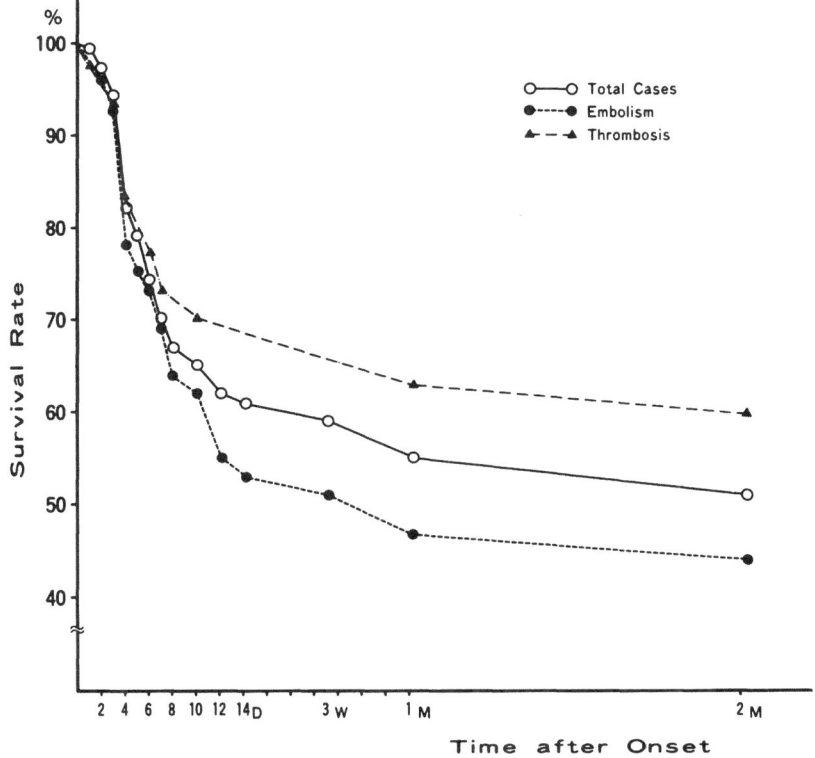

Fig. 7-13. Sequential change in the survival rate in cases of ICA occlusion

7.2.3 Occlusion of the Middle Cerebral Artery[8]

It has been mentioned above that intracranial lesions are particularly numerous among Japanese patients with cerebral infarction, but a high incidence of MCA lesions is also characteristic of the Japanese population and contrasts sharply with the pattern in the West.

7.2.3.1 MCA Occlusion and Prognosis

Among the 188 cases of MCA occlusion, the outcome two months from onset was excellent in 16 (8.5%), good in 23 (12.2%), fair in 49 (26.1%), poor in 74 (39.4%) and dead in 26 (13.8%).

Unlike the ICA cases, the death rate was low, but the rate of return to normal social life was also low—indicating a poor functional prognosis.

Fourteen of the 26 deaths were due to complications and 12 were due directly to the cerebral infarction, seven of which were cases of hemorrhagic

deaths among the thrombosis cases were due to complications, rather than directly due to the infarction. Hemorrhagic infarction was seen in 49 of the MCA cases, and the prognosis of similar cases was significantly worse than that of non-hemorrhagic cases. Hemorrhagic infarction was found in

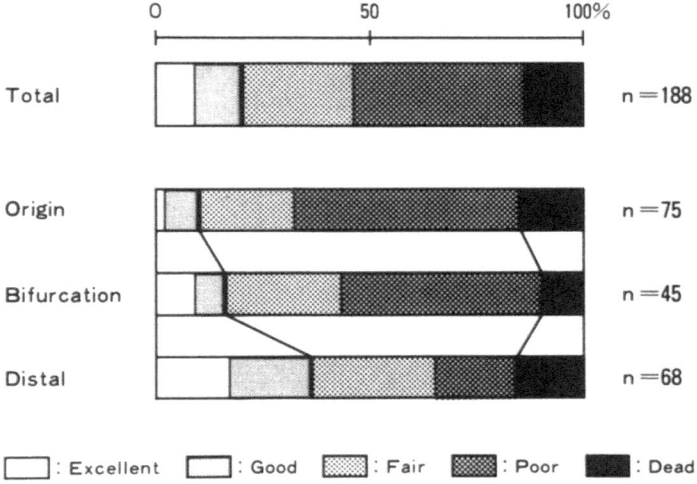

Fig. 7-14. Site of vascular lesions and prognosis at two months after the onset in cases of MCA occlusion

infarction. As noted earlier, these figures indicate that death due to cerebral infarction without cerebral hemorrhage among the MCA cases is exceptional.

The prognosis was worst for MCA cases with occlusion at the origin of the MCA and more favorable the more distal the occlusion—with a statistically significant difference found for distal and proximal locations (**Fig. 7-14**).

There were 90 emoblism cases, 44 thrombosis cases and 54 were uncertain—the prognosis of embolism being significantly worse than that for thrombosis (**Fig. 7-15**). All of the

33 of the 90 embolism cases (37%), but in only six of the 44 thrombosis cases (14%) (**Fig. 7-16**).

7.2.3.2 Acute Stage Disturbances of Consciousness and Prognosis

Nearly one half of the patients who were alert (Grade 0) at the time of admission were found to have favorable outcomes with a return to social living (excellent or good outcomes). In contrast, the more serious the disturbance of consciousness at onset, the lower the recovery rate, with no patients judged as Grade II or worse subsequently

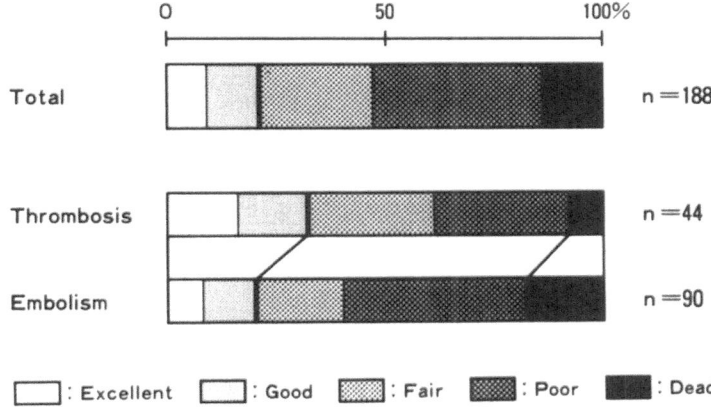

Fig. 7-15. Type of occlusion and prognosis at two months after the onset in cases of MCA occlusion

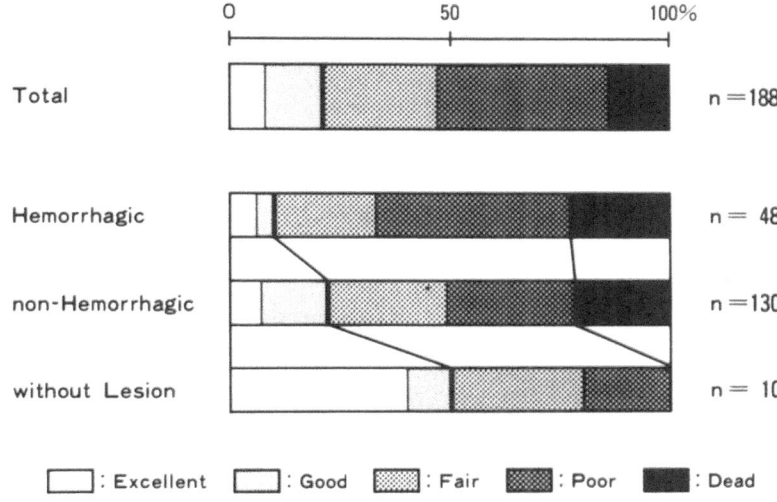

Fig. 7-16. Hemorrhagic infarction and prognosis at two months after the onset in cases of MCA occlusion

returning to social life. Nearly all of the patients who were comatose on admission (Grade III) died. In general, there was a close correlation between the state of consciousness on admission and the rate of survival and functional recovery (**Fig. 7-17**).

Disturbances of consciousness were significantly severer among the embolism than among the thrombosis cases— a finding which is thought to support the fact that the prognosis is poor both for embolism cases and for those with disturbances of consciousness (**Fig. 7-18**).

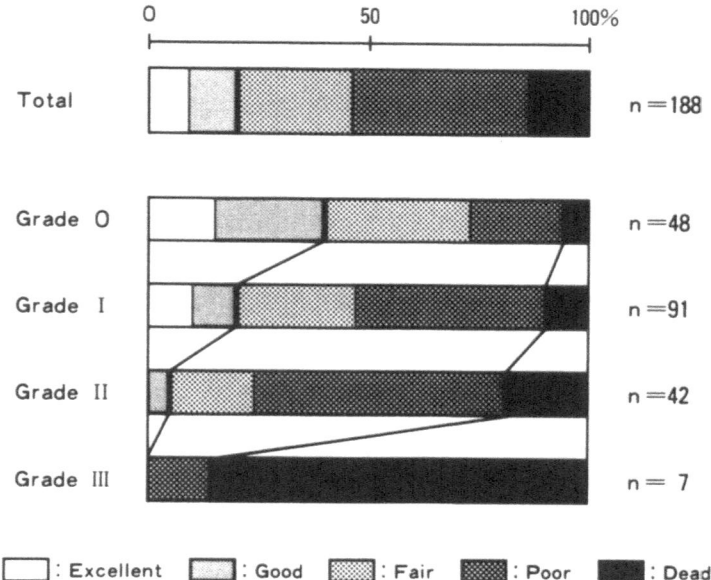

Fig. 7-17. Level of consciouness on admission and prognosis at two months after the onset in casesof MCA occlusion

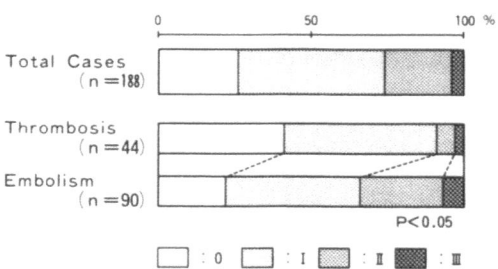

Fig. 7-18. Type of occlusion and level of consciousness on admission in cases of MCA occlusion

7.2.3.3 Acute Stage Motor Deficits and Prognosis

Although there was an even distribution of cases with mild and severe motor paresis among the fatalities, the prognosis of survivors was less favorable to severer the paresis on admission. There was functional recovery in only 10% of the cases with Grade 2 or worse motor deficits on admission (**Fig. 7-19**).

In order to study the recovery of neurological deficits, comparison was made between the severity of uppen limb motor paresis on admission and that after two months among 162 patients. Among those with paresis graded as 5, 4 or 3, approximately 80% were found after two months to have practical use of the arms (Grade 5–3) whereas only 30–40% of the patients with severe paresis on admission made such good recovery. Particularly among the patients with total paresis (Grade 0 on admission, about half showed no subsequent recovery. In contrast among the patients who showed good recovery despite having had severe paresis at the time of admission, the recovery tended to occur at an extremely early stage following onset (**Fig. 7-20**).

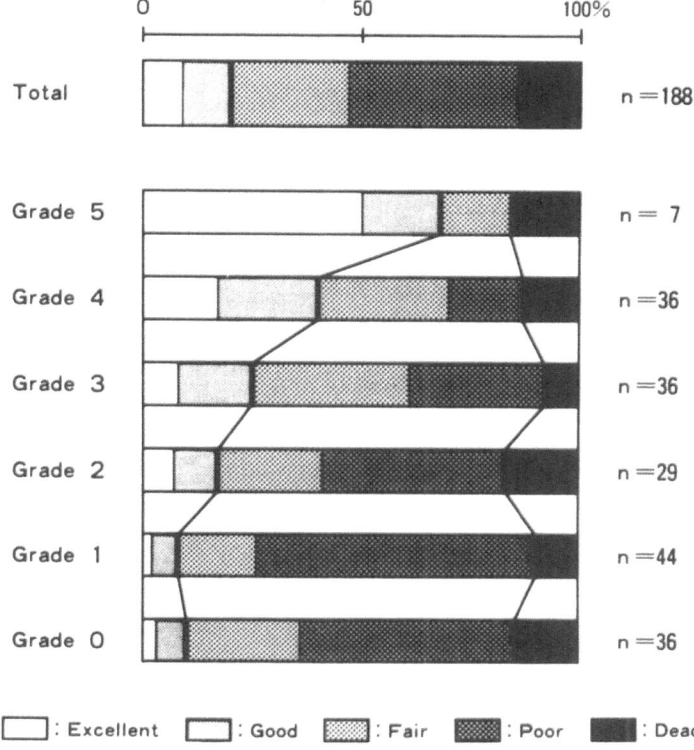

Fig. 7-19. Motor disturbance of upper limb on admission and prognosis at two months after the onset in cases of MCA occlusion

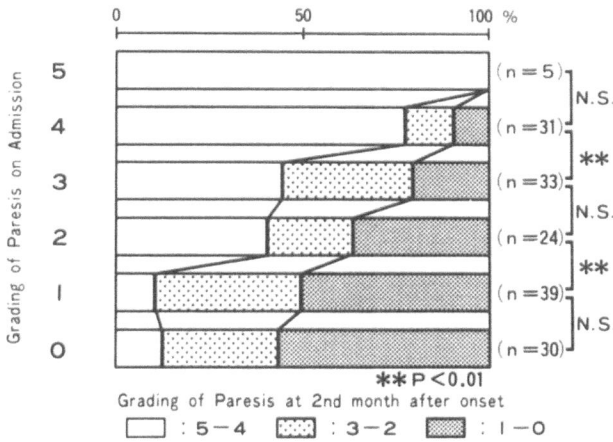

Fig. 7-20. Motor disturbance on admission and at two months after the onset in cases of MCA occlusion

7.2.4 Ischemic Cerebrovascular Disease Without Positive Angiographical Findings [17]

7.2.4.1 Relationship to Prognosis

No vascular lesion could be identified in 284 cases—approximately 35% of those cases for which angiography was performed. The outcome after two months in these cases was as follows: 105 in excellent condition (37%), 87 in good condition (30%), 53 in fair con-

were fatal, while 8% of the stenosis case and only 2% of the cases without identified lesions were. In other words, the six deaths among those without an identified vascular lesion were exceptional—two being due directly to the cerebral infarction and four to other complications. Both of the direct deaths

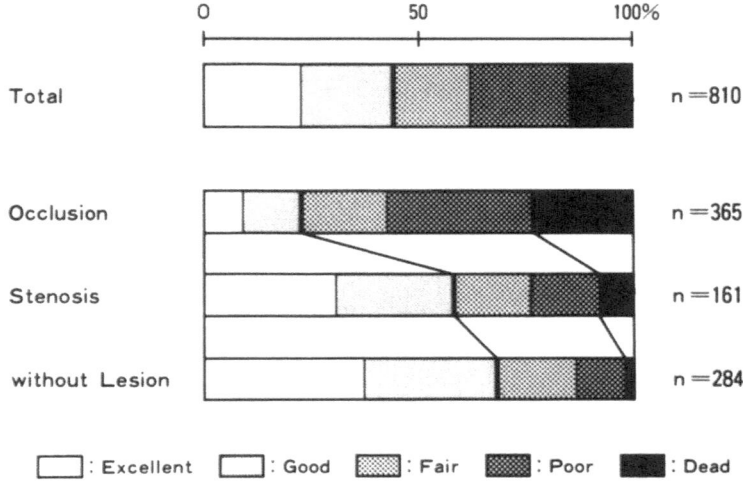

Fig. 7-21. Type of vascular lesion and prognosis at two months after the onset—from an analysis of 1,000 ischemic CVD cases

dition (19%), 33 in poor condition (12%) and six dead (2%). These results are significantly better than the results in patients in whom vascular stenosis or occlusion was identified angiographically. Specifically, recovery to an excellent or good condition was found in somewhat more than 20% of the occlusion cases, in slightly less than 60% of the stenosis cases and in slightly less than 70% of the cases with no identified vascular lesion (Fig. 7-21).

About 25% of the occlusion cases

were cases of deep coma immediately following onset in which there was no subsequent recovery, whereas all of the indirect deaths were relatively late (2–3 weeks from onset) and the causes on death were DIC in two, pneumonia in one and heart disease in one.

As we found for the entire series on 1000 cases, the recovery rate was highen for the younger patients, with abou 90% of the patients less than 40 years on age returning to useful social lives (Fig. 7-22).

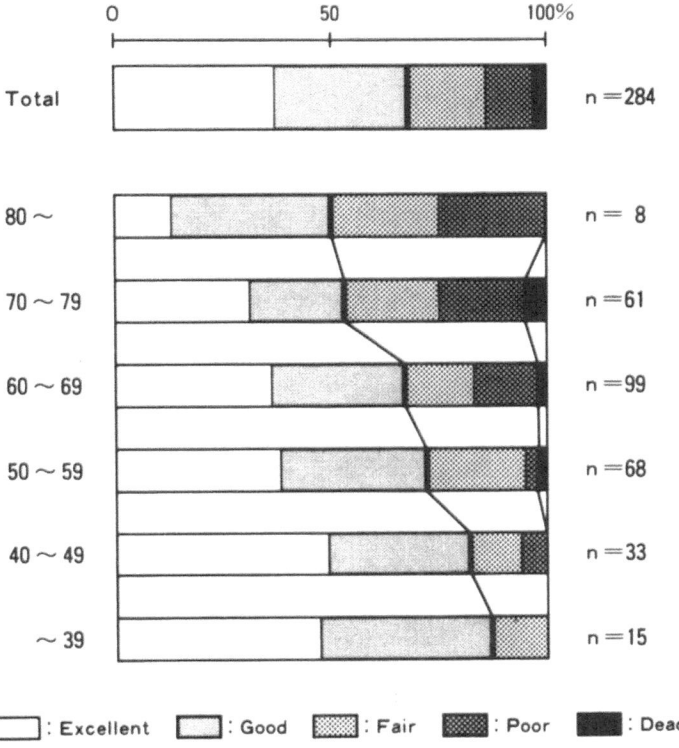

Fig. 7-22. Age distribution and prognosis at two months after the onset in cases of ischemic CVD without positive angiographical findings

Ten percent of the cases without angiographically identified lesions were embolism, 31% were determined as thrombosis and in 59% the diagnosis was uncertain. Although differential diagnosis was made following the criteria published by the Research Organs of the Japanese Ministry of Culture, it is uncertain how each individual decision at the various clinics was made. In light of the fact that no vascular lesion could be found angiographically among these cases, a differential diagnosis would be particularly difficult—as can be inferred from the relatively large number of cases for which a definitive diagnosis could not be made.

It is also notworthy that, in contrast to a ratio of five embolism cases for every three thrombosis cases among those with angiographically identified vascular occlusion, there was a ratio of one of three in both the group with identified stenosis and the group with no identified lesion. In light of the fact that about one half of these cases had cerebral angiography performed after the third day from onset and, as discussed below, in light of the fact that there were several cases of MCA embolism whose CT findings indicated spontaneous recanalization, it is thought that, among the cases for which differential diagnosis was not possible, there were likely several embolism cases with subsequent recanalization.

7.2.4.2 Acute Stage Neurological Deficits and Prognosis

In both the group of 1000 cases and in the groups with lesions on specific vessels, a strong correlation was found

identifiable vascular lesions. However, even some cases from the latter group with severe motor paresis during the acute stage showed good recovery and by two months from onset, one third had returned to normal social lives. The relatively good recovery of such severe cases was characteristic of this group **(Fig. 7-24)**.

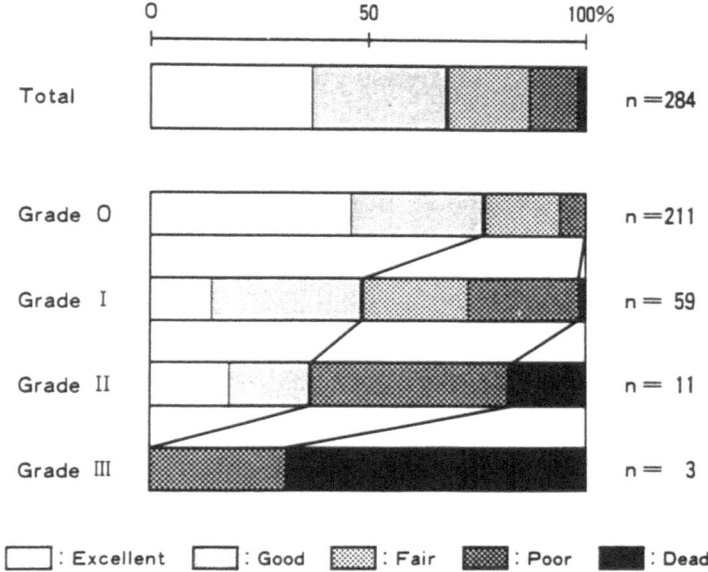

Fig. 7-23. Level of consciousness on admission and prognosis at two months after the onset in cases of ischemic CVD without positive angiographical findings

between the severity of the disturbances of consciousness on admission and the final outcome. The vast majority of the cases without angiographically identifiable lesions were alert or had mild deficits of consciousness on admission (270 out of 284 cases, or 95%) **(Fig. 7-23)**.

There was a similarly strong correlation between motor deficits and prognosis in the entire group of 1000 patients and in the 284 cases without

7.2.4.3 CT Findings

One hundred and fourteen cases (40%) did not show LDAs, while 105 (37%) showed LDAs of the basal ganglia and 50 (17.6%) showed LDAs in the region of the MCA. A smaller number of cases had LDAs in the posterior fossa and in the region of the PCA, regions of several of the trunk arteries, and regions of the ACA. These cases without LDAs are thought to

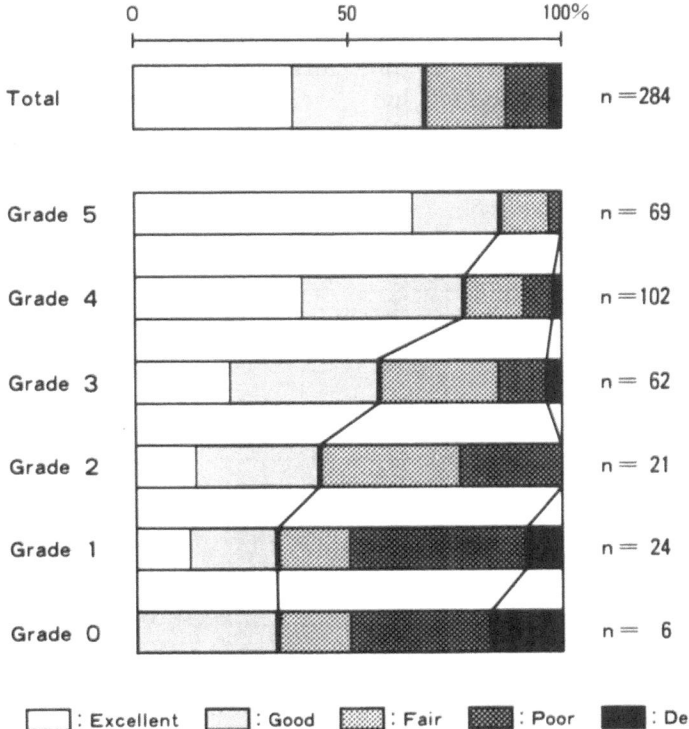

Fig. 7-24. Motor disturbance of upper limb on admission and prognosis at two months after the onset in cases of ischemic CVD without positive angiographical findings

include some with foci of infarction smaller than the resolution of the CT equipment and some with spontaneous recanalization of occluded vessels that occurred prior to the development of irreversible cerebral infarction.

The majority of the cases with distinct LDAs had such areas near the basal ganglia or in the vicinity of the MCA. It is thought, however, that the LDAs near the basal ganglia were probably cases of cerebral infarction due to occlusion of the perforating arteries emanating from the trunk arteries at the base of the brain. In patients with LDAs in the region of the MCA, it is thought that there were likely some in

which occlusion of cortical branches of the MCA did not show up clearly in angiograms and some cases of MCA embolism in which spontaneous recanalization occurred. The mechanisms of the onset of cerebral infarction in this group are thought to involve spontaneous recanalization in an early period following onset of occluded vessels primarily in embolism cases and the development of cerebral infarction in regions of perforating arteries due to occlusion of the perforating arteries.

With regard to the relationship between prognosis and the appearance of LDAs, it was found that 84% of the patients without LDAs returned to normal social lives, while only 60% of those with LDAs of the basal ganglia and 50% of those with LDAs of the MCA region did. Despite the fact that the LDAs of the basal ganglia were small, a relatively low proportion of such patients showed functional recovery. This result is due to the fact that, although disturbances of consciousness were mild, motor deficits tended to remain (**Fig. 7-25**).

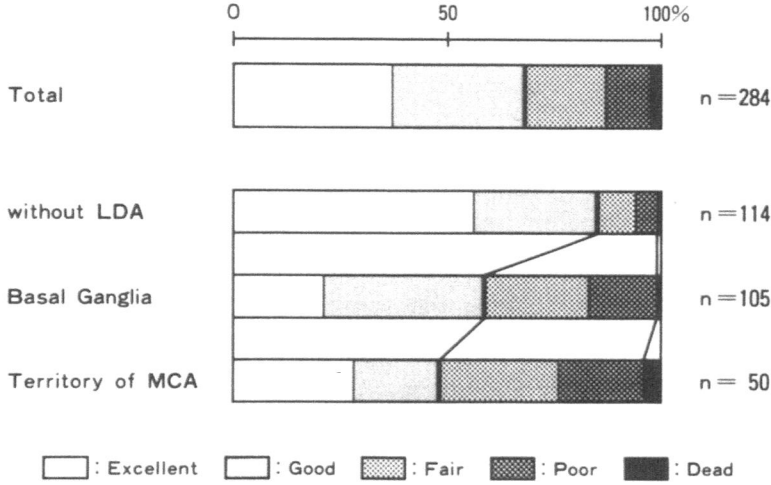

Fig. 7-25. Low density area on CT and prognosis at two months after the onset in cases of ischemic CVD without positive angiographical findings

8. DIAGNOSTIC TECHNIQUES

8.1 Introduction

Various diagnostic techniques are currently in use for the diagnosis of brain infarction and other intracranial diseases, and a wide variety of information concerning the metabolic and hemodynamic condition of the brain is thereby obtainable. However, in the short history of the development and establishment of these techniques, the appearance of X-ray computerized tomography (CT) has brought about the most dramatic improvements in the diagnosis of both the site and type of lesion in cases of cerebrovascular disease. For this reason, CT scanning must be regarded as the beginning of a new area of visual imaging of the brain.

Moreover, although the positron emission CT (PET) scanning techniques which have been developed in recent years remain somewhat inferior to those of conventional CT scanning in terms of their spatial resolution, these newer techniques have already provided a wealth of data on the dynamic changes in cerebral circulation, oxygen and glucose metabolism, etc., and represent another significant improvement in imaging of the brain. Nuclear magnetic resonance CT (NMR-CT) has not only provided morphological information with a spatial resolution

equal to or better than that of X-ray CT scanning, but now also promises to allow chemical imaging and the gathering of detailed information concerning changes in blood flow and high-energy phosphate compounds during cerebral ischemia.

There have also been important developments with regard to techniques for cerebral angiography—with a transition from direct puncture to the transfemoral Seldinger's method. More recently, there has been further evolution to angiographical techniques which allow for still clearer images using methods for infusion of contrast media which are simpler and less invasive. These include intraarterial and intravenous infusion techniques using digital subtraction angiography (DSA).

Among all intracranial lesions, determination of cerebral blood flow is most frequently done to elucidates the pathology in cases of cerebral infarction. Such measurements must be made while the patient is at rest, and are done to determine the CO_2 responsivity and autoregulation of cerebral vessels in response to hypertensive and hypotensive loads.

Here as well, significant advances have been made in the three dimensional imaging of regional cerebral blood flow (rCBF). Using techniques such as the dynamic CT, stable xenon-enhanced CT and singel photon emission CT (SPECT), it has become possible to obtain visual images of cerebral hemodynamics at deep portions of the brain. With the advent of PET-CT, it is now possible to image not only blood flow, but also glucose metabolism, metabolic rate and the consumption of oxygen in brain tissue.

In addition to these new techniques, traditional EEG techniques have been computerized—allowing for imaging using EEG topographic systems. Finally, other techniques have been developed for measuring the compliance of vascular walls, vessel diameter, flow velocity and flow rate in extracranial vessels using the echo and Doppler methods.

8.2 Techniques for Obtaining Information Concerning the Cerebral Vessels

Although we will here primarily discuss cerebral angiography, an important distinction must be made at the outset between the recently developed digital subtraction angiographical (DSA) method, on the one hand, and the transfemoral Seldinger and direct ICA puncture methods, which have been used for many years. The techniques differ both in terms of the angiographical method and the infusion of contrast medium. It is noteworthy,

however, that the older techniques are still superior to DSA both with regard to spatial resolution and with regard to the information obtainable concerning cerebrovascular lesions. In this section, therefore, cerebral angiography in cases of cerebral infarction using the older techniques will primarily be discussed and the methodology and unique aspects of DSA will be dealt with only briefly.

8.2.1 Cerebral Angiography Using the Direct Puncture and Seldinger's Methods

By means of cerebral angiography, it is possible to diagnose those lesions which are the direct causes of cerebral ischemia, i.e., stenosis or occlusion of extra—or intracranial vessels. Unfortunately, the site and severity of the vascular lesions and the severity and extent of the ischemia thus produced are heavily influenced by the rate of progression of the lesion and the presence of collateral pathways.

It is generally thought that infarctic attacks are caused by an embolism due to a thrombus with adherent atheroma plaque or to the plaque itself. It is, however, known that the thrombus itself can move, break into smaller pieces or dissolve in a matter of minutes in the case of white thrombi and in a matter of a relatively short period of 1–2 hours in the case of red thrombi. As a consequence, there are sequential changes in cerebral angiograms, so that already by the time of first angiography, recanalization may have occurred and there may be no obvious cause for the cerebral ischemia. Indeed, in a recent Japanese Cooperative Study on Cerebral Infarction (1984), no notable lesion was found in fully 35.1% of 810 cases in which cerebral angiography had been performed. In such cases it is necessary to consider the possibility of emoblism due, not only to cardiogenic thrombi, but also to thrombi adherent to atheroma plaques from the aorta, innominate artery, pulmonary arteries, etc.

In comparison with cases of thrombosis due to arteriosclerotic lesions, the symptoms are severer and the prognosis less favorable in embolism cases. Among such cases, it is rare that a concaved column head [Taveras[43]] is found at the site of occlusion—which is characteristic of the cerebral angiograms obtained in cases of embolism. Usually when vascular occlusion is seen at many branches of cerebral vessels, embolism should be considered the most likely diagnosis.

With regard to the incidence of occlusive lesions at various sites on cerebral vessels, it is generally thought that the highest incidence is found at the initial portion of the cervical internal carotid artery (ICA) and at the initial portion of the vertebrobasilar artery (VBA). In the above-mentioned Japanese Cooperative Study, however, the incidence of lesions of the middle cerebral artery (MCA) was relatively high among the 1000 infarction cases. Among the 365 cases in which cerebrovascular occlusion was seen in angiograms, 51.5% were occlusion of the MCA, 28.5% were occlusion of the ICA and 10.1% were occlusion of the VBA system. Among the 161 cases showing stenosis, 49.1% were of the ICA, 25.5% of the MCA and 13.0% of the VBA.

In addition to the characteristic stenosis or occlusion seen in angiograms, typical angiographical findings in cerebral infarction include local slowing of circulation or early venous filling, the capillary blush and indications of space occupying lesions[43], but it is important to note that these findings typically show sequential changes.

8.2.2 Digital Subtraction Angiography (DSA)

Digital subtraction angiography (DSA) is a technique which is easier, safer and less invasive than the direct puncture and Seldinger's methods described above, and it is now thought to be effective particularly as a screening technique[25, 39–42]. By means of computerized manipulation of the X-ray image, it allows for real-time visualization of the cerebrovascular system using a low concentration contrast medium.

In DSA, an intra-arterial catheter is used for infusion of contrast media (IADSA) or the media is injected intravenously into the cubital vein (IVDSA). It has been hoped that the latter method in particular would find wide usage for out-patient screening, but the spatial resolution has remained

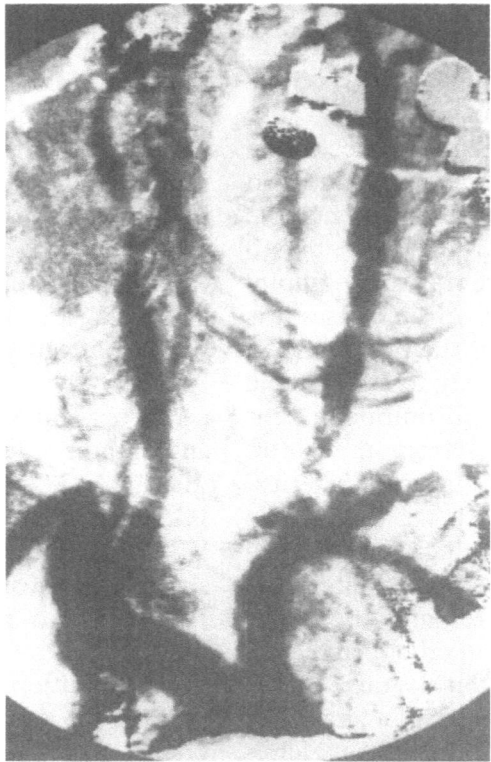

Fig. 8-1. Arteriogram of cervical vessels by
IVDSA. Severe stenosis of the right carotid
artery is demonstrated. However, precise diagnosis of the location or the degree of the
pathological lesion is not possible

poor (approximately 1 mm) and further
technical developments are yet needed.
As a consequence, the IVDSA technique does not yet have sufficient resolution for general use in most cases
expect those with lesions of relatively
large arteries, such as the ICA (**Figs.
8-1, 8-2, 8-3**).

It has been reported that diagnosis
using IVDSA is possible in some 30–
50% of patients with lesions of cerebral
vessels larger than about 3 mm in diameter (including the ICA, ACA, MCA,
PCA and VBA). It is also possible to

visualize relatively large pathways of
collateral circulation, but it remains
difficult to evaluate narrow collateral
pathways, mild plaque formation or
ulceration.

In contrast, the IADSA technique
can be used in either selective cath-
eterization or non-selective cath-
eterization using a diluted concentration of contrast medium, which would
not be possible in conventional cerebral
angiography. In order to minimize invasion due to the medium and in patients for whom a large volume of

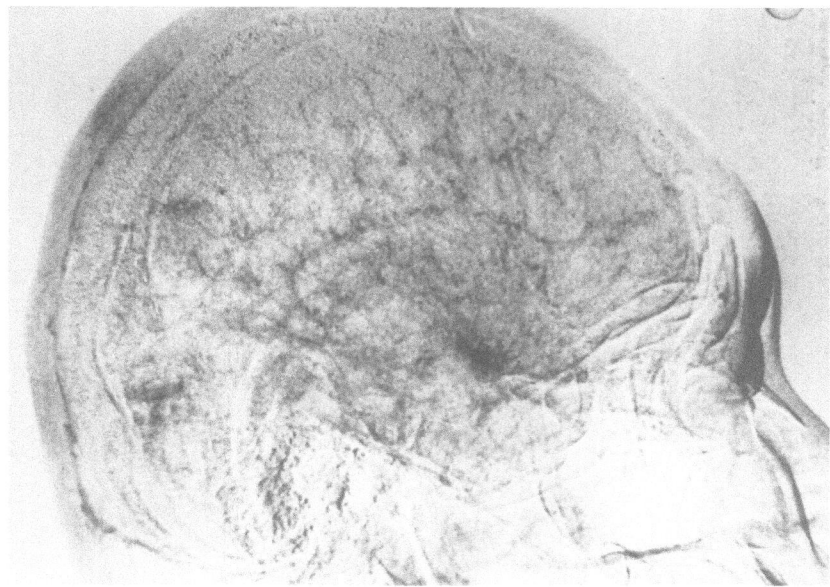

Fig. 8-2. Arteriogram of Moyamoya disease by IVDSA. Basal moyamoya vessels can be seen faintly. Demonstration of intracranial small vessels is almost impossible by IVDSA with the exception of these particular cases

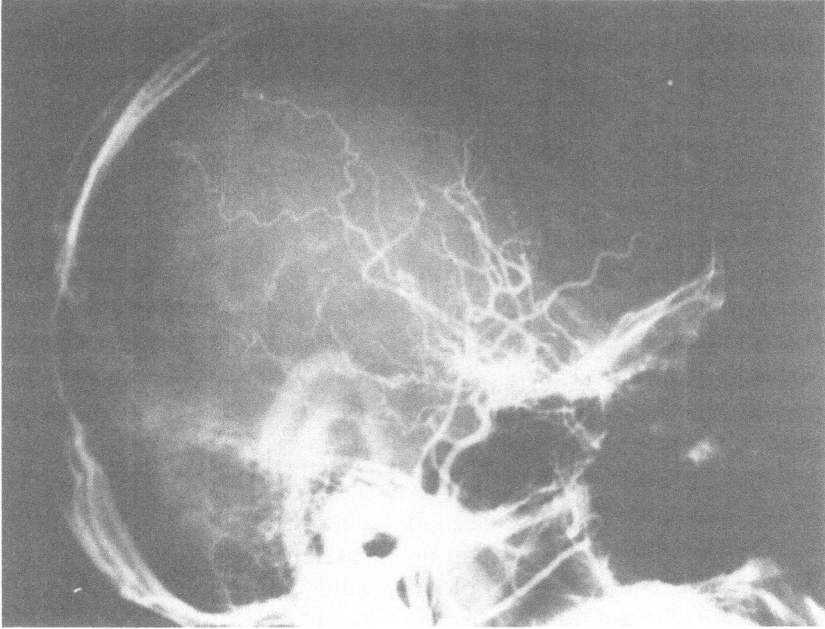

Fig. 8-3. Cerebral angiography by conventional Selginger's method. The same case of Moyamoya disease in Fig. 8-2

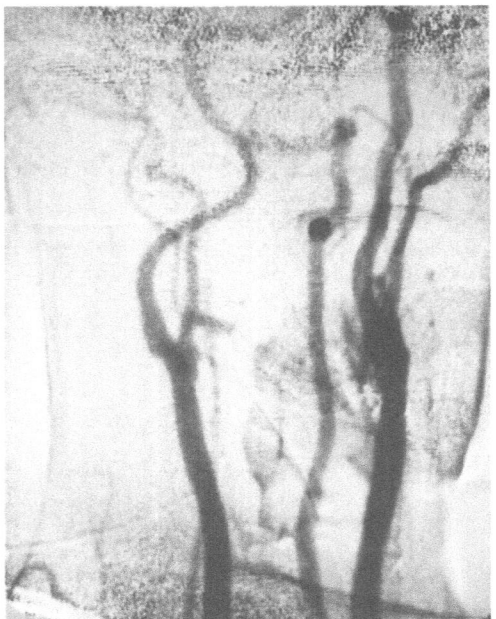

Fig. 8-4. Arteriogram of cervical vessels by IADSA (normal case). The image is much clearer than by IVDSA (compare Fig. 8-1)

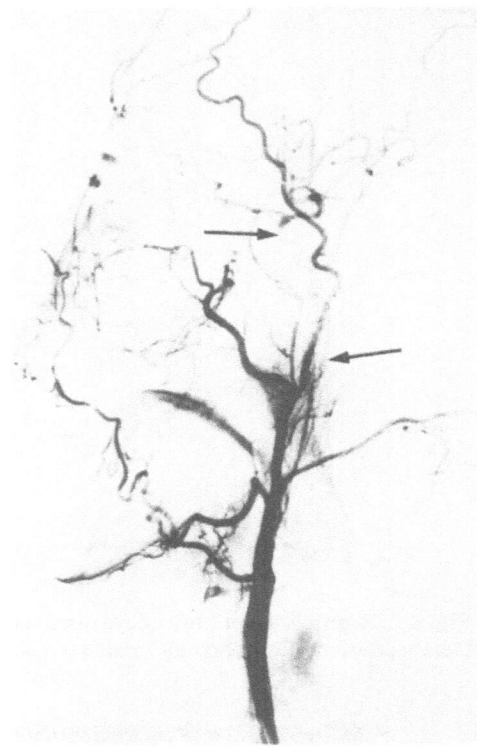

Fig. 8-5. Arteriogram of cervical and intracranial vessels by IADSA. Severe stenosis of the cervical portion of ICA is demonstrated (lower arrow). Intracranial ICA is scarcely seen (upper arrow)

medium cannot be used, IADSA is very effective. Provided that the vessels have a diameter larger than 1 mm, information comparable to that obtained using conventional angiography can be obtained (**Figs. 8-4, 8-5**).

The usefulness of DSA techniques will undoubtedly increase further due to their real-time imaging capability. Particularly when angioplasty or artificial embolization is undertaken, the ability to obtain angiographical information during the operation without the delay of photographic development makes it possible to evaluate the operation in mild-course and proceed to the next stage accordingly.

8.3 Techniques for Obtaining Morphological Information

In a broad sense, topics discussed in Section **IV** (techniques for obtaining information concerning cerebral blood flow and metabolism) could be discussed within the framework of the present section concerned with morphological data. Strictly speaking, in the present section the only technique which provides morphological information, X-ray CT scanning, should be discussed, but here we will review both the X-ray CT and the NMR-CT techniques.

8.3.1 X-ray Computerized Tomography (CT)

X-ray computerized tomography is a non-invasive technique for observing and diagnosing the infarctic brain. The images obtained in X-ray CTs are thought to reflect closely the histopathological changes of the brain during cerebral ischemia [3, 12, 15, 17, 20, 26, 27]. Moreover, the X-ray CT has proven to be effective in the differential diagnosis of ischemic infarction and hemorrhagic infarction [13, 45] and in determining the age of lesions (interval from onset) [26]. It must be pointed out, however, that accurate diagnoses of cerebrovascular lesions and information concerning cerebral hemodynamics cannot be obtained solely from X-ray CT findings. In this regard, Takahashi *et al.* reported that 244 out of 339 cases (72%), diagnosed clinically as cerebral infarction, showed positive findings on CT.

They found that, among their cases examined within 24 hours of onset, the incidence at which no findings of cerebral infarction were obtained was 60%. Moreover, they noted that, in contrast to findings of infarction in 81% of their supratentorial cases, the percentage of positive findings in infratentorial cases was quite low at 37%—due either to a large number of small infarctic foci or related to anatomical site (bone artifact, etc.).

It is often stated that CT findings in cerebral infarction include: (i) low density areas, (ii) mass signs due to cerebral edema, (iii) brain atrophy, (iv) contrast enhancement, and (v) high density areas (hemorrhagic infarction), but it must be pointed out that the most unique feature of CT scanning is the temporal changes occurring in the above findings following onset. The nature of these changes can be briefly summarized as follows.

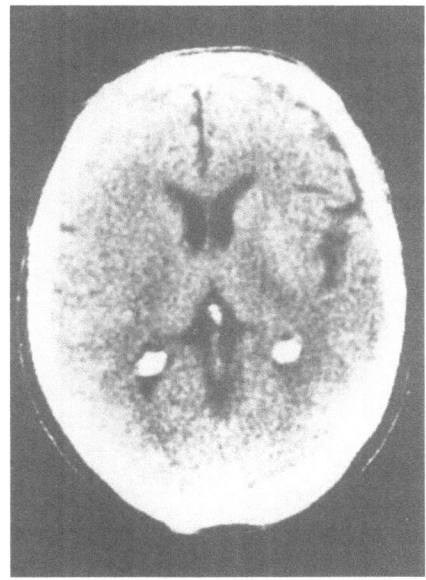

Fig. 8-6. Occlusion of the left internal carotid artery four hours from the onset (X-CT scan). Almost no abnormal findings are demonstrated

8.3.1.1 Low Density Areas

Low density areas, which are the principal finding in cases of ischemic infarction, are thought to be due to increases in the water content and decreases in the blood content of the focus, as well as to tissue necrosis [30]. The spread of the low density areas directly shows the size of infarctic focus and the occluded arteries can be identified from the distribution of the low density areas. Low density is not seen in the extremely early stage following onset (**Fig. 8-6**). Although the type and quality of CT equipment will have some influence, low density areas are not normally found in clinical patients for some 6–10 hours following onset, but will be seen in follow-up CT scans.

tendency for the infarctic regions to contract and become smaller than during the early period after onset. There are also many cases in which the density of the low density areas in-

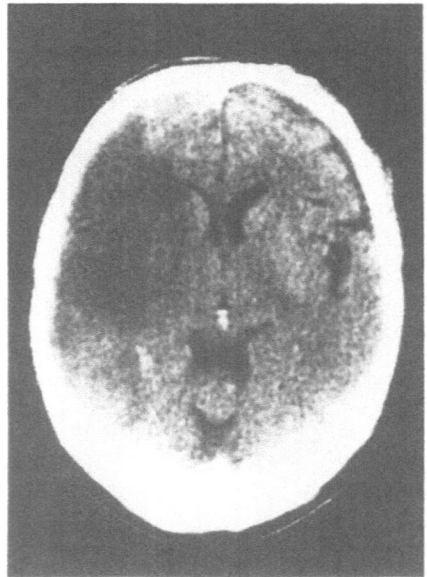

Fig. 8-7. The same case as in Fig. 8-6; 24 hours from the onset. Low density area in the left hemisphere and compression of the anterior horn are demonstrated

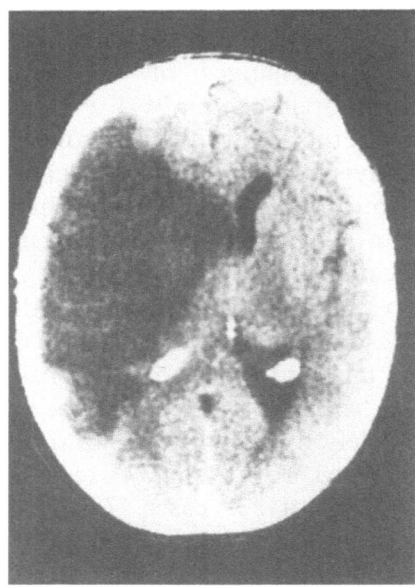

Fig. 8-8. The same case as in Fig. 8-6; three days after the onset. The size of the low density area and the mass effect have increased to the maximum degree

According to the previously-mentioned study by Takahashi *et al.*, low density areas were found in 40% of their patients studied within 24 hours of onset, and were first apparent at the earliest between 6 and 12 hours following onset. During the 2nd–7th days, 80% had low density areas and the border with surrounding tissue became distinct with absorption values of 10–30 Hounsfield Units (HU) (**Figs. 8-7, 8-8**).

Several months after onset, infarctic foci are found to be regions of markedly low and homogenous density with distinct borders. There is, in addition, a

creases after 3–4 weeks, gradually approaching isodensity. This process is called the "fogging effect"[37, 38] (**Fig. 8-9**), and is seen in approximately 50% of patients. This region is the same region that shows positive contrast enhancement effects (described below), and it is one of the causes for failing to find lesions using the CT.

8.3.1.2 Mass Signs Due to Cerebral Edema

Mass signs due to cerebral edema emerge 6–24 hours after onset, before the emergence of low density areas, and

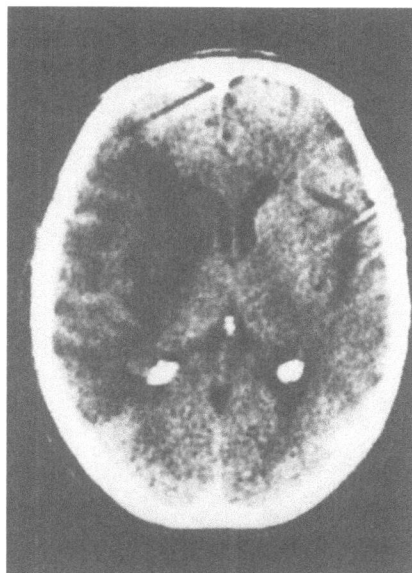

Fig. 8-9. The same case as in **Fig. 8-6**; 11 days from the onset. Irregular isodensity areas can already be seen in the low density lesion (fogging effect)

they are extremely important for diagnosis. Mass signs are usually manifest as lateralities due to pressure, deformation or narrowing of the sulci or cerebral ventricles or as shifting of micdline structures to one side (**Fig. 8-7**). Mass signs are thought to be due to a decrease in the X-ray density of an ischemic focus caused by cytotoxic edema. They are strongest on the 2nd–4th days of the disease (**Fig. 8-8**), start to subside after approximately one week (**Fig. 8-9**), and disappear after 1–2 months.

8.3.1.3 Cerebral Atrophy

Cerebral atrophy can often be severe in the vicinity of an infarctic focus in cases several months into the chronic stage, but there have also been many cases showing atrophy of an entire hemisphere. Near an infarctic focus, widening of cerebral sulci, ventricles and the cistern of the affected hemisphere are often seen (**Figs. 8-10C, D**).

8.3.1.4 Contrast Enhancement Effects [13, 38, 43, 48]

Again with reference to the work of Takashashi *et al.*, it can be said that contrast enhancement effects are found from about the third day from onset and persist for about two months. The effects are most frequently seen between two and five weeks from onset and are found in 60–70% of infarction patients. As shown in **Fig. 8-10B**, we have found a peak of contrast enhancement effects between the third and fourth week from onset. The enhancement is usually seen around the border of a low density area, and is frequently irregular in shape. As described above, if the sequential changes in low density areas are studied from onset, one portion often shows a fogging effect [37, 38] and in almost all such cases the area showing fogging also shows contrast enhancement.

Various suggestions have been put forward with regard to the mechanisms involved in contrast enhancement, among which the following two are noteworthy: (i) As a result of destruction of the blood-brain barrier, there is extravasation of the contrast medium. (ii) Pooling of contrast media occurs due to the proliferation of granulation tissue with the capillary neovascularization seen during the liquification of an infarctic focus.

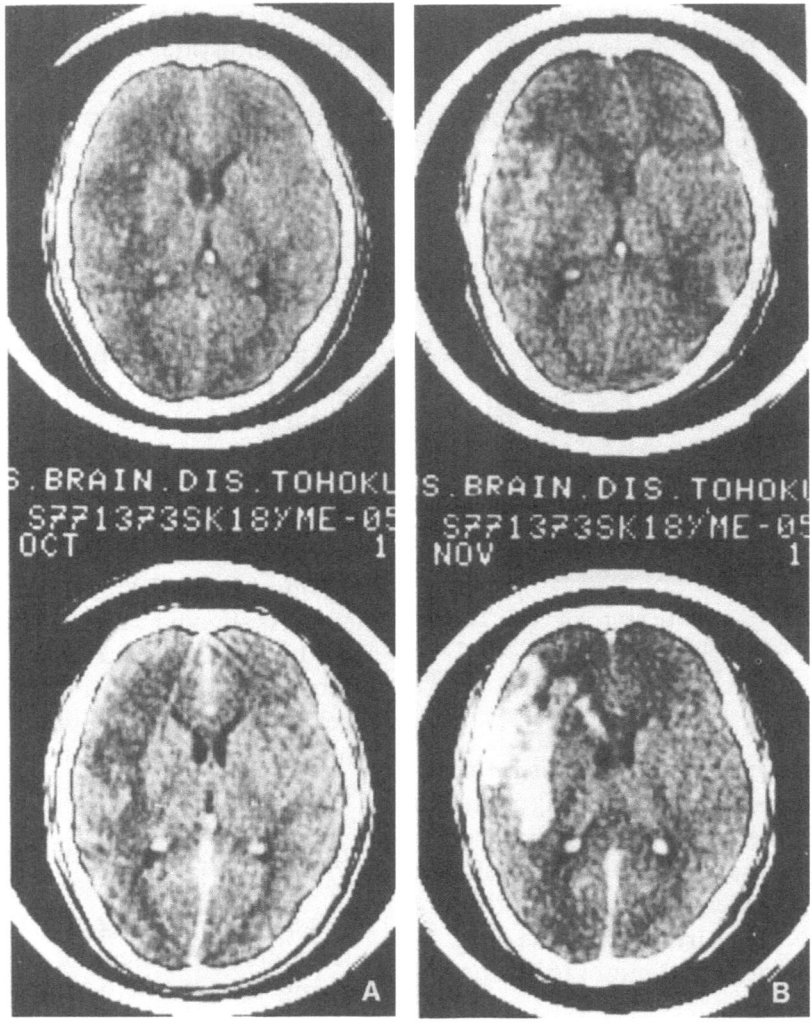

Fig. 8-10. X-CT image of the occlusion of the left internal carotid artery (upper: plain CT, lower enhanced CT). A) 28 hours, B) 16 days, C) 42 days, D) 99 days from the onset. Prominen enhancement effect is seen in B)

Contrast enhancement seen during the early stage following onset is thought to occur by the former mechanism, whereas the latter process is thought to account for the vast majority of contrast enhancement seen theraftre, in light of the following three facts: (i)

The period during which the abnorma accumulation of radioisotopes is seen ir scintigrams[14] (*i.e.*, after 3–4 weeks) i roughly the same as the peak of contras enhancement effects. (ii) The enhance ment is often seen in the periphery of ar infarctic focus. (iii) There is a stron

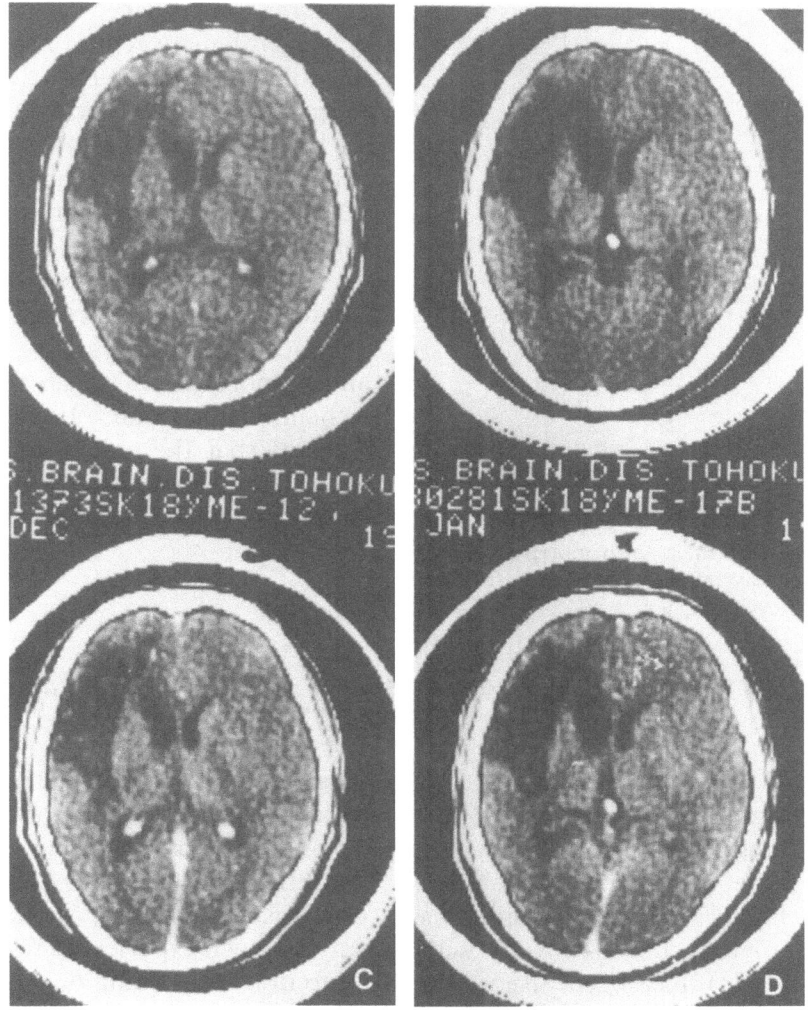

likelihood that the fogging effect (*i.e.*, the tendency toward isodensity) is related to the formation of granulation tissue.

8.3.1.5 High Density Areas

Within a low density infarctic focus which coincides with the area fed by branches of the main cerebral arteries, the intermingling of high density areas indicates hemorrhaging, and such cases are diagnosed as hemorrhagic infarction[13, 45]. The hemorrhage in such cases may take the form of hematomas or localized or diffuse petechial hemorrhage. The latter does not show up as high density region in CT scans and cannot be distinguished from ischemic infarction. Among hematoma cases, there are many in which the high density area is not as distinct as in the case of

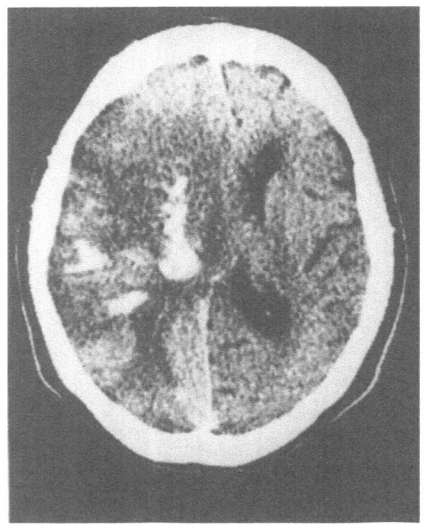

Fig. 8-11. X-CT image of hemorrhagic infarction. Three days after the onset of left middle cerebral artery occlusion

cerebral hemorrhage. Characteristically, a low density region is seen among irregular and uneven high density areas (**Fig. 8-11**). It has been reported that such findings are obtained in 8–10% of cerebral infarction cases and in some 40–50% in recanalization cases following the initial occlusion are included.

Contrast enhancement effects are found in nearly all such cases and the enhancement is seen at an earlier stage and is more distinct than that seen in ischemic infarction. It is generally thought that the enhancement effects are due to destruction of the blood-brain barrier in the recanalized vessels

8.3.2 Nuclear Magnetic Resonance Computed Tomography (NMR-CT)

NMR-CT is often referred to as simply magnetic resonance CT (MR-CT) or magnetic resonance imaging (MRI). Due to technical problems, current clinical NMR-CT imaging has been completely successful only when hydrogen (^1H) has been used. The imaging of ^{23}Na and ^{31}P and other nuclides has been only partially successful. The findings obtained with NMR-CT are not, however, confined solely to morphological information, but actually include significant chemical information. In the present section, we will confine the discussion primarily to the morphological findings in cerebral infarction using ^1H-NMR-CT.

Strictly speaking, when discussing the findings in clinical cases, an ac-

curate description of the image requires details concerning the procedural conditions (strength of the static magnetic field, the kind of pulse sequence, etc.) For this reason, here we will discuss only our own experiences with cerebral infarction cases using our machine.

The NMR-CT used in our clinic is a 0.15 Tasla resistive type, Model BNT 1000J from Bruker, Inc. of West Germany. The pulse sequence used was the Carr-Purcell-Meiboom-Gill (CPMG) method, which is an improvement on the spin echo method.

With regard to NMR detection of cerebral infarction in the acute stage several hours following onset[11, 36], data from our own and other departments[1, 35] remain equivocal. The main reason for this uncertainty is the small number

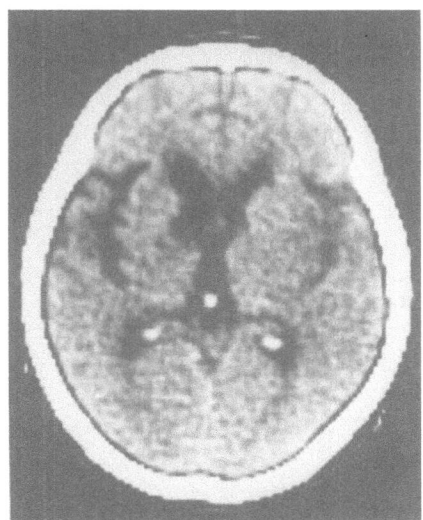

Fig. 8-12. Right middle cerebral artery occlusion 8 hours after the onset. No abnormal findings except slightly narrowed right Sylvian fissure

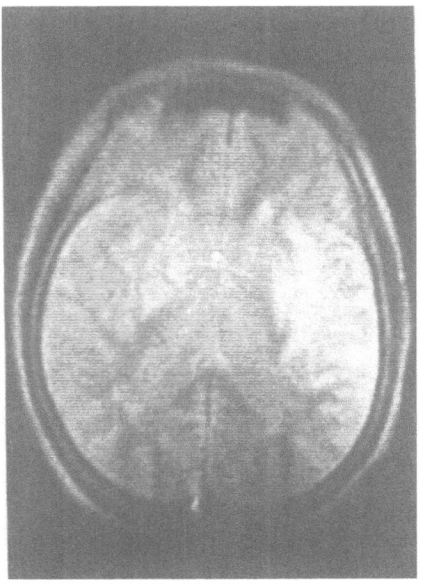

Fig. 8-13. NMR-CT image of the same case as in **Fig. 8-12** examined almost at the same time (proton density image). An increased proton density area is demonstrated in the right hemisphere

of such cases so far experienced. This is in turn due to problems concerning the time required for NMR examination of acute stage or severe cases with cerebrovascular disease, which have a high likelihood of exhibiting rapidly changing pathology.

Experimentally, Buonnano *et al.*[8, 9] have used carotid ligation in the rat and gerbil to study the effectiveness of NMR in cerebral ischemia. They found that there is prolongation of the relaxation time (T1) two hours after vascular ligation in animals later found to have infarctic foci. Seven and a half hours after ligation, the changes could be seen in the NMR images. In other words, prior to the detection of low density areas in conventional X-ray CT scans, changes in T1 and T2 occur and allow for earlier detection of the lesion with the NMR-CT.

In our department, the earliest NMR examination were performed ten hours following onset. Although slight narrowing of the Sylvian fissure (in comparison with that on the healthy side) was visible (*i.e.*, a slight mass sign) in X-ray CT scans (**Fig. 8-12**), the NMR-CT scan obtained at the same time showed the lesion as a distinct area of increased proton density (**Fig. 8-13**).

Sequential observations on such patients have shown further prolongation in T1 and T2[6, 7, 29, 35]. We have therefore made a comparison of such cases with the prolongation in hematoma cases and have found that the prolongation is considerably greater in infarction cases. At least from our own observations made over one year from onset, the prolongation gradually increases dur-

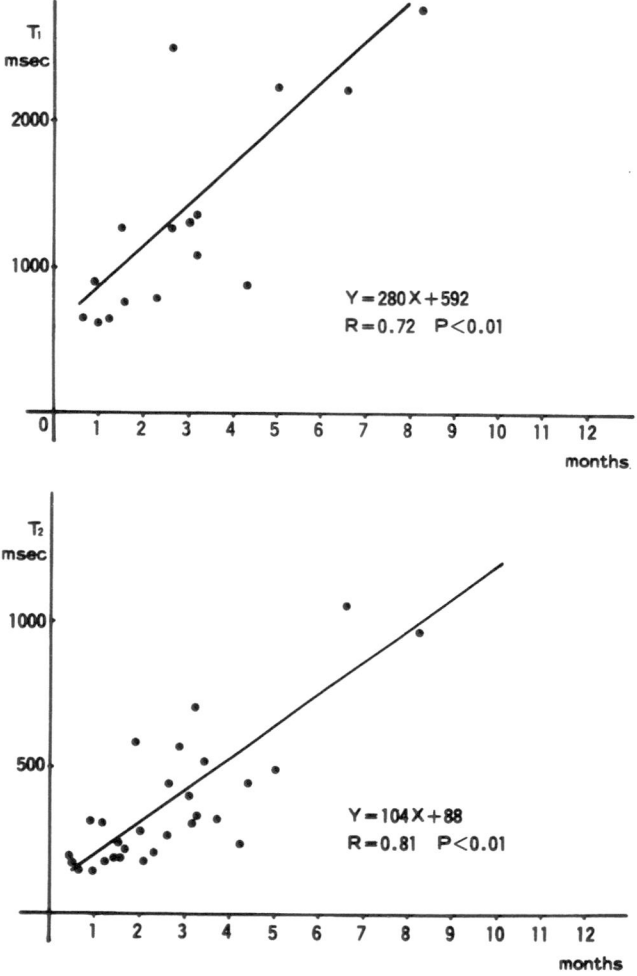

Fig. 8-14. Sequential changes in T1 and T2 in cerebral infarction

ing that period (**Fig. 8-14**). It is therefore possible to determine from the T1
and T2 values the period from onset
until examination and, in cases with
multiple infarct, to determine which is
the responsible lesion and how old the
other lesions are.

Among the strengths of the NMR-
CT technique, the following are
noteworthy: (i) there is a significant
degree of tissue specificity obtainable
from the NMR parameters such as
proton density, T1, T2, etc. so the
image provides easy identification of
mild or small lesions: (ii) images can be
obtained at any angle, even sagitally
without positioning of the patient, and
(iii) there are no bone artifacts to con-
tend with. For these reasons, the NMR
is particularly valuable for the depic-

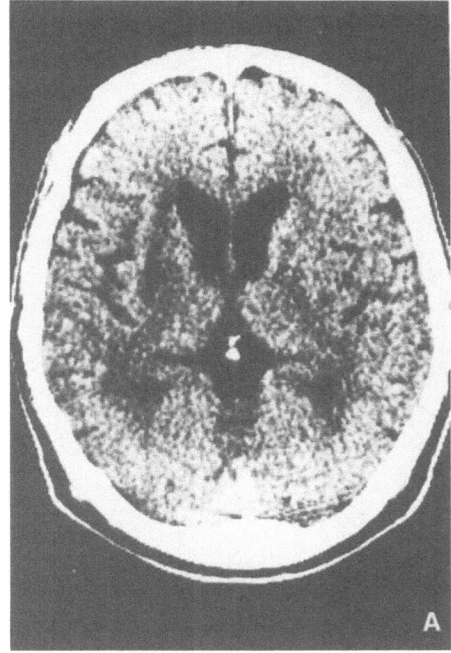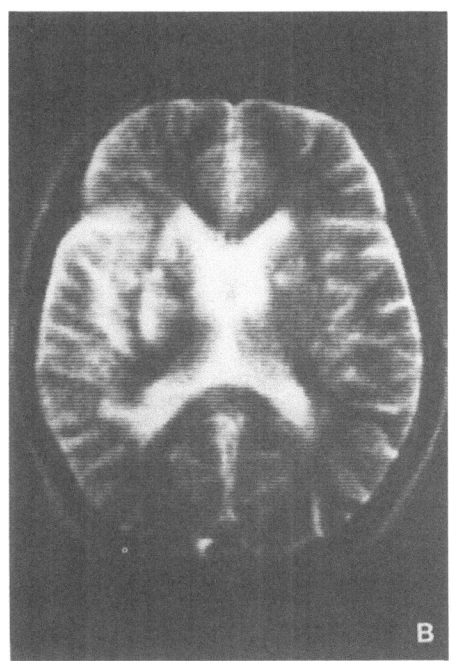

Fig. 8-15. A) Plain X-CT image B) NMR-CT image (Tr = 3.0 sec). As small pathological lesion in the right caudate nucleus is demonstrated by NMR, which is not seen in the X-CT image

tion of small lesions of the basal ganglia, brain stem or posterior fossa (**Figs. 8-15 A, B**).

Among infarction cases in the chronic stage, surrounding lesions of obsolete infarctions which cannot be detected using the X-ray CT can be detected as areas of slightly prolonged relaxation time using the NMR-CT (**Figs. 8-16 A, B**). The lesion depicted in **Fig. 8-16** shows a reduction in both the extent and signal intensity in NMR-CT scans following STA-MCA anastomosis. These findings were particularly valuable in suggesting that the ischemic lesion was extremely mild and possibly reversible. Since there was a strong likelihood that the lesion was a so-called ischemic penumbra, as defined by Astrup[2], these findings led to operative indication for bypass surgery in this chronic stage infarction case.

As is evident from the above, ^{1}H-NMR-CT is thought to be an effective technique for the diagnosis of cerebral infarction due to the close relationship between infarction and cerebral water metabolism. By means of NMR imaging of [23]Na, it is likely that the early stage changes due to ischemia will be observable, and by means of NMR imaging of the ^{31}P spectrum, it is likely that detection of changes in brain tissue pH and changes in high energy phosphorus compounds following ischemia will become possible. Clinical use of these techniques for treatment and for the evaluation of treatment is undoubtedly not far off.

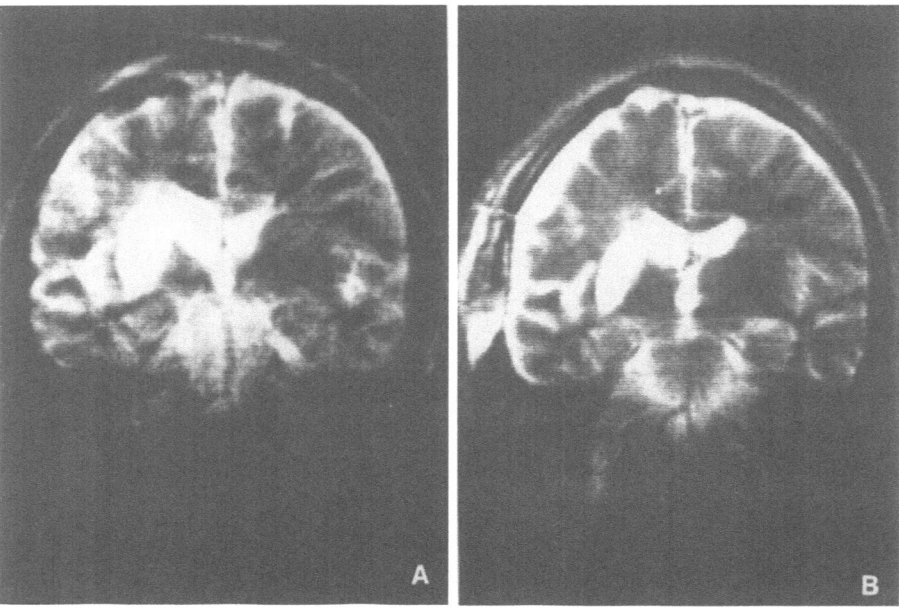

Fig. 8-16. A) NMR-CT (Tr = 3,0 sec) of cerebral infarction in the chronic stage. Two old infarctic lesions are seen in the left basal ganglia, from which a faint T2 elongated area continues to the cortex. B) After STA-MCA anastomosis. The faint T2 elongated area has decreased in size and degree

8.4 Techniques for Obtaining Information Concerning Cerebral Blood Flow and Metabolism

Unlike cerebral angiography and CT scanning, completely reliable techniques for measuring cerebral blood flow for the diagnosis of cerebral infarction have not yet been firmly established. Although such techniques are used as a supplementary diagnostic technique for determining the indication of anastomosis, etc., they are currently and primarily in the study of the pathophysiology of cerebral infarction. Due to developments in three-dimensional imaging techniques using PET-CT, it has become possible to compare simultaneously obtained images of cerebral blood flow and cerebral metabolism. Luxury perfu-sion, misery perfusion and the ischemic penumbra can be directly visualized. Such findings are thought to be useful in deciding on treatment, making clinical diagnoses and in determining the indication for anastomosis.

In the present section, a brief review of the two-dimensional measurement techniques will be carried out, various three-dimensional measurement techniques will be introduced, and finally discussion will be made of the pathophysiology of cerebral infarction as seen using techniques, primarily the PET scan, for determining cerebral blood flow and metabolism.

8.4.1 Two-dimensional Measurement Techniques

The first researchers to undertake measurements of the cerebral hemodynamics of the entire human brain were Kety and Schmidt, using the nitrous oxide (N_2O) method[19]. In contrast, rCBF measurements today are commonly carried out using the clearance method with a radio-inactive gas such as ^{85}Kr or ^{133}Xe. The Xe-clearance method in particular has been widely used, and intra-arterial infusion, inhalation[24, 31] and intravenous infusion methods have also been developed. The latter two techniques are thought to be suitable for use in cases of cerebral infarction, especially occlusive infarction.

8.4.2 Three-dimensional Measurement Techniques

The development of three-dimensional techniques has made it possible to measure and visualize the blood flow at locations deep within the brain. These techniques include: (i) dynamic CT, (ii) stable xenon-enhanced CT, (iii) single photon emission CT (SPECT) and (iv) positron-emission CT (PET). Not only does PET alone allow for measurement of

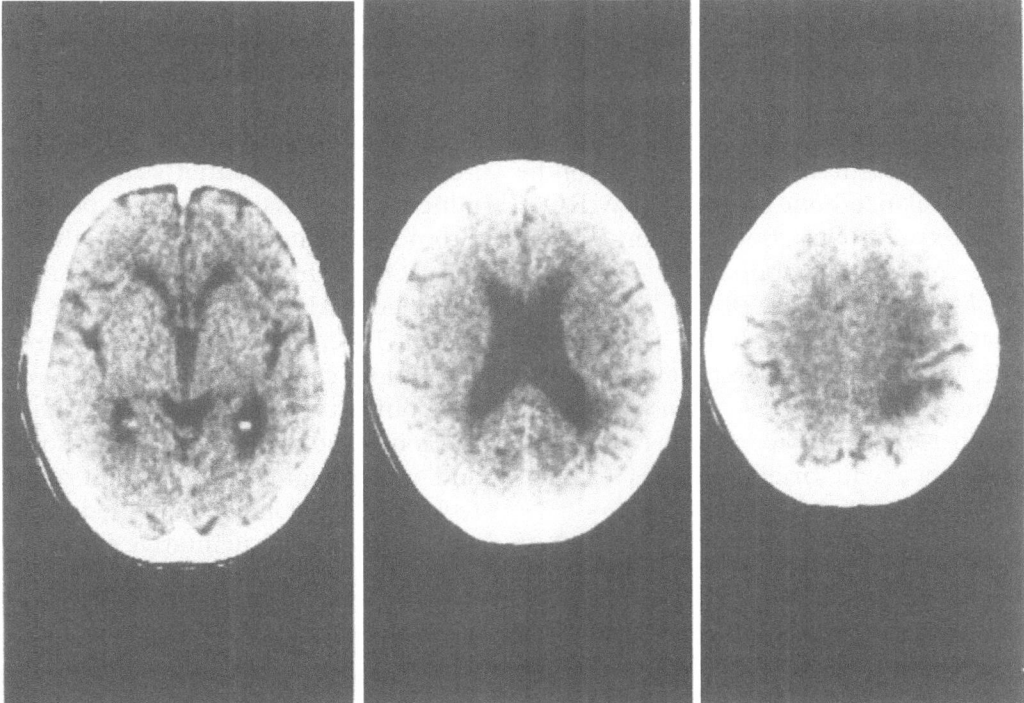

Fig. 8-17. X-CT of stenosis of the right internal carotid artery one month after the onset. Low density area is confined in the right parietal lobe (C)

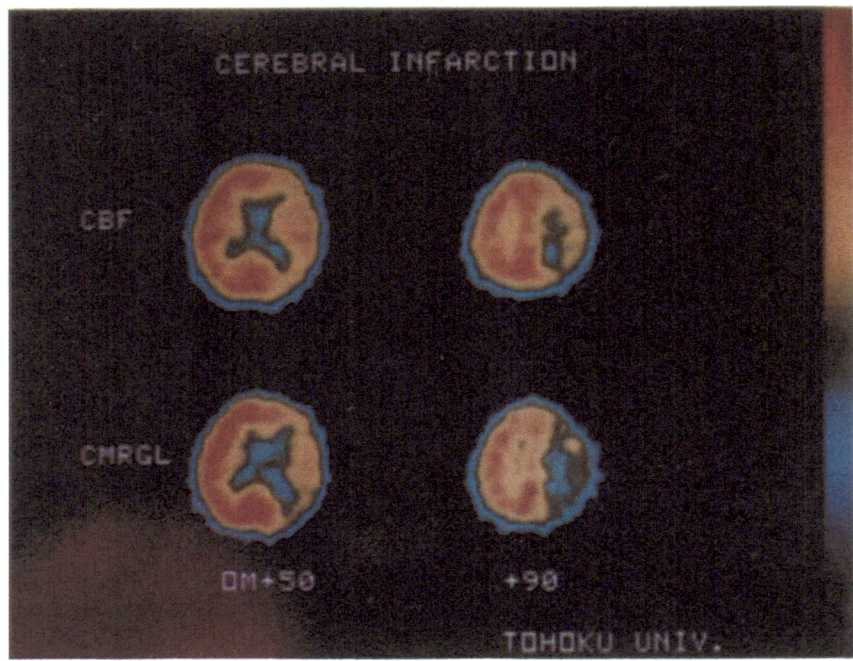

Fig. 8-18. PET image of the same case as in **Fig. 8-17.** Upper: CBF Lower: CMRGL. Left: a slice of OM + 50 mm. Right: a slice of OM 90 mm. The area of decreased CBF and CMRGL is much more extensive than the area demonstrated by the X-CT image

cerebral blood flow, but it also follows for the measurement and imaging of the cerebral metabolic rate of O_2 (CMRO$_2$), the oxygen extraction fraction (OEF), the cerebral metabolic rate of glucose (CMRGL). (**Figs. 8-17, 8-18**) and have been responsible for establishing the validity of several new concepts concerning cerebral hemodynamics, such as the luxury perfusion syndrome. Here we will briefly discuss each of these techniques.

8.4.2.1 Dynamic CT[1]

The principal advantage of the dynamic CT technique is that it uses the commonly available X-ray CT equipment and an iodine contrast medium. By means of *i.v.* infusion of 50 ml of the medium over 3–5 seconds, normally six sequential scans are obtained at the same "slice" position From the changes in the CT values of the ROI, a time-density curve can be obtained and the hemodynamics deduced. Due, however, to problems on resolution caused by recirculation on the contrast medium, extravasation on the iodine at the site of the pathology and a low signal to noise ratio, this technique has come to be valued primarily as a means of qualitative evaluation of cerebral blood flow.

8.4.2.2 The Stable Xenon-enhancec CT Technique[16, 18, 44]

The inert gas xenon (Xe), which is diffusible in brain tissue, and conventional X-ray CT equipment are usually used for stable Xenon-enhanced CT scanning. The xenon which becomes diffused within the brain brings about an enhancement effect in the X-ray CT. The changes in CT values are measured and the rCBF is calculated using the simultaneously recorded xenon concentration in arterial blood or expired gas. The advantages of this technique include the fact that only conventional CT equipment is needed and therefore the spatial resolution is good. Moreover, it is possible to make comparisons of the blood flow distribution with the morphological changes seen in the CT. The disadvantages include the high cost of xenon gas, the possibility that changes in blood flow are brought about by the anesthetic effect on xenon itself, the inability to make measurements due to the CT-image artifact, and a relatively low signal to noise ratio.

8.4.2.3 Single Photon Emission CT Scanning (SPECT)

Unlike PET scanning which requires the use of a large cyclotron, the SPECT technique is greatly facilitated by the use of γ-emitting isotopes such as ^{133}Xe and ^{123}I which are already widely available for other purposes. In comparison with PET scanning, however, the energy of the photons is only 1/3–1/5 the positron energy, leading to a reduced efficiency. Moreover, unlike the photons of positron emitters, there is no directionality of the photons in SPECT, so that rectification must be undertaken—leading to problems in quantification. Nevertheless, further improvements can be expected due to its fundamental simplicity, the availability of appropriate isotopes and developments in rectification techniques. Currently, the most-widely used techniques are that of Lassen et al.[22] using inhalation of ^{133}Xe, and that of Kuhl et al.[21] using N-isopropyl-p-[^{123}I]-iodoamphetamine (IMP) with the ^{123}I isotope.

8.4.2.4 PET Scanning[33, 34]

In comparison with the isotopes used in SPECT technique, those required for PET scanning have extremely short half-lives, and thus require synthesis in a cyclotron. For this reason, PET scanning is not as widely used, but the results obtained concerning cerebral blood flow, flow rate, metabolic rate, the distribution of receptors and other biochemical factors are truly outstanding. Although there are some limitations on the spatial resolution of the mapping of these parameters in transverse sections, PET scanning is in principle the same as the autoradiographic technique commonly used in animal experiments, and indeed has been referred to as "in vivo autoradiography". It can also be said that PET scanning has initiated a new field of study, namely, functional morphology.

The most widely used method for PET scanning employs continuous inhalation of $C^{15}O_2$ and $^{15}O_2$. Analysis of the emitted positrons allows conclusions to be drawn concerning cerebral flow (CBF), OEF and $CMRO_2$. Moreover, the cerebral blood volume (CBV) can be measured when $C^{15}O$ is inhaled.

Having briefly reviewed current techniques for measuring cerebral blood flow and metabolism, the hemo- dynamics of cerebral infarction, based primarily on PET scan findings, will now be discussed.

8.4.3 The Pathophysiology of Cerebral Infarction as Seen from Measurements of Cerebral Blood Flow and Metabolism

Using conventional two-dimensional imaging techniques, it has long been known that during the acute stage of cerebral infarction, there are decreases in cerebral blood flow and metabolism at the site of infarction and, moreover, decreases in blood flow at sites distant from the pathological focus (i.e., diaschisis)[28]. Due to developments in three-dimensional imaging techniques, the concepts of luxury perfusion[23] and misery perfusion[5] have also been established. Luxury perfusion, as first advocated by Lassen in 1966, is a state of hyperperfusion relative to the energy metabolism of the ischemic brain. Misery perfusion, as advocated by Baron et al. (1980), is the converse condition. In general, the clinical course of cases of cerebral infarction is thought to proceed typically from a stage of misery perfusion to one of luxury perfusion. Although it is often stated that every clinical case is different, the sequential PET scans in typical cases of cerebral infarction obtained by Uemura et al. are of interest.

In Uemura's case where both PET and X-ray CT scans were obtained at an early period following onset (9 hours), no signs of a low density area were found at the focus of pathology in CT scans. PET scans, however, showed an uncoupling of the hemodynamics and metabolism in the early period of ischemia—with marked decreases in CBF and $CMRO_2$, and marked decreases in CBF below the level of $CMRO_2$. Therefore, the OEF was highly elevated and a condition of misery perfusion was seen. On the 7th day from onset, low density areas were also apparent in the X-ray CT scans, but in PET scans it was seen that CBF had recovered, while $CMRO_2$ remained low and OEF was markedly decreased (i.e., luxury perfusion). One month after onset, CBF and $CMRO_2$ had decreased in parallel and OEF remained at approximately the same level as in other regions.

The contralateral mirror-image site of the infarction showed slight decreases in CBF and $CMRO_2$ from an early stage, and from around the 7th day the entire brain, including the mirror-site showed mild decreases in CBF and $CMRO_2$, while OEF showed slightly increased values.

In addition to direct damage to the brain due to ischemia, as discussed above, various secondary hemodynamic and metabolic disturbances (diaschisis)[28] located in the contralateral hemisphere or at sites distant from the pathology known since the time when the techniques of two-dimensional mapping alone were used. Due to developments in three-dimensional mapping, however, elucidation of the nature of the effects on the contralateral cerebellar hemisphere has been possible. Notable is the phenomenon of crossed cerebellar diaschisis[4], which is said to occur with a high incidence, although its clinical significance is yet uncertain.

Finally, with regard to autoregulation in cerebral infarction, it is known that various conditions may arise, including: (i) deficits in autoregulation of the entire brain, (ii) deficits confined to the infarctic hemisphere, (iii) deficits confined to the focus of infarction, and (iv) no deficits of autoregulation whatsoever. Factors such as the patient's age, the severity and site of the infarction, and the timing of measurements are all thought to have influence on the findings of autoregulation.

9A. MEDICAL TREATMENT OF CEREBRAL INFARCTION

9A.1 Introduction

In the treatment of cerebral infarction, it is of course important to distinguish between the prevention and the treatment of the underlying lesions which give rise to cerebral infarction, but in clinical practice the basis for treatment normally focuses on two problems: how to minimize cerebral edema and how to prevent progression of the lesion to a state of infarction. For these purposes, various drug therapies have been developed, but there are as yet few techniques which have been shown to be theoretically and experimentally sound and clinically effective. In practice, it is therefore necessary first to come to a sufficient understanding of the changing cerebral pathology of each patient, and then to undertake control of blood pressure, cerebral edema, etc., in order to maintain appropriate cerebral blood flow and control of the patient's systemic condition.

9A.2 Control of Blood Pressure

9A.2.1 Control of Blood Pressure in the Acute Stage

Control of blood pressure can be particularly important during the acute stage of cerebral infarction. Although it is generally the case that hypertension is found during the acute stage, since there is a loss of autoregulation of cerebral hemodynamics, imprudent use of hypotensive agents can lead to a sharp decrease in cerebral blood flow. As a rule, therefore, hypotensive drugs are not used during the acute stage of cerebral infarction.

It should be noted that, even when blood pressure is high immediately following onset, a gradual decrease is often observed thereafter. However, when the systolic pressure remains above 200 mmHg and the diastolic pressure above 120 mmHg for more than 24 hours, there is the danger of aggravation of cerebral edema, hemorrhagic infarction or the appearance of cardiovascular side effects[1, 15]. For these reasons, it is then necessary to regulate the blood pressure pharmacologically.

When administering hypotensives,

a systolic pressure between 160 and 180 mmHg and a distolic pressure below 100 mmHg are desirable, but it is advisable to avoid sharp fluctuations in blood pressure and to maintain pressure values slightly above those found prior to onset.

For reducing blood pressure in the acute stage, 250–500 mg or trimetaphan camsilate dissolved in 250 ml of physiological saline or in a 5% glucose solution should be administered by intravenous drip infusion. Intravenous or intramuscular injection of 0.5–2.0 mg of reserpine and 10–20 mg of hydralazine hydrochloride are also used. Nifedipine in a dose of 5–10 mg can be administered sublingually and nicardipine is also recommended. With all of these drugs, it is necessary to avoid sharp decreases in blood pressure.

In fact, when a hypotensive condition is encountered in an acute stage of cerebral infarction, then the need for bringing about increases in blood pressure must be considered. Even when at a normal level, the possibility of using hypertensive agents should be kept in mind, particularly when symptoms fluctuate. Indeed, among elderly patients with symptoms of stenosis of trunk arteries of the brain, an improvement in symptoms is sometimes seen following therapy with hypertensive agents. In such treatment, dopamine hydrochloride, dobutamine hydrochloride or metaraminol bitartrate is given by intravenous drip infusion and, while observing changes in heart rate and urine content blood pressure is maintained at 15–20% above the patient's normal level for 2–3 days. When no improvement in neurological symptoms is seen, therapy is halted, but a sudden decrease in blood pressure should also be avoided.

When heart failure is encountered, diuretics and digitalis should be administered and close observation of subsequent changes should be made.

9A.2.2 Control of Blood Pressure in the Chronic Stage

The primary purpose of controlling blood pressure in the chronic stage is to prevent the re-occurrence of cerebral infarction. For this reason, treatment of hypertension is most important, but there remains a problem of determining when the so-called chronic stage has in fact been entered and hypotensive therapy can begin. Hypotensive therapy should be avoided for at least one month following onset. In patients with a history of hypotension, once such therapy has begun, blood pressure should be maintained at a level somewhat higher than normal and sharp decreases should not be allowed to occur. The level at which the pressure should be maintained will vary depending upon the tension, but the diastolic pressure should be kept below 110 mmHg[4, 8].

9A.3 Treatment to Suppress Cerebral Edema

The most important drug therapy in the acute stage of cerebral infarction is the treatment of cerebral edema. Two kinds of edema in the acute stage can be distinguished—a transient edema appearing in the early stage of the disease and a more persistent edema appearing thereafter. It is thought that the principal causes of the transient edema are energy deficits, deficits in the extracellular release of Na^+, and water accumulation in cells due to changes in the hydrostatic pressure. In contrast, the subsequent edema is thought to be caused by progressive cellular damage due to ischemia, leading to increased vascular permeability and changes in colloid osmotic pressure. Most therapeutically difficult to treat is the latter form of edema, which becomes most severe between 3 and 5 days after the onset of the infarction. When cerebral edema occurs, pressure is exerted on the surrounding brain tissue—leading to a vicious cycle of secondary metabolic deficits and further disturbance of cerebral circulation. With further progression, the intracranial pressure increases and there is the danger of death due to herniation of the brain. For this reason, effective treatment of the edema is extremely important.

9A.3.1 Hypertonic Solutions

Ten percent glycerol or 20% mannitol is administered by intravenous drip infusion[20, 41]. Glycerol is more widely used than manntiol for the following reasons: (i) although it produces effects more slowly than mannitol, its actions are longer lasting, (ii) there is little rebound phenomenon, (iii) there is little stress placed on the kidneys and there is little fear of disturbing the electrolyte balance, (iv) glycerol is itself a source of energy for the brain, (v) it facilitates both cerebral circulation and metabolism[21], and (vi) because it has an antiketosis effect, it can also be used in diabetic patients.

As for side effects, glycerol can contribute to hypernatremia, acute renal failure and hemoglobinuria due to hemolysis, and there have been reports of non-ketotic hyperosmolar diabetic coma in elderly and diabetic patients.

Ten percent glycerol is usually administered by intravenous drip infusion over 1–2 hours 3–4 times a day at a dose of 200 ml or over 2–3 hours 1–3 times at a dose of 500 ml. When patients are in a poor condition, the glycerol is administered rapidly over a period of about 30 minutes.

Mannitol is known to be more effective than glycerol in reducing intracranial pressure and takes effect more rapidly, but its effectiveness is short-lived and the rebound phenomenon may occur. However, as discussed below, mannitol is known to protect the brain against the effects of ischemia[34, 35].

Twenty percent mannitol is administered by intravenous drip infusion over 30 minutes at a dose of 300–500 ml. It is given 2–4 times per day. From the second day onwards, if necessary, the drip can be administered slowly at a speed of 30 ml/hr. Since mannitol has a strong diuretic effect, care must be taken with regard to dehydration and the electrolyte balance.

9A.3.2 Corticosteroids

Corticosteroids are said to be effective against vasogenic edema, which occurs due to the leakage of serum components. There is continued controversy concerning the clinical usage of corticosteroids for cerebral infarction[9, 11, 15, 26, 27] and although there is currently a tendency towards decreased use, it has recently been re-evaluated as indeed having some protective effects on the brain[34].

Although administration of corticosteroids alone in the acute stage of cerebral infarction cannot be expected to be as effective as when administered for the edema accompanying brain tumors, it is effective when administered together with mannitol, glycerol or furosemide[17]. Among their side-effects, corticosteroids are known to lead to gastrointestinal bleeding and the aggravation of diabetes, and to facilitate the contraction of infectious diseases, so that their prolonged usage should be avoided.

Since it is thought that a relatively large dose is preferable as a first dose; 12 mg of dexamethasone or betamethasone is injected intravenously and thereafter 4 mg is injected intravenously or intramuscular every 6 hours. This regimen should be continued for 3–7 days, gradually decreased thereafter and all corticosteroid treatment should be halted by the 10th to 14th day. As will be described below, we have found intravenous drip infusion of corticosteroids together with mannitol and Vitamin E (the Sendai cocktail) to be effective, and recommend that method of administration.

9A.4 Antithrombotic Therapy

The formation of thrombi is thought to occur by, first, the adherence and aggregation of platelets to form a platelet thrombus, followed by the actions of coagulating mechanisms and the formation of fibrin thrombi. Antithrombotic therapy can therefore be directed at any of these stages—either as antiplatelet therapy[3, 5, 6, 10, 13, 14, 16, 19], anticoagulant therapy[2, 7, 11, 16, 23, 29, 40] or thrombolytic therapy[12, 16, 22, 30].

9A.4.1 Antiplatelet Agents

These drugs are not used in the acute stage of cerebral infarction, but are used in TIA patients and to prevent the recurrence of infarction in the chronic stage. In comparison with the anticoagulants, the administration and control of the antiplatelet agents is simpler and serious side-effects are uncommon. Among the drugs commonly used are aspirin[6, 10, 13, 33], dipyridamole[5], ticlopidine[14] and trapidil. Currently, the most widely used therapeutic techniques are (i) combined usage of 0.3 g of aspirin and 150 mg/kg of dipyridamole and (ii) 200–300 mg of ticlopidine per day. In either case, however, rapid drug effects cannot be expected*

9A.4.2 Anticoagulants

Heparin and warfarin are commonly used as anticoagulants. They are generally found to be effective in preventing cerebral infarction due to cardiac disorders such as atrial fibrillation[16]. However, since there is the danger of aggravating hemorrhagic infarction in the acute stage of cerebral infarction, it is generally agreed that anticoagulants are contra-indicated[15].

Fisher maintains that the use of anticoagulants in the acute stage of cerebral infarction prevents the formation of stagnation thrombus distal from the thrombotic occlusion and that it is suitable for a progressing infarction of the brainstem and pentrating branch (lacunar) syndromes[11]. Antiocoagulant therapy also reduces the incidence of or prevents embolism from the heart[11].

Treatment normally begins with intravenous injection of 5000–6000 units of heparin alone, and the coagulation time is adjusted to twice the normal level by administering 20,000–60,000 units per day. After 4–14 days of heparin treatment, heparin is replaced by warfarin[11]. At a maintenance dose of 2–10 mg per day of warfarin, the thrombin time is adjusted to 5–15 seconds and prothrombin time to 20–30%. Treatment is continued for 6–12 months and cessation of this drug treatment is undertaken gradually over a period of more than 6 weeks. There is a contraindication for cases of gastrointestinal ulcer, severe renal or hepatic disorders or severe hypertension.

9A.4.3 Thrombolytic Agents

Among the thrombolytic agents, urokinase (UK) is widely used[12, 30]. It is known to be an active plasminogen adherent to thrombi and to convert it to plasmin, which has lytic effects on the thrombi. Controversy continues, however, with regard to the applicability and effectiveness of UK in cases of cerebral infarction. Specifically, it has been virtually impossible to evaluate its effectiveness in the acute stage. Particularly in the light of the fact that no effects on the fibrinolytic system can be seen with small doses of less than 60,000 units, its effectiveness must be questioned. There is also the possibility of hemorrhagic infarction during reflow in the acute stage. As a consequence, it is suggested that thrombolytic drugs be used, not in order to allow recirculation through occluded vessels, but in order to prevent the formation of thrombi and improve cerebral microcirculation. In either case, it is thought that the indication for use of UK is limited to (i) the extremely early stage following onset and (ii) slowly progressive stroke.

It goes without saying that treatment with thrombolytic agents is not indicated in cases of hemorrhagic infarction, but neither should they be used in infarction cases where there is a fear of cerebral embolism or progression to hemorrhagic infarction.

The dosage of UK is normally 60,000–300,000 units per day and is often given together with low molecular weight dextran.

9A.5 Facilitators of Cerebral Blood Flow

9A.5.1 Cerebral Vasodilators

Cerebral vasodilators are not currently used in the acute stage of cerebral infarction for the following reasons: (i) in cases of the so-called luxury perfusion syndrome, there is the possibility of bringing about the so-called intracranial "steel" phenomenon; (ii) there is the danger of causing a rise in intracranial pressure due to an increase in the vascular bed; (iii) there is the fear of enlarging the focus of hemorrhage in cases of hemorrhagic infarction and (iv) there is the possibility of producing deficits in cerebral circulation due to a general lowering of blood pressure in patients already having systemic hypotension.

The mechanism of action of the cerebral vasodilators in cases of cerebral infarction showing the luxury perfusion syndrome is thought to be as follows. Although normal arteries become dilatated, those vessels around the lesion where autoregulation is lost do not dilatate. As a consequence, the blood flow increases in normal brain tissue, and the regions of hyperemia surrounding the lesion become ischemic (the "steal" phenomenon). Although it is sometimes noted that the luxury perfusion syndrome does not exceed 10–30% of the blood flow at the site of the cerebral infarction, since there is no easy way to determining the actual blood flow at this site in clinical cases, there is no reliable means for determining whether or not cerebral vasodilators should be used. Consequently, cerebral vasodilators are used primarily in the chronic stage of infarction.

In general, use of vasodilators is indicated in the following cases: (i) cases with mild symptoms and a small lesion or cases in which onset is extremely slow, (ii) immediately following onset in the acute stage, and (iii) 3 weeks after onset in cases where the vascular responsibility has returned to normal. It is worth noting, however, that even after 3 months a sharp decrease in blood pressure is not desirable since there are some patients still showing deficits of autoregulation in the chronic stage.

Cerebral vasodilators currently in use are listed in **Table 9 A-1**. Each of these drugs is known to have other actions, which must be fully appreciated before use.

In addition, inhalation of a 5% carbon dioxide gas is sometimes used for its cerebral vasodilatory effects.

Table 9 A-1. Classification of drugs for cerebral infarction based on their vasodilative actions

---

Relaxants for smooth muscle of vessel

    Derivatives of nicotinic acid
        inositol nicotinate
        nicotinyl tartrate
        tocopherol nicotinate
        nicametate citrate

    Drugs of papaverine like action
        papaverine hydrochloride
        cyclandelate
        cinnarizine
        bencyclane fumarate
        cinepazide maleate

    Xanthine derivarates
        pentoxifylline

    Calcium antagonists
        nicardipine hydrochloride
        flunarizine
        trapidil

    Vin alkaloids
        vinpocetine
        brovincamine

    Ergot alkaloids
        dihydroergotoxine mesitate
        lisuride hydrogen maleate

    Kinin
        kallidinogenase

    -adrenergic stimulants
        bemethan sulfate
        isoxsuprine hydrochloride
        nydidrin hydrochloride

    -adrenergic blockers
        tolazoline hydrochloride
        ifenprodil tartrate

9A.5.2 Low Molecular Weight Dextran (Dextran 40)

Low molecular weight dextran (dextran 40) is known to have a facilitating effect on microcirculation due to its capacity to reduce blood viscosity. It also prevents the aggregation of platelets and erythrocytes and is effective in cases of frequent transient ischemic attacks and progressive cerebral thrombosis. Attempts to confirm the overal beneficial effect of dextran, however have not been successful[15, 19].

Dextran is administered by intra

venous drip infusion 1–3 times per day at a dose of 500 ml. It should be used over as short a period as possible, however, and therapy should be completed within 5 days.

9A.5.3 Exsanguination

When cerebral infarction is considered to be due to an increased concentration of blood, improvements in symptoms can sometimes be obtained by diluting the blood by means of exsanguination and isovolemic infusion of isotonic saline[18].

9A.6 Activators of Cerebral Metabolism

There are a large number of drugs that are known as activators of cerebral metabolism, and their modes of action are varied. It is therefore particularly helpful in determining the indication and timing of therapy to consider which of these activators also act as cerebral vasodilators. Those that do not, as a rule, used in the acute stage of cerebral infarction. Only those activators of cerebral metabolism which do not have vasodilatatory effects are used in the acute stage following onset. Notable among drugs which can be used in the acute stage are meclofenoxate hydrochloride, citicolin, cytochrome C and ATP. These drugs are particularly widely used in cases of deficits of consciousness. As indicated in **Table 9 A-2**, most of the other activators of cerebral metabolism are used in the chronic stage of infarction. Such drugs include several which suppress the adherence and aggregation of blood platelets and which facilitate deformation of erythrocyte.

9A.7 Cerebral Protective Agents

In recent years, considerable attention has been given to drugs which have protective effects on the brain in the acute stage of cerebral infarction. In light of results in basic research, their clinical usage has begun. Drugs which are thought to have protective effects on the brain are listed in **Table 9 A-3**.

9A.7.1 Barbiturate Therapy

Since 1972, when Yatsu et al.[37] reported a remarkably good cerebral protective effect of barbiturate in animal experiments, several reports in man have been published[24, 28, 31, 32, 38]. For various reasons, however, barbiturate therapy has not been commonly used. The reasons for this are (i) the fact that a state of complete anesthesia is induced with high dosages, thus making evaluation of consciousness impossible, and (ii) strict control over respiratory

Table 9 A-2. Drugs with actions which improve cerebral metabolism and circulation

Name of drug	Dosage	Vasodilation
Meclofenoxate hydrochloride	PO 300–900 mg/day IV 250 mg × 1–3/day	–
Citicolin	IV/IM 100–500 mg × 1–2/day	–
Cytochrome C	IV 15–60 mg	–
ATP	PO 120–180 mg IV/IM/SC 10–40 mg	–
GABA	PO 3 g/day DIV 750–1000 mg	–
GABOB	PO 1.5 g/day IM/SC 50–100 mg	–
Extract from hemolysed	IM/IV 2 ml × 1–2/day	–
blood of young calves pyrithioxine	PO 300–600 mg/day	–
Hopatenate calcium	PO 1.5–3.0 g/day	–
Ifenprodil tartrate	PO 30–60 mg/day	+
Pentoxifylline	PO 300 mg/day	+
Bencyclane fumarate	PO 100–300 mg/day	+
Dihydroergotoxine mesitate	PO 3–6 mg/day	+
Cinepazide meleate	PO 600 mg/day	+
Vinpocetine	PO 15 mg/day	+
Dilazep dihydrochloride	PO 50–100 mg/day	+
Trapidil	PO 300 mg/day	+

Table 9 A-3. Brain protective substances

	Scavenge action of free radicals
1. Barbiturate	+
2. Vitamin E	+
3. Mannitol	+
4. Glycerol	+
5. Corticosteroid	+
6. DMSO	+
7. Imidazole derivarate	+
8. Naloxone	–
9. Phenytoin	–
10. Prostacycline	–

functions is needed. With regard to the effectiveness of barbiturate therapy, some questions also remain unanswered. Minematsu *et al.*[24] found such therapy to be effective in cases treated within several hours of onset, but all four of the patients treated by Rockoff *et al.*[28] died.

As for the mechanism of action through which barbiturates may protect the brain, several possibilities have been suggested, but final conclusions cannot be drawn at present. Barbiturates may work as suppressors of cerebral metabolism, as scavengers of free radicals, as inhibitors of cerebral edema, or as blockers of membrane Ca^{++} channels.

Pentobarbital sodium is administered by venous injection over 10–20 minutes in a dose of 3–5 mg/kg in the first instance, and 1–2 mg/kg every 1–2 hours thereafter. Mean arterial blood pressure is maintained at 60–90 mmHg and intracranial pressure is reduced by 15 mmHg[28].

9A.7.2 The Sendai Cocktail

The Sendai cocktail contains 20% mannitol, Vitamin E and dexamethasone in doses of 500 ml, 500 mg and 50 mg respectively. In animal research, these drugs have been shown to have protective effects on the brain by working as scavengers of free radicals and to be more effective in combating the effects of cerebral ischemia when administered together than when given individually. Still greater protective effects are found when the Sendai cocktail is administered together with perfluorochemicals or phenytoin[34]. Details concerning the actual usage of the Sendai cocktail are to be found in Chapter 6. It is worth noting that the three drugs contained in the Sendai cocktail are widely available and have been proven by lengthy clinical usage to be safe when used in human patients.

9A.7.3 Other Drugs with Protective Effects on the Brain

Recently, the imidazole derivarate, nizofenone (Y-9179), has been found to be effective in experimental studies and, while its hypotensive effects must be treated with care, it is thought likely that it will find clinical use in the near future[25, 36].

The opiate antagonist naloxone[3] and prostacycline[39] also have a possibility for clinical use, but further clinical studies are required.

9B. SURGICAL TREATMENT FOR CEREBRAL INFARCTION

9B.1 Introduction

Surgical treatment of ischemic brain disorders was first reported by Strully *et al.* in 1953[75] and already the most representative kind of neurosurgery, carotid endarterectomy, has a history of more than 30 years. Moreover, since Yaşargil[82] devised the superior temporal artery to middle cerebral artery (STA-MCA) anastomosis technique in 1967, a variety of vascular reconstruction methods have come into use. Together with developments in microsurgery, these techniques have been applied according to the sites of vascular occlusion and to recipient and donor vessels.

In the present chapter, those surgical therapeutic methods which have been used up to the present time will be critically reviewed.

9B.2 Surgical Treatment for Occlusive Diseases of the Extracranial Carotid Arteries

9B.2.1 Internal Carotid Endarterectomy (ICEA)

Stenotic changes in the cervical carotid arteries, particularly in arteriosclerotic pathology at the bifurcation of the internal carotid artery (ICA), is one of the major causes of ischemic brain disorders. It is said that, in the West, such lesions constitute 60–70% of all ischemic vascular lesions, and indeed many clinical studies on this topic have been carried out. The surgical technique used in such cases, carotid endarterectomy, already has a history of more than quarter of a century and various improvements have recently led to a mortality rate for good risk patients of less than 1–3%[23, 56]. Carotid endarterectomy is now one of the most widely used surgical methods in neurosurgery and vascular surgery.

Much controversy has surrounded the question of the surgical indication for ICEA, but the following conditions are widely acknowledged as suitable for such treatment[57, 64, 65]: (i) when the carotid artery on the side of the ischemia is stenotic to less than 1/2 its normal diameter, (ii) when the patient complains of incessant noises in the cervical region, (iii) when there is an attack of cerebral ischmia or progressing stroke followed by spontaneous dissection of the carotid artery which does not respond to drug therapy, (iv) for prevention of recurrence after a minor completed stroke caused by an embolism due to carotid plaque, and (v) when the carotid artery is thought to be the cause of amaurosis fugax.

In terms of classification according to the clinical symptoms of cerebral infarction, it is also thought that ICEA is often indicated for asymptomatic stenotic lesions or ulcerative lesions. Humphries et al.[40] reported that in a follow up study of cases of asymptomatic carotid stenosis where the carotid vessels were stenotic to less than 50% of their normal diameter, 15% later experienced attacks of cerebral schemia due to the stenosis. Moreover Durward et al.[22] reported that nearly all patients with similar lesions eventually experience transient ischemic attacks (TIA). Among operated cases, however, a follow up study by Nunn[59] showed favorable results in all 28 cases with asymptomatic carotid stenosis in which CEA was performed. Currently, it is generally acknowledged that, even among asymptomatic cases, CEA is indicated when the carotid vessel is stenotic to less than 20% of its normal size[66]. It goes without saying, however, that an experienced surgical technique approaching a mortality rate close to zero is required.

ICEA is also thought to be indicated in cases of TIA or reversible ischemic neurological deficit (RIND). According to a joint study reported by Fields and his colleagues[29], CEA is particularly effective in TIA cases. While only 38% of TIA cases undergoing CEA experienced further attacks, 54.5% of those undergoing medical therapy did so. Moreover, the appearance of other signs of cerebral infarction was noted during the follow up period in only 4% of the operated patients, but in 12.4% of the unoperated patients—further demonstrating the importance of CEA in TIA cases. It should be noted, however, that in a randomized trial comparing CEA cases with unoperated cases, placebo or aspirin, the best results were obtained in the CEA + aspirin group[27, 28]. These results thus suggest that the best method is not to rely solely on surgery, but to employ appropriate medical techniques as well.

With regard to completed stroke, the following two points are widely acknowledged. (i) In cases of major completed stroke in which there are widespread neurological deficits, improvements will not be found following CEA and the operation is therefore not indicated. (ii) In contrast, in cases of minor completed stroke, the danger of recurrence of cerebral infarction is similar to that in TIA patients and the CEA operation is thus indicated[26].

Final conclusions concerning the indication for CEA in the acute stage following onset have not yet been obtained. Whylie et al. report death due to massive hemorrhage in five of their nine CEA cases operated upon between 3 and 22 days after onset[80]. Blaisdell et al. experienced 21 deaths among 50 cases operated upon within two weeks[6] and Bauer et al. reported that 38% of their cases operated upon within two weeks died[35]. In contrast, Goldstone et al.[35] experienced no fatalities among their 26 cases of emergency CEA and Mentzer et al.[54] reported only one death among 24 such cases. Still, in Ojemann's report[60] of similar emergency CEA cases, he reported 4 deaths out of 9 cases.

As can be seen from the above, the results of acute stage CEA are varied, but the following view concerning the indication for CEA is widely held at the present time. That is, the operation is not indicated in cases of recent myocardial infarction, severe occlusive lung disease, or other lesions in which survival for more than 6 month is unlikely, in cases of total hemiparesis or complete aphasia, or in cases in which low density areas are already to be seen in CT scans[57].

Few studies dealing solely with CEA in progressing stroke are to be found in the literature, but Goldstone reported that all 18 of his cases showed dramatic recovery following CEA[34], suggesting that CEA is indicated in progressing stroke. It is, however, likely that one of the reasons for so few reports concerning progressing stroke is that confusion still remains concerning the definitions of progressing stroke, stroke-in-evolution and stroke-in-progression[33]. Some have argued that it is impossible to distinguish between progressing stroke and TIA within 24 hours of onset[55] and it is uncertain whether or not it would be possible, in a clinical setting, to apply the definition of progressing stroke suggested by Jones et al.[46], i.e., "a progressing stroke (stroke-in-evolution) is that temporal category in which there has been progression (increased severity of the neurological signs) within recent minutes". Although application of this definition is said to be valid for the aggravation of neurological deficits within 18–24 hours of onset in cases of ICA lesions and within 72 hours in vertebrobasilar artery (VBA) lesions, we and others[46, 47, 61] have experienced cases in which there was aggravation after those periods.

As is evident from the above, there is still controversy concerning the indication for CEA and these uncertainties remain not only about CEA, but, also about for the indication for surgery using other techniques in cases of ischemic cerebrovascular disorders, particularly in the acute stage after onset. The reasons for the continuing debate on these topics can be summarized as follows: (i) clinical diagnosis of TIA, RIND, progressing stroke and completed stroke can be made only retrospectively. (ii) As a consequence, it is difficult to make comparative studies of the effects of acute stage surgery versus non-surgical therapy for cases other than TIA or completed stroke. (iii) There are still no established criteria for surgical indications even when advanced physiological diagnostic techniques such as EEG, SEP, CBF, Xe-CT, NMR and PET scanning are used.

9B.2.2 External Carotid Endarterectomy (ECEA)

It is well known that the collateral pathways via the external carotid artery are extremely important in preventing the appearance of cerebral ischemic symptoms in cases of occlusive lesions of the ICA, but the role of occlusive lesions of the external carotid artery in the emergence of cerebral ischemia has only recently drawn significant attention. The principal collateral arteries in the brain from the external carotid artery (ECA) include the following[8, 14, 48]: (i) the ECA-ophthalmic artery system, (ii) the ECA-caroticotympanic artery system, (iii) the ECA-meningeal artery system, (iv) the ECA-ECA system, and (v) the ECA-vertebral artery system. For an understanding of the role of the ECA in preventing cerebral ischemia, it is important to have a proper understanding of these collateral pathways.

There is an indication for ECEA in cases where symptoms of TIA are found on the ipsilateral side to a unilateral ICA occlusion and where stenosis of the ECA is seen. There have been many reports of improvements in the symptoms of such cases following ispilateral or contralateral ECEA[13, 14, 17, 21, 37, 38]. ECEA is also indicated in cases for which extracranial-intracranial surgery, such as STA-MCA anastomosis, is to be performed and there is stenosis of the ECA[7].

9B.2.3 Stumpectomy

Following occlusion of the ICA, a short segment (several millimeters to two centimeters) of the proximal end of the ICA will remain as a vascular pouch. Due to the "turbulent flow" within such a pouch, the formation of thrombi is facilitated and in recent years the concept of a "stump syndrome", i.e., the induction of cerebral ischemia via the ECA due to such thrombi, has been advocated[4]. In cases where such a stump is present and thought to be the cause of TIA through the ECA, stumpectomy is indicated. In addition, when STA-MCA bypass surgery is performed, the possibility of induction of symptoms of cerebral ischemia by an embolus from the stump through the bypass has been point out, and requires careful study[16, 17].

9B.2.4 Tortuosity, Coiling and Kinking

There are reports that abnormal coursing of the cervical ICA may be seen in 35% of cerebral angiograms[81]. Bilateral abnormalities are frequently seen in children and they are thought likely to be congenital in nature[19, 49]. Such abnormal coursing as seen in angiograms is classified into three types: tortuosity, coiling and kinking[81]. S- or C-shaped coursing is called tortuosity, whereas coiling is when extreme an S-curve or helical coursing is observed. Kinking is less frequently seen, but is defined as a vessel showing an acute angular coursing with stenosis at this region.

Surgical treatment of such lesions includes correction of the coursing of the cervical arteries, shortening surgery together with end-to-end anastomosis, venous patch grafts. And these methods used in cases where the kinking or coiling is thought likely to be the cause of ischemic symptoms[15]. It must be said, however, that in many cases it is difficult or impossible to diagnose with certainty whether or not such lesions are the cause of cerebral ischemia. For this reason, it is essential that the patient be examined for the possibility of arteriosclerosis at the time of surgery and that the appropriate surgical steps be taken to prevent ischemia due to such arteriosclerosis.

9B.3 Surgical Therapy for Occlusive Lesions of the Extracranial Vertebral Artery

In comparison with occlusive lesions of the carotid arteries, surgical treatment of occlusive lesions of the VBA system has been carried out only reluctantly. This is because the pathophysiology of vertebrobasilar insufficiency is not well understood, because in general it is more difficult to pinpoint the clinical symptoms of such lesions, and because the incidence at which such cases progress to completed stroke is low[1, 9, 31]. Nevertheless, according to a recent report by Jones et al.[46], a large percentage of TIA cases of the VBA system progress to stroke within one year. Moreover, due to developments in neuroradiological diagnosis, such as digital subtraction angiography (DSA), it has become easier to discover lesions of this region, and various surgical approaches have been devised.

The first report of vascular reconstruction in the VBA system was the endarterectomy of vertebral artery performed by Crawford et al. in 1958[18]. The transthoracic approach used by them, however, is today little used and nearly all approaches at the VA are now made using extrathoracic approaches. Currently used surgical techniques include the following: (i) subclavian-vertebral endarterectomy[58], (ii) removal of osteophyte spur[35], (iii) ECA to distal VA anastomosis[12], (iv) VA to CCA transposition[32], (v) subclavian to CCA transposition[53], (vi) CCA to subclavian bypass graft[58], (vii) subclavian to CCA bypass graft[52], (viii) subclavian to subclavian bypass graft[30], (ix) axillo-axillary bypass graft[44]. There have also been reports of subclavian to VA bypass graft, correction of the coursing of the VA and perivascular sympathectomies of VA.

It is known that occlusive disorders of the VA manifest themselves in a variety of clinical pictures and the most appropriate therapeutic method will differ depending upon the cause of the lesion[10]. Particularly since the main cause of VBA lesions, atherosclerosis, is known to be a systemic disorder, multiple lesions are frequently found. Not infrequently, a combination of the above surgical techniques must therefore be used[73].

Indication for surgery in such lesions is thought to be similar to that for CEA in ICA cases. Due to the current developments in three-dimensional diagnostic tools, such as XeCT, SPECT, PET and NMR scanning, however, it can be expected that determination of the indication for surgery—which has until now been based solely upon clinical symptoms and cerebral angiography—will come to have a much firmer and theoretically more sound foundation.

9B.4 Intracranial Vascular Reconstruction

9B.4.1 Vascular Reconstruction of the Anterior Circulation

Following the demonstration by Jacobson et al.[45] in 1960 that anastomosis of the small vessels is possible, the first to apply such a technique in a human patient with an ischemic brain disorder was Yaşargil[82]. Since then, STA-MCA anastomosis had become the most widely performed bypass surgery.

Cases thought suitable for STA-MCA anastomosis on the basis of cerebral angiograms include the following: (i) Cases with insufficient collateral circulation to the intracranial region due to stenosis or occlusion of the cervical ICA, (ii) Cases with stenosis or occlusion of the intracranial ICA, (iii) Cases with stenosis or occlusion of the MCA, (iv) Cases in which it is necessary to occlude the ICA due to the presence of a giant aneurysm or CCF, and (v) Cases with a combination of the above lesions.

Clinical symptoms which would indicate surgery are similar to those indicative of CEA, and it is thought that there is good indication in cases of TIA and RIND. Several authors have reported that 85–95% of STA-MCA patients do not show TIA attacks over follow-up periods of 1–9 years[62, 63, 76]. Moreover, whereas reports of the natural history following minor completed strokes indicate that the incidence of new strokes during the subsequent two years is fully 67%[41], the incidence of recurrence following STA-MCA anastomosis is only 1.7%[63, 67, 69, 76].

In contrast, a randomized study of 1377 cases of TIA or minor completed stroke from 71 institutes world-wide by the EC/IC International Cooperative Study Group (headed by Barnett)[78] was done to compare the effects of aspirin treatment alone with bypass plus aspirin. As reported in 1985[79], despite the fact that the incidence of patency of the bypass surgery was 96% among the operated cases, significant differences were not found between the two groups. There is no doubt such results will have profound effects on the future of vascular reconstructive surgery.

It is noteworthy, however, that contrary results have been obtained in a recent PET study[70], which demonstrated increases in both CBF and $CMRO_2$ following bypass surgery without changes in oxygen extraction fraction (OEF) in cases with minor completed stroke. Such results suggest that the decreases in cerebral metabolic functions due to long-term circulatory insufficiency may recover following vascular reconstruction. In contrast to the above-mentioned cooperative study, in which the recurrence of TIA or stroke was mainly studied, henceforth it may become necessary to establish the indication for vascular reconstructive surgery on the basis of the possibility of improving cerebral functions for each individual patient.

Other kinds of vascular reconstruction for the anterior circulation when the STA cannot be used as the donor artery, include occipital artery (OA)-MCA anastomosis[71] and, for treatment of vascular insufficiency of the region of the anterior cerebral artery (ACA) there is ACA-ACA anastomosis[43].

9B.4.2 Vascular Reconstruction of the Posterior Circulation

Vascular reconstruction of the posterior circulation is thought to be indicated in the following cases: (i) Unilateral VA hypoplasia or stenosis or occlusion of the VA due to arteriosclerosis, when stenosis or occlusion is also seen contralaterally, (ii) Stenosis or occlusion of the basilar artery, and (iii) Stenosis or occlusion of branches of the vertebral or basilar arteries. Based upon clinical symptoms, there is good indication for surgery in cases of TIA or RIND, but unanimity of opinion has not been reached with regard to surgery in the acute stage—and the indication is thought to be similar to that for STA-MCA anastomosis.

Symptoms of occlusive disorders of the posterior circulation are known to vary depending upon the site of occlusion. Furthermore, since branches of the pial arteries go only to the surface of the cerebellum, it is necessary to choose the recipient artery according to the site of the occlusion. That is, in bypass surgery for the lower brainstem, OA-PICA anastomosis[2, 50, 77] is used when the occlusion is proximal to the posterior inferior cerebellar artery (PICA) and OA-anterior inferior cerebellar artery (AICA) anastomosis[2] is used when the site of occlusion is distal from the PICA and proximal to the AICA. When, however, the surgery is for lesions of the upper brainstem, OA-PCA anastomosis[39], STA-PCA anastomosis[39], OA-SCA anastomosis[39] and STA-SCA anastomosis[3] can be used.

9B.4.3 Vein Grafts

When a suitable donor artery cannot be obtained, or when a large bypass blood flow is desirable postoperatively, vein grafts are frequently used. The advantages of vein grafts include the following points: (i) Since any length of vessel can be obtained, any desired reconstruction can be done; and (ii) it is possible to allow for a large volume of blood to flow in the reconstructed vessels. At the same time, however, there are disadvantages: (i)

The anastomosis itself can be difficult due to large differences in the diameters of the recipient artery and donor vein, (ii) there is a danger of too large a perfusion pressure due to a large flow of blood through the graft, (iii) since there is no autoregulation in the vein graft, there is the possibility of increases in intracranial pressure and the occurrence of hemorrhagic infarction, (iv) the rate of patency of such grafts is known to be lower than in cases of arterial-arterial anastomosis.

Vein grafts are currently used in the following bypass procedures: CCA-ICA[51], STA-MCA[20, 68], STA-ACA[42], OA-MCA[69], CCA-ICA[72], CCA-MCA[68, 74], ECA-MCA[20, 24, 25], suclavian-MCA[68], innominate-MCA[68], ECA-PCA[68]. Radical artery grafts have also been used in extracranial-intracranial bypass operations[11], but this technique is not commonly used.

9C. REVASCULARIZATION IN ACUTE STAGE

9C.1 Introduction

In cases of cerebral infarction, it was once thought reasonable to remove the stenotic or occluded vessel in the acute stage following onset and to artificially construct collateral vessels—thereby allowing the recirculation of blocked blood flow. It was subsequently found, however, that, the surgical results of those cases undergoing vascular reconstruction in the acute period were in fact poor. Unfavorable results were due to the fact that, in the acute stage of cerebral infarction when vascular reconstruction permitted the flow of blood to a focus where already irreversible changes had occurred, a worsening of the cerebral edema, the occurrence of hemorrhagic infarction and further deterioration of histological damage were brought about.

It is noteworthy, however, that in a recent cooperative study of 1000 conservatively treated cases of ischemic cerebrovascular disease (CVD) (see Chapter 7 for further discussion), it was found that among 292 cases with confirmed occlusion of the internal carotid artery (ICA, 104 cases) or middle cerebral artery (MCA, 188 cases) only 17% were able to return to useful social lives within two months of onset. One fourth died and 57% had neurological deficits which prevented a return to normal life. These figures indicate that once cerebral infarction has set in and irreversible neuronal changes have occurred, all such therapeutic steps are fruitless and the recovery of neurological deficits is impossible. In light of this pessimistic situation, recent efforts have been made at several institutes to perform vascular reconstruction in the acute stage of cerebral infarction prior to the development of much of the pathology of cerebral infarction.

In the present chapter, we describe the current state of surgical techniques for vascular reconstruction in acute cerebral infarction in the internal carotid artery (ICA) system.

9C.2 Surgical Therapy in the Acute Stage of Major Stroke Cases

Ever since a report by Wylie in 1964[50], in which he noted that among 179 cases of carotid endoarterectomy there were 5 deaths among 9 cases operated upon between 2 and 21 days after onset, vascular reconstruction in the acute stage has been considered to be contraindicated. Among studies reported after the introduction of the superficial temporal artery—middle cerebral artery (STA-MCA) anastomosis technique, the report by Crowell in 1977[4] is noteworthy. He conducted a questionnaire study of 26 neurosurgeons and collected data on 12 cases from 6 institutes. He found that, among 12 patients undergoing STA-MCA anastomosis between 4 and 96 hours after onset, 8 survived and 3 of the 4 deaths were due to cerebral edema following vascular reconstruction. Only one case showed marked improvement in neurological deficits in an early stage following the surgery. He thus concluded that by-pass surgery in the acute period was contraindicated.

In a more recent study, as well, Diaz reported that, among two cases of ICA occlusion and two cases of MCA occlusion (all of which showed stable neurological deficits) STA-MCA anastomosis performed between 4 and 12 hours following onset resulted in one full recovery, and three cases with, mild, moderate or severe neurological deficits, respectively[5]. Tanaka et al. reported that among 16 major stroke cases in which vascular reconstruction was performed between 3 and 11 hours from onset, full recovery was seen in five cases, four were capable of independent home life, five returned to dependent home life and only two died. They noted that surgery had no therapeutic benefits for the severe cases with virtually no residual blood flow in the ischemic areas, but described the effectiveness of the preoperative dynamic CT scan for determining the suitability of surgical therapy. In all of these studies, the prognosis for patients with severe neurological deficits was poor[14].

Although no reviews of vascular reconstruction for ischemic CVD in patients in barbiturate coma have yet appeared, favorable results have been sporadically reported. The benefits of surgery in the acute stage under the administration of barbiturates[28] are thought to be similar to those obtained by administration of the "Sendai cocktail" and perfluorochemicals. That is, administration of these brain protective substances prolongs the time limit for vascular occlusion during reconstructive surgery and broadens the range of conditions to which acute stage surgery can be applied. As a consequence, these drugs allow for improvements in the prognosis and are thought to indicate the direction in which further developments can be expected in acute stage vascular reconstruction.

9C.3 Acute Stage Vascular Reconstruction Using the "Sendai Cocktail"—Our Series of Patients

As we have already reported in Chapter 6, we have developed new brain protective substance—the so-called "Sendai cocktail": 20% mannitol 500 ml, vitamine E 500 mg and phenytoin 500 mg/or dexamethasone 50 mg)—. Having performed considerable experimental testing of these drugs [13, 15–18, 22, 23, 27, 29–31, 33–36, 38, 39, 42–49], we have applied them in acute stage therapy of cerebral infarction [32, 37, 40, 41]. Specifically, we currently use perfluorochemicals (20% Fluosol DA 1000 ml) together with the Sendai cocktail as early after onset as possible in acute stage major stroke victims. While the progression of ischemic pathology is being suppressed in this way, we perform vascular reconstruction.

9C.3.1 Treatment and Indication

In patients with cerebral infarction, the Sendai cocktail is administered by *i.v.* drip for 60 minutes as soon after onset as possible. Perfluorochemicals are followed immediately by *i.v.* drip of the Sendai cocktail for two hours, during which time CT scans and cerebral angiography are performed. In cases for which surgery is indicated, vascular reconstruction (STA-MCA by-pass or embolectomy) is performed. During surgery, the patient is given oxygen in such a manner that PaO_2 is maintained at 200 mmHg, and the Sendai cocktail is administered at two hour intervals until recirculation is achieved. Using this technique, care must be taken to prevent dehydration, and urine and serum components must be regularly measured and appropriate supplements given (**Fig. 9C-1**).

Currently this therapeutic method is considered applicable to patients less them 70 years of age in whom the

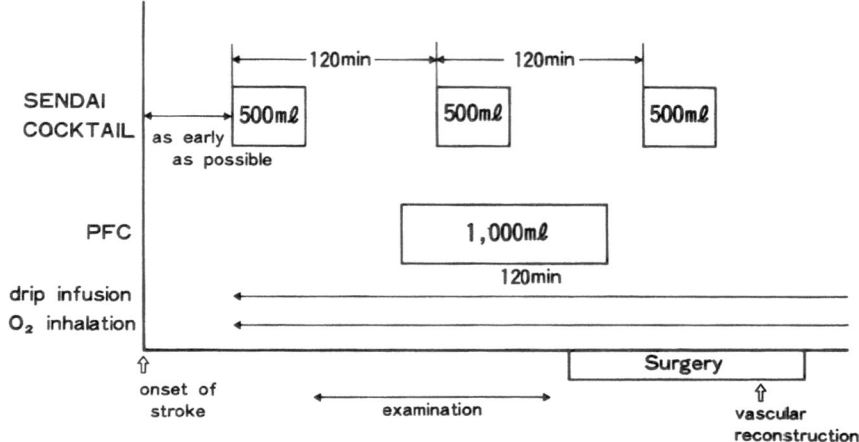

Fig. 9C-1. Programm of treatment: see the text for detail

responsible lesion has been identified by cerebral angiography, but in whom no LDAs are yet evident in CT scans, consciousness is stuporous or better (*i.e.,* better than Grade II using the 3-3-9 system)[24], and motor paresis is severe or worsening.

Among such patients, there are some whose motor paresis, improves following administration of the Sendai cocktail and perfluorochemicals. In most such cases, the vascular stenosis or occlusion is distal of the bifurcation of the MCA and the prognosis is generally good.

9C.3.2 Materials

Vascular reconstruction in the acute stage was performed using this method in 20 cases between November, 1980 and February, 1983. Ages ranged from 20 to 67 years; 17 were male and three were female. The vascular lesion was occlusion in all cases—eight of the ICA and 12 of the MCA. Vascular reconstruction was completed between 6 and 24 hours of after onset of the disease. Evaluation of motor paresis and the state of consciousness was done pre- and postoperatively and two months following surgery (**Table 9C-1**).

9C.3.3 Results

9C.3.3.1 Overall Results

Two months postoperatively, ten patients were in excellent condition (50%), three in good condition (15%), four in fair condition (20%), none in poor condition and three were dead (15%) (**Fig. 9C-2**)- These results can

Surgical treated cases- 20cases

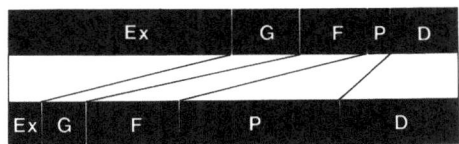

Conservatively treated cases - 292cases (Cooperative study)

Fig. 9 C-2. Prognosis of IC and MC occlusion (Nov. 1981–Feb. 1983). Upper: acute stage vascular reconstruction (20 cases), lower: conservative treatment (292 cases)

be compared with those of the 292 cases of angiographically confirmed occlusion of the ICA (104 cases) or MCA (188 cases) from a recent Japanese Cooperative Study which examined the results of 1000 conservatively treated cases. Two months from onset, those conservatively treated cases included: 21 in excellent condition (7%), 29 in good condition (10%), 60 in fair condition (21%), 105 in poor condition (36%) and 77 deaths (26%). When compared with the results for the 292 cases receiving medicinal treatment only, the results for our 20 cases were significantly better.

9C.3.3.2 Results of Cases of Thrombosis and Embolism

From a retrospective study of cerebral angiograms and findings on the spontaneous reopening of occluded vessels and embolus movement, it was

Table 9 C-1. Vascular reconstruction in

Case	Age	Sex	Occluded vessel	Pathology	Preoperative neurological condition		
						grading of paresis	
					cons-ciousness	arm	leg
1.	57 Y	M	M_1	thrombosis	30	1	1
2.	53 Y	M	M	embolism	30	1	1
3.	59 Y	F	M_1	embolism	3	2	2
4.	32 Y	F	C_3	embolism	1	2	0
5.	49 Y	F	Neck IC	embolism	100	1	2
6.	20 Y	M	Neck IC	undefined	0	0	1
7.	54 Y	M	M_2	embolism	1	0	1
8.	60 Y	M	M_1	thrombosis	1	2	4
9.	60 Y	M	M_1	thrombosis	1	3	3
10.	55 Y	M	M_2	thrombosis	3	2	3
11.	54 Y	M	Neck IC	undefined	1	1	1
12.	55 Y	M	Neck IC	thrombosis	3	2	2
13.	31 Y	M	M_1	embolism	1	3	3
14.	40 Y	M	$C_1M_1A_1$	embolism	30	2	2
15.	63 Y	M	M_1	embolism	2	1	3
16.	67 Y	M	$M_2(M_1)$	thrombosis	10	2	2
17.	47 Y	M	$C_1M_1A_1$	embolism	10	2	2
18.	68 Y	M	M_1	embolism	2	2	2
19.	45 Y	M	M_1	embolism	2	2	2
20.	64 Y	M	$C_1M_1A_1$	embolism	3	2	2

determined that there were six cases of thrombosis (one of ICA and five of MCA occlusive thrombosis), 12 cases of embolism (five of ICA and seven of MCA embolism), and two cases for which a choice between embolism and thrombosis could not be made.

a) Thrombosis. With the exception of one patient in fair condition, all had favourable outcomes without neurological deficits (**Fig. 9C-3**). The case with a fair result was one of ICA thrombosis in which therapy began 19 hours from onset and admission to our clinic and vascular reconstruction did not occur until 24 hours from onset. For the other

Surgical treated cases - 6 cases

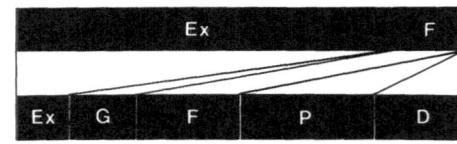

Conservatively treated cases - 74 cases (Cooperative study)

Fig. 9 C-3. Prognosis of IC and MC occlusior caused by thrombosis (Nov. 1981–Feb. 1983) Upper: acute stage vascular reconstruction (6 cases), lower: conservative treatment (74 cases)

five cases, the thrombosis was on the MCA, therapy began within six hours of onset and recirculation was achievec within 11 hours.

the acute stage of cerebral infarction

Surgery	Interval between onset and administration of drugs (hrs)	Onset to reflow (hrs)	Results	
			at discharge	at follow up
EC–IC	5	11	Ex	Ex
EC–IC	5	13	F	F
EC–IC	6	13	Ex	Ex
EC–IC	3	10	Ex	Ex
EC–IC	10	15	P	F
EC–IC	16	20	G	G
EC–IC	3	10	G	G
EC–IC	6	11	Ex	Ex
EC–IC	1	9	Ex	Ex
EC–IC	4	7	Ex	Ex
EC–IC	5	13	G	G
EC–IC	19	24	F	F
EC–IC	2.5	7	Ex	Ex
EMBOLECTOMY	2	9	F	F
EMBOLECTOMY	2	6	Ex	Ex
EC–IC	2	9	Ex	Ex
EMBOLECTOMY	3	9.5	D	—
EMBOLECTOMY	1	8	D	—
EC–IC	6	15.5	Ex	Ex
EMBOLECTOMY	1.5	7	D	—

Surgical treated cases - 7 cases

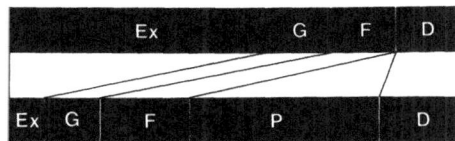

Conservatively treated cases - 90 cases (Cooperative study)

Fig. 9 C-4. Prognosis of MC occlusion caused by emoblism (Nov. 1981–Feb. 1983). Upper: acute stage vascular reconstruction (7 cases), lower: conservative treatment (90 cases)

b) MCA embolism. The results in these cases were excellent in four, good in one, fair in one and there was one death (**Fig. 9C-4**). The death was due to myocardial infarction one week post-operatively. The one fair case was a patient early in this series, and problems in the dose of protective drugs and the duration of administration are thought likely to have been responsible for the unfavourable result. Therapy began within six hours of onset and recirculation was achieved within 15 and 1/2 hours.

c) ICA embolism. The results in ICA embolism cases were excellent in one, fair in two and dead in two (**Fig. 9C-5**). The case with excellent recovery had collateral circulation via the circle of Willis. One of the fair cases was poor on discharge, but improved to a fair condition within two months; at the time of the start of therapy ten hours from

Surgical treated cases - 5 cases

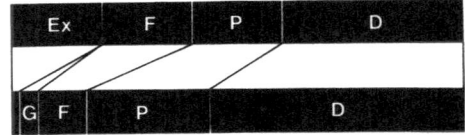

Conservatively treated cases - 55 cases (Cooperative study)

Fig. 9 C-5. Prognosis of IC occlusion caused by embolism (Nov. 1981–Feb. 1983). Upper: acute stage vascular reconstruction (5 cases), lower: conservative treatment (55 cases)

onset, consciousness was severely impaired (Grade III-100). In the remaining three cases, the terminal portion of the ICA was occluded including the ACA and MCA. Two of these cases showed postoperative deterioration, leading to death. Autopsy revealed cerebral edema in one case and hemorrhagic infarction in the other. The start of therapy and the completion of vascular reconstruction in all three cases was in fact relatively early after onset—being within three hours and within 9 an 1/2 hours, respectively.

9C.3.4 Conclusion

In comparison with previous reports, the therapeutic results obtained by us are quite good and we believe that our favorable results are due to the administration of brain protective substances during the acute stage of the disease. Those drugs, the Sendai cocktail and perfluorochemicals, suppress the appearance of cerebral infarction and ischemic changes and, moreover, prolong the time limit within which vascular reconstruction can be done. As a result, favorable prognoses have been obtained.

In light of the results of the present series, we conclude that a favorable outcome can be obtained in cases of occlusion of the MCA due either to thrombosis or to embolism if treatment is begun within six hours of onset and recirculation is achieved within 12 hours. In contrast, when there is occlusion of the ICA, the prognosis is good for cases with a limited region of occlusion and abundant collateral pathways via the PComA or the AComA, but in

embolism cases where the embolus is located between the ICA and the MCA or ACA and where there is a relatively large region of occlusion, no technique including ours will produce a favorable outcome. For this reason, surgical therapy is contraindicated in cases of ICA occlusion due to embolism in which the occlusion extends to the MCA or ACA and in which collateral pathways are meager.

With regard to the selection of the surgical technique for vascular reconstruction, it must be decided whether to choose EC-IC anastomosis or embolectomy and, if EC-IC anastomosis is chosen, whether STA-MCA anastomosis would be appropriate or a vein graft—which allows for a large increase in blood flow—would be suitable. Judging from previous clinical studies, the prognosis is poor when high flow by-pass[5] procedures are used for acute stage vascular reconstruction and such techniques have not been thought suitable during the acute period.

However, sufficient study of this problem has not been done and further work is yet needed. Despite the fact that distinguishing between thrombosis and embolism in an early period following onset is important for determining the prognosis and even for choosing the surgical method, no diagnostic technique has yet been established and previous studies have not examined this issue closely.

As mentioned above, the number of papers dealing with vascular reconstruction in the acute period for major stroke cases is yet small and further study is required. Specifically, in order to establish a therapeutic method for such patients, elucidation of the acute stage natural course of ischemic cerebrovascular disease must be done and various problems must be investigated, including the indications for surgery, the therapeutic program or surgical technique and surgical methods.

9C.4 Acute Vascular Reconstruction for Progressing Stroke

We have already discussed the issue of vascular reconstruction in the acute stage of major stroke cases which, among cerebral infarction cases, exhibit a rapid onset. The onset of cerebral infarction is, however, varied. The neurological symptoms of so-called "progressing stroke" progress gradually or stepwise and cases of crescendo TIAs show TIA which become more frequent and more severe.

The concept of progressing stroke was first presented by Millikan et al.[19, 20] and has also been referred to as stroke-in-evolution and stroke-in-progression. The incidence of progressing stroke has been reported to vary from 3% to 26–28% of that of ischemic cerebrovascular disease[2, 7, 12, 25]. It is known that the prognosis of progressing stroke is poor. In the natural course of the disease, some 64–69% of patients are left with severe neurological deficits and about 14% die[8, 21].

In marked contrast are the surgical results reported by Goldstein et al. All 18 of their patients receiving emergency carotid endarterectomy for extracranial ICA lesions had complete disappearance of neurological symptoms soon after surgery[9, 10]. Subsequent to the introduction of EC-IC by-pass techniques, several additional studies have been reported. An early study by Gratzl reported five deaths out by seven cases with neither survivor showing improvements[11], but most other studies have found such surgery to be effective in about 60% of cases. These include effective treatment in 58% of Engel's 21 patients[6], 56% among the 25 cases from the literature, as reported by Samson[26], 60% reported by Chater[3] and 60% reported by Andrews[1]. With regard to crescendo TIAs, a recent study has found favorable outcomes in all ten such cases, only one of which had mild neurological deficits remaining[5].

9C.5 The Present Series of Progressing Stroke Cases

We have performed acute vascular reconstruction in 12 cases of progressing stroke. Eleven were male and one was female. Ages ranged from 29 to 71 years (mean 55.8). The responsible lesion was an ICA in five cases (three occlusive, two stenotic) and an MCA in seven cases (four occlusive, three stenotic) (**Table 9 C-2**).

toms were found (improvement in 83%). The cause of death in the one fatality was pneumonia.

As can be seen from the above, the results of acute vascular reconstruction in our series of patients were relatively good, as compared to the reports of others, and again indicate the effectiveness of administration of the Sendai

Table 9 C-2. Acute vascular reconstruction for progressing stroke

Case	Angio-graphical findings	Preoperative most severe paresis		Speech distur-bance	Onset to reflow	Result
		arm	leg			
1. 29 F	rt. IC occl	3	4	−	4 days	Ex
2. 51 M	rt. IC sten	3	4	−	5 days	Ex
3. 50 M	lt. IC sten	2	3	+	5 days	Ex
4. 58 M	lt. IC occl	4	2	+	2 days	Ex
5. 64 M	lt. M_3 sten	4	4	+	4 days	Ex
6. 58 M	lt. M_1 occl	2	2	−	2 days	G
7. 61 M	lt. M_1 sten	4	4	+	6 days	G
8. 57 M	rt. M_{1-2} sten	1	4	−	3 days	G
9. 61 M	lt. M_1 occl	2	3	+	1 day	G
10. 51 M	lt. M_1 occl	2	2	+	2 days	F
11. 71 M	rt. M_1 sten	4	4	+	6 days	F
12. 59 M	lt. IC occl	2	3	+	8 days	D

As a rule, therapy included administration of the Sendai cocktail soon after admission and in some cases induced hypertension. CT scanning and cerebral angiography were then performed. If there was no marked improvement in the neurological deficits, EC-IC by-pass surgery was performed. The prognosis at the time of discharge from hospital was excellent in five cases, good in four, fair in two and dead in one. Except for Cases 11 and 12, improvements in neurological symp-

cocktail as a means of protecting the brain against the effects of cerebral ischemia. It must be said, however, that although favorable results were obtained for progressing stroke cases with occlusion of the ICA, cases with MCA occlusion retained neurological symptoms even when some improvement was seen. In CT scans, small foci of infarction in the vicinity of the perforating arteries leaving the MCA could be seen. This finding is thought to indicate that, even when the site of occlusion in an MCA occlusion is limited, a portion of the perforating arteries will also become occluded and this will have a deleterious effect upon the prognosis.

As seen from these cases, improvement in neurological symptoms due to acute vascular reconstruction is found in most cases of progressing stroke or crescendo TIA. Surgical therapy is thought to be indicated. Further study is still required, however, with regard to a more precise definition of surgical indication in these cases and the time limit for surgical therapy.

Part III
Appendixes

10. TEMPORARY OCCLUSION OF TRUNK ARTERIES OF THE BRAIN DURING SURGERY

10.1 Introduction

Together with increases in the use of microsurgery in recent years, there have even been reports that temporary clipping of afferent and efferent vessels during radical operations on cerebral aneurysms is not required[5, 15]. It is, however, indisputable that the incidence of rupture during peracute surgery is high and in order to obtain improved results from such early period surgery, temporary clipping is mandatory. That is, during surgery an immediately following the rupture of a cerebral aneurysm, the aneurysm may well rupture again even if blood pressure is controlled. Even when attempts to place temporary clips on a hemorrhaging vessel are then made, secondary damage to the substance of the brain or other vessels can easily be brought about in the ensuing struggle to prevent massive bleeding. If, however, temporary clips had previously been placed on the afferent artery prior to dissection of the aneurysm, then safe and certain treatment of the aneurysm neck would have been possible. Moreover, it is worth noting that not a small percentage of aneurysm cases present problems of various kinds. If, for example, the neck region is not completely dissected and neck clipping or neck ligation is performed, a daughter aneurysm present on the posterior side of the aneurysm may go unnoticed, kinking of afferent or efferent arteries may be brought about, or a branch of the artery leaving the back of the aneurysm may be overlooked and clipping might be done on the neck and artery together. For this reason, we believe that, prior to the start of aneurysm dissection, it is better to bring the internal pressure of the aneurysm to zero using a temporary clip, to collapse the aneurysm, and then to perform a rapid and certain treatment of the neck region.

The two prinicpal disadvantages of temporary clipping, as often mentioned in the literature, are: (i) the production of ischemia peripheral from the site of vascular occlusion, and (ii) the incurring of damage to the arterial wall due to the clip itself. Indeed, when using temporary clips, these two points must always be kept in mind. In the present chapter, problems concerning these two points and measures to prevent them will be discussed.

10.2 Decrease in Cerebral Blood Flow Due to Temporary Clipping and Countermeasures—the Development of Methods to Prolong the Safe Time Limit for Vascular Occlusion

No clinical studies have been reported that dealing specifically with the degree of reduction of blood flow peripheral to occlusion of a cerebral artery of with the degree of symptoms produced by such an occlusion. The absence of such research is due primarily to the fact that symptoms are varied because of large differences in the degree of collateral circulation caused by differences in the structure of the circle of Willis, the patient's age, the site and duration of occlusion. Needless to say, since various symptoms are known to arise after occlusion of cerebral arteries, some type of countermeasure must be taken. The first such countermeasure to be used widely in the clinic was hypothermia.

Lougheed et al.[4] were the first to make use of hypothermia in the field of neurosurgery. During operations in which four vessels (the bilateral carotid arteries and the bilateral vertebral arteries) were exposed and temporary occlusion was required, hypothermia was used to prolong the permissible time of occlusion. Since the metabolic rate of the brain is reduced to 50% when the brain temperature is reduced to 30 °C and to 30% at 25 °C, the ischemic brain is able to withstand prolonged periods of hypothermia. Subsequently, Botterell and Lougheed[2] reported that these four vessels can be occluded for up to 6 minutes when hypothermia to 30 °C is employed and for up to 8 minutes when 28 °C hypo-

thermia is used in both young and middle-aged patients. Moreover, Pool[7] found that bilateral occlusion of the anterior cerebral artery (ACA) for 20 minutes is safe when 28 °C hypothermia is used.

Subsequently, together with improvements in the surgical technique, it became known that somewhat longer periods of occlusion could be safely performed. Suzuki et al.[9] reported that, among 215 cerebral aneurysm cases operated upon at 22–30 °C (the majority having been operated upon at moderate hypothermia of 25–28 °C) and in which temporary vascular occlusion was undertaken, sequelae were not present in 177 cases (82.3%), were present in 16 cases (7.4%) and were fatal in 22 cases (10.2%). Among the fatalities and cases with sequelae, the vascular occlusion was thought to be the cause in 2 cases and the possibility of the same was noted in 6 other cases. The remaining 30 cases were fatalities or showed postoperative sequelae because of vasospasm, infection or other complications unrelated to the occlusion.

Study of the duration of vascular occlusion according to the site of the occlusion among cases without sequelae showed the following maximum occlusion times. Bilateral occlusion of the A_1 portion at 26 °C was done for 48.5 minutes and at 28 °C for 42 minutes. Occlusion of the A_1 portion on the dominant side only was done for 82 minutes at 26 °C and for 63 minutes at

28 °C. Unilateral occlusion of the M_1 portion at 26 °C was done for 30 minutes and at 30 °C for 40 minutes.

As is evident from the above data, it was found that hypothermia is a safe and certain means for prolonging the permissible time for occlusion of cerebral blood flow, and this technique has been effectively used for temporary occlusion—primarily in radical surgery on cerebral aneurysms. Subsequently, hypothermia has been used in cases of head injury and cerebral infarction[5], but the results in these latter conditions have not always been favorable. Moreover, even in the application of hypothermia in radical operations on cerebral aneurysms, the incidence of associated disorders due to the hypothermia, such as cardiac arrhythmia and ventricular fibrillation caused by myocardial damage, were found to increase together with improvements in the intracranial surgical results. Because the cooling and heating procedures involved in hypothermia require considerable time and because of related procedural problems, this technique has come to be used less frequently. Although the problem of associated complications remains, it is nonetheless true that hypothermia is still acknowledged as a reliable means for prolonging vascular occlusion time. The permissible duration for such occlusion in the case of 4-vessel occlusion is only 4 minutes at 38 °C, but this is doubled at 30 °C, quadrupled at 22 °C and is greater than 30 minutes at 16 °C[5, 11].

Just at the time when Suzuki *et al.* (1969) became aware of the limitations of the hypothermic method, they performed surgery on a cerebral aneurysm patient under normothermia who had associated cardiac disease and was consequently unsuitable for the hypothermic procedure. Unfortunately, the aneurysm ruptured during surgery and a lengthy period of vascular occlusion was unavoidable. Severe neurological deficits were of course thought likely to appear, but the patient showed no abnormalities and was discharged in an excellent condition. Surmising that this unexpected result might have been due to the administration of mannitol prior to vascular occlusion, they subsequently tested and confirmed the effectiveness of mannitol in protecting the brain in a series of animal experiments[6, 13, 16] and now perform radical aneurysm surgery at normothermia and normotension under the preadministration of manntiol.

Suzuki and Yoshimoto[12] have since reported a series of 384 cases of cerebral aneurysm which were operated on at normothermia and normotension and under the intravenous drip administration of 500–1000 ml of 20% mannitol over a period of 30–60 minutes prior to surgery. Temporary vascular occlusion was performed in all of these 384 cases; no sequelae were found in 330 cases (85.9%), sequelae were found in 41 cases (10.7%) and 13 were fatalities (3.4%). Among the 54 cases with unfavorable outcomes, only 4 (1.0%) were cases in which the possibility of deleterious effects of the vascular occlusion could not be denied. It was concluded that the administration of mannitol alone at normothermia and normotension is effective in prolonging the permissible occlusion time.

Among these patients, there were 60 cases undergoing bilateral occlusion of the A$_1$ portion, 52 cases undergoing internal carotid artery (ICA) occlusion, and 94 cases undergoing middle cerebral artery (MCA) occlusion. The incidence of death and sequelae in these three groups of patients is shown in **Figs. 10-1, 2, and 3** for, respectively,

were undertaken to clarify the mechanisms by which mannitol makes possible the prolongation of vascular occlusion and it has been demonstrated that it works as a scavenger of the free radicals produced in the ischemic brain. Experimental testing of other drugs which are known to have a scavenging effect similar to that of mannitol was then undertaken. Brain protective effects were demonstrated with vitamin E and steroids and a surprisingly strong

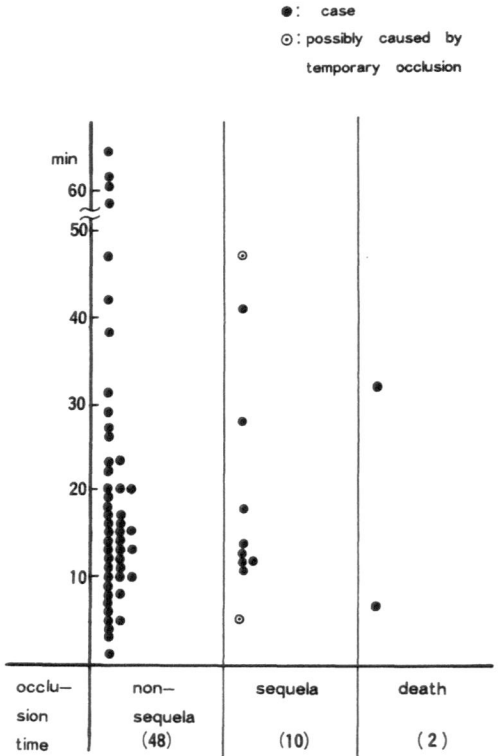

Fig. 10-1. Occlusion time of bilateral A$_1$ portion as related to sequelae or death in the cases with preoperative administration of mannitol

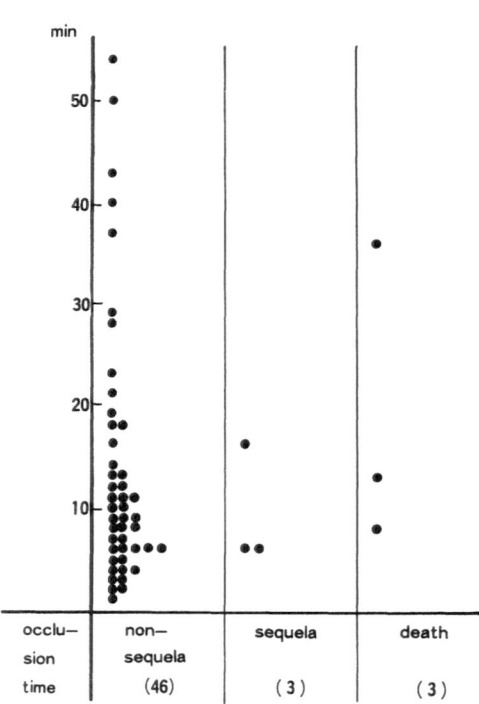

Fig. 10-2. Occlusion time of intracranial portion of internal carotid artery as related to sequelae or death in the cases with preoperative administration of mannitol

the A$_1$, ICA and MCA cases. These results indicate that vascular occlusion for up to 40 minutes is possible for any of the trunk arteries within the cranium.

Subsequently, animal experiments

effect was attained with the combined administration of all three drugs[14]. Since June of 1982, in place of the administration of mannitol alone prior to temporary vascular occlusion, com-

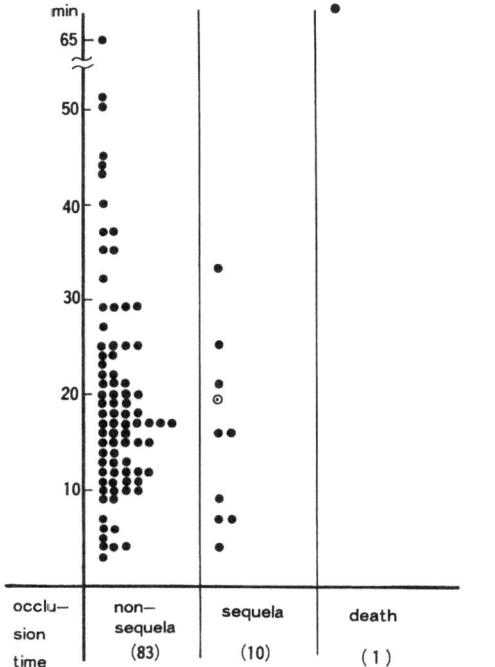

Fig. 10-3. Occlusion time of middle cerebral artery as related to sequelae or death in the cases with preoperative administration of mannitol

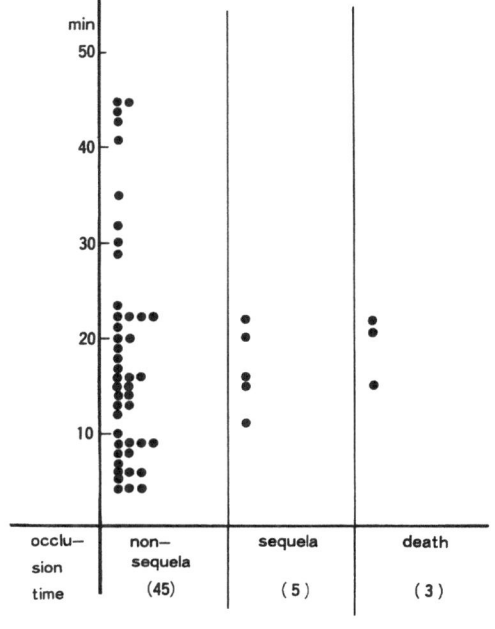

Fig. 10-4. Occlusion time of bilateral A_1 portion and sequelae in the cases with preoperative administration of Sendai cocktail

bined administration of 500 ml mannitol, 500 mg vitamin E and 50 mg dexamethasone has been used. This combination of drugs is now known as the Sendai cocktail.

Among the 148 cases operated on under the pre-administration of the Sendai cocktail until October of 1984, no sequelae were found in 122 cases (82.4%), 17 had sequelae (11,5%) and 9 have died (6,1%). Among the deaths and sequelae cases, however only 4 (2.7%) were thought to be neurological deficits attributable to the effects of vascular occlusion. **Figs. 10-4, 5,** and **6** illustrate the relationship between sequelae or death and the duration of the vascular occlusion among, respectively,

the 53 cases of bilateral A_1 occlusion, the 47 cases of ICA occlusion and the 41 cases of MCA occlusion.

It is worth noting that due to the recent increase in neurosurgical units in Northeastern Japan, there has been a tendency for our clinic to admit and treat only the more difficult cases. Simultaneously, there has also been a tendency to prolong the duration of occlusion during surgery. Consequently, among the earlier group administered only mannitol, only 69 of the 206 cases undergoing bilateral A_1 occlusion, ICA occlusion or MCA occlusion had the vascular occlusion for more than 20 minutes (33.5%). In contrast, among the more recent group of 141 cases

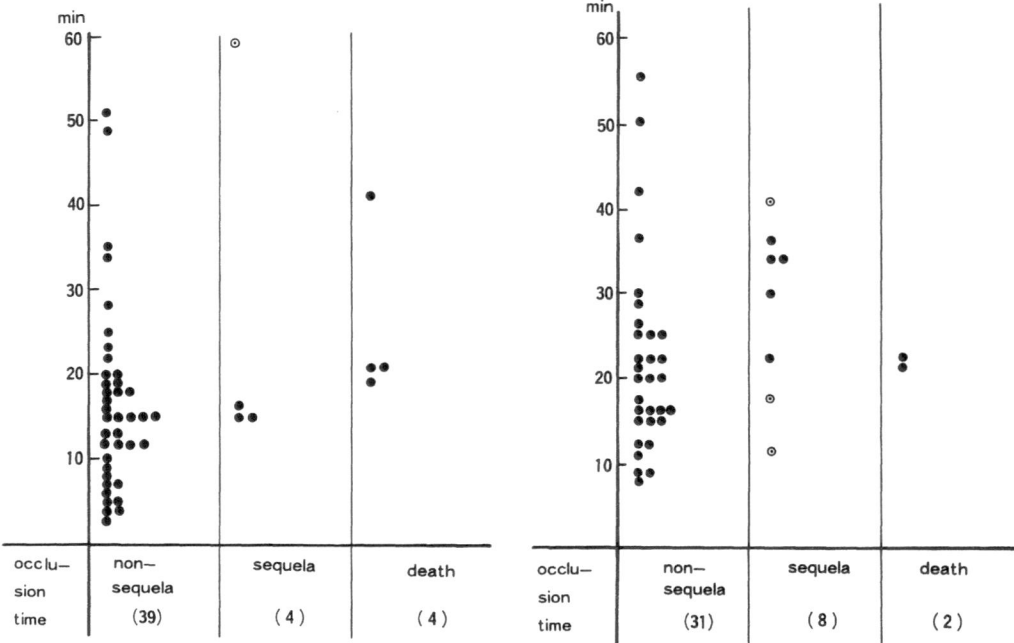

Fig. 10-5. Occlusion time of internal carotid artery and sequelae in the cases with preoperative administration of Sendai cocktail

Fig. 10-6. Occlusion time of middle cerebral artery and sequelae in the cases with preoperative administration of Sendai cocktail

administered the Sendai cocktail, a somewhat larger percentage (41.8% or 59 cases) of similar patients had vascular occlusion for more than 20 minutes. Since few patients undergo vascular occlusion for a period approaching the allowable limit, significant differences in the outcome of the cases administered manntiol alone or the Sendai cocktail have not been found, but it is nonetheless our impression that the Sendai cocktail is somewhat more effecitve in protecting the ischemic brain from permanent neurological deficits.

Both in the mannitol—only group and in the Sendai cocktail group, among the patients whose conditions became aggravated subsequent to tem-

porary clipping, the number of cases in which the duration of the occlusion is thought to have been a problem has decreased further and the majority with sequelae had occlusion of less than 20 minutes. The cause of the sequelae in these cases is in fact thought likely to have been endothelial damage to the vessels caused by the clip, rather than the duration of the occlusion.

With regard to actual administration of the Sendai cocktail, the following points should be noted. Since the period during which manntiol has been found to be effective in protecting brain tissue is approximately two hours, once 100 minutes have elapsed from the time of administration of the Sendai cocktail, readministration is necessary

During this 100 minute period, 40 minute occlusion of any cerebral vessel is safe. If still longer occlusion is required, 5 minutes of reflow is allowed, followed by up to another 40 minutes of occlusion—provided that it is within 100 minutes of administration of the Sendai cocktail. It should also be noted that, although steroids are effective in prolonging the permissible time of occlusion, they can also produce side-effects and must be used with caution and in smaller doses in patients with diabetes mellitus or in aged patients. As mentioned in other chapters, phenitoin is used as a new combination of the "Sendai cocktail" instead of steroids recently. Needless to say, even under the administration of the Sendai cocktail, every effort should be made during surgery to minimize of the vascular occlusion.

10.3 Damage to the Vascular Wall Due to the Temporary Clip

Application of a temporary clip can result in mechanical damage to the vascular endothel. At the site of such damage, a thrombus can easily form and subsequently be released as an atheroma plaque—ultimately causing occlusion of a peripheral vessel.

Experimentally, Dujouny *et al.*[3] have placed various kinds of temporary clips of the MCA of the dog for 45 minutes and measured the changes 20 minutes after release. They found laceration of the endothel and accumulation of fibrin, thrombocytes and erythrocytes at the site of laceration. The cause of such changes was due to the degree of pressure applied on the vessel by the clip end. Sugita *et al.*[8] have also studied the changes in vascular walls and vascular occlusion using a variety of clips. They maintain that blade tips with a pressure of less than 80 grams should be used for the temporary clipping of the parent arteries. When clips with a scissor-like action or in which the blade tip has an extremely high pressure are used, damage to the vascular wall can extend beyond the

endothel to the medial layer and, although rare, the site of such damage can give rise to a cerebral aneurysm[1].

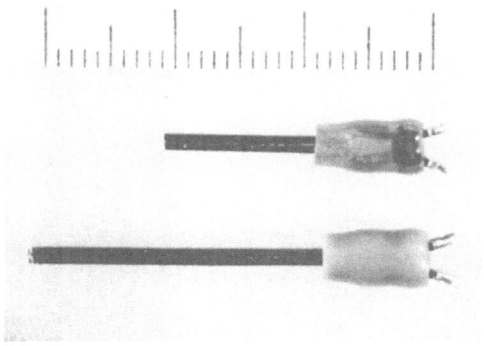

Fig. 10-7. Jiro's clip for the temporary occlusion of cerebral arteries

In light of these findings, Suzuki *et al.*[10] have developed an aneurysm clip suitable for neck clipping with a blade pressure of 100 grams and one suitable for temporary vessel clipping with a blade tip pressure of 40–50 grams (**Fig. 10-7**). It is important to note, however, that it is not enough merely to know the characteristics of the clip to be used, but also sufficient care must be taken with regard to the vessel to be clipped. Particularly at regions of severe arteriosclerosis, an atheroma plaque can be released by application of a temporary clip or, when the clip must nonetheless be applied to such a region, the sclerosis of the vessel wall can prevent complete occlusion of the vascular lumen and unexpected problems during aneurysm treatment can arise. As a rule, it is wisest to choose a vessel site for temporary clipping where arteriosclerosis is not evident.

11. THE PATHOLOGY OF CEREBRAL VASOSPASMS AND ITS TREATMENT

11.1 Introduction

It has long been known that autopsy studies will often reveal ischemic lesions in cases of death due to subarachnoid hemorrhage following the rupture of a cerebral aneurysm[59, 77]. Ischemic lesions suggest the possibility of stenosis due to vasospasm, but not until Ecker and Riemenschneider[13] demonstrated the reality of cerebral vasospasms in angiograms in 1951 did serious research on the genesis, pathology and treatment of cerebral vasospasms begin.

Since 1961, we have experienced more than 2000 cases of radical surgery on cerebral aneurysms. Study and review of the first 1000 cases experienced before 1976 revealed certain findings concerning the nature of cerebral vasospasm, which we have published previously[67]. Based upon those findings, we have undertaken basic and clinical research to further elucidate the mechanisms involved and have devised improved methods for the prevention and treatment of such vasospasm. In the present chapter, these techniques will be discussed.

11.2 The Pathology of Cerebral Vasospasm

According to several reports in the neurosurgical literature, the incidence of cerebral vasospasm due to subarachnoid hemorrhage varies considerably from 21 to 66% and varies also with regard to severity[43]. Although factors such as the timing of angiography following the onset of the spasms are also involved, a particularly important factor which influences the apparent incidence of cerebral vasospasm is the frequency of other lesions which are found to have similar morphology in angiograms[43].

In our own study, we have carefully examined factors such as the stenosis due to arteriosclerosis, as distinct from the stenosis due to vasospasm, compression due to hematoma formation, vascular narrowing caused by increases in intracranial pressure, and vascular change caused by congenital vascular hypoplasia. Without reference to the interval from the time of subarachnoid hemorrhage, among our cases fully 23.7% showed vasospasm[43]. It is noteworthy, however, that in angiograms vasospasm was extremely rare within three days of the first attack, more common following the fourth day and reached a peak incidence between 10 and 17 days from onset. Thereafter, a gradual decline in incidence was seen and vasospasm was extremely rare after 6 weeks. Fully 49.1% of the cases showed vasospasm between 10 and 17 days. It is therefore concluded that at least one half of all cases of a single attack of subarachnoid hemorrhage will experience vasospasm (**Fig. 11-1**).

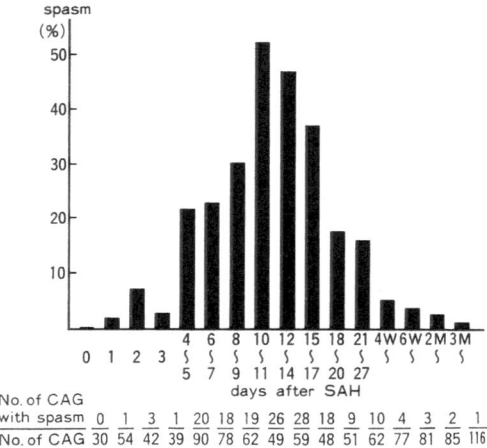

No. of CAG
with spasm 0 1 3 1 20 18 19 26 28 18 9 10 4 3 2 1
No. of CAG 30 54 42 39 90 78 62 49 59 48 51 62 77 81 85 116

Fig. 11-1. The incidence of vasospasm per number of days from subarachnoid hemorrhage (SAH) until carotid angiography (CAG) based on 1,023 carotid angiograms in 530 patients who experienced only one subarachnoid hemorrhage

It is not, however, the case that all cases with angiographical indication of cerebral vasospasm will show symptoms of cerebral ischemia. Various factors such as the abundance of collateral pathways, are thought to be involved in the development of such symptoms, but, in general, it is well-known that the severer the narrowing[20, 47] or the more diffuse the narrowing[48], the greater the likelihood that ischemic symptoms will develop.

In our recent statistical analysis, we found that among acute stage cases of aneurysm rupture, 32% showed ischemic symptoms—either transient symptoms or persistent neurological sequelae. About one half of these patients had irreversible neurological deficits. We therefore conclude that the incidence of permanent deficits is considerably lower than the incidence of cerebral vasospasm, as seen angiographically.

The development of an ischemic focus due to cerebral vasospasm differs according to the site of the ruptured aneurysm. In cases of ruptured aneurysms of the middle cerebral artery (MCA), the vasospasm is frequently within the region normally fed by the parent artery[22, 49], whereas in the case of internal carotid artery (ICA) or anterion communicating artery (AComA) aneurysms, it is often said that bilateral vasospasms are observed[12]. In our analysis, however, a consistent pattern was not found.

Among our 1000 aneurysm cases, death clearly due to vasospasm was found in six cases among operated cases and in two unoperated cases. Autopsy study of these eight cases revealed that the development of the ischemic focus did not coincide with the region of distribution of the parent artery[35] Angiographical study of the site of the vasospasm revealed similar results[47] There were few cases in which the vasospasm was restricted to the parent artery, and indeed there were many cases with vasospasm of the three main branches of the circle of Willis (the ACA, MCA and ICA).

In the recent years, since it has become possible using CT scans to identify accurately the severity and extent of the hematoma following subarachnoid hemorrhage, it has been found that there is a close relationship between the onset and location of cerebral vasospasm, on the one hand, and the severity and spread of the hematoma as seen in CT scans on the

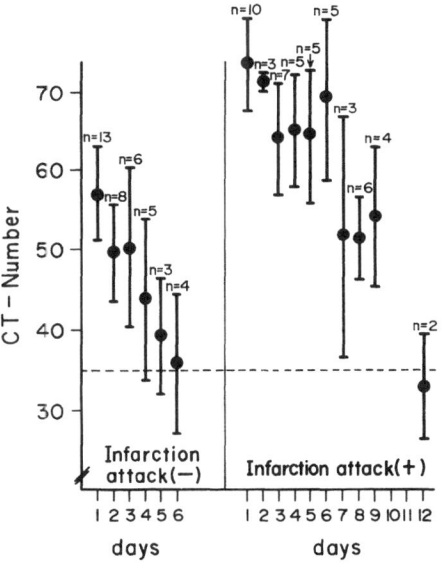

Fig. 11-2. Correlation between change in the highest density area and the occurrence of cerebral vasospasm. (◆: Hounsfield number of normal thalamus)

other[19, 40]. Specifically, among cases showing ischemic symptoms due to cerebral vasospasm, CT scans taken within four days of onset revealed high density areas ($>60 \mathrm{HU}$) in the subarachnoid space, whereas among cases without such symptoms, there were significantly lower CT values (**Fig. 11-2**). It was also found that the site of the

ischemic focus was often at regions normally fed by arteries where the highest CT values were found in the subarachnoid space.

Subsequent to these clinical findings, we have been able to predict with satisfactory precision the onset of cerebral vasospasm leading to ischemic symptoms solely from the severity of the high density regions seen in CT scans.

It has been found that, among ischemic symptoms due to cerebral vasospasm, the incidence of focal deficits, such as motor paresis, sensory deficits, aphasia, disturbances of consciousness or mental symptoms, is relatively high. Convulsions and electrolyte disturbances are sometimes found, but the incidence of headache is low[59]—a fact which is useful for distinguishing between ischemia and rerupture during the subsequent course of the disease.

Ischemic symptoms are found primarily between the 4th and the 15th days following the previous subarachnoid hemorrhage, regardless of the size of the hemorrhage (**Fig. 11-3**). Very few cases of symptoms before the 4th day or after the 15th day are experienced[39]. Although cerebral vasospasm is a completely different phenomenon from rerupture (which can occur anytime after the original subarachnoid hemorrhage), both phenomena produce similar symptoms and can be misdiagnosed because of this.

It is however, known that there are differences with regard to the timing of rerupture or vasospasm depending upon the severity of the attack. That is, in major attacks where consciousness is

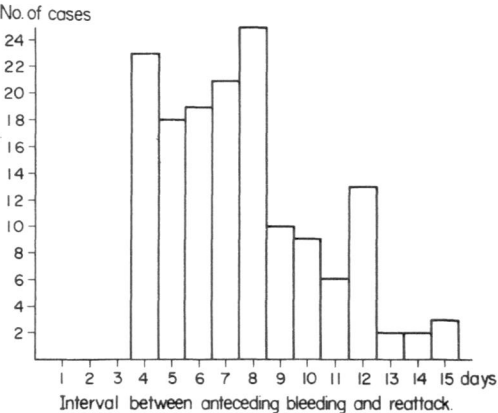

Fig. 11-3. Infarctic symptoms due to vasospasm occurred between the fourth and 15th day after the last subarachnoid hemorrhage

lost for more than one hour, rerupture rarely occurs during the 4th to 15th days from onset when the ischemic symptoms due to vasospasm are commonly observed, but in minor attacks where consciousness is not lost, rerupture can occur any time regardless of the interval from the attack. Even between the 4th and 14th days, there is a tendency for attacks to be more numerous than vasospasm. In patients with moderate attacks where consciousness is lost for less than one hour, both vasospasm and rerupture are likely to occur between the 4th and 14th days[39] (Fig. 11-4).

Cerebral vasospasms are thought to be due essentially to the continuous contraction of vascular smooth muscle. In fact, from the 18th day following a hemorrhage attack, vasospasm are seen to occur in angiograms less and less frequently. It must, however, also be said that the continous contraction of vascular smooth muscle over a period of several hours to several days is difficult

to explain in terms of usual physiological mechanisms. As a consequence, the suspicion arises that the vasospasm seen in clinical cases may be partly due to narrowing caused by organic changes in the vascular walls. This view is supported by the fact that in autopsy studies of clinical cases, organic changes in the vascular walls have been found. Among such organic changes, the following have been noted: deciduation and proliferation of the vascular endothelium, adherence of blood-born cells, formation of thrombi within the walls, and myonecrosis of the tunica media[1, 10, 18, 32, 63, 75].

Experimentally, we have confirmed these findings of pathological changes due to vasospasm[41, 50]. Moreover, from sequential observations of such pathology, we have found that changes occur in the endothelium and there is the adherence of blood-born cells several hours after induction of vasospasm, followed by myonecrosis of the tunica media[41, 50]. It is not yet clear to what degree such organic changes are involved in the pathology of cerebral vasospasm, but the fact that organic changes are found already within several hours of the start of the spasms suggests that early steps to prevent vasospasm from developing in clinical cases is extremely important.

In the preceding paragraphs we have discussed only cerebral vasospasms brought about by the rupture of cerebral aneurysms, but it is well known that spasm can occur due to other causes, including trauma[82] and inflammatory lesions. Unfortunately few studies on the nature of vasospasm in such conditions have been reported

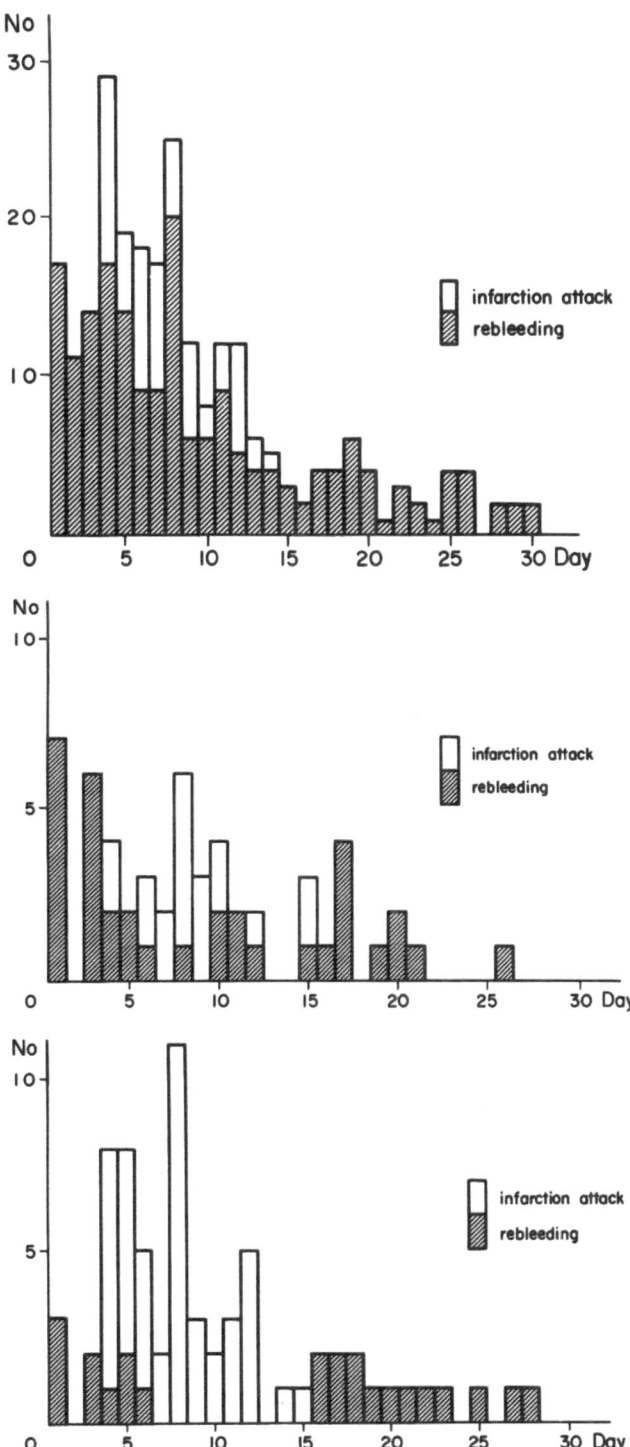

Fig. 11-4. The second attack after subarachnoid hemorrhage. In a minor attack (upper), rebleeding is more likely to occur than vasospasm. In a moderate attack (middle), both rebleeding and vasospasm occur in almost the same proportion between the fourth and 15th day. In a major attack (lower), rebleeding seldom occurs but vasospasm frequently occurs between the fourth and 15th day

and details concerning the similarity or dissimilarity with vasospasms occurring due to aneurysm rupture are not available.

11.3 Mechanisms of Onset and Causal Factors of Vasospasm

Despite numerous experimental and clinical studies of cerebral vasospasm, many unanswered questions remain concerning its genesis and the responsible biochemical substances (Table 11-1). Recently, there have are thought to bring about changes in the normal responsiveness of the vascular smooth muscles or to work only under unusual conditions. Among these conditions are: changes in the intra-hematoma pH[32] or potassium

Table 11-1. Substances which have been reported to constrict cerebral arteries. (From White RP)

1) Proteinaceous agents or peptides	3) Prostanoids
angiotensin II	arachidonic acid
bradykinin	PGA_1, A_2
FDP	PGB_1, B_2
hemoglobin	PGD_2
platelet factor	PGE_1, E_2
RBC ghosts	PGF_1, F_2
thrombin	PGH_2
vasopressin	PGI_2
	6-keto-PGF_1
2) Amines	TxA_2, B_2
	15-HPAA and other
acetylcholine	lipid peroxides
dopamine	
epinephrine	4) Inorganic agents
norepinephrine	Ca^{++}
isoproterenol	K^+
histamine	Mg^{++} deficiency
melatonine	$FeCl_2$
serotonin	$BaCl_2$
kynurenine	hydrogen peroxide
tryptamine	

been several studies on the substances contained within the hematoma. Some unknown (or known) substances which are produced within the hematoma are thought to act directly on blood vessels. Among the known substances are: the prostaglandins[5, 15, 27, 37, 53], hemoglobin[28, 34, 50, 64, 76, 77, 79, 84] and serotonin[3, 23, 31, 36, 58]. Other substances concentration[85]. Several studies on the mechanical factors involved in vasospasms have also been carried out[4, 55].

We have viewed the role of free oxyhemoglobin present around blood vessels as an important causal substance and have also viewed the distribution of catecholamines and sympathetic nerves to the vascular wall as

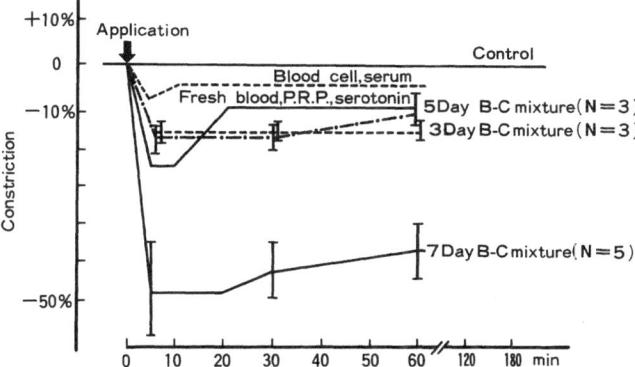

Fig. 11-5. Vasoconstriction of the basilar artery in cats. *PRP*: Platelet-rich plasma. B–C mixture: Supernatents of blood and cerebrospinal fluid mixture

important in the genesis of cerebral vasospasms. Among several experimental studies relevant to this topic[64], we have found that, using the **in situ** trunk arteries of the cat, the contact and adherence of fresh autologous blood produces transient vascular constriction, but the constriction is not severe. In contrast, application of blood mixed with cerebrospinal fluid which has been incubated for seven days at 37 °C produces continuous and severe constriction of the basilar artery. If incubated for 15 days, however, this phenomenon is not observed[64] (**Fig. 11-5**).

Biochemical analysis of the components of the blood after incubation of 7 or 15 days has revealed some interesting results. The substance found to have powerful constrictive actions after 7 days was oxyhemoglobin, but by 15 days it had been converted to methemoglobin. Application of oxyhemoglobin alone to the feline basilar artery produced continuous and powerful vasoconstriction, but application of a similar dose of methemoglobin produced only mild contraction, thus con-

firming that the active substance is oxyhemoglobin[64] (**Fig. 11-6**).

From these experimental findings, we have concluded that oxyhemoglobin is indeed an important causal factor in the genesis of vasospasm, but it is yet uncertain how it brings about such

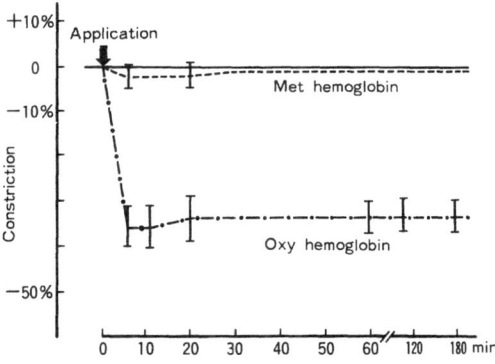

Fig. 11-6. Vasoconstriction of the basilar artery in cats. See the remarkable difference between the contractile activity of oxyhemoglobin and methemoglobin

effects. Since it is said that active oxygens are produced in the process of autoxidation of oxyhemoglobin[7, 44], we have investigated the role of such active

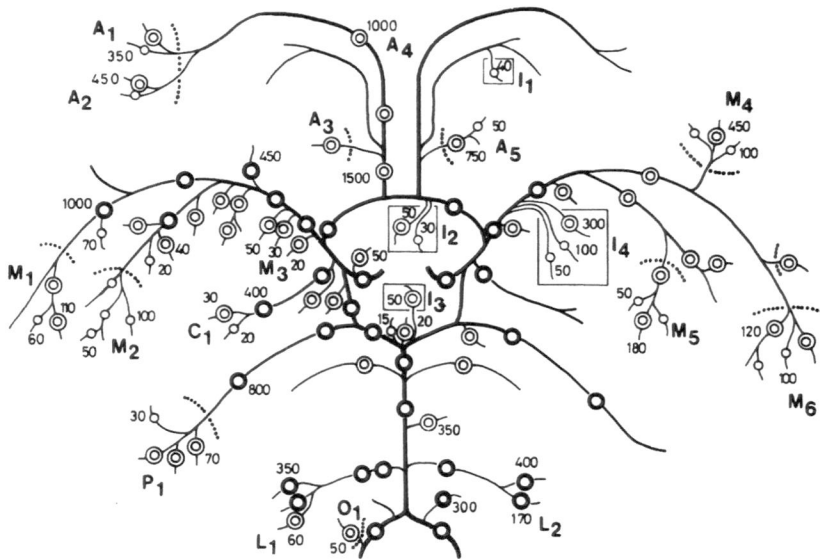

Fig. 11-7. Diagram showing the nerve supply to the intracranial arteries of the human brain. The circle of Willis is located in the center of the diagram. Arteries distal to the dotted lines are pial arteries. Arteries within the brain tissue are enclosed in squares ($I1–4$). The number beside an artery is the approximate size (in μm) of its outer diameter. A single circle indicates the total absence of vascular nerve fibers; a double circle indicates the presence of unmyelinated nerve fibers, and a dark circle indicates the presence of both myelinated and unmyelinated nerve fibers. $A1–5$, anterior cerebral artery: $M1–6$ middle cerebral artery

oxygens in vasospasm. Experimentally, it has been found that the administration of free radical scavengers together with oxyhemoglobin suppresses its constrictive effects and that the administration of active oxygens (or precursors of the same) with methemoglobin increases the constrictive effects of the methemoglobin. These findings strongly suggest the involvement of oxyhemoglobin and active oxygens in the genesis of vasospasm[34].

Since it has been reported by Sato *et al.* of our department that, contrary to widespread belief, the large cerebral vessels receive and abundant supply of nerve endings which contain amine secretory granules[61] **(Fig. 11-7)**, several studies have been made on the relationship between cerebral vasospasms and the sympathetic innervation of such vessels. On the basis of clinical experiences with cervical sympathectomy[51, 52, 68], we have previously surmised that sympathetic innervation plays an important role in its genesis. Moreover, in an electron microscopic study, we have found that, although a blood/CSF mixture incubated for 7 days induces spasms of the feline basilar artery, small cored vesicles thought to be amine secretory granules disappear 6 hours after induction of the vasospasm **(Fig. 11-8)**. This finding suggests that oxyhemoglobin brings about the release of catecholamines from nerve terminals[17, 50]. Furthermore, it was found that the degree

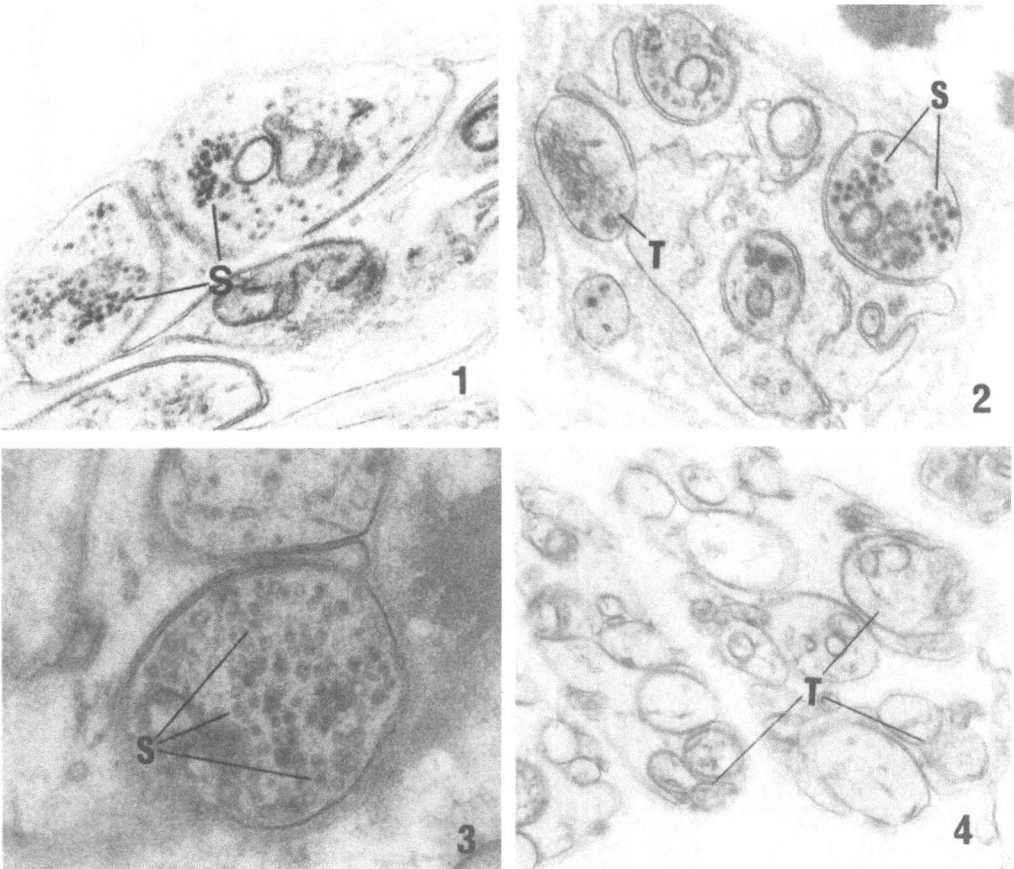

Fig. 11-8. No. 1–4 show the changes in small cored vesicles in the axon of the basilar artery with vasospasm induced by an incubated blood-CSF mixture. Small cored vesicles transformed, decreased and disappeared gradually in the course of time after the development of the vasospasm. Permanganate fixation. *1*, Control; abundant small cored vesicles (*S*) are observed. × 30,000. *2*, Immediately after the application of the blood and CSF mixture; small cored vesicles and small clear vesicles (*T*) are observed as in control cases. × 24,000. *3*, 20 minutes after the application; transformation of small cored vesicles (*S*) are observable. Dense components of vesicles have become small, shifted toward the vesicle membrane and disappeared in some vesicles. × 43,200. *4*, 10 hours after the application; small cored vesicles are not seen; only small clear vesicles (*T*) are observed. × 22,400

of vasospasm induced in this way differed according to the distance between the nerve ending and the smooth muscle tissue of the vascular wall. That is, the distance was greater at regions of mild constriction, and vice versa[17] (**Fig. 11-9**).

It has also been shown experimentally that cervical sympathectomy reduces the severity of the vasospasm induced by application of a blood/CSF mixture which has been incubated for 7 days[16] (**Fig. 11-10**). The origin of the sympathetic nerves terminating in

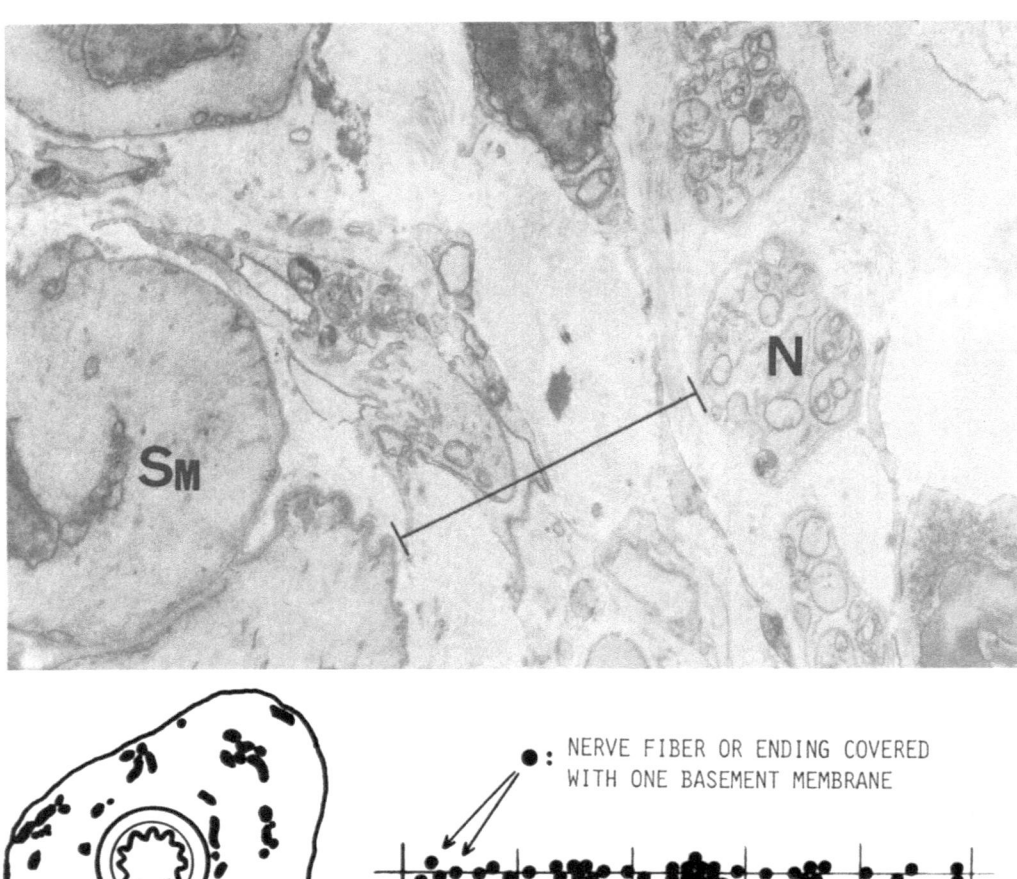

Fig. 11-9. To investigate the distribution of the nerves at the arterial wall, the shortest distance between the outer edge of the media and the nerve fibers and endings in the cross sections was measured (upper). *SM*: smooth muscle; *N*: unmyelinated nerve fibers. Relationship between vasospasm and the distribution of nerves at the arterial wall (lower)

these cerebral vessels has been found to be, with the exception of the anterior cerebral artery and the basilar artery (which receive bilateral input), from the ipsilateral superior cervical ganglion[62] (**Fig. 11-11**), and this finding is thought to confirm the effects found clinically[68].

There still remain "missing links" in the oxyhemoglobin-active oxygen-sympathetic innervation system, and further experimental work is yet required.

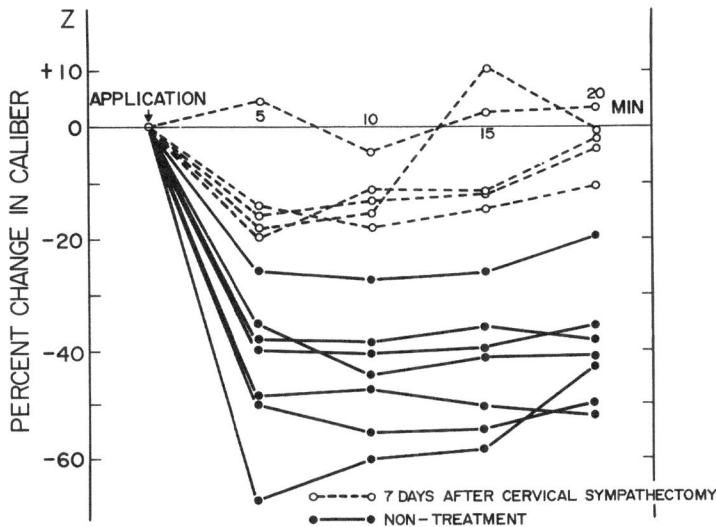

Fig. 11-10. The effect of cervical sympathectomy on vasospasm. In 5 cases undergoing cervical sympathectomy 7 days before vasospasm induced by the blood and CSF mixture (incubated for 7 days), vasospasm was definitely lighter than in nontreated cases

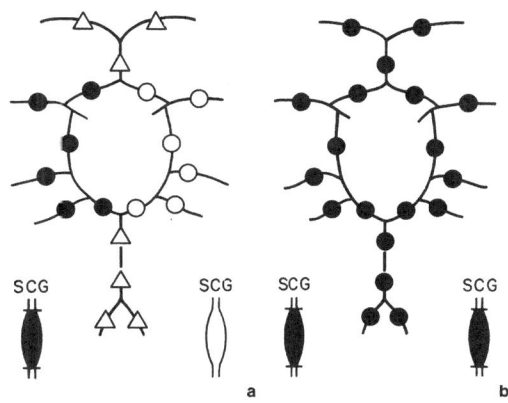

Fig. 11-11. Maps of ultrastructural changes in the cerebral nervi vasorum after superior cervical ganglionectomy. Schema of the degeneration of cerebral arterial nerve fibers following superior cervical ganglionectomy. (SCG) in the dog. a) Unilateral sup. cervical ganglionectomy. b) Bilateral sup. cervical ganglionectomy. ● Degenerated nerve fibers, ○ non-degenerated nerve fibers, △ combined: with and without degenerated nerve fibers

11.4 Treatment of Cerebral Vasospasm

11.4.1 Drug Therapy

A large number of drugs have been used to attempt to release cerebral vasospasm and it is probably not an exaggeration to say that all drugs with vasodilatatory effects have been tried. Since the effects of new drugs are being tested year in and year out, it would be virtually impossible to provide a comprehensive review of their modes of action in the present chapter. Fortunately, Wilkins[83] reviewed this topic in 1979, and the reader may be referred to his work for further details. It is, however, true that continued examination of such a large number of drugs testifies to the fact that as yet no entirely effective drug has been found.

Drugs which are intended to release vasospasm by means of direct application to cerebral vessels during surgery or by means of spinal injection include: local anesthetics[25, 54], Ca antagonists[14, 80], papaverin[2, 29, 86], α-blockers, haptoglobin[45, 66], vitamin C[48] and steroids[9].

The advantage of these means of administration is that a relatively high concentration of the drug can be applied directly to the spastic vessels. The procedure itself however, is complicated, there is the danger of infection and the different characteristics and doses required for each drug must be individually considered. Moreover, since there are progressive organic changes in the vascular walls due to continued vasospasm, the effects of drug administration after the vasospasm has been at work for several hours may already be minimal.

Under the assumption that one of the major causal factors of vasospasm is oxyhemoglobin and knowing that the oxidation of oxyhemoglobin to methemoglobin markedly reduces its vasospastic capabilities, we have adopted a policy of direct application of $NaNO_2$ (sodium nitrite) during surgery to convert the oxyhemoglobin to methemoglobin. It is known that, being a nitrogen compound, $NaNO_2$ itself has

vasodilatory effects, but by means of the above—mentioned conversion, it is thought likely to prevent the further genesis or progression of vasospasms[64].

The effects of the systemic administration of various vasodilatators have been reported in the literature. The prostaglandins have received particular attention recently. Since the suggestion of Boullin et al. that vasospasm may be the result of an imbalance between thromboxane A_2 (with its powerful vasoconstrictive effects) and PGI_2, which is synthesized in the vascular walls, (with its vasodilatory effects[5]), considerable interest in and investigation of the prostaglandins has been undertaken.

Experimentally, we have demonstrated the partial release of vasospasms, which had been induced using oxyhemoglobin, by the direct application of PGI_2[57]. Unfortunately, the fact that PGI_2 is an extremely unstable substance and maintains its activity only within a pH 10.0 buffer solution rules out its clinical application. A PGI_2 agonist which is stable and easily used is therefore much needed.

A second experimental approach has been concerned with the blockage of thromboxane A_2 synthesis[60, 72, 74], but definitive experimental results are yet to be obtained. Investigations have also been made of the relationship of the antiphlogistics and the prostaglandins[8], but results here are also inconclusive—perhaps due to the suppressive effects of the antiphlogistics on PGI_2 synthesis.

Having become aware of the possible involvement of sympathetic nerve activity in cerebral vasospasm, we have

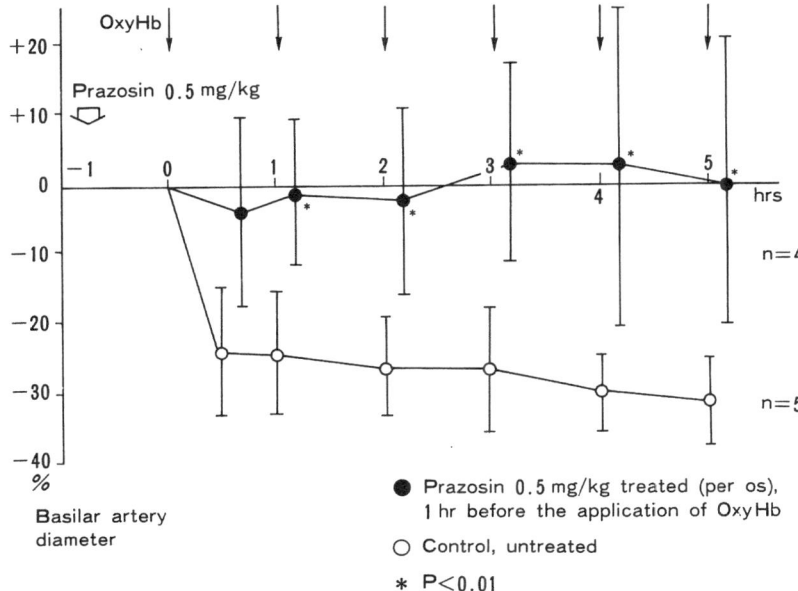

Fig. 11-12. Changes in the basilar artery diameter during vasospasm induced by hourly application of oxyhemoglobin solution. In the prazosin-treated group, vasospasm was not clear

undertaken experimental and clinical study of this problem as well. Since 1971, the classical sympatholytic agent, phenoxybenzamine, has been used in clinical cases and its effectiveness studied[11,21,26,81], but both positive[11] and negative[26] results have been reported. We have used the recently developed selective blocker of α1-receptors, prazosin hydrochloride, in an experimental study[33]. It is well-known that there are two varieties of α-receptor: α1, which is a postsynaptic receptor and α2, which is a pre-synaptic receptor. It is also known that the α2 receptor functions primarily for negative feedback control of the release of catecholamines. In effect, the α1 and α2 receptors have contrary effects. Prazosin is said to selectively block only the α1 receptor.

Prazosin was found to have pre-ventive effects by suppressing the vasospasm which would be induced by direct application of oxyhemoglobin to the feline basilar artery. That is, intragastric administration of 0.1–0.5 mg/kg of prazosin one hour before vasospasm induction resulted in a dose dependent suppression of the oxyhemoglobin-induced vasospasm (Fig. 11-12 and 11-13). It was also found that even if administered after the induction of vasospasm, prazosin was capable of partial release of the vasospasm. In light of these experimental results, we have used prazosin in clinical cases in order to prevent the occurrence of cerebral vasospasms.

On the other hand, cerebral vasospasms can be understood as one aspect of the pathology of cerebral ischemia. Consequently, as with cerebral infarction, it is necessary to maintain the

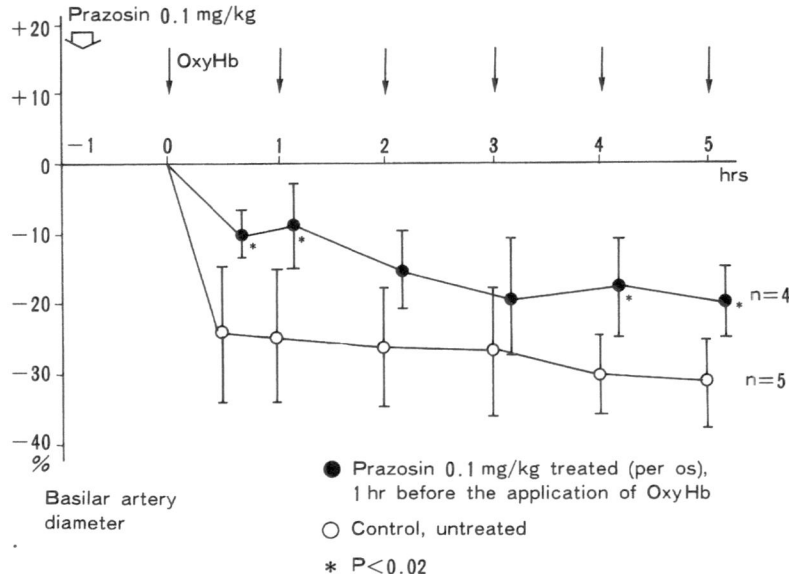

Fig. 11-13. Change in the basilar artery diameter. In the prazosintreated group, there was only partial vasospasm

perfusion pressure of the brain, improve microcirculation, prevent the formation of capillary thrombi and prevent the necrosis of brain tissue due to ischemia—regardless of the detailed mechanisms of vasospasm itself. It is known that the autoregulation of vasospastic vessels is disturbed, and blood flow within them will increase or decrease in relation to changes in blood pressure. Making use of this characteristic of vasospasm, hypervolemia-hypertension therapy is frequently used clinically[30]. Dopamine is widely used as a hypertensive agent[6, 73], but we have achieved relatively good results using Angiotensin II (Ciba). Unfortunately, Angiotensin II is not available commercially, and there remain some problems with regard to its widespread usage.

As hypervolemic agents, whole blood transfusion and plasma ex-panders are commonly used[42, 56], but they must be used with care to prevent abnormalities in the cardiovascular system. Moreover, there is the danger of hyperviscosity and a tendency toward hemorrhaging, so that care in these respects must also be taken.

The effects of various substances thought to have protective effects on the brain tissue during ischemia have been reported in the literature. We currently use the so-called "Sendai cocktail"[24] which contains 20% mannitol, vitamin . E and steroids. There are, however, time limitations on the effectiveness of all such protective drugs. In the case of the Sendai cocktail, one dose has been found to be effective for two hours[24]. During this two hour period, it is necessary to take steps—such as performing cervical sympathectomy—to improve the blood flow of the brain.

11.4.2 Surgical Treatment

We have come to view the role of the hematomas which surrounds blood vessels and the oxyhemoglobin contained therein as important causal factors in the genesis of cerebral vasospasm. Under the premise that sufficient removal of such hematomas would minimize the danger of continued vasospasm, we have actively pursued a program of acute stage radical surgery to extirpate the hematomas from around the cerebral vessels. We have found that such surgery has in fact reduced the incidence of cerebral vasospasms and reduced their severity. We therefore believe that acute stage surgery is particularly significant with regard to the prevention of cerebral vasospasm [38, 46, 70]. It must be noted, however, that there remain many technical difficulties in acute stage surgery and a proper understanding of such problems is prerequisite to such surgery. Unnecessary damage to the brain or the cutting of cerebral veins will have significant deleterious effects on the patient's prognosis.

The significance of both the completeness of hematoma extirpation and the use of continuous ventricular drainage must be emphasized. There are currently several hypotheses concerning the changes in intracranial pressure following subarachnoid hemorrhage, but it is known that the onset of hydrocephalus or cerebral edema due to vasospasm will bring about a decrease in the perfusion pressure, facilitate the appearance of cerebral ischemia due to vasospasm and contribute significantly to a poor prognosis. As a consequence, it is thought that, for acute stage radical surgery, continuous ventricular drainage should be instituted and the intracranial pressure regulated thereby [71].

As to surgical therapy in cases where ischemic symptoms due to vasospasm have already appeared, cervical sympathectomy should be considered. Such surgery has long been used in cases with circulatory deficits of the four limbs, but uncertainties concerning the distribution of sympathetic nerves to the cerebral vessels have produced doubts concerning its effectiveness. In an electron microscope study, however, we have demonstrated abundant distribution of sympathetic terminals to the major cerebral vessels [61]. The origin of this innervation has been found to be the superior cervical sympathetic ganglion, which distributes nerves to the ipsilateral vessels (expect the basilar artery and the anterior cerebral artery which receive bilateral input) [62].

In light of these findings, we have performed cervical sympathectomy in cases of cerebral vasospasm and have found this technique to be effective in some cases. Such surgery is now thought to be the final choice in cases of cerebral vasospasm. Since, however, sympathectomy is ineffective if too long a period has elapsed from the onset of symptoms of cerebral ischemia, it is necessary to perform the operation at as early a period as possible. Hypotension must be prevented before, during or after the surgery. In cases where ischemic symptoms due to vasospasm are already severe, no therapeutic technique has been found to be effective.

As a supplementary surgical technique in such cases, institution of drainage in the subarachnoid space has received considerable attention, but it must be said that its ultimate value remains uncertain.

11.4.3 Our Procedure for Prevention and Treatment

Based upon the above-mentioned basic and clinical research carried out in our department, we have adopted the following policy for the prevention and treatment of cerebral vasospasm (Table 11-2). As a rule, surgery is performed in the acute stage (particularly within 48 hours of onset). While taking care not to incur brain damage during the surgery, the subarachnoid hematoma is extirpated as completely as possible and continuous ventricular drainage is instituted. Postoperatively, great care is taken to maintain a normal systemic condition, and prazosin administration (8–16 mg/day) is begun no later than the third day from onset to prevent the development of symptoms due to vasospasm. If symptoms of cerebral ischemia are nonetheless encountered, the Sendai cocktail is administered and hypertension induced. If improvements are not soon apparent, cervical sympathectomy is performed without delay[68].

When radical surgery is performed during the period of onset of cerebral vasospasm, $NaNO_2$ (10 mM–100 mM) is applied during the surgery in order to convert the oxyhemoglobin in the cerebral vessels to methemoglobin. Following the radical surgery, the possibility of cervical sympathectomy should be considered. Using these procedures, we have recently obtained quite favorable results.

Table 11-2. Our methods for prevention and treatment of vasospasm following subarachnoid hemorrhage

---

1) To prevent vasospasm, early operation should be done, if possible.

    i)   removal of subarachnoid clot
   ii)   if nessesary, $NaNO_2$ should be used
  iii)   continuous ventricular drainage

2) Administration of prazosin should be begun on the third day.
    8–16 mg/day, every 3 hours

3) If ischemic symptoms appear,

    i)   administration of cerebral protective agents
        "Sendai cocktail"
   ii)   hypertensive therapy
        Angiotensin II
  iii)   cervical sympathectomy
        should be done.

---

12. SURGICAL THERAPY FOR MOYAMOYA DISEASE

12.1 Introduction

Moyamoya disease is a cerebrovascular disease in which vascular stenosis or occlusion of an unknown origin is seen in cerebral angiograms to extend from the termination of the bilateral internal carotid arteries (ICA) to the origin of the anterior and middle cerebral arteries (ACA and MCA). An abnormal, fine vascular network is also found at the base of the brain[16, 43, 47, 49, 53, 54]. Although there are cases in which basal moyamoya vessels are found in association with lesions of known etiology[25, 35, 36, 42, 60], such as von Recklinghausen's disease[58], trauma[54], fibromuscular dysplasia (FMD)[54], or atherosclerosis[34], when the etiology is known, it is not diagnosed as Moyamoya disease[54]. Similarly, cases presenting only unilateral basal moyamoya vessels are not considered to be true Moyamoya disease, but are referred to as quasi-Moyamoya disease.

The causal factors involved in the onset of this disease are still unknown and definitive therapy is yet to be established. As a consequence, it is inevitably the case that all therapeutic techniques are aimed at specific symptoms rather than complete cure. Since nearly all cases with intracranial hemorrhage at onset show intraventricular bleeding, it is essential to institute ventricular drainage surgically and then to extirpate the hematoma. In recent years, vascular reconstruction has been undertaken in ischemic cases[1, 3, 4, 8–10, 14, 15, 22, 26, 27, 29–31, 62]. It has also been argued that methods to improve the ischemia also prevent hemorrhage[30], but further study of this point is still required.

For patients presenting cerebral oligemia-like symptoms, however, various methods for cerebral revascularization have been attempted in order to bring about improvements in cerebral blood flow. Starting with our own technique for perivascular sympathectomy (PVS) and superior cervical ganglionectomy (SCG)[33, 39, 50], various other techniques have been devised[10, 14, 29, 31, 62]. These include STA-MCA anastomosis[1, 3, 4, 9, 22], encephalo-myo-synangiosis (EMS)[8, 15, 57], and encephalo-duro-arterio-synangiosis (EDAS)[24, 26, 27].

In the present chapter, we will describe primarily our own experiences with revascularizative operations.

12.2 The Characteristic Pathophysiology of Moyamoya Disease

In Moyamoya disease, onset in juvenile cases is usually in the form of cerebral ischemia and in adult cases usually in the form of intracerebral or intraventricular hemorrhage[6]. An understanding of these differences is essential for the selection of appropriate therapeutic methods.

In juvenile cases, the extent and severity of the cerebral ischemia is known to change with time. By means of sequential angiographical observations in many such cases, we have classified the dynamic course of juvenile Moyamoya disease into six stages-classification of basal moyamoya[5, 44–47, 53, 56]. These can be summarized as follows: (i) Narrowing of the carotid fork; only carotid fork stenosis is observed. (ii) Initiation of basal moyamoya; all the main cerebral arteries are dilated. (iii) Intensification of moyamoya; remarkable moyamoya vessels at the base of brain. Angiographical disappearance of the middle and anterior cerebral arteries is observed. (iv) Minimization of moyamoya; angiographical disappearance of the posterior cerebral artery is observed. (v) Reduction of moyamoya; all the main cerebral arteries are missing. (iv) Disappearance of moyamoya; cerebral blood flow supplied only from the external carotid artery (ECA) and vertebro-basilar artery (VBA)[21, 48, 49] (**Fig. 12-1**). These changes are seen to occur over a period of several months to several years, and ultimately most cases end up with cerebral circulation which is maintained solely via the ECA and VBA systems[56].

Fully 85% of juvenile cases exhibit symptoms thought to be due to cerebral ischemia, such as motor deficits and mental retardation[6], and it is thought that such symptoms are rooted in the chronic, progressive cerebral ischemia[51]. What is particularly noteworthy in the juvenile cases is the fact that there is apparently a robust capacity for the generation of collateral pathways[16, 17]. Two kinds of collateral pathways, ethmoidal[48, 49] and vault moyamoya[21], can develop in a surprisingly short period of time[56].

In contrast, in nearly all adult cases there are no changes in cerebral angiograms even when long-term follow-up is undertaken, and individual cases may be at any one of several stages of the disease, as seen agiographically. In many cases where it is apparent that the disorder has progressed to a certain stage during childhood and the patient has become an adult while the disorder has not progressed further, it is found that the disease is in its terminal stage. As a consequence, there is the possibility that among adult cases, some are cases due to unknown causes different from those in juvenile cases, but these unknown causes produce abnormal net-like vessels bilaterally at the base of brain.

Among all adult Moyamoya cases, 37% show cerebral ischemia-like symptoms and 43% have unmistakable attacks of intracranial hemorrhage[6]. Two mechanisms have been proposed

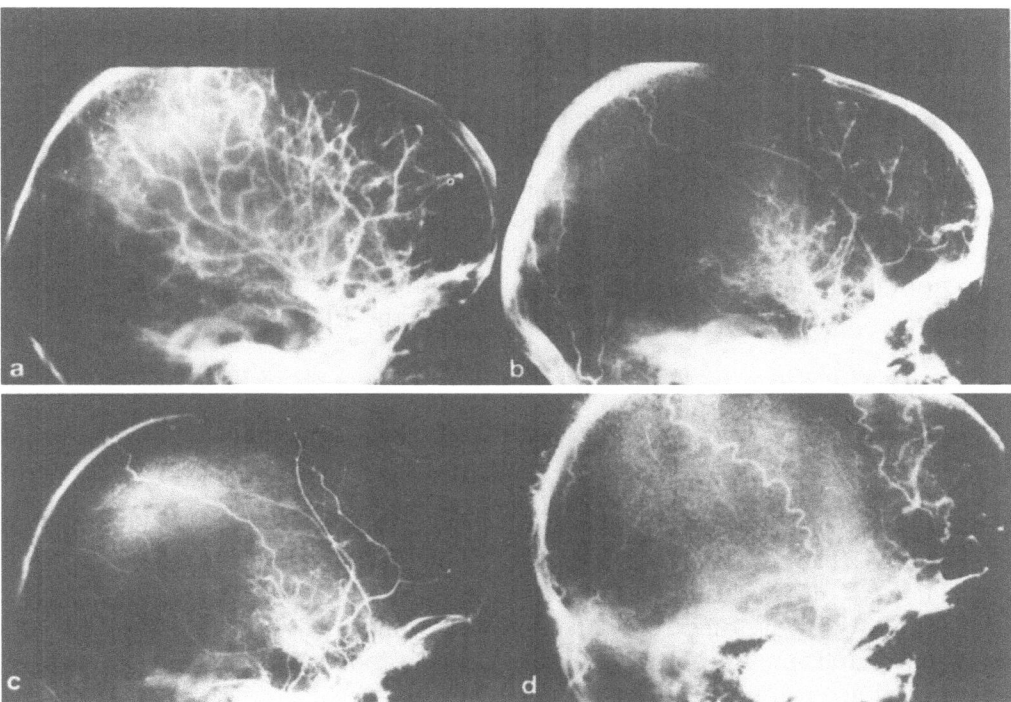

Fig. 12-1. Angiographical progression of Moyamoya disease. a) 4-year-old boy at the time of initial symptom. Note dilatation of the cortical arteries and the beginning of basal moyamoya (stage II). b) Follow-up angiogram 1 year and 9 months after initial study. Note the appearance of definite basal moyamoya (stage III). c) Follow-up angiogram 2 year and 6 months after initial study. Note the decreased size and intensity of basal moyamoya and the faint filling of the branches of the middle cerebral artery (stage IV). d) Follow-up angiogram 18 years and 3 months after initial study. Note disappearance of basal moyamoya. Vault moyamoya in the frontal region is recognized (stage VI)

to account for the strong correlation between decreased blood flow to the region fed by the ICA and intracranial hemorrhage. The first is that there may be an increase in blood flow in the VBA system together with the decrease in the ICA system, so that it tends to produce an aneurysm. In fact, the incidence of aneurysm in the VBA system is higher in Moyamoya cases than non Moyamoya cases. Although it is known that aneurysms on the VBA system can develop and rupture—causing subarachnoid hemorrhage, the incidence of the same is low[28]. The second mechanism is that there may be rupture of a small artery in the vicinity of the rostral lateral wall of the lateral ventricles, which are in a state of chronic oligemia[18, 19, 52]. As a consequence, it may be possible to justify some attempt at treating the decreased blood flow to the ICA or the consequent symptoms of chronic oligemia.

As is evident from the above, the chronic oligemia seen bilaterally in the region fed by the ICA is deeply involved in the onset of symptoms. At the current stage where the causes of Moyamoya disease are yet unknown, many surgical techniques have been tried to alleviate the ischemia. We have previously reported from the results of various experimental studies that cervical sympathetic nerves are involved in Moyamoya disease[6, 13, 63], and we have obtained favorable results from superior cervical ganglionectomy (SCG) and cervical perivascular sympathectomy (PVS)[51, 53].

Recently, anticipating the immediate effects obtained from surgery on the cervical region, we have also performed encephalo-duro-arterio-synangiosis (EDAS)[26, 27] in order to obtain long-term inprovements by the proliferation of collateral pathways from the ECA system as well.

12.3 Surgical Therapy for Ischemic Moyamoya Disease

12.3.1 Durapexia

In 1964, Tsubokawa et al.[59] opened on a 6 year old girl thought to have ischemic Moyamoya disease by placing a dural patch including the middle meningeal artery on the surface of the brain. Postoperatively, proliferation of collateral pathways was seen in angiograms, and there was an improvement in symptoms. Unfortunately, no further reports of this kind have been made.

12.3.2 Cervical Perivascular Sympathectomy and Superior Cervical Ganglionectomy (to be discussed below)

12.3.3 Superficial Temporal Artery-Cortical Branch of the Middle Cerebral Artery Anastomosis (STA-MCA Anastomosis) and Encephalo-myo-synangiosis (EMS)

The STA-MCA anastomosis technique developed in 1967 by Yaşargil and Donaghy and others for cases of ischemic cerebral lesions was first applied in Moyamoya cases in 1974 by Krayenbühl[22], Karasawa and Kikuchi[9]. Unlike most vascular disorders, however, there are few appropriate recipient arteries at the surface of the brain in Moyamoya disease, and there are consequently many difficulties in perform-ing the anastomosis. In cases where no suitable recipient artery can be found, Kikuchi et al.[8] used their encephalo-myo-synangiosis (EMS) technique in which temporal muscle is attached to the brain surface. Subsequently, STA-MCA anastomosis, EMS or a combination of such techniques have been used for treatment of Moyamoya disease[8, 9, 12, 14]. In postoperative external carotid angiograms, in addition to visualization of the arteries at the surface of the brain, an extremely complex vascular pattern, including spontaneous anastomosis of the other meningeal arteries is seen. In many cases, the whole MCA branches can be visualized three months postoperatively. After 6–12 months have elapsed, there is the formation of an abnormal vascular network near the site of the anastomosis, together with occlusion of the original anastomosis site. Cortical arteries can then be visualized through the abnormal network. In contrast, in internal carotid angiograms, there is contraction of the basal moyamoya vessels and in some cases it cannot be visualized at all[11, 15].

In a follow-up study of 29 cases, such surgery was found to have no effect upon symptoms due to ischemia at locations unrelated to the MCA (such as visual acuity, visual field and swallowing disturbances), but notable improvements were found in motor and somatosensory deficits, speech disturbances and involuntary movements[12].

Others have reported similarly good therapeutic results[1, 3, 4, 15, 29–31, 57] and this form of surgery is thought to be the most frequently used therapeutic technique for ischemic Moyamoya disease in Japan.

There have also been, however reports of postoperative aggravation of symptoms and complications[29, 40] and we have also experienced cases, operated upon at other clinics, showing postoperative bilateral motor deficits, and speech and swallowing disturbances. Problems associated with this method included, the fact that: (i) the craniotomy is large, (ii) there is the possibility of sacrificing collateral pathways already present due to STA transection, (iii) there is the possibility of producing an epileptic focus due to the widespread adhesion of the temporal muscle to the surface of the brain, and (iv) there is the possibility of destroying neovascularization due to trauma. For these reasons, a comprehensive long-term follow-up study of this method is still needed.

12.3.4 Transplantation of Omentum

This surgical technique involves partial resection of the omentum majus, including the gastroepiploic artery and vein, and after anastomosis of the artery and vein to the respective superficial temporal artery and vein, the omentum patch is placed on the surface of the brain[61]. Karasawa *et al.*[10] and

Yonekawa *et al.*[62] have used this technique in cases of Moyamoya disease. In comparison with temporal muscle, the omentum majus can be applied over a wider area, but this technique has not been widely used and accurate evaluation of its clinical effects is not yet possible.

12.3.5 Encephalo-Duro-Arterio-Synangiosis (EDAS)

In 1980, Matsushima *et al.*[26] devised a technique which involves first exposing scalp arteries (the superficial temporal artery and/or the occipital artery).

Next, through a linear craniotomy, the scalp artery with a strip of galea is suture-fixed to the linear dural opening edge without cutting the distal end. The technique is simple and any trans-dural anastomosis present prior to the operation can be preserved. Moreover, the invasion of the brain itself is minimal—that is, there is no compression of the brain, no temporary occlusion of cerebral arteries and no disturbances of cerebral circulation due to decreases in the buoyancy of the brain caused by aspiration of CSF. It is therefore thought likely that most of the complications often found with STA-MCA anastomosis, EMS and other kinds of similar operations will not arise

with EDAS. Another advantage of this technique is that donor arteries from almost all over the cranium can be used (including the anterior and posterior superficial temporal artery and the occipital artery).

According to a report by Matsu-shima et al.[27], revascularization of vessels at the surface of the brain was seen in angiograms obtained on 13 sides in 9 cases of juvenile Moyamoya disease, together with contraction of the basal moyamoya vessels. Spetzler et al.[41] and Lesoin et al.[24] have reported having used this technique in other chronic occlusive cerebral lesions, and had favorable results.

12.3.6 Other Techniques for Improving Cerebral Blood Flow

In addition to the surgical techniques noted above, there have been reports of an encephalo-arterio-synangiosis (EAS) technique in which the cut end of the STA is ligated and placed on the surface of the brain, and a technique combining both EMS and EAS, the so-called encephalo-myo-arterio-synangiosis (EMAS) method[29, 31]. Improvements in collateral circulation have been thus obtained. Some

believe that, since the brain in Moyamoya disease is in a state of chronic oligemia, it will demand blood from any source available[2, 41], and these surgical techniques are simply means for facilitating the generation of collateral pathways. It must be said, however, that regardless of which of these surgical techniques is used, the above-mentioned risks involved in performing craniotomy exist.

12.4 Our Own Therapeutic Technique: Cervical Perivascular Sympathectomy (PVS) and Superior Cervical Ganglionectomy (SCG)

In an electron microscopical study, we have observed sympathetic nerve trunks in cerebral vascular walls in arterioles as small as 30–50 microns in diameter and have found them to be

particularly abundant in the vicinity of the circle of Willis[37]. Furthermore, experimentally we have shown that the origin of the sympathetic nerves within the vascular walls of the intracranial

ICA system was the superior cervical ganglion[38]. In light of these factors, we concluded that—similar to the lumbar sympathetic ganglionectomy performed in cases of Buerger's disease of the lower extremities—ganglionectomy in Moamoya disease could lead to an increase in cerebral blood flow and facilitate the development of collateral pathways. In this way, ganglionectomy could be a means of treating cerebral ischemia. PVS and SCG are currently used by us in order to improve symptoms of cerebral infarction due to vasospasm following aneurysm rupture[33], [39] and they are known to be effective with but few undesired side effects.

Thus far we have experienced a total of 119 cases of Moyamoya disease, among which this technique was used in 40 juvenile and 20 adult cases pre-

senting primarily symptoms of cerebral ischemia (**Table 12-1**). Among our early patients, there were some undergoing only PVS, but recently we have performed bilateral PVS together with SCG as a rule.

Table 12-1. Moyamoya disease

(Tohoku Univ. 1963~1984.5)

	Male	Female	Total
Children	18	38	56
Adults	28	35	63
Total	46	73	119

In the present section we will discuss the effects of such surgery on these patients after follow-up periods varying from three weeks to 14 years.

12.4.1 The Surgical Method

As shown in **Fig. 12-2**, the head is turned slightly to the side contralateral from the lesion and the cervical region is extended. A 6 cm skin incision is made from slightly above the angle of the lower jaw proceeding caudally along the anteromedial border of the sterno-cleido-mastoid muscle (**Fig. 12-2A**). After retracting this muscle laterally, about 10 ml of physiological saline is injected above and below the muscle through a small hole in the fascia cervicalis using an injection syringe without the needle attached. This procedure greatly facilitates the subsequent dissection of the cervical region. The CCA and ICA are then fully exposed. Dissection of the nerve plexi and these

arteries is done using a fine saline injection between the outer and the inner membrane. The PVS should be performed over the entire region extending 1 cm caudal from the bifurcation of the CCA and along about 1 cm of the ICA (**Fig. 12-2B**). The superior cervical ganglion lies slightly rostral and posteromedialy from the bifurcation of the common carotid artery and is upon the transverse process of the cervical spine, but discovery of the ganglion can sometimes be difficult. Since, however, the sympathetic nerve trunk can be discovered by patiently searching in this region with a clear idea of the local anatomy in mind, the superior cervical ganglion can be found by tracing this trunk rostrally. While severing all branches of this ganglion, cuts should be made as rostrally as possible (**Fig. 12-2 C**). During this procedure, it is inevitable that the ICA will be compressed somewhat, but care should be taken not to cause temporary occlusion of this vessel. Damage to the vagal and hypoglossal nerves should also be avoided. Bilateral surgery of this kind should not be attempted in one stage, but should be completed with at least two week interval between surgery on one side and that on the other.

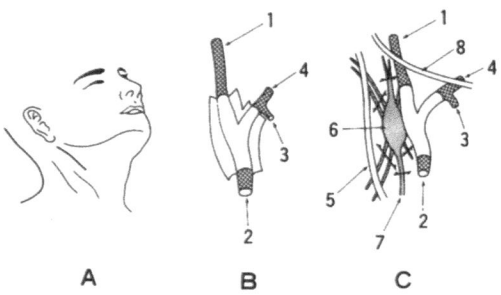

A   **B**   **C**

Fig. 12-2. The surgical method of perivascular sympathectomy and superior cervical ganglionectomy. A) A skin incision is made from above the angle of the lower jaw proceeding along the sternocleidomastoid muscle. B) The perivascular sympathectomy should be performed over the entire region extending 1 cm caudal from the bifurcation of the CCA and along about 1 cm of the ICA. C) The superior cervical ganglionectomy. The ganglion can be found by tracing the sympathetic nerve trunk rostrally. While serving all branches of this ganglion, cuts should be made as rostrally as possible. *1* internal carotid artery, *2* common carotid artery, *3* superior thyroid artery, *4* external carotid artery, *5* vagal nerve, *6* superior cervical ganglion, *7* sympathetic nerve trunk, *8* hypoglossal nerve

12.4.2 Follow-up Clinical Symptoms

Follow-up of clinical symptoms was done primarily by collecting records of the patients' conditions as noted by parents or family members (**Table 12-2, 12-3**). Improvements were seen in 28 of 40 juvenile cases (70%) and in 10 of 20 adult cases (50%). Nine of the juvenile cases remained unchanged (22.5%) and 3 (7.5%) showed further deterioration. Among these latter 3 cases, 2 suffered compression of the ICA and one showed a decrease in cerebral blood flow due to hypotension during the operation. These cases serve to emphasize the importance of care during the surgical operation. Ten of the adult cases (50%) remained unchanged following surgery and none showed further deterioration.

Study of the symptoms involved showed that, among the juvenile cases, TIA previously seen in 35 of the cases disappeared in 24 cases and the incidence decreased in another 8. The involuntary movements seen in 8 patients disappeared in all. Mental retardation was found in 17 cases. Among 11 mild cases of the same, one showed recovery to a normal IQ and 8 showed improvements, but among the severely

Table 12-2. Surgical procedures for children

Operation	Results			
	Improved	Unchanged	Aggravated	
Bilateral PVS	1	2	1	4
Unilateral PVS & Contralateral PVS+SCG	3	3	0	6
Bilateral PVS+SCG	24	4	2	30
Total	28(70.0%)	9 (22.5%)	3 (7.5%)	40

PVS : perivascular sympathectomy
SCG : superior cervical ganglionectomy

Table 12-3. Surgical procedures for adults

Operation	Results			
	Improved	Unchanged	Aggravated	
Bilateral PVS	4	0	0	4
Unilateral PVS & Contralateral PVS+SCG	1	2	0	3
Bilateral PVS+SCG	5	8	0	13
Total	10(50.0%)	10(50.0%)	0 (0 %)	20

Table 12-4. Results of surgical treatment in children

Symptom	Presurgery	Follow up
Episode of T.I.A.	35 Cases	24 No episode / 8 Improved / 3 Worsened
Involuntary movement	8	8 Cured
Mental handicap	17	
Educationable (50<IQ<80)	11	1 Normal range / 8 Improved / 2 Stationary
Uneducationable (IQ<50)	6	6 Stationary

Table 12-5. Results of surgical treatment in adults

Symptom	Presurgery	Follow up
Motor disturbance	11 Cases	6 Improved / 5 Stationary
Sensory disturbance	4	1 Improved / 3 Stationary
Speech disturbance	4	3 Improved / 1 Stationary
Visual disturbance	4	2 Improved / 2 Stationary
Mental handicap	9	4 Improved / 5 Stationary

retarded cases with IQs below 50, no postoperative changes were apparent (Table 12-4). Among the adult cases, there were no cases of deterioration. The improvements were found in the following (Table 12-5): 6 of 11 cases with motor deficits, 3 of 4 cases with speech disturbances, 2 of 4 cases with deficits of visual acuity, and 4 of 9 cases with psychic or intellectual impairments. Moreover, such improvements were seen in the early period following surgery and were maintained thereafter.

As seen from the above, this surgical procedure is reasonable effective and, since it does not require an intracranial procedure, complications are uncommon. We believe that this operation should be considered as the first choice for surgical therapy in Moyamoya disease.

12.4.3 Angiographical Follow-up

A follow-up angiographical study was made of some of the juvenile Moyamoya cases undergoing PVS and SCG[50, 56]. Carotid angiograms were taken of 13 sides in 9 cases within two months of surgery and follow-up was made of 20 sides in 10 cases for 6–19 years starting from at least 6 months after the operation. The angiograms showed improvements in 8 sides of the sides there was deterioration—specifically, progression to further stages of the disease as defined by Suzuki's 6-stage classification[5, 47, 53].

Table 12-6. Results of postoperative short-term follow-up angiography

			within 2 mos (9cases 13sides)	more than 6 mos (6cases 10sides)
1) TS*	R CAG		○	×
	L CAG		○	×
2) CA	R CAG		×	×
	L CAG		○	×
3) HY	R CAG		○	×
	L CAG		○	×
4) HS*	R CAG		×	△
	L CAG		○	△
5) NK*	R CAG			
	L CAG		△	×
6) CS	R CAG			×
	L CAG			
7) YK	R CAG			△
	L CAG			
8) HI	R CAG	△		
	L CAG	○		
9) YM*	R CAG			
	L CAG	○		

○ : improved, △ : unchanged, × : deteriorated
R : right, L : left, **CAG** : carotid angiography
* indicates the same case with long-term follow-up

cases studied within two months of surgery. Specifically, there was dilatation of basal moyamoya vessels, the appearance of trunk arteries and dilatation of the terminal portion of the ICA (**Table 12-6**). In three sides, however, there were no changes and in two

Among the 20 sides in 10 patients studied for 6–19 years postoperatively, however, the results were completely different from those obtained within two months[56] (**Table 12-7**). That is, progression of the disease to the next 1–4 stages was seen in 19 of the 20 sides, and in the final angiogram obtained in 14 of the sides the disease had progressed to one of the terminal stages (5 or 6). Only one side remained unchanged and none showed improvements. Among these cases, 13 sides showed the proliferation of collateral pathways together with clinical improvements, such as ethmoidal moyamoya[48, 49] and vault moyamoya[21] (the so-called transdural anastomosis or rete mirabile anastomosis) **(Fig. 12-3)**. In contrast, a reduction in collateral pathways was seen on 5 sides in 5 cases, and all but one case presented with mental retardation.

It is consequently believed that the

Table **12-7**. Results of postoperative long-term follow-up angiography

CASE	SIDE	AGE (year)					
		0	5	10	15	20	25
1) YS	R			3-③---------4			
	L			2-③---------5			
2) TK	R			④---------------------6			
	L			②--------------------3			
3) TI	R			⑤----ᵣ-----6			
	L			⑤----------6			
4) TS	R		2-3-④4-----------5----------------6				
	L		2-3-④4-----------5----------------6				
5) HH	R		④---------------------5				
	L		③---------------------5				
6) NS	R		5--------⑤-----------6				
	L		5--------⑤-----------6				
7) YM	R		②2----------------6				
	L		⑤5----------------5				
8) HS	R	③3-------4---4----------------4					
	L	③3-------4---4----------------4					
9) NK	R			3-③-------------4			
	L			②2---------------3			
10) HI	R		4------⑤---------6				
	L		4------⑤---------6				

number indicates the angiographical stage
circle indicates the operation (PVS, SCG)
R : right,   L : left

progression of Moyamoya disease, as seen in cerebral angiograms, can be halted for only a short period of about two months. Thereafter, although progression of the disease as seen on angiograms, and continued improvement in postoperative clinical symptoms appear to be contradictory, it is thought likely that the PVS and SCG surgery allow for the proliferation of collateral pathways other than basal moyamoya vessels from the ECA and VBA systems, thus preventing the aggravation of symptoms.

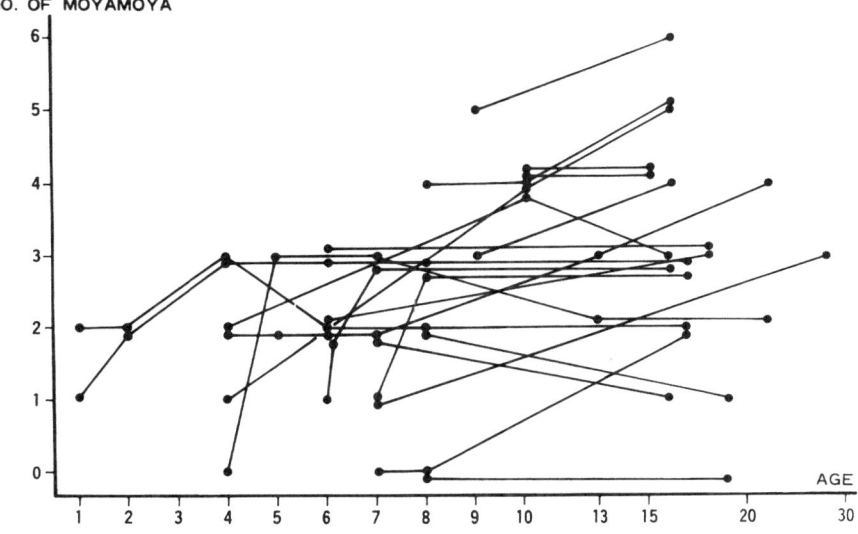

Fig. 12-3. Follow-up results on the changes in collateral circulation; vault and ethmoidal moyamoya

12.4.4 Electroencephalographical Follow-up

Follow-up EEG studies were made in a total of 62 sides—51 sides among juvenile cases and 11 sides among adult cases (**Table 12-8**). Improvements in electrical activity 3–4 weeks postoperatively were seen in 31 sides of ju-

Table 12-8. Evaluation of effects of surgical treatment by EEG

	Improved	Unchanged	Aggravated	Total
Children	31 sides	15	5	51
Adults	9	2	0	11
Total	40(64.5%)	17(27.4%)	5 (8.1%)	62

venile cases (60.8%), and in 9 sides of adult cases (81.8%). Signs of improvement were most commonly seen as a reduction in slow waves (posterior and centro-temporal slow waves), which are typically found in juvenile cases of

Moyamoya disease. There were also cases showing a disappearance of epileptic discharges and the reduction or disappearance of re-build up (Table 12-9).

Table 12-9. Improvements in EEG findings

	Children	Adults
1. Improving of basic patterns	4 sides	1
2. Decreasing of slow waves	19	4
3. Diminishing of epileptic discharges	5	1
4. Diminishing of HVS bursts	4	
5. Suppressing of re-build ups	4	
6. Others	11	4

12.4.5 The Effects of PVS and SCG, as Monitored During Surgery

Based upon the above results, we can now anticipate immediate therapeutic effects following PVS and SCG. In order to obtain long-term increases in cerebral blood flow, in our more recent cases, we have also performed EDAS. A study has been made of the cerebral hemodynamics and metabolism during surgery in 7 juvenile and 4 adult cases (a total of 18 sides in 11 patients). As a rule, surgery is performed on only one side at a time and for monitoring the cerebral condition throughout the surgery, various physiological measurements are made. These include: systemic blood pressure, cortical cerebral blood flow (using the Laser-Doppler method), cortical tissue

oxygen pressure and CO_2 pressure and CCA blood flow (using an electro magnetic flow meter).

The operation begins with SCG, followed by PVS after about a 20–30 minute interval. No changes in systemic blood pressure have been found due to SCG. At the completion of PVS, the blood flow at the surface of the brain and in the CCA has been found to have increased an average of 8% and 13% respectively. Although no changes in cortical tissue oxygen pressure have been found, there has been a tendency for a slight decrease in CO_2 pressure (Table 12-10). The immediate improvements in cerebral blood flow due to this surgical method are evident. We have used this method also even in acute stage Moyamoya disease cases with frequent transient ischemic attacks. Thus far, we have obtained good results which we believe are due to the rapid improvements in blood flow.

Table 12-10. Sequential changes in CBF, EMF, PtO_2, $PtCO_2$, BP after SCG and PVS

	SCG	PVS
CBF (n=10)	105.4±12.1 (%)	107.9±11.8 (%)
EMF (n=15)	113.9±22.7	113.1±16.9
PtO_2 (n=12)	99.7±13.1	99.5±15.5
$PtCO_2$ (n=11)	95.8± 7.5	94.2±12.6
BP (n=14)	101.1±14.8	100.9±15.9

CBF : cortical cerebral blood flow by laser doppler
EMF : electromagnetic blood flow of common carotid artery
PtO_2 : tissue oxygen pressure
$PtCO_2$: tissue carbon dioxide pressure
BP : systemic blood pressure
     each values are presented mean±SD compared with preoperative values

12.5 Anesthesia, Pre- and Postoperative Management in Moyamoya Disease

Together with the increasing applicability of surgical techniques for therapy of Moyamoya disease, there has been an growing recognition of the importance of not only the problems in the surgical method itself, but also of appropriate pre- and postoperative and anesthetic management. Even in Moyamoya patients who do not show any clinical symptoms of ischemia, the cerebral blood flow is thought to be near the critical level, so that particularly attentive pre- and postoperative care should be required for these patients. Next, we will discuss our recent study of these problems, based upon these considerations.

12.5.1 Study of the "Re-build Up" Phenomenon Using EEG, Angiography and PET

It is well-known that a state of cerebral ischemia can be induced by hyperventilation in Moyamoya disease, particularly among juvenile cases. Having noticed that after completion of hyperventilation there is a reduction or

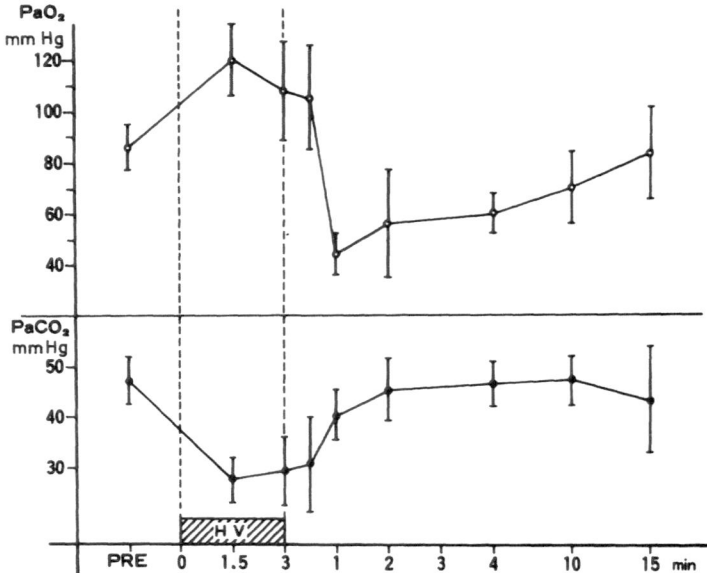

Fig. 12-4. Sequential changes in PaO_2 and $PaCO_2$ (Hyperventilation in room air: HV). Note decrease in $PaCO_2$ during hyperventilation and decrease in PaO_2 after hyperventilation

disappearance of the build up seen during hyperventilation, followed by a recurrence of slow-wave activity, we have labeled this phenomenon "re-build up" and have reported several studies on it[7, 20, 32, 55]. Since this phenomenon is particularly characteristic in Moyamoya disease, it has been shown to be effective in screening for this disorder[20]. Since re-build up is particularly common in juvenile cases with frequent TIAs and in patients immediately following an ischemic attack, it is thought to provide important information concerning pre- and postoperative care and anesthesia. In order to study this problem, positron emission tomography (PET) scans, EEG recording and cerebral angiograms were obtained in several cases before and after hyperventilation. Due to the hyperventilation load, changes in PaO_2 and $PaCO_2$ in juvenile Moyamoya cases and normal children were found to be as follows. During hyperventilation PaO_2 rose and $PaCO_2$ fell. After hyperventilation, however, $PaCO_2$ immediately returned to normal, while there was a marked decrease in PaO_2 due to the suppression of the ventilation. Approximately 15 minutes were required for a return to normal[32, 55] (**Fig. 12-4**).

Build-up in the EEG was seen during the hyperventilation which caused decrease in $PaCO_2$, and re-build up was characteristically seen appearing together with the prolonged decrease in the PaO_2 after the completion of hyperventilation[32]. Study was also made of the effects of hyperventilation of 100% O_2 or air base 8% CO_2. The re-build up was suppressed when breathing 100% O_2, and both build up an

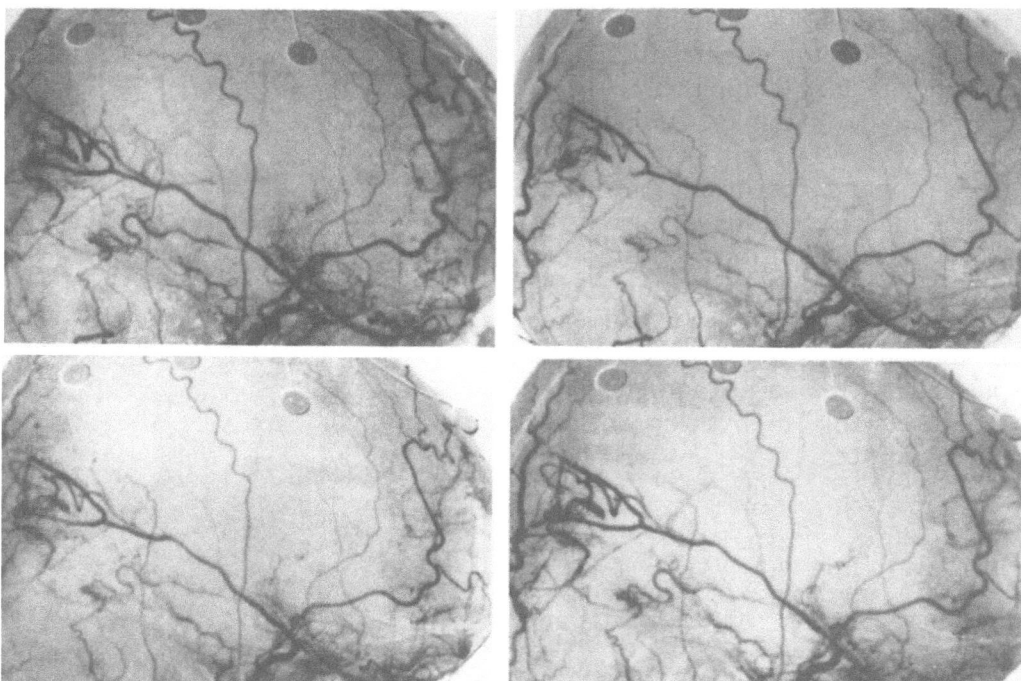

Fig. 12-5. Angiograms before and after hyperventilation. Upper left: pre-hyperventilation, upper right: post-hyperventilation 1 min, lower left: post-hyperventilation 7.5 min, lower right: post-hyperventilation 15 min. Note significant decrease in the size of the basal moyamoya and in the diameter of the cortical arteries in the frontal and occipital areas. These changes are most prominent in the early period of post-hyperventilation (upper right), and gradually return to the pre-hyperventilation state (lower left to right)

re-build up were suppressed when breathing 8% CO_2[32].

In patients showing the re-build up on EEG, cerebral angiograms showed that immediately following the completion of hyperventilation, there was contraction of the basal moyamoya vessels and a reduction in the diameter of cortical arteries. Furthermore, these changes were found to persist even following the recovery of $PaCO_2$ (**Fig. 12-5**). In contrast, no changes in pre- and post-hyperventilation cerebral angiograms were apparent when 8% CO_2 was used[55] (**Fig. 12-6**). In the PET study, a fall in CBF was seen at the time of re-build up and an even more marked decrease was seen in $CMRO_2$. Moreover, the fall in CBF was significantly greater in whichever hemisphere had a more advanced stage lesion, as determined angiographically[7] (**Fig. 12-7**).

In light of the above findings, it is thought that ischemia in a brain which is already at a critical level of CBF is brought about due to a decrease in $PaCO_2$ caused by hyperventilation. Moreover, due to hypoxia caused by a decrease in PaO_2 following hyperventilation, there then occurs the re-build up. EEG build up is generally thought to originate deep within the brain since it is monorhythmic and diffuse, whereas the re-build up is thought to be of cortical or subcortical origin since it is polymorphous with changes in its wave shapes and site with time. For these reasons, the re-build up characteristically seen in Moyamoya disease is thought to indicate that such brains, particularly the cortex and subcortex, are susceptible to hypoxia with decreased CBF.

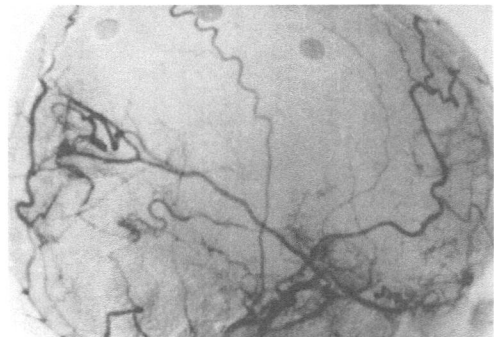

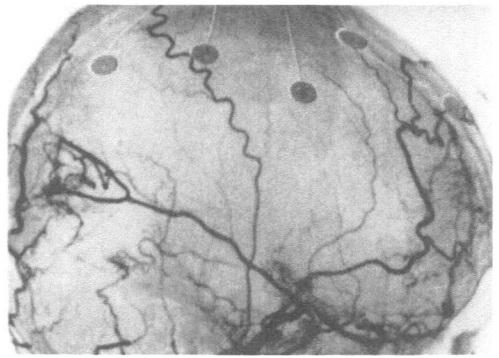

Fig. 12-6. Angiogram before and after hyperventilation (same case as in **Fig. 12-5**) hyperventilation during 8% CO_2 inhalation. Upper: pre-hyperventilation, lower: post-hyperventilation. Note no significant changes in spite of hyperventilation

12.5.2 Anesthesia, Pre- and Postoperative Management

It is fully evident from the above-mentioned study that care must be taken in Moyamoya cases to prevent hyperventilation and hypoxia. We normally prohibit hyperventilation (such as strenuous exercise, the playing of wind instruments) before surgery and during the acute period after surgery in patients who show re-build up. In cases in the acute stage of an ischemic attack or in cases with frequent TIAs, we perform PVS and SCG while administering a gas mixture of 8% CO_2 and 92% O_2 before, during and after surgery. In juvenile cases, it is essential to carry out CT and angiographical follow-up examinations and

to check the patient's EEG at regular intervals postoperatively. The fact that decrease in or disappearance of the re-build up is closely correlated with an improvement in clinical symptoms is particularly useful. Finally, care to prevent hypotension in everyday life must be taken, and for this purpose facilitators of cerebral circulation, vasodilators and stabilizers of the vascular wall should be administered.

With regard to anesthesia, it is essential to avoid hypocapnia and systemic hypotension. A study by Kuru et al.[23] is of some relevance. They investigated the relationship among $PaCO_2$, systemic blood pressure and

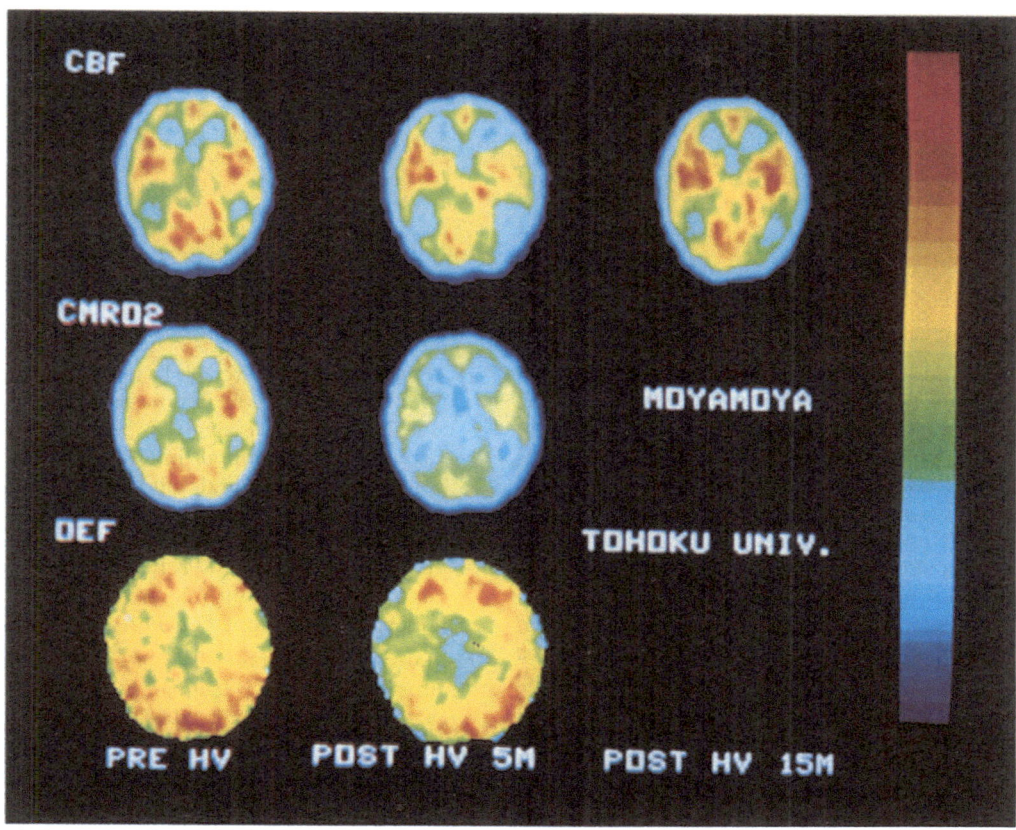

Fig. 12-7. Positron emission tomograms before and after hyperventilation. Decreased CBF in a 13-year-old boy was observed mainly in the right hemisphere where the angiographic stage was more advanced in a posthyperventilation 5-minute scan. In contrast, the decrease in $CMRO_2$ was much more than that in CBF in the whole brain region

the oxygen pressure of the internal jugular vein (PjO_2) as an indicator of CBF. They found that, although there were some differences according to the anesthetic agent used, there was close correlation between $PaCO_2$ and CBF. Using mNLA anesthesia, CBF began to fall when $PaCO_2$ fell below 30 mmHg and, using GOF anesthesia, it began to fall when $PaCO_2$ fell below 25 mmHg.

In either case, it is suggested that $PaCO_2$ should be maintained at around 45 mmHg. Moreover, GOF anesthesia had strong effects on autoregulation, indicating that particular care should be taken to avoid hypotension. During surgery, we usually prevent a fall in $PaCO_2$ by making frequent checks of blood gas analyses.

References

CHAPTER 1

1. Branston NM, Symon L, Crockard HA *et al* (1974) Relationship between the cortical evoked potential and local cortical blood flow following acute cerebral artery occlusion in the baboon. Exp. Neurol 45: 195–208

2. Crowell RM, Olsson Y, Klatzo I *et al* (1970) Temporary occlusion of the MCA in monkey, clinical and pathological observation. Stroke 1: 439–448

3. Diaz FG, Mastri AR, Ausmann JI *et al* (1979) Acute cerebral revascularization: Part I. Cerebral ischemia experimental animal model. Surg Neurol 12: 353–362

4. Eklöf B, Seisjö BK (1972) The effects of bilateral carotid artery ligation upon the blood flow and the energy state of the rat brain. Acta Physiol Scand 86: 155–165

5. Fujishima M, Sugi T, Morotomi Y *et al* (1975) Effects of bilateral carotid artery ligation on brain lactate and pyruvate concentrations in normotensive and spontaneously hypertensive rats. Stroke 6: 62–66

6. Ginsberg MD, Welsh FA, Budd WW *et al* (1980) Deleterious effect of glucose pretreatment on recovery from diffuse cerebral ischemia in cat. Stroke 11: 347–354

7. Hossmann KA, Zimmermann V (1974) Resuscitatin of the monkey brain after 1 h complete ischemia. I. Physiological and morphological observations. Brain Res 81: 59–74

8. Hossmann KA, Schuier FJ (1980) Experimental brain infarcts in cats. I. Pathophysiological observations. Stroke 11: 583–592

9. Hossman V, Hossman KA, Takagi S (1980) Effect of intravascular platelet aggregation on blood recirculation following prolonged ischemia of the cat brain. J Neurol 222: 159–170

10. Hudgins WR, Garcia JH (1970) Transorbital approach to the middle cerebral artery of the squirrel monkey. A technique for experimental cerebral infarction applicable to ultrastructural studies. Stroke 1: 107–111

11. Ito U, Spatz M, Walker JT Jr *et al* (1975) Experimental cerebral ischemia in mongolian gerbils. I. Light microscopic observations. Acta Neuropathol (Berl) 32: 209–223

12. Jarrott DM, Domer FR (1980) A gerbil model of cerebral ischemia suitable for drug evaluation. Stroke 11: 203–209

13. Kameyama M, Shirane R, Suzuki J *et al* (1985) A new model of bilateral hemispheric ischemia in the rat—three vessel occlusion model. Stroke 16: 489–493

14. Kayama T, Mizoi K, Suzuki J (1981) A canine model of a completely ischemic brain regulated with the perfusion method. Surg Neurol 16: 167–172

15. Kogure K, Busto R, Scheinberg P *et al* (1974) Energy metabolites and water content in rat brain during the early stage of development of cerebral infarction. Brain 97: 103–114

16. Kudo M, Aoyama A, Ichimori S *et al* (1982) An animal model of cerebral infarction. Homologous blood clot emboli in rats. Stroke 13: 505–508

17. Laha RK, Israeli J, Dujouny M *et al* (1980) Low molecular weight dextran in experimental embolectomy. Stroke 11: 59–63

18. Lavyne MH, Hariri RJ, Tankosic T *et al* (1983) Effect of low dose γ-butyrolactone therapy on forebrain neuronal ischemia in the unanesthetized, awake rat. Neurosurgery 12: 430–434

19. Meyer JS (1958) Circulatory changes following occlusion of the middle cerebral artery and their relation to function. J Neurosurg 15: 653–673

20. Miller CL, Lampard DG, Alexander K et al (1980) Local cerebral blood flow following transient cerebral ischemia. I. Onset of impaired reperfusion within the first hour following global ischemia. Stroke 1: 534–541

21. Molinali GF (1970) Experimental cerebral infarction I: Selective segmental occlusion of intracranial arteries in the dog. Stroke 1: 224–231

22. Morawetz RB, DeGirolami U, Ojeman RG et al (1978) Cerebral blood flow determined by hydrogen clearance during middle cerebral artery occlusion in unanesthetized monkeys. Stroke 9: 143–149

23. Nakagawa Y, Yamamoto YL, Meyer E et al (1981) Effects of hypercapnia on enhancement of decreaced perfusion flow in noninfarcted brain tissues. Stroke 12: 86–92

24. Okada Y, Shima T, Yokoyama N et al (1983) Comparison of middle cerebral artery trunk occlusion by silicone cylinder embolization and by trapping. J Neurosurg 58: 492–499

25. Okamoto K, Yamori Y, Nagaoka A (1974) Establishment of the stroke-prone spontaneously hypertensive rats (SHR). Circ Rec SL 34: 143–153

26. Osterholm JL, Alderman JB, Triolo AJ et al (1983) Severe cerebral ischemia treatment by ventriculosubarachnoid perfusion with an oxygenated fluorocarbon emulsion. Neurosurgery 13: 381–387

27. Pulsinelli WA, Brierly JB (1979) A new model of bilateral hemispheric ischemia in the unanesthetized rat. Stroke 10: 267–272

28. Sakamoto T, Tanaka S, Yoshimoto T et al (1978) Experimental cerebral infarction. Part 2: Electroencephalographic changes produced by experimental thalamic infarction in dogs. Stroke 9: 214–216

29. Sakurada O, Kennedy C, Jehle J et al (1978) Measurement of local cerebral blood flow with iodo[14C]antipyrine. Am J Physiol 234: H 59–H 66

30. Shibata S, Hodge CP, Pappius HM (1974) Effect of experimental ischemia on cerebral water and electrolyte. J Neurosurg 41: 146–159

31. Steen PA, Newberg LA, Milde JH et al (1983) Nimodipine improves cerebral blood flow and neurologic recovery after complete cerebral ischemia in the dog. J Cereb Blood Flow Metab: 3: 38–43

32. Sundt TM Jr, Waltz AG (1966) Experimental cerebral infarction, retro-orbital extradural approach for occluding the middle cerebral artery. Mayo Clin Proc 41: 159–168

33. Sundt TM, Grant WC, Garcia JH (1969) Restoration of middle cerebral artery flow in experimental infarction. J Neurosurg 31: 311–322

34. Suzuki J, Yoshimoto T, Tanaka S et al (1980) Production of various models of cerebral infarction in the dog by means of occlusion of intracranial trunk arteries. Stroke 11: 337–341

35. Tamura A, Asano T, Sano K (1980) Correlation between rCBF and histological changes following temporary middle cerebral artery occlusion. Stroke 11: 487–493

36. Tamura A, Graham DI, McCulloch J et al (1981) Focal cerebral ischemia in the rat: 1. Description of technique and early neuropathological consequences following middle cerebral artery occlusion. J Cereb Blood Flow Metab 1: 53–60

37. Todd MM, Dunlop BJ, Shapiro HM et al (1981) Ventricular fibrillation in the cat. A model for global cerebral ischemia. Stroke 12: 808–815

38. Welch FA, O'Connor MJ, Marcy VR et al (1982) Factors limiting regeneration of ATP following temporary ischemia in cat brain. Stroke 13: 234–242

39. Wexler BC (1980) Comparative effects of unilateral and bilateral carotid artery ligation in the spontaneously hypertensive rat. Stroke 11: 72–78

40. Yamada K, Hayakawa T, Yoshimine T et al (1984) A new model of transient hind brain ischemia in gerbils. J Neurosurg 60: 1054–1058

41. Yonas H, Walfson SK Jr, Dujovny M et al (1981) Selective lenticulostriate occlusion in the primate. A high focal cerebral ischemia model. Stroke 12: 567–572

42. Yoshimine T, Yanagihara T (1983) Regional cerebral ischemia by occlusion of the posterior communicating artery and the

middle cerebral artery in gerbils. J Neurosurg 58: 362–367

43. Yoshimoto T, Sakamoto T, Suzuki J (1978)

Experimental cerebral infarction. Part 1: Production of thalamic infarction in dogs. Stroke 9: 211–214

CHAPTER 2

1. Ames A, Wright RL, Kowada M et al (1968) Cerebral ischemia. II. The no-reflow phenomenon. Am J Pathol 52: 437–453

2. Arsénio-Nunes ML, Hossmann KA, Farkas-Bargeton E (1973) Ultrastructural and histochemical investigation of the cerebral cortex of cat during and after complete ischemia. Acta Neuropathol (Berl) 26: 329–344

3. Brown AW, Brierley JB (1966) Evidence for early anoxic-ischemic cell damage in the rat brain. Experientia (Basel) 22: 546–547

4. Brown AW, Brierley JB (1968) The nature, distribution and earliest stages of anoxic-ischemic nerve cell damage in the rat brain as defined by the optical microscope. Br J Exp Pathol 49: 87–106

5. Brown AW, Brierley JB (1972) Anoxic-ischemic cell change in the rat brain. Light microscopic and fine structural observations. J Neurol Sci 16: 59–84

6. Brown AW, Brierley JB (1973) The earliest alterations in rat neurons and astrocytes after anoxia-ischemia. Acta Neuropathol (Berl) 23: 9–22

7. Brown AW (1977) Structural abnormalities in neurons. J Clin Pathol [Suppl] 11: 155–169

8. Calhoun CL, Mottaz JH (1966) Capillary bed of rat cerebral cortex. The fine structure in experimental cerebral infarction. Arch Neurol (Chicago) 15: 320–328

9. Cammermeyer J (1961) The importance of avoiding dark neurons in the experimental neuropathology. Acta Neuropathol 1: 245–270

10. Cammermeyer J (1973) Ischemic neuronal disease of Spielmeyer. A reevaluation. Arch Neurol 29: 390–393

11. Dodson RF, Kawamura Y, Aoyagi M et al (1973) A comparative evaluation of the ultrastructural changes following induced cerebral infarction in the squirrel monkey and baboon. Cytobios 8: 175–182

12. Fernand DA, Lau JK (1978) An electron microscope study of the effects of acute ischemia in the brain. Acta Anat (Basel) 100: 241–249

13. Fischer EG, Ames A III, Hedley-Whyte ET et al (1977) Reassessment of cerebral capillary changes in acute global ischemia and their relationship to the "no-reflow phenomenon". Stroke 8: 36–39

14. Furlow TW, Martin FM, Harrison LE (1984) Simultaneous measurement of local glucose utilization and blood flow in the rat brain. An autoradiographic method using two tracers labelled with carbon-14. J Cereb Blood Flow Metab 3: 62–66

15. Garcia JH, Kamijyo Y (1974) Cerebral infarction. Evaluation of histopathological changes after occlusion of a middle cerebral artery in primates. J Neuropath Exp Neurol 33: 408–421

16. Garcia JH, Trump BF (1975) Cerebral ischemia. The early structural changes and correlation of dynamic abnormalities. In: Whisnant J, Sandok BA (eds) Ninth Conference. Cerebral vascular disease. Grune and Stratton, New York, pp 313–323

17. Garcia JH, Lossinsky AS, Kauffman FC et al (1978) Neuronal ischemic injury: light microscopy, ultrastructure and biochemistry. Acta Neuropathol (Berl) 43: 85–95

18. Gildea EF, Cobb S (1930) Effects of anemia on the cerebral cortex of cat. Arch Neurol Psychiat 23: 876–903

19. Hills CP (1964) Ultrastructural changes in the capillary bed of the rat cerebral cortex in anoxic ischemic brain lesions. Am J Pathol 44: 531–551

20. Hossmann KA, Sato K (1975) Recovery of neuronal function after prolonged cerebral ischemia. Science 168: 375–376

21. Hossmann KA, Kleihues P (1973) Reversibility of ischemic brain damage. Arch Neurol 29: 375–384

22. Hounthoff KJ, Go KG (1980) Endogenous versus exogenous protein tracer passage in blood-brain barrier damage. In: Cervos-Navarro J, Ferszt R (eds) Brain edema. Adv Neurology 28. Raven Press, New York, pp 78–81

23. Ito U, Spatz M, Walker JT Jr *et al* (1975) Experimental cerebral ischemia in mongolian gerbils. I. Light microscopic observations. Acta Neuropathol (Berl) 32: 209–223

24. Jenkins LW, Povlishock JT, Becker DP *et al* (1979) Complete cerebral ischemia. An ultrastructural study. Acta Neuropathol (Berl) 48: 113–125

25. Johansen FF, Jorgensen MB, Diemer N (1983) Resistance of hippocampal CA-1 interneurons to 20 min of transient cerebral ischemia in the rat. Acta Neuropathol 61: 135–140

26. Kalimo H, Garcia JH, Kamijyo Y *et al* (1977) The ultrastructure of „brain death". II. Electron microscopy of feline cortex after complete ischemia. Virchows Archiv (Cell Pathol) 25: 207–220

27. Kalimo H, Rehncrona S, Söderfeldt B *et al* (1981) Brain lactic acidosis and ischemic cell damage: 2. Histopathology. J Cereb Blood Flow Metab 1: 313–327

28. Kirino T, Sano K (1980) Changes in the contralateral dentate gyrus in Mongolian gerbils subjected to unilateral cerebral ischemia. Acta Neuropathol (Berl) 50: 121–129

29. Kirino T (1980) Degeneration and repair of the brain after cerebral infarction—Changes in the neuropil of the contralateral hemisphere in Mongolian Gerbils. No To Shinkei 32: 1071–1079 (Eng Abstr)

30. Kirino T, Sano K (1984) Fine structural nature of delayed neuronal death following ischemia in the gerbil hippocamps. Acta Neuropathol (Berl) 62: 209–218

31. Koshu K, Yoshimoto T, Suzuki J (1980) Experimental study on hemorrhagic infarction following recirculation in thalamic ischemic lesion. Neurol Med Chir (Tokyo) 20: 935–938 (Eng Abstr)

32. Kowada M, Ames A III, Majno G *et al* (1968) Cerebral ischemia. I. An improved experimental method for study; cardiovascular effects and demonstration of an early

vascular lesion in the rabbit. J Neurosurg 28: 150–157

33. Lear JL, Jones SC, Greenberg JH *et al* (1981) Use of ^{123}I and ^{14}C in a double radionuclide autoradiographic technique for simultaneous measurement of LCBF and LCMRgl. Stroke 12: 589–597

34. Levine S, Payan H (1966) Effects of ischemia and other procedures on the brain and retina of the gerbil (Meriones unguiculatus). Exp Neurol 16: 255–262

35. Levine S (1960) Anoxic-ischemic encephalopathy in rats. Am J Pathol 36: 1–17

36. Lindberg R (1955) Compression of brain arteries as pathogenetic factor for tissue necrosis and their areas of pedilection. J Neuropathol Exp Neurol 14: 223–243

37. Lindberg R (1956) Morphometric and morphostatic necrobiosis. Am J Pathol 23: 1147–1177

38. Little JR, Kerr FW, Sundt TM Jr (1974) The role of lysosomes in production of ischemic nerve cell changes. Arch Neurol 30: 448–455

39. Little JR, Kerr FW, Sundt TM Jr (1974) Significance of neuronal alterations in developing cortical infarction. Mayo Clin Proc 49: 827–837

40. Little JR, Sundt TM Jr, Kerr FW (1974). Neuronal alterations in developing cortical infarction. An experimental study in monkeys. J Neurosurg 40: 186–198

41. Little JR, Kerr FW, Sundt TM Jr (1975) Microcirculatory observation in focal cerebral ischemia. An electron microscopic investigation in monkeys. Stroke 7: 25–30

42. Matakas F, Cervos-Navarro J, Schneider H (1973) Experimental brain death. 1. Morphology and fine structure of the brain. J Neurol Neurosurg Psychiatry 36: 497–508

43. McGee-Russell SM, Brown AW, Brierley JB (1970) A combined light and electron microscope study of early anoxic-ischemic cell change in rat brain. Brain Res 20: 193–200

44. Meyer JS (1958) Importance of ischemic damage to small vessels in experimental cerebral infarction. J Neuropathol Exp Neurol 17: 571–585

45. Mies G, Niebuhr I, Hossmann KA (1981) Simultaneous measurement of blood flow

and glucose metabolism by autoradiographic techniques. Stroke 12: 581–588

46. Mizoi K, Ogawa A, Seki H *et al* (1980) Changes of tissue pH in dog during ischemia—Alkaline shift. No To Shinkei 32: 265–268 (Eng Abstr)

47. Nishijima M, Tanaka S, Watanabe T *et al* (1981) Sequential changes in nerve cells during complete ischemia and the preventive effects of various drugs on cerebral infarction. No To Shinkei 33: 291–299 (Eng Abstr)

48. Ohishi H, Koshu K, Yoshimoto T *et al* (1981) Contrast enhancement on computed tomography in experimental cerebral infarction in dog. Jpn J Stroke 3: 23–26 (Eng Abstr)

49. Ohishi H, Watanabe T, Seki H *et al* (1983) Sequential changes of experimental cerebral infarction—CT and histological study. No To Shinkei 35: 983–988 (Eng Abstr)

50. Ohishi H, Nishijima M, Ogawa A *et al* (1984) Protective effect of mannitol in cerebral infarction—CT findings and physiological observation in experimental cerebral infarction in dogs. No Shinkei Geka 12: 153–158 (Eng Abstr)

51. Petito CK (1979) Platelet thrombi in experimental cerebral infarction. Stroke 10: 192–196

52. Petito CK, Babiak T (1982) Early proliferative changes in astrocytes in postischemic non infarcted rat brain. Ann Neurol 11: 510–518

53. Pulsinelli WA, Brierley JB (1979) A new method of bilateral hemispheric ischemia in the unanesthetized rat. Stroke 10: 267–272

54. Pulsinelli WA, Brierley JB, Plum F (1982) Temporal profile of neuronal damage in a model of transient forebrain ischemia. Ann Neurol 11: 491–498

55. Schneider H, Dralle J (1973) Ultrastructural changes in the rat spinal cord after temporary occlusion of the thoracic aorta. Acta Neuropathol (Berl) 26: 301–315

56. Scholtz W (1959) The contribution of pathoanatomical research to problem of epilepsy. Epilepsia 1: 36–55

57. Seki H, Ogawa A, Tanaka S *et al* (1980) Correlation between the thalamus EEG and rCBF in the thalamus infarction in the dogs. No To Shinkei 32: 1065–1069 (Eng Abstr)

58. Spielmeyer W (1922) Histopathologie des Nervensystems. Springer, Berlin

59. Suzuki M, Iwasaki Y, Yamamoto T *et al* (1984) Disintegration of orthogonal arrays in perivascular astrocytic processes as an early event in acute global ischemia. Brain Res 300: 141–145

60. Tureen LL (1936) Effect of experimental temporary vascular occlusion on the spinal cord: correlation between structural and functional changes. Arch Neurol Psychiat 35: 789–807

61. Watanabe T, Yoshimoto T, Tanaka S *et al* (1979) Ultrastructural observation of infarction changes of cerebral tissue in dog—I. Neuronal alterations. Neurol Med Chir (Tokyo) 19: 279–285

62. Watanabe T, Yoshimoto T, Koshu K *et al* (1979) Ultrastructural observation on the infarctic sequential changes in the cerebral tissue of dog—II. changes in small vessels. Neurol Med Chir (Tokyo) 19: 811–816 (Eng Abstr)

63. Watanabe T, Suzuki M, Yoshimoto T *et al* (1985) Recirculation in ischemic focus in the acute stage—electron microscopical examination. Neurol Med Chir (Tokyo) 25: 81–88 (Eng Abstr)

64. Welsh FA, Rieder W (1978) Evaluation of in situ freezing of cat brain by NADH fluorescence. J Neurochem 31: 299–309

65. Yoshimine T, Morimoto K, Yanagihara T (1982) Immunohistochemical investigation on cerebral ischemia. Stroke 13: 119

66. Yoshimine T, Yanagihara T (1983) Regional cerebral ischemia by occlusion of the posterior communicating artery and of the middle cerebral artery in gerbils. J Neurosurg 58: 362–367

67. Yoshimoto T, Sakamoto T, Suzuki J (1978) Experimental cerebral infarction. Part I. Production of thalamic infarction in dogs. Stroke 9: 211–214

CHAPTER 3

1. Ames A, Wright RL, Kowada M *et al* (1968) Cerebral ischemia II. The no-reflow phenomenon. Am J Pathol 52: 437–453

2. Astrup J, Siesjo BK, Symon L (1981) Thresholds in cerebral ischemia—the ischemic penumbra—. Stroke 12: 723–725

3. Aukland K, Bower BF, Berliner RW (1964) Measurement of local blood flow with hydrogen gas. Circ Res 14: 164–187

4. Branston NM, Symon L, Crockard HA *et al* (1974) Relationship between the cortical evoked potential and local cortical blood flow following acute middle cerebral artery occlusion in the baboon. Exp Neurol 45: 195–208

5. Davis KR, Ackerman RH, Kistler JP *et al* (1977) Computed tomography of cerebral infarction: hemorrhagic, contrast enhancement, and time of appearance. Comput Tomogr 1: 71–86

6. Doyle TF, Martins AN, Kobrine AI (1975) Estimating total cerebral blood flow from the initial slope of hydrogen washout curve. Stroke 6: 149–152

7. Fieschi C, Bozzao L, Agnol A (1965) Regional clearance of hydrogen as a measure of cerebral blood flow. Acta Neurol Scand [Suppl] 14: 46–52

8. Halsey JH Jr, Capra NF, McFarland RS (1977) Use of hydrogen for measurement of regional cerebral blood flow: problem of intercompartmental diffusion. Stroke 8: 351–357

9. Hasuo M, Furuse M, Brock M (1978) Brain tissue pressure gradients and cerebral blood flow dynamics. No Shinkei Geka 6: 789–794 (Eng Abstr)

10. Heiss WD, Hayakawa T, Waltz AG (1976) Cortical neuronal function during ischemia. Effects of occlusion of one middle cerebral artery on single-unit activity in cats. Arch Neurol 33: 813–820

11. Heiss WD (1983) Flow thresholds of functional and morphologic damage of brain tissue. Stroke 14: 329–331

12. Hossmann KA, Lechtape-Gruter H, Hossmann V (1973) The role of cerebral blood flow for the recovery of the brain after prolonged ischemia. Z Neurol 204: 281–299

13. Irino T (1978) Review of clinical reports about recanalization of occluded cerebral artery. No To Shinkei 30: 135–151 (Eng Abstr)

14. Ito U, Ohno K, Tomita H *et al* (1976) Cerebral changes during recirculation following temporary ischemia in mongolian gerbils. Neurol Med Chir (Tokyo) (16) 2: 313–322 (Eng Abstr)

15. Jones TH, Morawetz RB, Crowell RM *et al* (1981) Thresholds of focal cerebral ischemia in awake monkeys. J Neurosurg 54: 773–782

16. Kogure K, Fujishima M, Scheinberg P *et al* (1969) Effects of changes in carbon dioxide pressure and arterial pressure on blood flow in ischemic regions of the brain in dogs. Circ Res 24: 557–565

17. Koshu K, Seki H, Yoshimoto T *et al* (1981) Experimental hemorrhagic thalamic infarction in the dog. Surg Neurol 16: 274–279

18. Lassen NA (1966) The luxury-perfusion syndrome and its possible relationship to acute metabolic acidosis localized within the brain. Lancet 2: 1113–1115

19. Lim RSK, Lim CN, Moffitt RL (eds) (1960) A stereotaxic atlas of the dog's brain. Charles C Thomas, Springfield

20. Mabe H, Uemura S, Yoshida T *et al* (1982) Correlation between local cerebral blood flow and EEG in experimental cerebral ischemia. No To Shinkei 34: 585–589 (Eng Abstr)

21. Marmarou A, Poll W, Shapiro K *et al* (1976) The influence of brain tissue pressure gradience upon local cerebral blood flow in vasogenic edema. In: Becks JWF, Bosch DA, Brock M (eds) Intracranial pressure, vol 3. Springer, Berlin Heidelberg New York, pp 10–13

22. Meyer JS, Fand HC, Denny-Brown D (1954) Polarographic study of cerebral collateral circulation. Arch Neurol Psychiat 72: 296–312

23. Meyer JS, Gotoh F, Tazaki Y (1962) Circulation and metabolism following experimental cerebral embolism. J Neuropath Exp Neurol 21: 4–24

24. Miller JD, Garibi J, North JB *et al* (1975) Effects of increased arterial pressure on blood flow in the damaged brain. J Neurol Neurosurg Psychiatry 38: 657–665

25. Morawetz RB, DeGirolami U, Ojeman RG et al (1978) Cerebral blood flow determined by hydrogen clearance during middle cerebral artery occlusion in unasthetized monkeys. Stroke 9: 143–149

26. Nemoto EM, Snyder JV, Carroll Rg et al (1965) Global ischemia in dogs—cerebrovascular CO_2 reactivity and autoregulation—. Stroke 6: 425–431

27. Ogawa A, Seki H, Mizoi K et al (1979) Circulatory dynamics in experimental focal cerebral infarction. Acta Neurol Scand [Suppl] 72: 284–285

28. Ogawa A, Seki H, Yoshimoto T et al (1982) Experimental focal cerebral infarction. Part 1: Hemodynamics at the center of the focal cerebral infarction following recirculation. Jpn J Stroke 4: 1–9 (Eng Abstr)

29. Ogawa A, Seki H, Yoshimoto T et al (1982) Experimental focal cerebral infarction. Part 2: Hemodynamics in and around the focal cerebral infarction following recirculation. Jpn J Stroke 4: 10–17 (Eng Abstr)

30. Olsen TS, Larsen B, Herning M et al (1983) Blood flow and vascular reactivity in collaterally perfused brain tissue. Evidence of an ischemic penumbra in patients in acute stroke. Stroke 14: 332–341

31. Sakamoto T, Tanaka S, Yoshimoto T et al (1978) Experimental cerebral infarction. Part 2: Electroencephalographic changes produced by experimental thalamic infarction in dogs. Stroke 9: 214–216

32. Sato H, Seki H, Ogawa A et al (1985) Sequential changes of autoregulation in and around focal ischemia. In: Spetzler RF, Carter LP, Selman WR et al (eds) Cerebral revascularization for stroke, pp 48–52

33. Seki H, Ogawa A, Tanaka S et al (1980) Correlation between the the thalamus EEG and rCBF in the thalamus infarction in the dogs. No To Shinkei 32: 1065–1069 (Eng Abstr)

34. Seki H, Yoshimoto T, Ogawa A et al (1984). The CO_2 response in focal cerebral ischemia—sequential changes following recirculation. Stroke 15: 699–704

35. Seki H, Yoshimoto T, Ogawa A et al (1983) Effect of mannitol on rCBF in canine thalamic ischemia—an experimental study—. Stroke 14: 46–50

36. Seki H, Yoshimoto T, Ogawa A et al (1985) Hemodynamics in hemorrhagic infarction—an experimental study—. Stroke 16: 647–651

37. Strong AJ, Venables GS, Gibson G (1983) The cortical ischemic penumbra associated with occlusion of the middle cerebral artery in the cat. 1. Topography of changes in blood flow, potassium ion acitivity, and EEG. J Cereb Blood Flow Metab 3: 86–96

38. Sundt TM, Michenfelder JD (1972) Focal transient cerebral ischemia in the squirrel monkey. Effect on brain adenosin triphosphate and lactate levels with electrocorticographic and pathologic correlation. Circ Res 30: 703–712

39. Suzuki H, Kimura S, Watanabe M et al (1979) The EEG changes during the extracorporeal circulation. Rinsho Noha 21: 1–9 (Eng Abstr)

40. Symon L, Pasztor E, Branston NM (1974) The distribution and density of reduced cerebral blood flow following acute middle cerebral artery occlusion—An experimental study by the technique of hydrogen clearance in baboons. Stroke 5: 355–364

41. Symon L, Branston NM, Strong AJ (1976) Autoregulation in acute focal ischemia—an experimental study—. Stroke 7: 547–554

42. Tamura A, Sano K (1978) The temporary occlusion of the middle cerebral artery in cat—The correlation between the rCBF and the histological changes. No To Shinkei 31: 1005–1014 (Eng Abstr)

43. Traupe H, Kruse E, Heiss WD (1982) Reperfusion of focal ischemia of varing duration. Postischemic hyper- and hypoperfusion. Stroke 13: 615–622

44. Waltz AG, Sundt TM Jr (1967) The microvasculature and microciruclation of the cerebral cortex after arterial occlusion. Brain 90: 681–691

45. Waltz AG (1970) Effect of $PaCO_2$ on blood flow and microvasculature of ischemic and nonischemic cerebral cortex. Stroke 1: 27–37

46. Waltz AG (1974) Effect of blood pressure on blood flow in ischemic and nonischemic cerebral cortex—the phenomena of autoregulation and luxury perfusion—. Neurology 41: 755–757

47. Watanabe T, Yoshimoto T, Koshu K *et al* (1976) Ultrastructural observation on the infarctic sequential changes in the cerebral tissue of dog. Part 2. Changes in small vessels. Neurol Med Chir (Tokyo) 19: 811–816 (Eng Abstr)

48. Watanabe T, Suzuki M, Yoshimoto T *et al.* (1985) Recirculation in ischemic focus in the acute stage—electron microscopical observation. Neurol Med Chir (Tokyo) 25: 81–88 (Eng Abstr)

49. Yoshimoto T, Sakamoto T, Suzuki J (1978) Experimental cerebral infarction. Part 1: Production of thalamic infarction in dogs. Stroke 9: 211–214

CHAPTER 4

1. Arai H, Suzuki M, Yoshimoto T *et al* (1985) The early permeability change to macromolecules after transient global ischemia. In: Inaba Y, Klatzo I, Spatz M (eds) Brain edema. Springer, Berlin Heidelberg New York Tokyo, pp 149–154

2. Bergström M, Ericson K (1979) Compartment analysis of contrast enhancement in brain infarctions. J Comput Assist Tomogr 3: 234–240

3. Bradbury M (1979) The conept of a blood brain barrier. John Wiley, New York Brisbane Toronto

4. Bralet J, Beley P, Bralet AM *et al* (1983) Comparison of the effects of hypertonic glycerol and urea on brain edema, energy metabolism and blood flow following cerebral microembolism in the rat. Deleterious effect of glycerol treatment. Stroke 14: 597–604

5. Bremer AW, West CR, Yamada K (1978) Alteration in the distribution of water, sodium, potassium, and chloride in brain during the evolution of ischemic cerebral edema. Neurosurgery 3: 187–195

6. Brightmann MW, Hori M, Rapoport SI *et al* (1973) Osmotic opening of tight junctions in cerebral endothelium. J Comp Neurol 152: 317–325

7. Buchnill JC, Tuke DH (1874) A manual of psychological medicine, 3rd edn Churchill, London, p 587

8. Chan FH, Fishman RA (1984) Phospholipid degradation and the early release of polyunsaturated fatty acida in the revolution of brain edema. In: Go KG *et al* (eds) Recent progress in the study and therapy of brain edema. Plenum, New York London, pp 193–203

9. Cohadon F, Rigoulet M, Averet N (1982) Alterations of membrane-bound enzymes in vasogenic edema. In: Go G, Beathmann A (eds) Recent progress in the study and therapy of brain edema. Plenum, New York London, pp 223–231

10. Deane Br, Greendwood J, Lantos PL *et al* (1984) The vasculature of experimental brain tumors Part 4: The quanitification of vascular permeability. J Neurol Sci 65: 59–68

11. Fenske A, Fischer M, Regli F *et al* (1979) The response of focal ischemic cerebral edema to dexamethasone. J Neurol 220: 199–209

12. Fujimoto T, Walker JT, Spatz W Jr *et al* (1976) Pathophysiologic aspects of ischemic edema. In: Pappius HM, Feindel W (eds), dynamics of brain edema, Springer, Berlin Heidelberg New York, pp 171–180

13. Galicich JH, French LA, Melby JC (1961) Use of dexamethasone in treatment of cerebral edema associated with brain tumors. Lancet 81: 46–53

14. Gobiet W (1977) Monitoring of intracranial pressure in patients with severe head injury: a review of 100 cases. Acta Neurochir (Wien) 20: 35–47

15. Greenfield JG (1939) The histology of cerebral edema associated with intracranial tumors (with special reference to changes in the nerve fibers of the centrum ovales). Brain 62: 129–151

16. Higashi K, Maza T (1976) The effect of osmotic pressure on the production of cerebrospinal bluid. Neurol Med Chir (Tokyo) 16: 5–13 (Eng Abstr)

17. Hirano A (1980) Fine structure of edematous encephalopathy. Adv Neurology 28: 83–97

18. Hossmann KA, Lechtape Yruter H, Hossmann V (1973) The role of cerebral blood flow for the recovery of the brain after

prolonged ischemia. J Neurol 304: 281–299

19. Hutton M, Rhodes RS, Chapman G (1980) the lowering of postischemic compartment pressures with mannitol. J Surg Res 32: 239–242

20. Ito U, Go KG, Walker JT *et al.* (1976) Experimental cerebral ischemia. II Behavior of blood-brain barrier. Acta Neuropathol (Berl) 34: 106

21. Ito U, Suganuma Y, Suzuki K *et al* (1980) Effect of steroid on ischemic brain edema. Analysis of cytotoxic and vasogenic edema occuring during ischemia and after restoration of blood flow. Stroke 11: 166–167

22. Jerusalem C, Polder T, Kubat K *et al* (1982) Brain edema in cerebral malaria: a comparative clinical and experimental, ultrastructural and histochemical study. In: Go G, Beathmann A (eds) Recent progress in the study and therapy of brain edema. Plenum Press, New York London, pp 126–127

23. Katzman R, Passius HM (1973) Brain electrolytes and fluid metabolism. Williams and Wilkins, Baltimore

24. Kayama T, Sakurai Y, Fujimoto S *et al* (1985) Experimental study of ischemic brain edema by differential scanning calorimetry and NMR spin analyser. In: Inaba Y, Klatzo I, Spatz M (eds) Brain edema. Springer, Berlin Heidelberg New York Tokyo, pp 383–391

25. Klatzo I (1972) Neurpathological aspects of brain edema. J Neuropath Exp Neurol 26: 1–15

26. Klatzo I (1983) Disturbance of the blood-brain barrier in cerebrovascular disorders. Acta Neuropathol [Suppl] (Berl) 8: 31–38

27. Klatzo I, Suzuki R, Orzi F *et al* (1984) Pathomechanisms of ischemic brain edema. In: Go KG *et al* (eds) Recent progress in the study and therapy of brain edema. Plenum, New York London, pp 1–10

28. Kogure K, Busto R, Scheinberg P (1981) The role of hydrostatic pressure in ischemic brain edema. Ann Neurol 9: 273–282

29. Kostron H, Fischer J (1983) Regional, cellular and subcellular distribution of [³H]dexamethasone in rat. Surg Neurol 20: 48–54

30. Kuroiwa T, Ting P, Suzuki R *et al* (1982) The relationship of the blood-brain barrier (BBB) opening to the thresholds of regional cerebral blood flow (rCBF) in cerebral ischemia. J Neuropath Exp Neurol 34: 352

31. Langfitt TW (1961) Possible mechanisms of action of hypertonic urea in reducing intracranial pressure. Neurology (Mineap) 11: 196–209

32. Laursen H, Westergaard E (1978) The permeability of the blood-brain barrier und cell membranes to horseradish peroxidase in hyperammonaemia. Acta Neuropathol 54: 293–299

33. Little JR (1978) Modification of acute focal ischemia by treatment with mannitol. Stroke 9: 4–9

34. Little JR (1978) Modification of acute focal ischemia by treatment with mannitol and dexamethasone. J Neurosurg 49: 517–524

35. Lorenzo AV, Heley Whyte ET, Eisenberg HM (1975) Increased penetration of horseradish peroxidase across the blood-brain barrier induced by metrazol seizures. Brain Res 88: 136–140

36. Maier-Hauff K, Lange M, Shcuerer L *et al* (1984) Glutamate and free fatty acid concentrations in extracellular vasogenic edema fluid. In: Go KG *et al* (eds) Recent progress in the study and therapy of brain edema. Plenum, New York London, pp 183–192

37. Marmarrow A, Nakamura T, Tanaka K *et al* (1982) The time course and distribution of water in the resolution phase of infusion edema. In: Go G *et al* (eds) Recent progress in the study and therapy of brain edema. Plenum, New York London, pp 37–44

38. Marshal LF, Smith RW, Rauscher LA *et al* (1978) Mannitol dose requirements in brain-infured patients. J Neurosurg 48: 169–172

39. McGraw CP, Alexander E, Howard G (1978) Effect of dose and dose schedule on the response of intracranial pressure to mannitol. Surg Neurol 10: 127–130

40. McQueen JD, Jeanes LD (1964) Dehydration and rehydration of the brain with hypertonic urea and mannitol. J Neurosurg 21: 118–128

41. Mening G, Reulen HJ, Simon RS *et al* (1980) Clinical, chemical and CT avaluation of short-term and long-term antiedema therapy with dexamethasone and diuretics. In: Cervos-Navaro J, Ferzi R (eds) Brain edema. Raven Press, New York, pp 471–489

42. Meyer JS, Fukuchi Y, Shimazu K et al (1972) Effect of intravenous infusion of glycerol on hemispheric blood flow and metabolism in patients with acute cerebral infarction. Stroke 3: 168–180

43. Meyer JS, Itoh Y, Okamoto S et al (1978) Effect of dose and dose schedule on the response of intracranial pressure to mannitol. Surg Neurol 10: 127–130

44. Mrsulija BB, Kurcic BM, Cvejic V et al (1980) Biochemistry of experimental ischemic brain edema. In: Cervos-Navarro J et al (eds) Brain edema. Advances in neurology, Vol 28. Raven, New York, pp 217–230

45. Muizelaar JP, Wei EP, Kontos HA et al (1983) Mannitol causes compensatory cerebral vasoconstriction and vasodilatation in response to blood viscosity changes. J Neurosurg 59: 822–828

46. Node Y, Yajima K, Nakazawa S (1983) A study of mannitol and glycerol on the reduction of raised in pressure and on their rebound phenomenon. No Shinkei Geka 11: 259–267 (Eng Abstr)

47. Norton GA, Kishore PR, Lin J (1978) CT contrast enhancement in cerebral infarction. JAR 131: 881–885

48. Oldenhoff (1975) Permeability of the blood-brain barrier. In: Brady RO (ed) The nervous system, Vol 1. Raven Press, New York, pp 4–63

49. Pappius HM, Dayer LA (1964) Hypertonic urea; its effect on the distribution of water and electrolytes in normal and edematous brain tissued. Arch Neurol 13: 395–402

50. Parsson LI, Johanson BB, Hanston HS (1978) Ultrastructural studies in blood-brain barrier dysfunction after cerebral air embolism in the rat. Acta Neuropathol (Berl) 44: 53–56

51. Petito CK (1979) Early and late mechanisms of increased vascular permeability following experimental cerebral infarction. J Neuropathol Exp Neurol 38: 224–234

52. Pollay M, Fullenwider C, Roberts A et al (1983) Effect of mannitol and furosemide on blood-brain osmotic gradient and intracranial pressure. J Neurosurg 59: 945–950

53. Ransohoff J (1979) Membrane perturbations in central nervous system injury: Theoretical basis for free radical damage and a review of the experimental data. In: Popp AJ et al (eds) Neural trauma. Raven Press, New York, pp 63–78

54. Rapoport SI, Hori M, Klatzo I (1972) Testing of a hypothesis for osmotic opening of the blood-brain barrier. Am J Physiol 223: 323–331

55. Reichardt M (1985) Zur Entsuchung des Hirndrück bei Hirngeschwülsten und deren Hirn-Krankheiten und über eine bei diesen zu beobachtende besondere Art der Hirnschwellung. Dtsch Z Nervenheilk 28: 306–355

56. Reulen HJ, Tsuyumu M, Tack A et al (1978) Clearance of edema fluid into cerebrospinal fluid. J Neurosurg 48: 754–769

57. Rob CG (1969) Operation for acute complete stroke due to thrombosis of the internal carotid artery. Surgery 65: 862–863

58. Schuier FJ, Hossmann KA (1980) Experimental brain infarcts in cats. II. Ischemic brain edema. Stroke 11: 593–601

59. Selbach H (1952) Physikalischchemische Untersuchungen zur Grage der Hirnvolumsvermehrung (Hirnschwellung und Hirnoedem). Arch Psychiat 112: 409–440

60. Shaller CA, Jacques S, Shelen CH (1980) The pathophysiology of stroke: a review with molecular considerations. Surg Neurol 14: 433–443

61. Sheinker IM (1947) Cerebral swelling; histology, classification and clinical significance of brain edema. J Neurosurg 4: 255–275

62. Siesjö BK (1979) Ischemia. In: Siesjö BK (ed) Brain energy metabolism. Williams and Wilkins, Baltimore, pp 453–526

63. Siesjö BK (1981) Cell damage in the brain: a speculative synthesis. J Cereb Blood Flow Metab 1: 155–185

64. Spatz H (1929) Die Bedeutung der „Symptomatischen" Hirnschwellung für die Hirntumoren und für andere raumbeengende Prozesse in der Schaedelgrube. Arch Psychiat 88: 790–794

65. Spatz M, Go GK, Klatzo I (1974) The effect of ischemia on the brain uptake of 14C glucose analogues and 14C sucrose. In: Cervos-Navarro J (ed) Pathology of cerebral microcirculation. De Gruyter, Berlin, pp 361–366

66. Spatz M, Go KG, Klatzo I (1980) Monoamine neurotransmitter change vascular permeability in cerebral ischemia. Ann Neurol 10: 273–279
67. Suzuki J, Kayama T, Yoshimoto T et al (1984) Suppression of brain swelling with mannitol and perfluorochemicals: an experimental study. In: Go KG, Beathman A (eds) Progress in the study and therapy of brain edema. Plenum, New York London, pp 711–718
68. Suzuki R, Yamaguchi T, Kirino T et al (1983) The effects of 5 minute ischemia in mongolian gerbils: I. Blood-brain barrier, cerebral blood flow, and local cerebral glucose utilization changes. Acta Neuropathol (Berl) 60: 207–216
69. Tomita M, Gotoh F, Sato T et al (1979) Determination of the osmotic potential for swelling of cat brain in vitro. Exp Neurol 65: 66–77
70. Unterberg A, Maier-Hauff K, Wahl M et al (1984) Cerebral uptake and consumption of plasma-kininogens in vasogenic brain edema. Pathomechanisms of ischemic brain edema. In: Go KG et al (eds) Recent Progress in the study and therapy of brain edema. Plenum, New York London, pp 175–182
71. Weed LH, McKiben PS (1919) Pressure changes in cerebrospinal fluid following intravenous infection of solutions of various concentrations. Am J Physiol 48: 512–530
72. Westergaard E, Go G, Klatzo I et al (1976) Increased permiability of cerebral vessels to horseradish peroxidase induced by ischemia in mongolian gerbil. Acta Neuropathol (Berl) 35: 307–325
73. Wise BL, Chater N (1962) The value of hypertonic mannitol solution in decreasing brain mass and lowering cerebrospinal fluid pressure. J Neurosurg 19: 1038–1043
74. Zülich KJ (1962) Hirnödem, Hirnschwellung, Hirndruck. Zbl Neurochir 12: 174–186

CHAPTER 5

1. Abdel-Halim MS, Von Holst H, Meyerson B et al (1980) Prostaglandin profiles in tissue and blood vessels from human brain. J Neurochem 34: 1331–1333
2. Abe K, Yoshida S, Watson BD et al (1983) Alpha-Tocopherol and ubiquinones in rat brain subjected to decapitation ischemia. Brain Res 273: 166–169
3. Aldrete JA, Remo-Salas F, Jankovsky L et al (1979) Effect of pretreatment with thiopental and phenytoin on postischemic brain damage in rabbits. Crit Care Med 7: 466–470
4. Aldrete JA, Romo-Salas F, Mazzia VDB et al (1981) Phenytoin for brain resuscitation after cardiac arrest: an uncontrolled clinical trial. Crit Care Med 9: 474–477
5. Anderson BR, Lint TF, Brenzel AM (1978) Chemically shifted singlet oxygen spectrum. Biochim Biophys Acta 542: 527–536
6. Artru AA, Michenfelder JD (1980) Cerebral protective metabolic and vascular effects of phenytoin. Stroke 2: 377–382
7. Artru AA, Michenfelder JD (1981) Anoxic cerebral potassium accumulation reduced by phenytoin: mechanism of cerebral protection? Anesth Analg 60: 41–45
8. Aust SD, Svingen BA (1982) The role of iron in enzymatic lipid peroxidation. In: Pryor WA (ed) Free radicals in biology, vol 4. Academic Press, New York, pp 1–28
9. Baker PF (1978) The regulation of intracellular calcium in giant axons of loligo and myxicola. Ann NY Acad Sci 307: 250–268
10. Baker PF, Blaustein MP, Hodgkin AL et al (1969) The influence of calcium on sodium efflux in squid axons. J Physiol 200: 431–458
11. Baldy-Moulinier M (1971/72) Cerebral blood flow and membrane ionic pump, cerebral blood flow and intracranial pressure. Proc 5th Int, Symp, Roma-Siena 1971 Part I. Eur Neurol 6: 107–113
12. Barb WG, Baxendale JH, Gorge P et al (1951) Reactions of ferrous and ferric ions with hydrogen peroxide. Part II: the ferric ion reaction. Trans Faraday Soc 47: 591–616

13. Barrit GJ (1981) Calcium transport across cell membranes: progress toward molecular mechanism. Trends in Biochem Sci 6: 322–325

14. Bartschat DK, Cyr DL, Lindenmayer GE (1980) Depolarization induced calcium uptake by vesicle in a highly enriched sarcolemma preparation from canine ventricle. J Biol Chem 255: 10044–10047

15. Bateman L (1954) Olefin oxidation. Quart Rev 8: 147–167

16. Baulieu E-E (1978) Cell membrane, a target for steroid hormones. Mol Cell Endocrinol 12: 247–254

17. Bażan NG Jr (1970) Effects of ischemia and electroconvulsive shock on free fatty acid pool in the brain. Biochim Biophys Acta 218: 1–10

18. Bażan NG, Rodriguez de Turco EB (1980) Membrane lipids in the pathogenesis of brain edema; phospholipids and arachidonic acid, the earliest membrane components changed at the onset of ischemia. In: Cervós-Navarro J, Ferszt R (eds) Brain edema. Advances in neurology, vol 28. Raven Press, New York, pp 197–205

19. Bażan NG Jr, de Bazan HEP, Kennedy WG et al (1971) Regional distribution and rate of production of free fatty acids in rat brain. J Neurochem 18: 1387–1393

20. Bażan NG Jr (1976) Free arachidonic acid and other lipids in the nervous system during early ischemia and after electroshock. In: Porcellati G, Amaducci L, Galli C (eds) Function and metabolism of phospholipids in the central and peripheral nervous systems. Plenum Press, New York, pp 317–335

21. Berliner LJ (1976) Spin labeling theory and applications. Academic Press, New York

22. Berridge MJ (1979) Modulation of nervous activity by cyclic nucleotides and calcium. In: Schmit FO, Worden FG (eds) The neurosciences: fourth study program. MIT Press, Cambridge, Mass, pp 873–889

23. Berridge MJ (1982) A novel cellular signaling system based on the integration of phospholipid and calcium metabolism. In: Cheung WY (ed) Calcium and cell function, vol III. Academic Press, New York, pp 1–36

24. Berridge MJ (1983) Rapid accumulation of inositol triphosphate reveals that agonists hydrolyze polyphosphoinositides instead of phosphatidylinositol. Biochem J 212: 849–858

25. Berridge MJ (1984) Inositol triphosphate and diacylglycerol as second messengers. Biochem J 220: 345–360

26. Blackwell GJ, Flower RJ, Nijkamp FP et al (1978) Phospholipase A_2 activity of guinea-pig isolated perfused lungs: stimulation and inhibition by anti-inflammatory steroids. Br J Pharmacol 62: 79–89

27. Boehme DH, Koseki R, Carson S et al (1977) Lipidperoxidation in human and rat brain tissue: Developmental and regional studies. Brain Res 136: 11–21

28. Bolland JL (1949) Kinetics of olefin oxidation. Quart Rev 3: 1–21

29. Bothe HW, Bodsch W, Hossmann KA (1984) Relationship between specific gravity, water content and serum protein extravasation in various types of vasogenic brain edema. Acta Neuropathol 64: 37–42

30. Boveris A, Chance B, Filipkowski M et al (1978) Enhancement of the chemiluminescence of perfused rat liver and isolated mitochondria and microsomes by hydroperoxides. In: Scarpa A, Dutton PL, Leigh JS (eds) Frontiers to biological energetics, electron to tissues, vol 2. Academic Press, New York, pp 975–984

31. Boveris A, Cadenas E, Reiter R et al (1980) Organ chemiluminescence: Noninvasive assay for oxidation radical reactions. Proc Natl Acad Sci USA 77: 347–351

32. Boveris A, Cadenas E, Chance B (1981) Ultraweak chemiluminescence: A sensitive assay for oxidative radical reactions. Fed Proc 40: 195–198

33. Braughler JM (1985) Lipid peroxidation-induced inhibition of γ-aminobutyric acid uptake in rat brain synaptosomes: protection by glucocorticoids. J Neurochem 44: 1282–1288

34. Bremer AM, Yamada K, West CR (1980) Ischemic cerebral edema in primates: effect of acetazolamide, phenytoin, sorbitol, dexamethasone, and methylprednisolone on brain water and electrolytes. Neurosurgery 6: 149–154

35. Broddle WD, Nelson SR (1968) The effect of diphenylhydantoin on energy reserve levels in brain. Fed Proc 27: 751

36. Bunting S, Moncada S, Vane JR (1983) The prostacyclin-thromboxane A$_2$ balance: pathophysiological and therapeutic implications. Br Med Bull 39: 271–276

37. Cadenas E, Arad ID, Boveris A et al (1980) Partial spectral analysis of the hydroperoxide-induced chemiluminescence of the perfused lung. FEBS Lett 111: 413–418

38. Cadenas E, Arad ID, Fisher AB et al (1980) Hydroperoxide induced chemiluminescene of the perfused lung. Biochem J 192: 303–309

39. Cadenas E, Boveris A, Chance B (1985) Low-level chemiluminescence of hydroperoxide-supplemented cytochrome c. Biochem J 187: 131–140

40. Cadenas E, Boveris A, Chance B (1980) Spectral analysis of low level chemiluminescence of hydrogen peroxide supplemented ferricytochrome c. FEBS Lett 112: 285–288

41. Cadenas E, Boveris A, Chance B (1980) Low-level chemiluminescence of bovine heart submitochondrial particles. Biochem J 186: 659–667

42. Cadenas E, Boveris A, Chance B (1980) Chemiluminescence of lipid vesicles supplemented with cytochrome c and hydroperoxide. Biochem J 188: 577–583

43. Capaldi RA, Vanderkooi G (1972) The low polarity of many membrane proteins. Proc Natl Acad Sci USA 69: 930–932

44. Capaldi RA (1982) Arrangement of proteins in the mitochondrial inner membrane. Biochim Biophys Acta 694: 291–306

45. Carafoli E, Crompton M (1978) The regulation of intracellular calcium. Curr Top Membr Transp 10: 151–216

46. Chan PH, Fishman RA (1978) Brain Edema: Induction in cortical slices by polyunsaturated fatty acid. Science 201: 358–360

47. Chan PH, Fishman RA (1980) Transient formation of superoxide radicals in polyunsaturated fatty acid-induced brain swelling. J Neurochem 35: 1004–1007

48. Chandrabose KA, Lapetina EG, Schmitges CJ et al (1978) Action of corticosteroids in regulation of prostaglandin biosynthesis in cultured fibroblasts. Proc Natl Acad Sci USA 75: 214–217

49. Cheung Y (1980) Calmodulin plays a pivotal role in cellular regulation. Science 207: 19–27

50. Chien KR, Abrams J, Serroni A et al (1978) Accelerated phospholipid degeneration and associated membrane dysfunction in irreversible ischemic liver cell injury. J Biol Chem 253: 4809–4817

51. Cooper AJL, Pulsinelli WA, Duffy TE (1980) Glutathione and ascorbate during and postischemic reperfusion in rat brain. J Neurochem 35: 1242–1245

52. Crabtree GR, Munck A, Smith KA (1979) Glucocorticoids inhibit expression of Fc receptors on the human granulocytic cell line HL-60. Nature 279: 338–339

53. Crane P, Swanson PD (1970) Diphenylhydantoin and the cations and phosphates of electrically stimulated brain slices. Neurology (Minneap) 20: 1119–1123

54. Cullen JP, Aldrete JA, Jankovsky L et al (1979) Protective action of phenytoin in cerebral ischemia. Anesth Analg 58: 165–169

55. Dahle LK, Hill EG, Holman RT (1962) The thiobarbituric acid reaction and the autoxidation of polyunsaturated fatty acid methyl esters. Arch Biochem Biophys 98: 253–261

56. Dahlen SE, Björk J, Hedquist P et al (1981) Leukotriens promote plasma leakage and leukocyte adhesion in post capillary vessels: In vivo effects with relevance to the acute inflammatory response. Proc Natl Acad Sci USA 78: 3887–3891

57. Deneke CF, Krinksy NI (1977) Inhibition and enhancement of singlet oxygen ($^1\Delta$g) dimol chemiluminescence. Photochem Photobiol 25: 299–304

58. Dannenberg JrAM (1979) The antiinflammatory effects of glucocorticosteroids. A brief review of the literature. Inflammation 3: 329–343

59. De Medio GE, Goracci G, Horrocks LA et al (1980) The effect of transient ischemia on fatty acid and lipid metabolism in the gerbil brain. Ital J Biochem 29: 412–432

60. DeVries GH, Norton WT (1974) The lipid composition of axons from bovine brain. J Neurochem 22: 259–264

61. Dembinska-Kiec A, Korbut R, Zmuda A *et al* (1984) Formation of lipoxygenase and cyclooxygenase metabolites of arachidonic acid by brain tissue. Biomed Biochim Acta 43: 222–226

62. Demopoulos HB, Flamm ES, Seligman ML *et al* (1977) Antioxidant effects of barbiturates in model membranes undergoing free radical damage. Acta Neurol Scand [Suppl] 56: 152–153

63. Demopoulos HB, Landgraf W, Duke PS *et al* (1966) Light-induced alterations in melanoma-related free radicals, and the consequences on respiration and growth. Lab Invest 15: 1652–1658

64. Demopoulos HB, Milvy P, Kakari S *et al* (1972) Molecular aspects of membrane structure in cerebral edema. In: Reulen HJ, Schurmann K (eds) Steroid and brain edema. Springer, Berlin Heidelberg New York, pp 20–39

65. Demopoulos HB (1973) The basis of free radical pathology. Fed Proc 32: 1859–1861

66. Demopoulos HB, Flamm ES, Seligman ML *et al* (1979) Membrane pertubations in central nervous system injury; theoretical basis for free radical damage and a review of the experimental data. In: Popp AJ *et al* (eds) Neural trauma. Raven Press, New York, pp 63–78

67. Dillard CJ, Kunert KJ, Tappel AL (1982) Effects of vitamin E, ascorbic acid and mannitol on alloxan-induced lipid peroxidation in rats. Arch Biochem Biophys 216: 204–212

68. Diplock AT, Baum H, Lucy JA (1971) The effect of vitamin E on the oxidation state of selenium in rat liver. Biochem J 123: 721–729

69. Diplock AT, Lucy JA (1973) The biochemical mode of action of vitamin E and selenium: a hypothesis. FEBS Lett 29: 205–210

70. Diplock AT (1974) Possible stabilizing effect of vitamin E on microsomal, membrane-bound, selenide-containing proteins and drug metabolizing enzyme systems. Am J Clin Nutr 27: 995–1004

71. Elkes J (1970) Psychopharmacology: on beginning in a new science. In: Ayd FJ Jr, Blackwell B (eds) Discoveries in biological psychiatry. Lippincott, Philadelphia, pp 30–52

72. Enseleit WH, Domer FR, Jarrott DM *et al.* (1984) Cerebral phospholipid content and Na^+-K^+ ATPase activity during ischemia and post ischemic reperfusion in the mongolian gerbil. J Neurochem 43: 320–327

73. Erecinska M, Nelson D, Wilson DF *et al.* (1984) Neurotransmitter amino acid in the CNS. I. Regional change in amino acid levels in rat brain during ischemia and reperfusion. Brain Res 304: 9–22

74. Escueta AV, Appel SH (1972) Brain Synapses: An *in vitro* model for the study of seizures. Arch Intern Med 129: 333–344

75. Fertziger AP, Liuzzi SE, Dunham PB (1971) Diphenylhydantoin (Dilantin) stimulation of potassium influx in lobster axons. Brain Res 33: 592–596

76. Festoff BW, Appel SH (1968) Effect of diphenylhydantoin on synaptosome soduim potassium ATPase. J Clin Invest 47: 2752–2758

77. Finkelstein E, Rosen GM, Raukman EJ (1985) Spin trapping of superoxide and hydroxy radical: practical aspects. Arch Biochem Biophys 200: 1–16

78. Flamm ES, Demopoulos HB, Myron MD *et al* (1978) Free radicals in cerebral ischemia. Stroke 9: 445–447

79. Franklin ML, Horlick G, Malmstadt HV (1969) Basic and practical consideration in utilizing photon counting for quantitative spectrochemical methods. Anal Chem 7: 2–10

80. Fong KL, McCay PB, Poyer JL (1973) Evidence that peroxidation of lysosomal membranes is initiated by hydroxy radicals produced during flavin enzyme activity. J Biol Chem 248: 7792–7797

81. Foots CS, Ching TY, Geller GG (1974) Chemistry of singlet oxygen-XVIII: Rates of reaction and quenching of α-tocopherol and singlet oxygen. Photochem Photobiol 20: 511–513

82. Foots CS (1976) Photosensitization oxidation and singlet oxygen; consequences in biological systems. In: Pryor WA (ed) Free radicals in biology, vol 2. Academic Press, New York, pp 85–133

83. Foots CS, Shook FC, Abakerli RA (1980) Chemistry of superoxide ion 4. Singlet oxygen is not a major product of dismutation. J Am Chem Soc 102: 2503–2504

84. Forman HJ, Kennedy J (1976) Dihydroorotate-dependent superoxide production in rat brain and liver: a function of the primary dehydrogenase. Arch Biochem Biophys 173: 219–224

85. Fourcroy AF (1789) Sur l'existence de la matière albumineuse dans les végétaux. Ann Chimi (Paris) 3: 252–261

86. Fox CF (1972) The structure of cell membranes. Sci Am 226: 30–38

87. Fujita Y, Shingu T, Kohno M (1983) Radical reactions in the ischemic brain damage—basic studies with ESR trap method. In: Asano T (ed) Brain ischemia and free radicals. Neuron Pub, Tokyo, pp 101–110

88. Fukuda A, Tabuse H, Ihara N et al (1983) Effect of phenytoin on regional cerebral blood flow, electroencephalogram, and electrolyte contents in cerebral blood and cerebral cortex following total cerebral ischemia in dogs. Circ Shock 10: 341–350

89. Galli C, Spagnuolo C (1976) The release of brain free fatty acids during ischemia in essential fatty acid-deficient rats. J Neurochem 26: 401–404

90. Gaudet RJ, Alam I, Levine L (1980) Accumulation of cyclooxygenase products of arachidonic acid metabolism in gerbil brain during reperfusion after bilateral common carotid artery occlusion. J Neurochem 35: 653–658

91. Gilman AG, Goodman LS, Gilman A (1980) Hydantoins. In: Gilman AG, Goodman LS, et al (eds). The pharmacological basis of therapeutics. Macmillan, pp 452–456

92. Ginsberg MD, Watson BD, Yoshida S et al (1983) Aspects of tissue injury in cerebral ischemia. In: Reivich M, Hurtig HI, et al (eds) Cerebrovascular diseases. 13th Princeton Conference. Raven Press, New York, pp 237–247

93. Goldberg WJ, Watson BD, Busto R et al (1984) Concurrent measurement of (Na^+, K^+)-ATPase activity and lipid peroxides in rat brain following reversible global ischemia. Neurochem Res 9: 1737–1747

94. Gonzales RA, Crews FT (1985) Cholinergic- and adrenergic-stimulated inositide hydrolysis in brain: interaction, regional distribution, and coupling mechanisms. J Neurochem 45: 1076–1084

95. Guidotti G (1972) Membrane proteins. Ann Rev Biochem 41: 731–752

96. Haber F, Weiss J (1934) The catalytic decomposition of hydrogen peroxide by iron saits. Proc R Soc Lond (Biol) Ser A 147: 332–351, 1934

97. Hafeman DG, Hoekstra WG (1977) Lipid peroxidation *in vivo* during vitamin E and selenium deficiency in rat as monitored by ethane evolution. J Nutr 107: 666–672

98. Haglund L, Kohler C, Haaparanta T *et al* (1984) Presence of NADPH-cytochrome P450 reductase in central cathecholaminergic neurons. Nature 307: 259–262

99. Hall ED, Braughler JM (1981) Acute effects of intravenous glucocorticoid pretreatment on the *in vitro* peroxidation of cat spinal cord tissue. Exp Neurol 73: 321–324

100. Halliwell B (1976) An attempt to demonstrate reaction between superoxide and hydrogen peroxide. FEBS Lett 72: 8–10

101. Hammarstrom S, Hamberg M, Duell EA *et al* (1977) Glucocorticoid in inflammatory proliferative skin disease reduces arachidonic and hydroxyeicosatetraenoic acids. Science 197: 994–996

102. Hansen AJ (1978) The extracellular potassium concentration in brain cortex following ischemia in hypo- and hyperglycemic rats. Acta Physiol Scand 102: 324–329

103. Hansen AJ (1981) Extracellular ion concentration in cerebral ischemia. In: Zeuthen (ed) The application of ion selective microelectrode. Elsevier/North-Holland Biomedical Press, Amsterdam, pp 239–254

104. Harbour JR, Chew V, Bolton JR (1974) An electron spin resonance study of the spin adducts of OH and OH_2 radicals with nitrones in the ultraviolet photolysis of aqueous hydrogen peroxide solutions. Car J Chem 52: 3549–3553

105. Harris RJ, Simon L, Branston NM *et a* (1981) Changes in extracellular calcium activity in cerebral ischemia. J Cereb Blood Flow Metab 2: 203–211

106. Hirata F (1981) The regulation of lipomo·

dulin, a phospholipase inhibitory protein, in rabbit neutrophils by phosphorylation. J Biol chem 256: 7780–7783

107. Holman RT, Burr GO (1946) Spectrophotometric studies of the oxidation of fats. VI. Oxygen absorption and chromophore production in fatty esters. J Am Chem Soc 68: 562–566

108. Holton FA, Holton P (1954) The capillary dilator substances in dry powders of spinal roots. A possible role of adenosine triphosphate in chemical transmission from nerve endings. J Physiol 126: 124–140

109. Holub BJ, Kuksis A, Thompson W (1970) Molecular species of mono-, di- and triphosphoinositides of bovine brain. J Lipid Res 11: 558–564

110. Hong SCL, Levine L (1976) Inhibition of arachidonic acid release from cells as the biochemical action of anti-inflammatory corticosteroids. Proc Natl Acad Sci USA 73: 1730–1734

111. Igarashi O, Matsukawa H, Inagaki C (1976) Reactivity of α-tocopherol with hydroperoxide of methyl linoleate. J Nutr Sci Vitaminol (Tokyo) 22: 267–270

112. Imaizumi S, Kayama T, Suzuki J (1984) Chemiluminescence in hypoxic brain—the first report: correlation between energy metabolism and free radical reaction. Stroke 15: 1061–1065

113. Imaizumi S, Kayama T, Suzuki J (1985) Chemiluminescence in hypoxic brain—the 2nd report: effects of free radical scavengers. No To Shinkei 37: 161–168 (Eng Abstr)

114. Imaizumi S, Suzuki J, Tominaga T et al (1985) Effect of free radical scavengers on cerebral ischemia and hypoxia evaluated by chemiluminescence. In: Spetzler RF, Selman WR, et al (eds) Cerebral revascularization for stroke. Thieme-Stratton Inc, New York, pp 299–306

115. Inaba H, Shimizu Y, Tsuji Y et al (1979) Photon counting spectral analyzing system of extra-weak chemi- and bioluminescence for biochemical applications. Photochem Photobiol 30: 169–175

116. Inaba H, Yamagishi A, Takyu C et al (1982) Development of an ultra-high sensitive photon counting system and its application to biomedical measurements. Opt Lasers Engineering 3: 125–130

117. Ishikawa K, Hanaoka Y, Kondo Y et al (1977) Primary action of steroid hormone at the surface of amphibian oocyte in the induction of germinal vesicle breakdown. Mol Cell Endocrinol 9: 91–100

118. Janzen EG, Blackburn BJ (1969) Detection and identification of short-lived free radicals by electron spin resonance trapping techniques (Spin Trapping). Photolysis of organolead, -tin and -mercury compounds. J Am Chem Soc 91: 4481–4490

119. Janzen EG (1971) Spin trapping. Acc Chem Res 4: 31–40

120. Janzen EG (1980) A critical review of spin trapping in biological system. In: Pryor WA (ed) Free radicals in biology, vol IV. Academic Press, New York, pp 115–154

121. Johnston MV (1983) Neurotransmitter alterations in a model of perinatal hypoxic-ischemic brain injury. Ann Neurol 13: 511–518

122. Joseph SK, Thomas AP, Williams RJ et al (1984) Myo-inositol 1, 4, 5—triphosphate: a second messenger for the normal mobilization of intracellular Ca^{2+} in liver. J Biol Chem 259: 3077–3081

123. Jost PC, Griffith OH, Capaldi RA, et al (1973) Evidence for boundary lipid in membrane. Proc Natl Acad Sci USA 70: 480–484

124. Kahn AU, Kasha M (1970) Chemiluminescence arising from simultaneous transitions in pairs of singlet oxygen molecules. J Am Chem Soc 92: 3293–3300

125. Kakiuchi S, Rall TW (1968) Studies on adenosine 3'-5'-phosphate in rabbit cerebral cortex. Mol Pharmacol 4: 379–388

126. Kalynaraman B, Perez-Reyes E, Mason PP (1980) Spin-trapping and direct electron spin resonance investigations of the redox metabolism of quinone anticancer drugs. Biochim Biophys Acta 630: 119–130

127. Kameyama M, Suzuki J, Shirane R et al (1985) A new model of bilateral hemispheric ischemia—three vessel occlusion model. Stroke 16: 489–493

128. Kellog III, Fridovich I (1975) Superoxide, hydrogen peroxide, and singlet oxygen in lipid peroxidation by xanthine oxidase system. J Biol Chem 250: 8812–8817

129. Kiwak KJ, Moskowitz MA, Levine L (1985) Leukotriene production in gerbil brain after ischemic insult, subarachnoid hemorrhage, and concussive injury. J Neurosurg 62: 865–869

130. Klee CB, Crouch TH, Richman PG (1980) Calmodulin. Ann Rev Biochem 49: 489–515

131. Klee CB, Vanaman TC (1982) Calmodulin. Adv Protein Chem 35: 213–321

132. Kobayashi M, Lust WD, Passonneau JV (1977) Concentrations of energy metabolites and cyclic nucleotides during and after bilateral ischemia in the gerbil cerebral cortex. J Neurochem 29: 53–59

133. Kogure K, Scheinberg P, Kishikawa H et al. (1979) Adrenergic control of cerebral blood flow and energy metabolism in the rat. Stroke 10: 179–184

134. Kogure K, Watson BD, Busto R et al (1982) Potentiation of lipid peroxides by ischemia in rat brain. Neurochem Res 7: 437–454

135. Kovachich GB, Mishra OP (1980) Lipid peroxidation in rat brain cortical slices as measured by the thiobarbituric acid test. J Neurochem 35: 1449–1452

136. Kretsinger RH (1979) Calcium in neurology, a general theory of its function and evolution. In: Schmitt FO, Worden FG (eds) The neuroscience: fourth study program. MIT Press, Cambridge, Mass, pp 617–622

137. Krishtal OA, Marchenko SM, Pidoplichko VI (1983) Receptor for ATP in the membrane of mammalian sensory neurones. Neurosci Lett 35: 41–45

138. Lai C-S, Piette LH (1977) Hydroxy radical production in lipid peroxidation of rat liver microsome. Biochem Biophys Res Commun 78: 51–59

139. Lai C-S, Grover TA, Piette LH (1979) Hydroxy radical production in a purified NADPH-cytochrome c (P-450) reductase system. Arch Biochem Biophys 193: 373–378

140. Lazarewicz JW, Strosznajder J, Gromek A (1972) Effect of ischemia and exogeneous fatty acid on the energy metabolism in brain mitochondria. Bull Acad Pol Sci 20: 599–606

141. Lehninger AL, Carafoli E, Rossi CS (1967) Energy-linked ion movements in mitochondrial systems. Adv Enzymol 29: 259–320

142. Littel JR, O'Shaughnessy D (1979) Treatment of acute focal ischemia with continuous CSF drainage and mannitol. Stroke 10: 446–450

143. Lowry OH, Passonneau JV, Hasselberger FX et al (1964) Effect of ischemia on known substrates and cofactors of the glycolytic pathway in brain. J Biol Chem 239: 18–30

144. Lucy JA, Dingle JT (1964) Fat-soluble vitamins and biological membranes. Nature 204: 156–204

145. Lucy JA (1972) Functional and structural aspects of biological membranes: A suggested structural role for vitamin E in the control of membrane permeability and stability. Ann NY Acad Sci 203: 4–11

146. MacMillan V, Shankaran R (1984) Influence of lactate accumulation of Na^+-K^+ ATPase activity of ischemic and postischemic brain.

147. Majewska MR, Strosznajder J, Lazarewicz J (1978) Effect of ischemic anoxia and barbiturate anesthesia on free radical oxidation of mitochondrial phospholipids. Brain Res 158: 423–434

148. Marion J, Wolfe LS (1979) Origin of the arachidonic acid released post-mortem in rat forebrain. Biochim Biophys Acta (Amst) 574: 25–32.

149. Markelonis G, Garbus J (1975) Alterations of intracellular oxidative metabolism as stimuli evoking prostaglandin biosynthesis. Prostaglandins 10: 1087–1106

150. McCay PB, Noguchi T, Fong K-L et al (1980) Production of radicals from enzyme systems and the use of spin traps. In: Pryor WA (ed) Free radicals in biology, vol 4. Academic Press, New York, pp 155–186

151. McCord JM, Fridovich I (1969) Superoxide dismutase. An enzymatic function for erythrocuprein (hemocuprein). J Biol Chem 249: 6049–6055

152. McCord JM, Fridovich I (1973) Production of O_2^- in photolyzed water demonstrated through the use of superoxide dismutase. Photochem Photobiol 17: 115–121

153. Merritt HH, Putnam TJ (1938) A new series of anticonvulsant drugs tested by

experiments on animals. Arch Neurol Psychiat 39: 1003–1015

154. Means AR, Tash JS, Chafouleas JG (1982) Physiological implications of the presence, distribution and regulation of calmodulin in eukaryotic cells. Physiol Rev 62: 1–30

155. Mitchell P (1972) Chemiosmotic coupling in energy transduction: A logical development of biochemical knowledge. J Bioenerg 3: 5–24

156. Mitchell P (1973) Performance and conservation of osmotic work by proton-coupled solute porter systems. J Bioenerg 4: 63–91

157. Mizuno K, Hata S, Tomioka S (1970) Measurement of extra-weak chemiluminescence. Chem Pharm Bull (Tokyo) 18: 2588–2589

158. Moncada S, Vane JR (1979) Arachidonic acid metabolites and the interactions between platelets and blood-vessel walls. New Eng J Med 300: 1142–1147

159. Morton EG (1968) Photon counting. Appl Opt 7: 1–10

160. Moskowitz MA, Kiwak KJ, Hekimian K et al (1984) Synthesis of compounds with properties of leukotriene C_4 and D_4 in gerbil brain after ischemia and reperfusion. Science 224: 886–889

161. Mrsǔlja BB, Djuričić BM, Cvejić V et al (1980) Biochemistry of experimental ischemic brain edema: In: Cervós-Navarro J, Ferszt R (eds) Brain edema. Advances in neurology, vol 28. Raven Press, New York, pp 217–230

162. Mullins LJ (1979) The generation of electric currents in cardiac fibers by Na/Ca exchange. Am J Physiol 236: C 103–110

163. Murphy RC, Hammarström S, Samuelsson B et al (1979) Leukotrien C: a slow-reacting substance (SRS) from murine mastcytoma cells. Proc Natl Acad Sci USA 76: 4275–4279

164. Neifakh YA (1971) Free radical mechanism of ultraweak chemiluminescence coupled with peroxide oxidation of unsaturated fatty acids. Biofizica 16: 584–588

165. Nemoto EM, Bleyaert AL, Stezoski SW et al (1977) Global brain ischemia: a reproducible monkey model. Stroke 8: 558–564

166. Nemoto EM (1985) Brain ischemia. In: Lajtha A (ed) Handbook of neurochemistry 9. Plenum Press, New York, pp 553–588

167. Nordström CH, Siesjö BK (1978) Effects of phenobarbital in cerebral ischemia. Part I: Cerebral energy metabolism during pronounced incomplete ischemia. Stroke 9: 327–335

168. Nordström CH, Rehncrona S, Siesjö BK (1978) Effects of phenobarbital in cerebral ischemia. Part II: Restitution of cerebral energy state, as well as of glycolytic metabolities, citric ascid cycle intermediates and associated amino acids after pronounced incomplete ischemia. Stroke 9: 335–343

169. Nordström CH, Rehncrona S, Siesjö BK (1978) Restitution of cerebral energy state, as well as of glycolytic metabolism and associated amino acids after 30 min of complete ischemia in rats anesthetized with nitrous oxide or phenobarbital. J Neurochem 30: 479–486

170. Norton WT, Poduslo SE (1971) Neuronal perikarya and astroglia of rat brain: chemical composition during myelination. J Lipid Res 12: 84–90

171. O'Brien JS, Sampson EL (1965) Lipid composition of the normal human brain: gray matter, white matter, and myelin. J Lipid Res 6: 537–545

172. O'Brien JS (1965) Stability of the myelin membrane. Science 147: 1099–1107

173. Ozawa K, Seta K, Handa H (1969) Biochemical studies on brain swelling II. influence of brain swelling and ischemia on the formation of an endogenous inhibitor in mitochondria. J Biochem (Tokyo) 66: 361–367

174. Ozawa K, Seta K, Takeda H et al (1966) On the isolation of mitochondria with high respiratory control from rat brain. J Biochem (Tokyo) 59: 501–510

175. Ozawa K, Kitamura O, Ohsawa T et al (1966) Mitochondrial vulnerability and lipid metabolism. Folia Psychiatr Neurol 20: 73–84

176. Ozawa T, Hanaki A, Matsumoto S et al (1978) Electron spin resonance studies of radicals obtained by the reaction of alpha-tocopherol and its model compound with superoxide ion. Biochim Biophys Acta 531: 72–78

177. Pavlock GS, Southard JH, Litz MF *et al* (1981) Effect of mannitol and chlorpromazine pretreatment of rabbits on kidney mitochondria following *in vivo* ischemia and reflow. Life Sci 29: 2667–2672

178. Pederson TC, Aust SD (1975) The mechanism of liver microsomal lipid peroxidation. Biochim Biophys Acta 385: 232–241

179. Pincus JH, Grove I, Marino BB *et al* (1970) Studies on the mechanism of action of diphenylhydantoin. Arch Neurol 22: 566–571

180. Pincus JH (1972) Diphenylhydantoin and ion flux in lobster nerve. Arch Neurol 26: 4–10

181. Pincus JH, Rawson MD (1969) Diphenylhydantoin and intracellular sodium concentration. Neurology 19: 419–422

182. Pietronigro DD, Mignano JE, Demopoulos HB (1983) Direct quenching of adriamycin radicals by coenzyme Q_{10} and tetrazolium salts. Biochem Pharmacol 32: 1441–1444

183. Pietronigro DD, Hovsepian M, Demopoulos HB *et al* (1985) Reductive metabolism of ascorbic acid in the central nervous system. Brain Res 333: 161–164

184. Poduslo SE, Norton WT (1972) Isolation and some chemical properties of oligodendroglia from calf brain. J Neurochem 19: 727–736

185. Pollay M, Fullenwider C, Roberts A *et al* (1983) Effect of mannitol and furosemide on blood-brain osmotic gradient and intracranial pressure. J Neurosurg 59: 945–950

186. Porcellati G, De Medio GE, Fini C *et al* (1978) Phospholipid and its metabolism in ischemia. In: Neuhoff V (ed) Prov Europ Soc Neurochem, vol 1, pp 285–302

187. Poyer JL, McCay PB, Lai EK *et al* (1980) Confirmation of asignment of the trichloromethyl radical spin adduct detected by spin trapping ^{13}C-carbon tetrachloride metabolism *in vitro* and *in vivo*. Biochem Biophys Res Commun 94: 1154–1160

188. Prioleau GR, Fishman RA, Chan PH (1979) Induction of brain edema by fatty acids *in vivo*. Trans Am Neurol Ass 104: 147–150

189. Pryor WA (1973) Free radical reactions and their importance in biochemical systems. Fed Proc 32: 1862–1869

190. Pryor WA (1976) The role of free radical reactions in biological systems. In: WA Pryor (ed) Free radicals in biology, vol 1. Academic Press, New York, pp 1-49

191. Reeves JP, Sutro JL (1980) Sodium-calcium exchange activity generates a current in cardiac membrane vesicles. Science 208: 1461–1464

192. Rehncrona S, Mela L., Siesjö BK (1979) Recovery of brain mitochondrial function in the rat after complete and incomplete cerebral ischemia. Stroke 10: 437–446

193. Rehncrona S, Smith DS, Åkesson B *et al* (1980) Peroxidative changes in brain cortical fatty acids and phospholipids, as characterized during Fe^{2+}—and ascorbic acid-stimulated lipid peroxidation *in vitro*. J Neurochem 34: 1630–1638

194. Rehncrona S, Rosén I, Siesjö BK (1981) Brain lactic acidosis and ischemic cell damage: 1. Biochemistry and neurophysiology. J Cereb Blood Flow Metab 1: 297–313

195. Rehncrona S, Westerberg E, Åkesson B *et al* (1982) Brain cortical fatty acids and phospholipids during and following complete and severe incomplete ischemia. J Neurochem 38: 84–93

196. Rieley CA, Cohen C, Lieberman M (1974) Ethane evolution: a new index of lipid peroxidation. Science 183: 208–210

197. Rothman JE, Lenard J (1977) Membrane asymmetry. Science 195: 743–753

198. Russell GA (1957) Deuterium-isotope effects in the autoxidation of aralkyl hydrocarbons. Mechanism of the interaction of peroxy radicals. J Am Chem Soc 79: 3871–3877

199. Saprin AN, Piette LH (1977) Spin trapping and its application in the study of lipid peroxidation and free radical production with liver microsomes. Arch Biochem Biophys 180: 480–492

200. Schreier S, Polneszek CF, Smith IC (1978) Spin labels in membranes: problems in practice. Biochim Biophys Acta 515: 395–436

201. Seliger HH (1975) The origin of bioluminescence. Photochem Photobiol 21: 355–361

202. Seligman ML, Demopoulos HB (1973) Spin-probe analysis of membrane per-

turbations produced by clinical and physical agents. Ann NY Acad Sci 222: 640–667

203. Seligman ML, Mitamura J, Shera N et al (1979) Corticosteroid (methylprednisolone) modulation of photoperoxidation by ultraviolet light in liposomes. Photochem Photobiol 29: 549–558

204. Shimizu Y, Inaba H, Kumaki K et al (1973) Measuring methods for ultra-light intensity and their application to extraweak bioluminescence from living tissures. IEEE, Trans Inst Meas IM 22: 153–157

205. Shiu GK, Nemoto EM (1981) Barbiturate attenuation of brain free fatty acid liberation during global ischemia. J Neurochem 37: 1448–1456

206. Shiu GK, Nemmer JP, Nemoto EM (1983) Reassessment of brain free fatty acid liberation during global ischemia and its attenuation by barbiturate anesthesia. J Neurochem 40: 880–884

207. Shiu GK, Nemoto EM, Nemmer J (1983) Dose of thiopental, pentobarbital and phenytoin for maximal therapeutic effects in cerebral ischemic anoxia. Crit Care Med 11: 452–459

208. Shohami E, Rosenthal J, Lavy S (1982) The effect of incomplete cerebral ischemia on prostaglandin levels in rat brain. Stroke 13: 494–499

209. Siesjö BK (1978) Brain energy metabolism. John Wiley, New York

210. Siesjö BK (1981) Cell damage in the brain: a speculative synthesis. J Cereb Blood Flow Metab 1: 155–185

211. Singer SJ, Nicolson GL (1972) The fluid mosaic model of the structure of cell membranes. Science 175: 720–731

212. Sokoloff L (1977) Relation between physiological function and energy metabolism in the central nervous system. J Neurochem 29: 13–26

213. Spagnuolo C, Sautebin L, Galli G et al (1979) $PGF_{2\alpha}$, thromboxane B_2 and HETE levels in gerbil brain cortex after ligation of common carotid arteries and decapitation. Prostaglandins 18: 53–61

214. Stauff J, Schmidkunz H, Hartman G (1963) Weak chemiluminescence of oxidations reactions. Nature 198: 281–282

215. Sugioka K, Nakano M (1983) Mechanism of phospholipid peroxidation by ferric ion-ADP-adriamycin-coordination complex. Biochim Biophys Acta 713: 333–343

216. Sun GY, Manning R, Strosznajder J (1980) Effect of postdecapitative ischemia and hypoxia on the phosphoglyceride acyl groups of rat brain membranes. Neurochem Res 5: 1211–1219

217. Suzuki J, Yoshimoto T, Kodama N et al (1982) A new therapeutic method for acute brain infarction: revascularization following the administration of mannitol and perfluorochemicals—a preliminary report. Surg Neurol 17: 325–332

218. Suzuki J, Imaizumi S, Kayama T (1985) Chemiluminescence in hypoxic brain— The second report: cerebral protective effect of mannitol, vitamin E and glucocorticoid. Stroke 16: 695–700

219. Suzuki J, Ogawa A, Yoshimoto T et al (1985) Indications for surgery in the acute stage of cerebral infarction: The role of new cerebral protective drugs—"Sendai Cocktail" and perfluorochemicals. In: Spetzler RF, et al (eds) Cerebral revascularization for stroke. Thieme-Stratton Inc., New York, pp 392–396

220. Svingen BA, O'Neal FO, Aust SD (1978) The role of superoxide and singlet oxygen in lipid peroxidation. Photochem Photobiol 28: 803–809

221. Svingen BA, Buege JA, O'Neal FO et al (1979) The mechanism of NADPH-dependent lipid peroxidation. J Biol Chem 254: 5892–5899

222. Tappel AL (1962) Vitamin E as the biological lipid antioxidant. Vitamins Hormones 20: 493–510

223. Tappel AL (1973) Lipid peroxidation damage to cell components. Fed Proc 32: 1870–1874

224. Tappel AL (1972) Vitamin E and free radical peroxidation of lipids. Ann NY Acad Sci 203: 12–28

225. Tappel AL (1975) Lipid peroxidation and fluorescent molecular damage to membrane. In: Trump BF, et al (eds) Pathological aspects of cell membranes, vol 1. Academic Press, New York, pp 145–170

226. Tappel AL (1980) Measurement of and protection from in vivo lipid peroxidation. In: Pryor WA (ed) Free radicals in biology, vol 4, Academic Press, New York, pp 1–47

227. Tarusov BN, Polidova AI, Zhuravlev AI et al (1962) Ultraweak luminescence of animal tissues. Tsitologiia 4: 696–699

228. Taylor GW, Morris HR (1983) Lipoxygenase pathways. Br Med Bull 39: 219–222

229. Thudichum JLW (1901) Die Chemische Konstitution des Gehirns und der Tiere. Franz Pietzcher, Tübingen

230. Tien M, Svingen BA, Aust SD (1981) Initiation of lipid peroxidation by perferryl complexes. In: Rodgers MAJ, Powers EL (eds) Oxygen and oxy-radicals in chemistry and biology. Academic Press, New York, pp 147–152

231. Tominaga T, Imaizumi S, Yoshimoto T et al (1985) Protective effects of radical scavengers on cerebral infarction— experimental study utilizing spin trapping method of ESR. No To Shinkei 37: 555–560 (Eng Abstr)

232. Tominaga T, Imaizumi S, Yoshimoto T et al (1986) Detection of free radicals generated in NADPH-dependent lipid peroxidation of rat brain homogenate— application of spin trapping technique. No To Shinkei 38: 169–175 (Eng Abstr)

233. Vassil'ev RF, Vichutinskii AA (1962) Chemiluminescence and oxidation. Nature 194: 1276–1277

234. Walling C (1975) Fenton's reagent revisited. Acc Chem Res 8: 125–131

235. Watanabe T, Yoshimoto T, Suzuki M et al (1984) Supression effect of mannitol upon cerebral infarct formation—an electron microscopical investigation. Recent progress in the study and therapy of brain edema. Plenum, New York, 551–559

236. Watson BD, Busto R, Goldberg WJ et al (1984) Lipid peroxidation in vivo induced by reversible global ischemia in rat brain. J Neurochem 42: 268–274

237. Westerberg E, Åkesson B, Rehncrona S et al (1979) Lipid peroxidation in brain tissue in vitro: Effects on phospholipids and fatty acids. Acta Physiol Scand 105: 524–526

238. Westerberg E, Friberg M, Åkesson B (1981) Assay of brain tocopherol using high performance liquid chromatography. J Lipid Chromatogr 4: 109–121

239. Wieloch T, Siesjö BK (1982) Ischemic brain injury: the importance of calcium, lipolytic activity and free fatty acid. Pathol Biol 30: 269–277

240. Woelk H, Goraci G, Gatti A et al (1973) Phospholipase A_1 and A_2 activities of neuronal and glial cells of the rabbit brain. Hoppe-Seyler's Z Physiol chem 354: 729–736

241. Woelk H, Rubly N, Arienti G et al (1981) Occurrence and properties of phospholipase A_1 of plasma membrane prepared from neuronal and glial enriched fraction of the rabbit cerebral cortex. J Neurochem 36: 875–880

242. Wolfe LS, Coceani F (1979) The role of prostaglandins in the central nervous system. Annu Rev Physiol 41: 669–684

243. Woodbury DM, Timiras PS, Vernadakis A (1957) Modification of adrenocortical function by centrally acting drugs and the influence of such modification on the central response to these drugs. In: Hoagland H (ed) Hormones, brain function, and behavior. Academic Press, New York, pp 38–50

244. Woodbury DM (1955) Effect of diphenylhydantoin on electrolytes and radiosodium turnover in brain and other tissues of normal, hyponatremic and postictal rats. J Pharmacol Exp Ther 115: 74–95

245. Yagi K (1976) A simple fluorometric assay for lipidperoxide in blood plasma. Biochem Med 15: 212

246. Yamada K, Saito Y, Matsuoka N et al (1978) Studies on adrenaline-induced lipolysis in adrenalectomized rats. Endocrinol Jpn 25: 315–320

247. Yoshida S, Abe K, Busto R et al (1982) Influence of transient ischemia on lipid-soluble antioxidants, free fatty acids and energy metabolites in rat brain. Brain Res 245: 307–316

248. Yoshida S, Inoh S, Asano T et al (1980) Effect of transient ischemia on free fatty acids and phospholipids in the gerbil brain: Lipid peroxidation as possible cause of postischemic injury. J Neurosurg 53: 323–331

249. Yoshida S, Harik SI, Busto R et al (1984) Free fatty acids and energy metabolites in ischemic cerebral cortex with noradrenaline depletion. J. Neurochem 42: 711–717

CHAPTER 6

1. Adlard BPF (1974) Ascorbic acid in developing brain—possible function as an inhibition of lipid peroxidation. Biochem Soc Trans 2: 281–284

2. Aldrete JA, Romo-Salas F, Jankovsky L et al (1979) Effect of pretreatment with thiopental and phenytoin on postischemic brain damage in rabbits. Crit Care Med 7: 466–470

3. Anderson DC, Cranford RE (1979) Corticosteroids in ischemic stroke. Stroke 10: 68–71

4. Arnfred I, Secher O (1962) Anoxia and barbiturates, tolerance to anoxia in mice influenced by barbiturates. Arch Int Pharmacodyn Ther 139: 67–73

5. Artru AA, Michenfelder JD (1980) Cerebral protective, metabolic, and vascular effects of phenytoin. Stroke 11: 377–382

6. Artru AA, Michenfelder JD (1981) Anoxic cerebral potassium accumulation reduced by phenytoin—mechanism of cerebral protection? Anesth Analg 60: 41–45

7. Bandaranayake NM, Nemoto EM, Stezoski SW (1978) Rat brain osmolality during barbiturate anesthesia and global brain ischemia. Stroke 9: 249–254

8. Baskin DS, Hosobuchi Y (1981) Naloxone reversal of ischemic deficits in man. Lancet 8: 272–275

9. Baskin DS, Kieck CG, Hosobuchi Y (1982) Naloxone reversal of ischemic neurologic deficits in baboons is not mediated by systemic effects. Life Sci 31: 2201–2204

10. Baskin DS, Hosobuchi Y, Grevel JC (1986) Treatment of experimental stroke with opiate antagonists. Effects on neurological function, infarct size, and survival. J Neurosurg 64: 99–103

11. Black KL, Weidler DJ, Jalld NS et al (1978) Delayed pentobarbital therapy of acute focal cerebral ischemia. Stroke 9: 245–254

12. Bleyaert AL, Nemoto EM, Safar P et al (1978) Thiopental amelioration of brain damage after global ischemia in monkeys. Anesthesiology 49: 390–398

13. Bodannes RS, Chan PC (1979) Ascorbic acid as a scavenger of singlet oxygen. FEBS Lett 105: 195–196

14. Bolu RG, Plotkine M, Guenidu C et al (1982) Effect of indomethacin in experimental cerebral ischemia. Pathol Biol 30: 278

15. Branstron NM, Hope DT, Symon L (1979) Barbiturates in focal ischemia of primate cortex—effects on blood flow distribution, evoked potential and extracellular potasium. Stroke 10: 647–653

16. Broddle W, Nelson SR (1968) The effect of diphenylhydantoin on energy reserve levels in brain. Fed Proc 27: 751

17. Brown FD, Johns L, Jafer JJ et al (1979) Detailed monitoring of the effect of mannitol following experimental head injury. J Neurosurg 50: 423–432

18. Caplan LR, Skillman J, Ojemann R et al (1978) Intracerebral hemorrhage following carotid endoarterectomy. A hypertensive complication? Stroke 9: 457–460

19. Clark LC Jr, Gollan F (1966) Survival of mammals breathing organic liquids equilibrated with oxygen at atmospheric pressure. Science 152: 1755–1766

20. Corkill G, Chikovani OK, McLeish I et al (1976) Timing of pentobarbital administration for brain protection in experimental stroke. Surg Neurol 5: 147–149

21. Corkill G, Sivalingam S, Reitan JA et al (1978) Dose dependency of the post-insult protective effect of pentobarbital in the canine experimental stroke model. Stroke 9: 10–18

22. Demopoulos HB (1973) The basis of free radical pathology. Fed Proc 32: 1859–1861

23. Faden AI, Hallenbeck JM, Brown CQ (1982) Treatment of experimental stroke: comparison of naloxone and thyrotropin-releasing hormone. Neurology (Minneap) 32: 1083–1088

24. Faden AI, Jacobs TP, Smith MT (1984) Evaluation of the calcium channel antagonist nimodipine in experimental spinal cord ischemia. J Neurosurg 60: 796–799

25. Feustel PJ, Ingvar MC, Severinghaus JW (1981) Cerebral oxygen availability and

blood flow during middle cerebral artery occlusion: effects of pentobarbital. Stroke 12: 858–863

26. Flamm ES, Demopoulos HB, Seligman ML *et al* (1978) Free radicals in cerebral ischemia. Stroke 9: 445–447

27. Foote CS, Ching TY, Geller GG (1974) Chemistry of singlet oxygen-XVIII: rates of reaction and quenching of α-tocopherol and singlet oxygen. Photochem Photobiol 20: 511–513

28. Gaines C, Nehls DG, Suess DM *et al* (1981) Effect of naloxone on experimental stroke in awake monkeys. Neurosurgery 14: 308–314

29. Gaudet RJ, Levine L (1979) Transient cerebral ischemia and brain prostaglandins. Biochem Biophys Res Commun 86: 893–901

30. Gaudet RJ, Alam I, Levine A (1980) Accumulation of cyclooxygenase products of arachidonic acid metabolism in gerbil brain during reperfusion after bilateral common carotid artery. J Neurochem 35: 653–658

31. Goldstein A, Wells BA, Keats AS (1966) Increased tolerance to cerebral anoxia by pentobarbital. Arch Int Parmacodyn Thér 161: 138–143

32. Gratzl O, Schmiedek P, Spetzler R *et al* (1976) Clinical experience with extraintracranial arterial anastomosis in 65 cases. J Neurosurg 44: 313–324

33. Hallenbeck JM, Leitch DR, Dutka AJ *et al* (1982) Prostaglandin I$_2$, indomethacin and heparin promote postischemic neuronal recovery in dogs. Ann Neurol 12: 145–156

34. Harris RJ, Bayhan M, Branston NM *et al* (1982) Modulation of the pathophysiology of primate focal cerebral ischemia by indomethacin. Stroke 13: 17–24

35. Harris RJ, Branston NM, Symon L *et al* (1982) The effects of a calcium antagonist, nimodipine, upon physiological responses of the cerebral vasculature and its possible influence upon focal cerebral ischemia. Stroke 13: 759–766

36. Hoff JT, Smith AL, Hankinson HL *et al* (1975) Barbiturate protection from cerebral infarction in primates. Stroke 6: 28–33

37. Holaday JW, D'Amato RJ (1982) Naloxone or THR fails to improve neurological deficits in gerbils models of "stroke". Life Sci 31: 385–392

38. Hosobuchi Y, Baskin DS (1982) Reversal of induced ischemic neurologic deficit in gerbils by the opiate antagonist naloxone. Science 215: 69–71

39. Hossmann KA, Paschen W, Csiba L (1983) Relationship between calcium accumulation and recovery of cat brain after prolonged cerebral ischemia. J Cereb Blood Flow Metab 3: 346–353

40. Hubbard JL, Sundt TM (1983) Failure of naloxone to affect focal incomplete cerebral ischemia and collateral blood flow in cats. J Neurosurg 59: 237–244

41. Johnston IH, Harper AM (1973) The effect of mannitol on cerebral blood flow. An experimental study. J Neurosurg 38: 461–471

42. Kagawa S, Kosyu K, Yoshimoto T *et al* (1982) The protective effect of mannitol and perfluorochemicals on hemorrhagic infarction: an experimental study. Surg Neurol 17: 66–70

43. Karasawa A, Kumada Y, Yamada K *et al* (1982) Protective effect of flunarizine against hypoxic-anoxia in mice and rats. J Pharmacobiodyn 5: 295–300

44. Kassel NF, Baumann KW, Hitchon PW *et al* (1982) The effect of high dose mannitol on cerebral blood flow in dogs with normal intracranial pressure. Stroke 13: 59–61

45. Kayama T, Mizoi K, Suzuki J (1981) A canine model of a completely ischemic brain regulated with the perfusion method. Surg Neurol 16: 167–172

46. Kazda S, Hoffmeister F, Garthoff B *et al* (1973) Prevention of the postischemic impaired reperfusion of the brain by nimodipine (BAY e 9736). Acta Neurol Scand [Suppl] 60: 302–303

47. Kennedy C, Grave GD, Jehle JW *et al* (1972) The effect of diphenylhydantoin on local cerebral blood flow. Neurology (Abstr) 22: 451

48. Koreh K, Seligman ML, Flamm ES *et al* (1981) Lipid antioxidant properties of naloxone *in vivo*. Biochem Biophys Res Commun 102: 1317–1322

49. Lavyne MH, Hariri RJ, Tankosic T *et al* (1983) Effect of low dose γ-butyrolactone therapy on forebrain neuronal ischemia in the unrestrained, awake rat. Neurosurgery 12: 430–434

50. Levy DM, Brierley JB (1979) Delayed pentobarbital administration limits ischemic brain damage in gerbils. Ann Neurol 5: 59–64

51. Levy R, Feustel P, Severinghaus J *et al* (1982) Effect of naloxone on neurologic deficit and cortical blood flow during focal cerebral ischemia in cats. Life Sci 31: 2205–2208

52. Little JR (1978) Modification of acute focal ischemia by treatment with mannitol. Stroke 9: 4–9

53. Little JR (1979) Treatment of acute focal cerebral ischemia with intermittent low dose mannitol. Neurosurgery 5: 687–691

54. Majewska MD, Strosznajder J, Lazarewicz J (1978) Effect of ischemic anoxia and barbiturate anesthesia on free radical oxidation on mitochondrial phospholipids. Brain Res 158: 423–434

55. Meyer JS, Fukuuchi Y, Shimazu K *et al* (1982) Abnormal hemispheric blood flow and metabolism in cerebrovascular disease. Part 2: Therapeutic trials with 5% CO_2 inhalation, hyperventilation, and intravenous infusion of THAM and mannitol. Stroke 3: 157–167

56. Michenfelder JD, Theye RA (1970) The effects of anesthesia and hypothermia on canine cerebral ATP and lactate during anoxia produced by decapitation. Anesthesiology 33: 430–439

57. Michenfelder JD, Theye RA (1973) Cerebral protection by thiopental during hypoxic. Anesthesiology 39: 510–517

58. Michenfelder JD, Milde JH (1975) Influence of anesthetics on metabolic, functional and pathological responses to regional cerebral ischemia. Stroke 6: 405–410

59. Michenfelder JD, Milde JH, Sundt TM Jr (1976) Cerebral protection by barbiturate anesthesia. Use after middle cerebral artery occlusion in java monkeys. Arch Neurol 33: 345–350

60. Miller CL, Lampard DG, Alexander K *et al* (1980) Local cerebral blood flow following transient cerebral ischemia. I. Onset of impaired reperfusion within the first hour following global ischemia. Stroke 11: 534–541

61. Mitsuno T (1986) Studies on perfluorochemical emulsion as artificial blood. Surg Ther (Osaka) 34: 75–84 (Eng Abstr)

62. Mizoi K, Yoshimoto T, Suzuki J (1981) Experimental study of new cerebral protective substances—functional recovery of severe incomplete ischemic brain lesions, pretreated with mannitol and perfluorochemicals. Acta Neurochir (Wien) 56: 157–166

63. Moseley JI, Laurent JP, Molinari GF (1975) Barbiturate attenuation of clinical course and pathologic lesions in a primate stroke model. Neurology 25: 870–874

64. Naito R, Yokohama K (1978) Perfluorochemical blood substitutes. Fluosol-43, Fluosol-DA 20% and 35%. Technical Information Series No 5. The Green Cross Corporation, 1–177, June 30

65. Newberg LA, Steen PA, Milde JH *et al* (1984) Failure of flunarizine to improve cerebral blood flow or neurologic recovery in a canine model of complete cerebral ischemia. Stroke 15: 666–671

66. Nilsson L (1971) The influence of barbiturate anesthesia upon the energy state and upon acid-base parameters of the brain in arterial hypotension and asphyxia. Acta Neurol Scand 47: 233–253

67. Nishikimi M (1975) Oxidation of ascorbic acid with superoxide anion generated by the xanthine-xanthine oxidase system. Biochem Biophys Res Commun 63: 463–467

68. Nordström CH, Siesjö BK (1978) Effects of phenobarbital in cerebral ischemia. Part I: Cerebral energy metabolism during pronounced incomplete ischemia. Stroke 9: 327–335

69. Nordström CH, Rehncrona S, Siejö BK (1978) Effects of phenobarbital in cerebral ischemia. Part II: Restitution of cerebral energy state, as well as of glycolytic metabolites, citric acid cycle intermediates and associated amino acids after pronounced incomplete ischemia. Stroke 9: 335–343

70. Nugent M, Artru AA, Michenfelder JD (1982) Cerebral metabolic vascular and

protective effects of midazolam maleate. Anesthesiology 56: 172–176

71. Ochiai C, Asano T, Takakura K et al (1982) Mechanism of cerebral protection by pentobarbital and nizofenone correlated with the course of local cerebral blood flow changes. Stroke 13: 788–796

72. Ogawa A, Seki H, Mizoi K et al (1979) Circulatory dynamics in experimental focal cerebral infarction. Acta Neurol Scand [Suppl] 72: 284–285

73. Ogawa A, Seki H, Yoshimoto T et al (1982) Experimental focal cerebral infarction. Part 1: Hemodynamics at the center of the focal cerebral infarction following recirculation. Jpn J Stroke 4: 1–9 (Eng Abstr)

74. Pulsinelli WA, Plum F, Brierley JB (1979). Barbiturate exacerbation of ischemic brain damage following bilateral hemispheric ischemia in the rat. Ann Neurol 6: 156 (Abstr)

75. Raichel ME (1983) The pathophysiology of brain ischemia. Ann Neurol 13: 2–10

76. Redpath JL, Willson RL (1973) Reducing compounds in radioprotection and radiosensitization: model experiments using ascorbic acid. Int J Radiat Biol 23: 51–65

77. Reedy DP, Little JR, Capro JA et al (1983) Effects of verapamil on acute focal cerebral ischemia. Neurosurgery 12: 272–276

78. Roba J, Reuse-Blom S, Lambelin G (1976) In vivo antispasmodic activity of suloctidil. Arch Int Phamacodyn Ther 221: 54–59

79. Roba J, Roncucci R, Lambelin G (1977) Parmacological properties of suloctidil. Acta Clin Belg 32: 3–7

80. Roy MW, Dempsey RJ, Meyer KL et al (1985) Effect of verapamil and diltiazem on acute stroke in cats. J Neurosurg 63: 929–936

81. Sawynok J, Pinsky C, LaBella FS (1979) Mini review on the specificity of naloxone as an opiate antagonist. Life Sci 25: 1621–1632

82. Secher O, Wilhjem BI (1968) The protective action of anesthetics against hypoxia. Can Andes Soc J 15: 166–182

83. Seki H, Yoshimoto T, Ogawa A et al (1983) Effect of mannitol on rCBF in canine thalamic ischemia—an experimental study. Stroke 14: 46–50

84. Selman WR, Spetzler RF, Roessmann UR et al (1981) Barbiturate-induced coma therapy for focal cerebral ischemia. J Neurosurg 55: 220–226

85. Selman WR, Spetzler RF, Roski RA et al (1981) Regional cerebral blood flow following middle cerebral artery occlusion and barbiturate therapy in baboons. J Cereb Blood Flow Metab [Suppl 1] 1: 214–215

86. Selman WR, Spetzler RF, Roski RA et al (1982) Barbiturate coma in focal cerebral ischemia: relationship of protection of timing of therapy. J Neurosurg 56: 685–690

87. Shiu GK, Nemoto EM (1981) Barbiturate attenuation of brain free fatty acid liberation during global ischemia. J Neurochem 37: 1448–1456

88. Shiu GK, Nemoto EM, Nemmer J (1983) Dose of thiopental, pentobarbital and phenytoin for maximal therapeutic effects in cerebral ischemic anoxia. Crit Care Med 11: 452–459

89. Shohami E, Rosenthal T, Lavy S (1982) The effect of incomplete cerebral ischemia on prostaglandin levels in rat brain. Stroke 13: 494–499

90. Siesjö BK (ed) (1978) Brain energy metabolism. John Wiley and Sons, New York

91. Siesjö BK (1981) Cell damage in the brain—a speculative synthesis. J Cereb Blood Flow Metab 1: 155–185

92. Siesjö BK (1984) Cerebral circulation and metabolism. J Neurosurg 60: 883–908

93. Simeone FA, Frazer G, Lawner P (1979) Ischemic brain edema: comparative effects of barbiturates and hypothermia. Stroke 10: 8–12

94. Smith AL, Hoff JT, Nielsen SL et al (1974) Barbiturate protection in acute focal cerebral ischemia. Stroke 5: 1–7

95. Smith DS, Rehncrona S, Siesjö BK (1980) Inhibitor effects of different barbiturates on lipid peroxidation in brain tissue in vitro: comparison with the effects of promethazine and chlorpromazine. Anesthesiology 53: 186–194

96. Smith DS, Rehncrona S, Siesjö BK (1980) Barbiturates as protective agents in brain ischemia and as free radical scavenger in vitro. Acta Physiol Scand [Suppl] 492: 129–134

97. Spector RG (1972) Effects of formyl tetrahydrofolic acid and noradrenaline on the

oxygen consumption of rat brain synaptosome-mitochondrial preparation. Br J Pharmacol 44: 279–285

98. Steen PA, Michenfelder JD (1978) Cerebral protection with barbiturates. Relation to anesthetic effect. Stroke 9: 140–142

99. Steen PA, Milde JH, Michenfelder JD (1978) No barbiturate protection in a dog model of complete cerebral ischemia. Ann Neurol 5: 343–349

100. Steen PA, Michenfelder JD (1979) Barbiturate protection in tolerant and nontolerant hypoxic mice: comparison with hypothermic protection. Anesthesiology 50: 404–408

101. Steen PA, Newberg LA, Milde JH et al (1983) Nimodipine improves cerebral blood flow and neurologic recovery after complete cerebral ischemia in the dog. J Cereb Blood Flow Metab 3: 38–43

102. Suzuki O, Yagi K (1973) Formation of lipoperoxide in brain edema by cold injury. Experientia 30: 248

103. Suzuki J (1974) Method of prolongation of temporary stopping of the cerebral blood flow. Presidential address. 33rd Annual Meeting of Japan Neurosurgical Society. Sendai, Japan, Oct 22

104. Suzuki J, Tanaka S, Yoshimoto T et al (1980) Recirculation in the acute period of cerebral infarction. Experimental research on brain swelling and its suppression by using mannitol or glycerol. Acta Neurochir (Wien) 54: 219–231

105. Suzuki J, Yoshimoto T, Tanaka S et al (1980) Production of various models of cerebral infarction in the dog by means of occlusion of intracranial trunk arteries. Stroke 11: 337–341

106. Suzuki J, Tanaka S, Yoshimoto T (1981) Suppression of brain swelling with mannitol and perfluorochemicals—an experimental study. Acta Neurochir (Wien) 58: 149–160

107. Suzuki J, Fujimoto S, Mizoi K et al (1984) The protective effect of combined administration of anti-oxidants and perfluorochemicals on cerebral ischemia. Stroke 15: 672–679

108. Suzuki J, Imaizumi S, Kayama T (1985) chemiluminescence in hypoxic brain. The

2nd report: cerebral protective effect of mannitol, vitamin E and glucocorticoid. Stroke 16: 695–700

109. Tamura A, Asano T, Sano K et al (1979) Protection from cerebral ischemia by a new imidazole derivative (Y-9179) and pentobarbital. A comparative study in chronic middle cerebral artery occlusion in cats. Stroke 10: 126–134

110. Tappel AL (1973) Lipid peroxidation damage to cell components. Fed Proc 32: 1870–1874

111. Van Reempts J, Borgers M, Van Eyndhofen J et al (1982) Protective effects of etomidate in hypoxic-ischemic brain damage in the morphologic assessment. Exp Neurol 76: 181–195

112. Watanabe T, Yoshimoto T, Ogawa A et al (1979) The effect of mannitol in preventing the development of cerebral infarction. An electron microscopical investigation. No Shinkei Geka 7: 859–866 (Eng Abstr)

113. Watanabe T, Yoshimoto T, Suzuki J et al (1984) Suppressive effect of mannitol upon cerebral infarction formation. An electron microscopical investigation. In: Go KG, Bethmann A (eds) Recent progress in the study and therapy of brain edema. Plenum Press, New York, pp 551–554

114. White BC, Gadztnski DS, Hoehner PJ et al (1982e) Effect of flunarizine on canine cerebral cortical blood flow and vascular resistance post cardiac arrest. Ann Emerg Med 11: 119–126

115. Wiernsperger N, Gygay P, Hofmann A (1984) Calcium antagonist PY 108–068: demonstration of its efficacy in various types of experimental brain ischemia. Stroke 15: 679–685

116. Wolfe LS (1982) Eicosanoids: prostaglandins, thromboxanes, leukotrienes, and other derviatives of carbon-20 unsaturated fatty acids. J Neurochem 38: 1–14

117. Wylie EJ, Hein MF, Adams JE (1964) Intracranial hemorrhage following surgical revascularization for treatment of acute strokes. J Neurosurg 21: 212–215

118. Yasuda H, Shimada O, Nakajima A et al (1981) Cerebral protective effect and radical scavenging action. J Neurochem 37: 934–938

119. Yatsu FM, Diamond I, Graziano C et al

(1972) Experimental brain ischemia: protection from irreversible damage with a rapid-acting barbiturate (methohexital) Stroke 3: 726–732

120. Yoshida S, Inoh S, Asano T *et al* (1980) Effect of transient ischemia on free fatty acids and phospholipids in gerbil brain. Lipid peroxidation as a possible cause of postischemic injury. J Neurosurg 53: 323–331

121. Yoshimoto T, Sakamoto T, Suzuki J *et al* (1978) Experimental cerebral infarction.

Part 1: Production of thalamic infarction in dogs. Stroke 9: 211–214

122. Yoshimoto T, Sakamoto T, Watanabe T *et al* (1978) Experimental cerebral infarction. Part 3: Protective effect of mannitol in thalamic infarction in dogs. Stroke 9: 217–218

123. Zabramski JM, Spetzler RF, Selman WR *et al* (1984) Naloxone therapy during focal cerebral ischemia. Evaluation in a primate model. Stroke 15: 621–626

CHAPTER 7

1. Aoki N, Horibe H, Kasagi F (1984) International mortality statistics for all causes, cerebrovascular disease, ischemic heart disease and diabetes mellitus 1958–1978. Department of Epidemiology National Cardiovascular Center Research Institute. Osaka, Japan, pp 207–208

2. Baum HM, Goldstein M (1982) Cerebrovascular disease type specific mortality: 1968–1977. Stroke 13: 810–817

3. Berry RG, Alpers BJ (1957) Occlusion of the carotid circulation: Pathologic considerations. Neurology 7: 223–237

4. Bounds JV, Wiebers DO, Whisnant JP *et al* (1981) Mechanisms and timing of deaths from cerebral infarction. Stroke 12: 474–477

5. Brown M, Glassenberg M (1973) Mortality factors in patients with acute stroke. JAMA 224: 1493–1495

6. Castaigne P, Lhermitte F, Gautier J-C *et al* (1970) Interal carotid artery occlusion. A study of 61 instances in 50 patients with postmortem data. Brain 93: 231–258

7. DeJong RN (1976) The neurological examination. Evanston, New York, Harper and Row, London, pp 452–453

8. Fukada N, Ogawa A, Yoshimoto T *et al* (1985) Clinical course and prognosis of middle cerebral artery occlusion in the acute stage. Jpn J Stroke 7: 425–432 (Eng Abstr)

9. Greenwood J (1968) Acute brain infarction with high intracranial pressure: Surgical indications. Johns Hopkins Med J 122: 250–260

10. Hurwitz LJ, Groch SN, Wright IS *et al* (1959) Carotid artery occlusive syndrome. Arch Neurol 1: 491–501

11. Jones HR, Millikan CH (1976) Temporal profile (clinical course) of acute carotid system cerebral infarction. Stroke 7: 64–71

12. Katz *et al* (1966) Prognosis after stroke, part 2. Long term course of 159 patients. Medicine (Balt) 45: 236

13. Kogure T, Ogawa A, Seki H *et al* (1985) Clinical course and prognosis of internal carotid artery occlusion in the acute stage. Jpn J Stroke 7: 394–401 (Eng Abstr)

14. Larry KY, Nimmaitya J (1970) Massive cerebral infarction with severe brain swelling. Stroke 1: 158–163

15. Marquardsen J (1969) The natural history of acute cerebrovascular disease. A retrospective study of 769 patients. Munksgaard, Copenhagen, pp 44–45

16. Marshall J, Kaeser AC (1961) Survival after nonhemorrhagic cerebrovascular accidents. Br Med J 2: 73

17. Nakamura N, Kogure T, Ogawa A *et al* (1986) Prognosis of ischemic cerebrovascular disease without positive angiographical findings. Jpn J Stroke (in press) (Eng Abstr)

18. National public health, it's trend, health and welfare statistics, Health and Welfare Statistics Soc: 64 (1983)

19. Nishimaru K, Omae T (1972) Three years follow-up of survivors from acute ischemic stroke. Nippon Ronen 9: 42–49 (Eng Abstr)

20. Nishimaru K, Beppu M (1978) The actural circumstances of transitory ischemic apoplexy in Japan—the relationship of its etiology to cerebral infarct apoplexy. The Journal Clinical Science 14: 165–173

21. Ohta T, Waga S, Handa H *et al* (1974) New grading of level of disordered consciousness. No Shinkei Geka 2: 623–627 (Eng Abstr)
22. Okinaka S (1966) Epidemiological study of cerebral apoplexy—follow-up research in 17 municipalities throughout Japan during the last 3 years. Japanese Medical Journal (Tokyo) 2221: 19–28 (Eng Abstr)
23. Otomo E (1982) Easy treatable cerebral infarct—complete recuperative cerebral apoplexy (RIND: reversible ischemic neurological deficit). Nippon Rinsho 40: 2208–2212 (Eng Abstr)
24. Radue EW; Moseley IF (1978) Carotid artery occlusion and computed tomography. A clinicoradiological study. Neuroradiology 17: 7–12
25. Shaw CM, Alvord EC Jr, Berry RG (1959) Swelling of the brain following ischemic infarction with arterial occlusion. Arch Neurol 1: 161–177
26. Shenkin HA, Haft H, Somach FM (1965) Prognostic significance of arteriography in nonhemorrhagic strokes. JAMA 194: 612–616
27. Statistical analysis of movement of population by the Ministry of Health and Welfare (1983)
28. Takanashi A, Ogawa A, Sakurai Y *et al* (1982) CT findings of cerebral hemorrhagic infarction. Progress in CT (Tokyo) 4: f 153–158 (Eng Abstr)
29. Whisnant JP *et al* (1973) Transient cerebral ischemic attacks in a community; Rochester, Minnesota, 1955 through 1969. Mayo Clin Proc 48: 194–198
30. World health statistics annual, WHO, Geneva (1983)

CHAPTER 8

1. Alex L (1980): Cerebral blood flow determination by rapidsequence computed tomography. A theoretical analysis. Radiology 137: 679–686
2. Astrup J, Siesjö BK, Symon L (1981): Thresholds in cerebral ischemia—the ischemic penumbra. Stroke 12: 723–725
3. Aulich A, Wende S *et al* (1976): Diagnosis and follow up studies in cerebral infarcts. In: Lansk W, Kazner E (eds) Cranial computerized tomography. Springer, Berlin Heidelberg New York, pp 273–283
4. Baron JC, Bousser MG; Coman D *et al* (1981): "Crossed cerebellar diaschisis". A remote functional depression secondary to supratentorial infarction of man. J Cereb Blood Flow Metab [Suppl 1] 1: 500
5. Baron JC, Bousser MG, Rey A *et al* (1981): Reversal of focal misery-perfusion syndrome by extra-intracranial arterial bypass in hemodynamic cerebral ischemia. A case study with ^{15}O positron emission tomography. Stroke 12: 454–459
6. Bryan JT, Willcott MR, Schneiders NJ *et al* (1983): Nuclear magnetic resonance evaluation of stroke. Radiology 149: 189–192
7. Buonanno FS, Pykett IL, Brady TJ *et al* (1983): Proton NMR imaging in experimental ischemic infarction. Stroke 14: 178–183
8. Buonanno FS, Brady TJ, Pykett IL *et al* (1981): Proton NMR imaging in experimental ischemic cerebral infarction. Ann Neurol 10: 75
9. Buonanno FS, Pykett IL *et al* (1983): Cranial anatomy and detection of ischemic stroke in the cat by nuclear magnetic resonance imaging. Radiology 143: 187–193
10. Bydder GM, Steiner RE, Young IR *et al* (1982): Clinical NMR imaging of the brain: 140 cases. AJR 139: 215–236
11. Bydder GM, Steiner RE (1982): NMR imaging of the brain. Neuroradiology 23: 231–240
12. Davis KR, Taveras JM *et al* (1975): Cerebral infarction diagnosis by computed tomography: Analysis and evaluation of findings. AJR 124: 643–660
13. Davis KR, Ackerman JP (1977): Computed tomography of cerebral infarction: hemorrhagic, contrast enhancement, and time of appearance. Computerized Tomogr 1: 71–86
14. Di Chiro G, Timins EL *et al* (1974). Radionuclide scanning and microangiography of evolving and completed brain in-

farction: a correlative study in monkey. Neurology 24: 418–423

15. Drayer BP, Dujovny M *et al* (1977): The capacity of computed tomography diagnosis of cerebral infarction. Radiology 125: 393–402

16. Drayer BP, Wolfson SK Jr, Reinmuth OM *et al* (1978): Xenon enhanced CT for analysis of cerebral integrity, perfusion, and blood flow. Stroke 9: 123–130

17. Gacs G, Fox AJ, Barnett HJM (1983): CT visualization of intracranial arterial thromboembolism. Stroke 14: 756

18. Gur D, Wolfson SK Jr, Yonas H *et al* (1982). Progress in cerebrovascular desease: local cerebral blood flow by xenon enhanced CT. Stroke 13: 750–758

19. Kety SS, Schmidt CF (1945): The determination of cerebral blood flow in man by the use of nitrous oxide in low concentration. Am J Physiol 143: 53–66

20. Kjos BO, Brant-Zawadzki M, Young RG (1983). Early CT findings of global central nervous system hypoperfusion. AJR 14: 1277

21. Kuhl DE, Barrio JR, Haung SC *et al* (1982): Quantifying local cerebral blood flow by N-Isopropyl-p-(^{123}I) Iodoamphetamin (IMP) tomography. J Nucl Med 23: 196–203

22. Lassen NA, Henriksen L, Paulson O (1981): Regional cerebral blood flow in stroke by ^{133}Xenon inhalation and emission tomography. Stroke 12: 284–288

23. Lassen NA (1966): The luxury-perfusion syndrome and its possible relation to acute metabolic acidosis localized within the brain. Lancet 2: 1113–1115

24. Mallett BL, Veall N (1965): The measurement of regional cerebral clearance rates in man using xenon-133 inhalation and extracranial recording. Clin Sci 29: 179–191

25. Manelfe C, Ducos de Lahitte M, Marc Vergnes JP *et al* (1982): Investigation of major cerebral arteries by intravenous angiography: Report of 100 cases. AJNR 3: 287–293

26. Masdeu JC, Azar-Kia B, Rubino FA (1977): Evaluation of recent cerebral infarction by computerized tomography. Arch Neurology 34: 417–421

27. Masdeu JC (1983): Enhancing mass on CT: neoplasma or recent infarction? Neurology (NY) 33: 836–840

28. Meyer JS, Shinohara Y, Kanda T *et al* (1970): Diaschisis resulting from acute unilateral cerebral infarction. Quantitative evidence for man. Arch Neurol 23: 241–247

29. Mills CM, Crooks LE *et al* (1984): Cerebral abnormalities; use of calculated T_1 and T_1 magnetic resonance images for diagnosis. Radiology 150: 87–94

30. O'Brien MD, Waltz AG, Jordan MM (1974): Ischemic cerebral edema distribution of water in brain of cats after occlusion of the middle cerebral artery. Arch Neurol 30: 456–460

31. Obrist WD, Thompson HK Jr, King CH *et al* (1967): Determination of regional cerebral blood flow by inhalation of 133-Xenon. Circ Res 20: 124–135

32. Ogawa A, Suzuki J (1984): Prognosis of 1000 cases of acute stage. Japanisch-Deutsche Medizinische Berichte 29: 578–589 (Jpn)

33. Phelps ME (1982): Emission computed tomography. Semin Nucl Med 7: 337–365

34. Phelps ME (1982): The study of cerebral function with positron emission tomography. J Cereb Blood Flow Metab 2: 113–162

35. Sipponen JT, Kaste M, Ketonen L *et al* (1983): Serial magnetic resonance (NMR) imaging in patients with cerebral infarction. J Comput Assist Tomogr 7: 585–589

36. Sipponen JT (1984): Visualization of brain infarction with nuclear magnetic resonance imaging. Neuroradiology 26: 387–391

37. Striver EB, Olsen TS (1981): Transient disappearance of cerebral infarct on CT scan, the so-called fogging effect. Neuroradiology 22: 61

38. Skriver EB, Olsen TS (1982): Contrast enhancement of cerebral infarcts: Incedence and clinical value in different states of cerebral infarction. Neuroradiology 23: 259

39. Takahashi M, Hirota Y, Tsuchigame M *et al* (1982): Clinical value of digital fluoroscopic angiography. Journal of Medical Images 2: 660–672

40. Takahashi M, Bussaka H, Nonaka S *et al* (1983): Degital fluoroscopic angiography in the diagnosis of central nervous system

diseases. Neurol Med Chir (Tokyo) 23: 116–112 (Eng Abstr)

41. Takahashi M, Bussaka H, Nakagawa N (1984): Evaluation of cerebral vasculature by intraarterial DSA-with emphasis on *in vivo* resolution. Neuroradiology 26: 253–259

42. Takahashi M, Hirota Y, Bussaka H *et al* (1983): Evaluation of prototype equipment of digital subtraction angiography in diagnosing intracranial lesions. AJNR 4: 259–262

43. Taveras JM, Wood EH (1976): Diagnostic neuroradiology. Williams and Wilkins, Baltimore

44. Tomita M, Gotoh F (1981): Local cerebral blood flow values as estimated with diffusible tracers: validity of assumptions in normal and ischemic tissue. J Cereb Blood Flow Metab 1: 403–411

45. Vonfakos D, Artmann H (1983): T finding in hemorrhagic cerebral infarction. Comput Radiol 7: 75

46. Wing SD, Norman D *et al* (1976): Contrast enhancement of cerebral infarcts in computed tomography. Radiology 121: 89–92

47. Yamaguchi K, Uemura K (1979): Neuroradiology. Tokyo, Asakura-Shoten (Jpn), Tokyo, pp 147–190

48. Yock DJ, Marshall WH (1975): Recent cerebral infarcts at computed tomography: appearance pre-and-post contrast infusion. Radiology 117: 599–608

CHAPTER 9 A

1. Acheson J (1971): Factors affecting the natural history of focal cerebral vascular disease. Q J Med 40: 25–46

2. Baker RN, Broward JH, Fang HC *et al* (1962): Anticoagulant therapy in cerebral infarction: report of a co-operative study. Neurology (Minneap) 12: 823–835

3. Baskin DS, Hosobuchi Y (1981): Naloxone reversal of ischemic neurological deficits in man. Lancet 2: 272–275

4. Beevers DG, Fairman MJ, Hamilton M *et al* (1973): Antihypertensive treatment and the course of established cerebral disease. Lancet 1: 1407–1409

5. Bousser MG, Eschwege E, Hauenau M *et al* (1983): „AICLA" controlled trial of aspirin and dipyridamole in the secondary prevention of athero-thrombotic cerebral ischemia. Stroke 14: 5–14

6. Canadian Cooperative Study Group (1978): A randomized trial of aspirin and sulfinphrazone in threatened stroke. N Engl J Med 299: 53–59

7. Carter AB (1961): Anticoagulant treatment in progressing stroke. Br Med J 5244: 70–73

8. Carter AB (1970): Hypotensive therapy in stroke survivors. Lancet 1: 485–489

9. Dyken M, White PT (1956): Evaluation of cortisone in treatment of cerebral infarction. JAMA 162: 1531–1534

10. Fields W, Lemak NA, Frankowski RF *et al* (1977): Controlled trial of aspirin in cerebral ischemia. Stroke 8: 301–316

11. Fisher CM (1982): Management of occlusive cerebrovascular disease. In: Ropper AH, Kennedy SK *et al* (eds) Neurological and neurosurgical intensive care. University Park Press, Baltimore, pp 189–205

12. Fletcher AP, Alkjaersig N, Lewis M *et al* (1976): A pilot study of urokinase therapy in cerebral infarction. Stroke 7: 135–142

13. Harter HR, Burch, Majerus PW *et al* (1979): Prevention of thrombosis in patients on hemodialysis by low-dose aspirin. N Engl J Med 301: 577–579

14. Heck AF (1984): Medical management of TIAs and small strokes. In: Smith RR (ed): Stroke and the extracranial vessels. Raven Press, New York, pp 149–158

15. Hutchinson EC (1983): Management of cerebral infarction. In: Russel RWR (ed) Vascular disease of the central nervous system, edn 2, Churchill Livingstone, Edinburgh London Melbourne New York, pp 185–203

16. Genton E, Barnett HJM, Fields WS *et al* (1977): Cerebral ischemia: The role of thrombosis and of antithrombotic therapy. Stroke 8: 150–175

17. Katzman R, Clasen R, Klatzo I *et al* (1977): Brain edema in stroke. Stroke 8: 512–540

18. Kuriyama Y, Sawada T, Yamaguchi T *et al* (1980): Induced hemodilution in acute stage of cerebral infarction, its pathophysiological

and therapeutic consideration. Jpn J Stroke 2: 309–315 (Eng Abstr)

19. Matthews WB, Oxbury JM, Grainger KMR et al (1976): A blind controlled trial of dextran 40 in the treatment of ischemic stroke. Brain 99: 193–206

20. Meyer JS, Charney JZ, Mathew NT (1971): Treatment with glycerol of cerebral edema due to acute cerebral infarction. Lancet 2: 993–997

21. Meyer JS, Fukuuchi Y, Shimazu J et al (1972): Effect of intravenous infusion of glycerol on hemispheric blood flow and metabolism in patients with acute cerebral infarction. Stroke 3: 167–180

22. Meyer JS, Gilroy JG, Barnhart MI et al (1963): Therapeutic thrombosis in cerebral thromboembolism. Neurology 13: 927–937

23. Millikan CH (1971): Reassessment of anti-coagulant therapy in various types of occlusive cerebrovascular disease. Stroke 2: 201–208

24. Minematsu K, Yamaguchi T, Kuryama Y et al (1983): Clinical application of bar-biturate therapy for cerebral infarction: a preliminary report. Jpn J Stroke 5: 286–294 (Eng Abstr)

25. Mizoi M, Oba M, Fujimoto S et al (1984): An experimental study on the protective effect of nizofenone in cerebral ischemia. No To Shinkei 36: 1089–1093 (Eng Abstr)

26. Mulley G, Wilcox RG, Mitchell JRA (1978): Dexamethasone in acute stroke. Br Med J 2: 994–996

27. Patten BM, Mendell J, Bruun B et al (1972): Doubleblind study of the effects of de-xamethasone on acute stroke. Neurology 22: 337–383

28. Rockoff MA, Marshall LF, Shapiro HM (1979): high-dose barbiturate therapy in humans: a clinical review of 60 patients. Ann Neurol 6: 194–199

29. Sandok BA, Furlan AJ, Whisnant JP et al (1978): Guidelines for the management of transient ischemic attacks. Mayo Clin Proc 53: 665–674

30. Satake K, Kondo H, Fukase M (1979): A double-blind study on the clinical efficacy of urokinase in treatment of patients with cerebral infarction. In: Yamori et al (eds) Prophylactic approach to hypertensive disease. Perspective in Cardiovascular Reseach 4: 433–440

31. Smith AL, Hoff JT, Nielsen SL et al (1974) Barbiturate protection in acute focal cerebral ischemia. Stroke 5: 1–5

32. Smith AL (1977) Barbiturate protection in cerebral hypoxia. Anesthesiology 47: 285–293

33. Sorensen PS, Pedersen H, Marquardsen J et al (1983) Acetylsalicylic acid in the pre-vention of stroke in patients with reversible cerebral ischemic attacks: a Danich co-operative study. Stroke 14: 15–22

34. Suzuki J, Fujimoto S, Mizoi K et al (1984) The protective effect of combined admin-istration of anti-oxidants and perflu-orochemicals on cerebral ischemia. Stroke 15: 672–679

35. Suzuki J, Takahashi A, Ycshimoto T et al (1985) Use of balloon occlusion and sub-stances to protect ischemic brain during resection of posterior fossa AVM: case report. J Neurosurg 63: 626–629

36. Tamura A, Asano T, Sano K et al (1979) Protection from cerebral ischemia by a new imidazole derivative (Y-9179) and pento-barbital: a comparative study in chronic MCA occlusion in cats. Stroke 10: 126–134

37. Yatsu FM, Diamond I, Graziano C et al (1972) Experimental brain ischemia: pro-tection from irreversible damage with a rapid-acting barbiturate (methohexital) Stroke 3: 726–732

38. Yatsu FM (1975) Protective effects of bar-biturates on cerebral ischemia. In: McLaurin RL (ed) Head injuries, 2nd Chicago Symposium on Neural Trauma. Grune and Stratton, New York, pp 185–187

39. Yatsu FM (1984) Medical management of complete strokes. In: Smith RR (ed) Stroke and the extracranial vessels. Raven Press, New York, pp 159–166

40. Whisnant JP, Matsumoto N, Elveback LR (1973) The effect of anticoagulation therapy on the prognosis of patients with transient cerebral ischemic attacks in a community: Rochester, Minnesota 1955 through 1969. Mayo Clin Proc 48: 844–848

41. Wise BL, Chater N (1962) The value of hypertonic mannitol solution in decreasing brain mass and lowering cerebrospinal fluid pressure. J Neurosurg 19: 1038–1043

CHAPTER 9 B

1. Asplund K, Liliequist B, Fodstad H et al (1981) Long-term outcome in cerebrovascular disease in relation to findings at aortocervical angiography. A 12-year follow up. Stroke 12: 307–313

2. Ausman JI, Diaz FG, de los Reyes RA et al (1981) Anastomasis of occipital artery to anterior inferior cerebellar artery for vertebrobasilar junction stenosis. Surg Neurol 16: 99–102

3. Ausman JI, Lee MC, Chater N (1979) Superificial temporal artery to superior cerebellar artery anastomosis for distal basilar artery stenosis. Surg Neurol 12: 277–282

4. Barnett HJM, Peerless SJ, Kaufmann JCE (1978) "Stump" of internal carotid artery— A source for further cerebral embolic ischemia. Stroke 9: 448–456

5. Bauer RB, Meyer JS, Gothman JE, Gilroy J (1966) A controlled study of surgical treatment of cerebrovascular disease—42 months experience with 183 cases. In: Millikan C, Sickert R, Whisman JP (eds) Cerebral vascular diseases. Grune and Stratton, New York

6. Blaisdell WF, Clauss RH, Galbraith JG et al (1969) Joint study of extracranial arterial occlusion. JAMA 209: 1889–1895

7. Bogousslavsky J, Regli JF, Hungerdichler JP et al (1981) Transient ischemic attacks and external carotid artery. A retrospective study of 23 patients with an occlusion of the internal carotid artery. Stroke 12: 627–630

8. Bosnic MA (1964) Cervical arterial pathways associated with brachiocephalic occlusion disease. AJR 91: 1232–1244

9. Caplan LR (1979) Occlusion of the vertebral or basilar artery. Follow up analysis of some patients with benign outcome. Stroke 10: 277–282

10. Caplan LR (1981) Vertebrobasilar diseases. Time for a new strategy. Stroke 12: 111–114

11. Chater N (1978) Microsurgical vascular bypass for occlusive cerebrovascular disease. In: Rand RW (ed) Microneurosurgery. CV Mosby, St Louis, pp 71–92

12. Clark K, Perry MO (1966) Carotid vertebral anastomosis: an alternate technique for repair of the subclavian steal syndrome. Ann Surg 163: 414–416

13. Clayson KR, Edwards WH (1971) Importance of the external carotid artery in extracranial cerebrovascular occlusive disease. South Med J 79: 904–909

14. Conolly JE, Stemmer EA (1973) Endarterectomy of the external carotid artery: its importance in the surgical management of extracranial cerebrovascular occlusive disease. Arch Surg 106: 799–802

15. Correll JW, Quest DO, Darpenter DB (1985) Nonathermoatous lesions of the extracranial cerebral arteries. In: Smith RR (ed) Stroke and extracranial vessels. Raven Press, New York, pp 321–332

16. Countee RW, Vijayanathan T, Chavis P (1981) Recurrent retinal ischemia beyond cervical carotid occlusions. Clinical angiographic correlations and therapeutic implications. J Neurosurg 55: 532–542

17. Countee RW, Vijayanathan T (1979) External carotid artery in internal carotid artery occlusions. Angiographic, therapeutic, and prognostic considerations. Stroke 10: 450–460

18. Crawford ES, DeBakey ME, Fields WS (1958) Roentogenographic diagnosis and surgical treatment of basilar artery insufficiency. JAMA 168: 509–514

19. Culligan JA (1960) Backling and kinking of the carotid vessels. Minn Med 43: 678–733

20. DeBenedetti G, Marra A, Pozzi M (1982) Bypass with basillic autogenous venous graft because of insufficient caliber of the superficial temporal artery. Surg Neurol 18: 303–306

21. Diethrich EB, Liddicoat JE, McCutchen JJ et al (1968) Surgical significance of the external carotid artery in the treatment of cerebrovascular insufficiency. J Cardiovasc Surg 9: 213–223

22. Durward QJ, Ferguson GG, Barn HWK (1982) The natural history of asymptomatic carotid bifuraction plaques. Stroke 13: 459–464

23. Eaton DJ, Sherman DG (1977) Stroke and mortality rate in carotid endarterectomy: 228 consecutive operations. Stroke 8: 565–568

24. Eguchi T, Mayanagi Y, Funamura T (1982) Treatment of bilateral spontaneous carotid

cavernous fistula by Hamby's method combined with an extracranial-intracranial bypass procedure. Neurosurgery 11: 706–711

25. Eguchi T, Mayanagi Y, Iai S *et al* (1983) Extracranial-intracranial bypass to the proximal posterior cerebral artery and to the proximal middle cerebral artery for multiple occlusive cerebrovascular disease. Surg Neurol 19: 131–136

26. Fein JM (1985) Carotid endarterectomy. In: Fein JM (ed) Cerebrovascular surgery. Springer, New York Berlin Heidelberg, pp 399–427

27. Fields WS, Lamak NA, Frakowski RF *et al* (1977) Controlled trial of aspirin in cerebral ischemia. Stroke 8: 301–319

28. Fields WS, Lemak NA, Frakowski RF *et al* (1978) Controlled trial of aspirin in cerebral ischemia. Part II: Surgical group. Stroke 9: 309–319

29. Fields WS, Lemak NA (1976) Joint study of extracranial arterial occlusion. IX Transient ischemic attacks in the carotid artery. JAMA 235: 2608–2610

30. Finkenstein JM, Byer A, Rush BF Jr (1972) Subclavian-subclavian bypass for the subclavian steal syndrome. Surgery 71: 142–145

31. Fisher CM (1970) Occlusion of the vertebral arteries causing transient basilar symptoms. Arch Neurol 22: 13–19

32. Galbraith JG, McDowell HA Jr (1969) Stroke and occlusive cerebrovascular disease. Reviews and surgical results in 265 cases. J Med Assoc Stroke Ala 38: 1107–1111

33. Gautier JC (1985) Stroke-in-progression. Stroke 16: 729–733

34. Goldstone J, Moore WS (1976) Emergency carotid artery surgery in neurologically unstable patients. Arch Surg 111: 1284–1291

35. Goldstone J, Moore WS (1978) A new look at emergency carotid artery operations for the treatment of cerebral vascular insufficiency. Stroke 9: 599–602

36. Hardin CA, Williamson WP, Steegmann AT (1960) Vertebral artery insufficiency produced by cervical osteoarthritis spurs. Neurology 10: 855–858

37. Heros RC, Sekhar LN (1981) Diagnostic and therapeutic alternatives in patients with symptomatic "carotid occlusion" referred for extrancranial-intracranial bypass surgery. J Neurosurg 54: 790–796

38. Hodosh RM, Boone S (1981) Neurological manifestation of external carotid artery disease. Clin Neurosurg 28: 384–406

39. Hopkins LN, Bundy JC, Spetzler RF (1982) Revascularization of the rostral brain stem. Neurosurg 10: 364–369

40. Humphries AW, Young JR, Santilli PH *et al* (1976) Unoperated asymptomatic significant internal carotid artery stenosis. A review of 182 instances. Surgery 80: 695–698

41. Hutchinson EC, Acheson EJ (1975) Strokes; natural history, pathology and surgical treatment. Saunders, London

42. Ishii R, Koike T, Takeuchi S (1983) Anastomosis of the superficial temporal artery to the distal anterior cerebral artery with interposed cephalic vein graft. J Neurosurg 58: 425–429

43. Ito Z (1981) A new technique of intracranial anastomosis between distal anterior cerebral arteries (ACA) for ACA occlusion and its indication. Neurol Med Chir (Tokyo) 21: 931–939 (Eng Abstr)

44. Jacobson JH II, Mozersky DJ, Mitty HA *et al* (1973) Axillary-axillary bypass for the "subclavian steal" syndrome. Arch Surg 106: 24–27

45. Jacobson JH II, Suarez EL (1960) Microsurgery in anastomosis of small vessels. Surg Forum 11: 243–245

46. Jones HR, Millikan CH, Sandok BA (1980) Temporal profile of acute vertebro-basilar system cerebral infarction. Stroke 11: 173–177

47. Jones HR, Millikan CH (1976) Temporal profile (clinical course) of acute carotid system cerebral infarction. Stroke 7: 64–71

48. Kaplan HA (1961) Collateral circulation of the brain. Neurology 11: 9–15

49. Kelly AB (1925) Tortousy of the internal carotid artery in relation to the pharynx. J L Otology 40: 15–23

50. Khodadad G (1976) Occipital artery—posterior inferior cerebellar artery anastomosis. Surg Neurol 5: 225–227

51. Lawrie GM, Morris GC Jr, Chapman DW (1977) Patterns of patency of 596 vein grafts up to seven years after aorto-coronary bypass. J Thrac Cardiovasc Surg 73: 443–448

52. Lyons C, Galbraith G (1957) Surgical treatment of atherosclerotic occlusion of the internal carotid artery. Ann Surg 146: 487–498

53. Mehigan JT, Buch WS, Pipkin RD et al (1978) Subclavian-carotid transposition for the subclavian steal syndrome. Am J Surg 136 15–20

54. Mentzer RM Jr, Finkelmeier BA, Crosby IK et al (1981) Emergency carotid endarterectomy for fluctuating neurologic dificits. Surgery 89: 60–66

55. Millikan CH, McDowell FH (1981) Treatment of progressing stroke. Stroke 12: 397–409

56. Mohr JP (1982) Symptomatic carotid artery desease. Stroke 13: 431–433.

57. Murphey F, Miller JH (1959) Carotid insufficiency—diagnosis and surgical treatment. A report of twenty-one cases. J Neurosurg 16: 1–23

58. North RR, Fields WS, DeBakey ME et al (1962) Brachial-basilar insufficiency syndrome. Neurology 12: 810–820

59. Nunn DB (1975) Carotid endarterectomy: analysis of 234 operative cases. Ann Surg 182: 733–738

60. Ojemann RG, Crowell RM, Robertson GH et al (1974) Surgical treatment of extracranial carotid occlusive disease. Clin Neurosurg 22: 214–263

61. Patrick BK, Ramirez-Lassepas M, Synder BD (1980) Temporal profile of vertebrobasilar territory infarction. Prognostic implications. Stroke 11: 643–648

62. Reichman OH (1975) Extracranial-intracranial arterial anastomosis. In: Whisnant JP, Sandkok BA (eds) Cerebral vascular diseases. Grune and Stratton, New York, pp 175–185

63. Reichman OH (1976) Neurosurgical anastomosis for cerebral ischemia. Five year's experience. In: Sheinberg P (ed) Cerebrovascular diseases. Raven Press, New York, pp 311–330

64. Robertson JT, Auer NJ (1982) Extracranial occlusive disease of the carotid artery. In: Youmanns JR (ed) Neurological surgery, vol 3, Saunders, Philadelphia, pp 1559–1583

65. Robertson JT, Watridge CB (1979) Surgical management of extracranial-intracranial occlusive disease. Symposium on clinical neurology. Med Clin North Am 63: 681–693

66. Robertson JT (1985) The surgical candidates. In: Smith RD (ed) Stroke and extracranial vessels. Raven Press, New York, pp 167–174

67. Samson DS, Boones S (1978) Extracranial-intracranial (EC-IC) arterial bypass: past performance and current concepts. Neurosurg 3: 79–86

68. Samson DS, Gerwertz BL, Beyer CW Jr et al (1981) Saphenous vein interposition grafts in the microsurgical treatment of cerebral ischemia. Arch Surg 116: 1578–1582

69. Samson DS, Hodosh RM, Clark WK (1979) Microsurgical treatment of transient cerebral ischemia. Preliminary results in 50 cases. JAMA 241: 376–378

70. Samson Y, Baron JC, Boussen MG et al (1985) Effect of extra-intracranial arterial bypass on cerebral blood flow and oxygen metabolism in humans. Stroke 16: 609–616

71. Spetzler RF, Chater N (1974) Occipital artery-middle cerebral artery anastomosis for cerebral artery occlusive disease. Surg Neurol 2: 235–238

72. Story JL, Brown WE Jr, Eidelberg E (1979) Cerebral revascularization: cervical carotid—intracranial arterial long graft bypass. In: Marguth F, Brock M, Kazner E (eds) Advances in neurosurgery. Springer, Berlin Heidelberg New York, vol 7, pp 15–23

73. Story JL, Brown WE, Ansell LV et al (1985) The management of the vertebrobasilar insufficiency by extrancranial surgical procedure. In: Smith RR (ed) Stroke and extracranial vessels. Raven Press, New York, pp 227–254

74. Story JL, Brown WE Jr, Eidelberg E (1978) Cerebral revascularization: Common carotid to distal middle cerebral artery bypass. Neurosurg 2: 131–135

75. Strully KJ, Hurwitt ES, Blankenberg HW (1953) Thrombo-endarterectomy. For thrombosis of the internal carotid artery in the neck. J Neurosurg 10: 474–481

76. Sundt TM Jr, Siekert RG, Piepgras DG et al (1976) Bypass surgery for vascular disease of the carotid system. Mayo Clin Proc 51: 677–692

77. Sundt TM Jr, Piepgras DG (1978) Occipital to posterior inferior cerebellar artery bypass surgery. J Neurosurg 48: 916–928

78. The EC/IC Bypass Study Group (1985) The international cooperative study of extracranial/intracranial arterial anastomosis (EC/IC Bypass Study): Methodology and entry characteristics. Stroke 16: 397–406

79. The EC/IC Bypass study Group (1985) Failure of extracranial-intracranial arterial bypass to reduce the risk of ischemic stroke. Results of an international randomized trial. New Eng J Med 313: 1191–1200

80. Whylie EJ, Hein MF, Adam JE (1964) Intracranial hemorrhage following surgical revascularization for treatment of acute stroke. J Neurosurg 21: 212–215

81. Weibel J, Fields WS (1965) Tortuousity, coiling and kinking of the internal carotid artery. I. Etiology and radiographic anatomy. Neurology 15: 7–18

82. Yaşargil MG (1969) Anastomosis between the superficial temporal artery and a branch of the middle cerebral artery. In: Yaşargil MG (ed) Microsurgery applied to neurosurgery. Thieme, Stuttgart, pp 105–115

CHAPTER 9 C

1. Andrews BT, Weinstein PR, Chater NL (1985) Emergency extrcranial-intracranial arterial bypass for stroke in evolution. In: Spetzler RF, Carter LP, Selman WR, Martin NA (eds) Cerebral revascularization for stroke. Thieme-Stratton, New York, pp 542–547

2. Baker RN, Broward JA, Fang HC et al (1962) Anticoagulant therapy in cerebral infarction. Report on cooperative study. Neurology 12: 828–835

3. Chater IL, Popp J (1976) Microsurgical vascular bypass for occlusive cerebrovascular disease; review of 100 cases. Surg Neurol 6: 115–118

4. Crowell RM (1977) STA-MCA bypass for acute focal cerebral ischemia. In: Schmiedek P (ed) Microsurgery for stroke. Springer, Berlin Heidelberg New York, pp 244–250

5. Diaz FG, Ausman JI, Metha B et al (1985) Acute cerebral revascularization. J Neurosurg 63: 200–209

6. Engel LS, Ferraz H (1983) Indication and results of extra-intracranial arterial bypasses. Int Surg 68: 197–200

7. Fisher CM (1961) Anticoagulant therapy in cerebral thrombosis and cerebral embolism. A national cooperative study. Interim report. Neurology 11: 119–131

8. Gentou E, Barnett HJM, Fields WS et al (1977) XIV. Cerebral ischemia; the role of thrombosis and antithrombotic therapy. Study group on antithrombotic therapy. Stroke 8: 150–175

9. Goldstone J, Effeney D (1980) The role of carotid endarterectomy in the treatment of acute neurologic deficit. Prog Cardiovasc Dis 22: 6

10. Goldstone J, Moore WS (1978) A look at emergency carotid artery operations for the treatment of cerebrovascular insufficiency. Stroke 9: 599–602

11. Gratzl O, Schmidek P, Spetzler R et al (1976) Clinical experience with extraintracranial anastomosis in 65 cases. J Neurosurg 44: 313–324

12. Irino T, Watanabe M, Nishide M et al (1983) Angiographical analysis of acute cerebral infarction followed by "cascade"-like deterioration of minor neurological deficits. What is progressing stroke? Stroke 14: 363–368

13. Kagawa S, Koshu K, Yoshimoto T et al (1982) The protective effect of mannitol and perfluorochemicals on hemorrhage infarction: an experimental study. Surg Neurol 17: 66–70

14. Kayama T, Kaneko M, Muraki M et al (1985) Surgical results of emergent revasuclarization for the acute major stroke. Hemodynamic observation and clinical course. Cerebral revascularization for stroke. Thieme-Stratton, New York, pp 555–563

15. Kayama T, Mizoi K, Suzuki J (1981) A canine model of a completely ischemic brain regulated with the perfusion model. Surg Neurol 16: 167–172

16. Kayama T, Mizoi K, Yoshimoto T et al

(1980) Metabolic changes in complete ischemia using a canine model of complete ischemic brain regulated with a perfusion method. Analysis of the lipid peroxidate metabolites, energy metabolites and cortical electrical activity. In: Sano K, Handa H (eds) Cerebral ischemia and cell damage. Neuron Pub, Tokyo, pp 33–38

17. Kayama T, Mizoi K, Yoshimoto T *et al* (1980) Protective effect of mannitol and fluorocarbon in cerebral ischemia using a dog model of complete ischemia brain with perfusion model. Surg Ther (Osaka) 43: 700 (Eng Abstr)

18. Koshu K, Seki H, Yoshimoto T *et al* (1981) A canine model of a completely ischemic brain regulated with the perfusion method. Surg Neurol 16: 274–279

19. Millikan CH, Siekart RG (1955) Studies in cerebral vascular disease. 1. The syndrome of intermittent insufficiency of the basilar arterial system. Proc Staff Meet Mayo Clin 30: 61–68

20. Millikan CH, Siekart RG (1955) Studies in cerebral vascular disease. IV. The syndrome of intermittent insufficiency of the carotid. Proc Staff Meet Mayo Clin 30: 186–191

21. Millikan CH (1983) Clinical management of cerebral ischemia. In: McDowell FH, Brennan RW (eds) Cerebral vascular diseases, 8th Conference. Grune and Stratton, New York

22. Mizoi K, Yoshimoto TL, Suzuki J (1981) Experimental study of new cerebral protective substance-functional recovery of severe incomplete ischemic brain lesions, pretreated with mannitol and fluorocarbon emulsion. Acta Neurochir (Wien) 56: 157–166

23. Nishijima M, Tanaka S, Watanabe T *et al* (1981) Sequential changes in nerve cells during complete ischemia and the preventive effects of various drugs on cerebral infarction. No To Shinkei 33: 291–299 (Eng Abstr)

24. Ohta T, Waga S, Handa H *et al* (1974) New grading of level of disordered consciousness. No Shinkei Geka 2: 623–627 (Eng Abstr)

25. Royden Jones H, Millikan CH (1976) Temporal profile (clinical course) of acute carotid system cerebral infarction. Stroke 7: 64–71

26. Samson S, Boone S (1978) Extracranial-intracranial (EC-IC) arterial bypass: past performance and current concepts. Neurosurgery 3: 79–86

27. Seki H, Ogawa A, Yoshimoto T *et al* (1981) Effect of mannitol on rCBF in canine thalamus ischemia—an experimental study. No To Shinkei 33: 1101–1105 (Eng Abstr)

28. Spetzlar RG, Selman WR, Roski RA *et al* (1982) Cerebral revascularization during barbiturate coma in primates and humans. Surg Neurol 17: 111–115

29. Suzuki J (1974) A method of prolongation of temporary stopping of the cerebral blood flow. Presidential Address, 33rd Annual Meeting of the Japan Neurological Society, Sendai, Japan, October

30. Suzuki J, Fujimoto S, Mizoi K *et al* (1984) The protective effect of combined administration of anti-oxidants and perfluorochemicals on cerebral ischemia. Stroke (In press)

31. Suzuki J, Imaizumi S, Kayama T (1984) Chemiluminescence on hypoxic brain. The 2nd report; effect of free radical scavengers. Stroke (In press)

32. Suzuki J, Ogawa A, Yoshimoto T *et al* (1985) Indications for surgery in the acute stage of cerebral infarction. The role of new cerebral protective drugs—"Sendai cocktail" and perfluorochemicals. Cerebral revascularization for Stroke. Thieme-Stratton, New York, pp 392–396

33. Suzuki J, Onuma T, Yoshimoto T (1979) Results of early operation on cerebral aneurysms. Surg Neurol 11: 407–412

34. Suzuki J, Tanaka S, Yoshimoto T (1980) Recirculation in the acute period of cerebral infarction; brain swelling and its suppression using mannitol. Surg Neurol 14: 467–472

35. Suzuki J, Tanaka S, Yoshimoto T (1981) Suppression of brain swelling with mannitol and perfluorochemicals—an experimental study. Acta Neurochir (Wien) 58: 149–160

36. Suzuki J, Tanaka S, Yoshimoto T (1980) Recirculation in the acute period of cerebral infarction; experimental research on brain swelling and its suppression by using

mannitol or glycerol. Acta Neurochir (Wien) 54: 219–231

37. Suzuki J, Yoshimoto T (1981) A combined treatment of hypertonic solution and blood substitute for cerebral infarction. Medicine (Tokyo) 18: 1660–1661

38. Suzuki J, Yoshimoto T (1978) Indication and timing in the surgery of ruptured cerebral aneurysm. Phronesis 36: 34–48

39. Suzuki J, Yoshimoto T (1979) The effect of mannitol in prolongation of permissible occlusion time of cerebral artery—clinical data of aneurysm surgery. Neurosurg Rev 1: 13–19

40. Suzuki J, Yoshimoto T, Kadama N et al (1982) A new therapeutic method for acute brain infarction; revascularization following the administration of mannitol and perfluorochemicals. A preliminary report. Surg Neurol 17: 325–332

41. Suzuki J, Yoshimoto T, Mozoi K et al (1981) A new therapeutic method for the acute period of brain infarction. Jpn J Stroke 3: 130–131 (Eng Abstr)

42. Suzuki J, Yoshimoto T, Tanaka S et al (1981) Development of new methods in suppressing brain infarction—combined administration of mannitol and perfluorochemicals. No Shinkei Geka 91: 465–470 (Eng Abstr)

43. Suzuki J, Yoshimoto T, Tanaka S et al (1980) Production of various models of cerebral infarction in the dog by means of occlusion of intracranial trunk arteries. Stroke 11: 337–341

44. Tanaka S, Yoshimoto T, Seki H et al (1979) Production of infarction model in dogs by severe brain arteries occlusion. Part 1: Production of hemispheric non-blood supply model. No To Shinkei 31: 665–669 (Eng Abstr)

45. Yoshimoto T, Sakamoto T, Suzuki J (1978) Experimental cerebral infarction. Part 1: Production of thalamus infarction in dogs. Stroke 9: 211–214

46. Yoshimoto T, Sakamoto T, Watanabe T et al (1978) Experimental cerebral infarction. Part 3: Protective effect of mannitol in thalamic infarction in dogs. Stroke 9: 217–218

47. Yoshimoto T, Suzuki J (1976) Intracranial definitive aneurysm surgery under normothermia and normotension—utilizing temporary occlusion of brain artery and preoperative mannitol administration. No Shinkei Geka 4: 775–783 (Eng Abstr)

48. Yoshimoto T, Suzuki J (1979) Temporary clipping prolongation of the time of occlusion by mannitol. In: Pia HW, Langmaid C, Zierski J (eds) Cerebral aneurysms: advances in diagnosis and therapy. Springer, Berlin Heidelberg New York, pp 388–392

49. Watanabe T, Yoshimoto T, Suzuki J (1979) The effect of mannitol in preventing the development of cerebral infarction; an electron microscopical investigation. In: Suzuki J (ed) Cerebral aneurysms—experience with 1000 directly operated cases. Neuron Pub, Tokyo, pp 359–365

50. Wylie EJ, Hein MF, Akams JE (1964) Intracranial hemorrhage following surgical revascularization for treatment of acute stroke. J Neurosurg 21: 212–215

CHAPTER 10

1. Alexander E, Adams JE, Davis CH (1963) Complications in the use of temporary intracranial arterial clip. J Neurosurg 20: 810–811

2. Botterell EH, Lougheed WM, Morley TP et al (1958) Hypothermia in the surgical treatment of ruptured intracranial aneurysms. J Neurosurg 15: 4–18

3. Dujovny M, Osgood CP, Barrionuevo PJ et al (1978) SEM evaluation of endotherial damage following temporary middle cerebral artery occlusion in dogs. J Neurosurg 48: 42–48

4. Lougheed WM, Sweet WH, White JC et al (1955) The use of hypothermia in surgical treatment of cerebral vascular lesions. A preliminary report. J Neurosurg 12: 240–255

5. Michenfelder JD, Gronert GA, Rehder K (1982) Anesthesia. In: Youmans JR (ed) Neurological surgery, 2 edn. Saunders, Philadelphia, pp 1119–1135

6. Mizoi K, Yoshimoto T, Suzuki J (1981) Experimental study of new protective substances—functional recovery of severe incomplete ischemic brain lesion, pretreated with mannital and perfluorochemicals—. Acta Neurochir (Wien) 56: 157–166

7. Pool JL (1961) Aneurysms of the anterior communicating artery. Bifrontal craniotomy and routine use of temporary clips. J Neurosurg 18: 98–111

8. Sugita K, Hirota T, Iguchi I et al (1976) Comparative study of the pressure of various aneurysm clips. J Neurosurg 44: 723–727

9. Suzuki J, Fujimoto S, Mizoi K et al (1984) The protective effect of combined administration of anti-oxidants and perfluorochemicals on cerebral ischemia. Stroke 15: 672–679

10. Suzuki J, Komada N, Homma M (1979) New clip and flexible clip forceps for neurosurgery. In: Suzuki J (ed) Cerebral aneurysms. Neuron Pub, Tokyo, pp 386–388

11. Suzuki J, Kwak R, Okudaira Y (1979) The safe time limit of temporary clamping of cerebral arteries in the direct surgical treatment of intracranial aneurysm under moderate hypothermia. Tohoku J Exp Med 127: 1–7

12. Suzuki J, Onuma T, Yoshimoto T (1979) Results of early operation on cerebral aneurysms. Surg Neurol 11: 407–412

13. Suzuki J, Tanaka S, Yoshimoto T (1981) Suppression of brain swelling with mannitol and perfluorochemicals—an experimental study. Acta Neurochir (Wien) 58: 149–160

14. Suzuki J, Yoshimoto T (1979) The effect of mannitol in prolongation of permissible occlusion time of cerebral arteries. Clinical date of aneurysm surgery. In: Suzuki J (ed) Cerebral aneurysms. Neuron Pub, Tokyo, pp 330–337

15. Yaşargil MG, Fox JL (1975) The microsurgical approach to intracranial aneurysms. Surg Neurol 3: 7–14

16. Yoshimoto T, Sakamoto T, Watanabe T et al (1978) Experimental cerebral infarction. Part III: Protective effect of Mannitol in thalamic infarction in dogs. Stroke 9: 217–218

CHAPTER 11

1. Alksne JF, Greenhoot JM (1974) Experimental cetecholamine-induced chronic cerebral vasospasm. Myonecrosis in vessel wall. J Neurosurg 41: 440–445

2. Allcock JM, Drake CG (1965) Ruptured intracranial aneurysms—the role of arterial spasm. J Neurosurg 22: 21–29

3. Allen GS, Gold LHA, Chou SN et al (1974) Cerebral arterial spasm. Part 3: In vivo intracisternal production of spasm by serotonin and blood and its reversal by phenoxybenzamine. J Neurosurg 40: 451–458

4. Arutiunov AI, Baron MA, Majorova NA (1974) The role of mechanical factors in the pathogenesis of short-term and prolonged spasm of the cerebral arteries. J Neurosurg 40: 459–472

5. Boullin DJ, Bunting S, Blaso WP et al (1979) Responses of human and baboon arteries to prostaglandin endoperoxides and biologically generated and synthetic prostacyclin. Their relevance to cerebral arterial spasm in man. Br J Clin Pharmacol 7: 139–147

6. Brown FD, Hanlon K, Mullan S (1978) Treatment of aneurysmal hemiplegia with dopamine and mannitol. J Neurosurg 49: 525–529

7. Brunori M, Falcioni G, Fioretti E et al (1975) Formation of superoxide in the autoxidation of isolated a and b chains of human hemoglobin and its involvement in emichrome precipitation. Eur J Biochem 53: 99–104

8. Chyatte D, Rusch N, Sundt TM Jr (1983) Prevention of chronic experimental cerebral vasospasm with ibuprofen and high-dose methylpredonisolone. J Neurosurg 59: 925–932

9. Clarisse J, Jomin M, Andreussi L et al (1972) Prognostic significance of cerebral arterial spasm in the course of meningeal hemorrhage. Neuroradiology 3: 150–152

10. Conway LW, McDonald LW (1972) Structual changes of the intracranial arteries following subarachnoid hemorrhage. J Neurosurg 37: 715–723

11. Cummins BH, Griffith HB (1971) Intracarotid phenoxybenzamine for cerebral arterial spasm. Br Med J 1: 382–383

12. DeBouley G (1963) Distribution of spasm in the intracranial arteries after subarachnoid hemorrhage. Acta Radiol 1: 257–266

13. Ecker A, Riemenschneider PA (1951) Arteriographic demonstration of spasm of the intracranial arteries. With special reference to saccular arterial aneurysms. J Neurosurg 8: 660–667

14. Edvinson L, Brandt L, Anderson KE et al (1979) Effect of a calcium antagonist on experimental constriction of human brain vessels. Surg Neurol 11: 327–330

15. Ellis EF, Nies AS, Oates JA (1977) Cerebral arterial smooth muscle contraction by thromboxane A2. Stroke 8: 480–486

16. Endo S, Suzuki J (1977) Experimental cerebral vasospasm after subarachnoid hemorrhage. Development and degree of vasospasm. Stroke 8: 702–707

17. Endo S, Suzuki J (1979) Experimental cerebral vasospasm after subarachnoid hemorrhage. The participiation of adrenergic nerves of cerebral vessel walls. Stroke 10: 703–711

18. Fein JM, Flor WJ, Cohan SL et al (1974) Sequential changes of vascular ultrastructure in experimental cerebral vasospasm. Myonecrosis of subarachnoid arteries. J Neurosurg 41: 49–58

19. Fisher CM, Kistler JP, Davis JM (1980) Relation of cerebral vasospasm to subarachnoid hemorrhage visualized by computed tomographic scanning. Neurosurgery 6: 1–9

20. Fisher CM, Robertson GH, Ojeman RG (1977) Cerebral vasospasm after ruptured aneurysm. Stroke 8: 11

21. Flamm ES, Yasargil MG, Ransohoff J II (1972) Alteration of experimental cerebral vasospasm by adrenergic blockade. J Neurosurg 37: 294–301

22. Fletcher TM, Taveras JM, Pool JL (1959) Cerebral vasospasm in angiography for intracranial aneurysms. Incidence and significance in one hundred consecutive angiograms. Arch Neurol 1: 38–47

23. Fox JL (1983) Intracranial aneurysms. Springer, New York Berlin Heidelberg

24. Fujimoto S, Mizoi K, Oba M et al (1984) Experimental study of protective effect on cerebral ischemia of various antioxidants and other agents. No Shinkei Geka 12: 171–180 (Eng Abstr)

25. Giannotta SL, Kindt GW, Haar FL (1975) Topical lidocaine in treatment of cerebral vasospasm. Surg Neurol 4: 13–16

26. Handa H, Osaka K, Okamoto S (1979) Breakdown products of erythrocytes as a cause of cerebral vasospasm. In: Wilkins RH (ed) Cerebral arterial spasm. Williams and Wilkins, Baltimore, pp 158–165

27. Handa J, Matsuda M, Ohtsubo K et al (1973) Effect of intracarotid phenoxybenzamine on cerebral blood flow and vasospasm: a clinical study. Surg Neurol 1: 229–232

28. Handa J, Yoneda S, Matsuda M et al (1974) Effects of prostaglandins A_1, E_1, E_2 and $F_{2\alpha}$ on the basilar artery of cats. Surg Neurol 2: 251–255

29. Harris P, Udvarhelyi GB (1957) Aneurysms arising at the internal carotid-posterior communicating artery junction. J Neurosurg 14: 180–191

30. Heilburn MP, Olesen J, Lassen NA (1972) Regional cerebral blood flow studies in subarachnoid hemorrhage. J Neurosurg 37: 36–44

31. Heros RC, Zervas NT, Negoro M (1976) Cerebral vasospasm. Surg Neurol 5: 354–362

32. Hughes JT, Schianti PM (1978) Cerebral artery spasm. A histological study at necropsy of the blood vessels in cases of subarachnoid hemorrhage. J Neurosurg 48: 515–525

33. Ishibashi Y, Konda R, Yoshimoto T (1984) The effect of prazosin hydrochloride on cerebral vasospasm. No Shinkei Geka 12: 133–139 (Eng Abstr)

34. Kamiyama K, Okada H, Suzuki J (1981) The relation between cerebral vasospasm and superoxide. Neurol Med Chir (Tokyo) 21: 201–209 (Eng Abstr)

35. Kaneko U, Yoshimoto T, Suzuki J et al (1979) A histopathological study of patients

dying from cerebral vasospasm after rupture of cerebral aneurysm. In: Suzuki J (ed) Cerebral aneurysms. Sendai: Neuron Pub, Tokyo, pp 535–540

36. Kapp J, Mehaley MS Jr, Odom GL (1968) Cerebral arterial spasm. Part 2: Experimental evaluation of mechanical and humoral factors in pathogenesis. J Neurosurg 29: 339–349

37. Kapp JP, Roberston JT, White RP (1976) Spasmogenic qualities of prostaglandin $F_{2\alpha}$ in the cat. J Neurosurg 44: 173–175

38. Kassel NF, Drake CG (1982) Timing of aneurysm surgery. Neurosurgery 10: 514–519

39. Kodama N, Mizoi K, Sakurai Y et al (1979) Incidence and onset of vasospasm. In: Wilkins RH (ed) Cerebral arterial spasm. Williams and Wilkins, Baltimore, pp 361–365

40. Komatsu S, Sato T, Ogawa A et al (1981) Correlation between CT findings and subsequent development of cerebral infarction due to vasospasm in subarachnoid hemorrhage cases. Neurol Med Chir (Tokyo) 21: 373–377 (Eng Abstr)

41. Konda R, Ishibashi Y, Okada H et al (1984) Sequential changes of vascular intimal ultrastructure in experimental cerebral vasospasm induced by oxyhemoglobin. No To Shinkei 36: 275–283 (Eng Abstr)

42. Kosnik EJ, Hunt WE (1976) Postoperative hypertension in the management of patients with intracranial arterial aneurysms. J Neurosurg 45: 148–154

43. Kwak R, Niizuma H, Ohi T et al (1979) Angiographic study of cerebral vasospasm following rupture of intracranial aneurysms. Part I: Time of the appearance. Surg Neurol 11: 257–262

44. Misra HP, Fridovich I (1972) The generation of superoxide radical during the autoxidation of hemoglobin. J Biol Chem 247: 2960–2962

45. Miyaoka M, Nonaka T, Watanabe H et al (1976) Etiology and treatment of prolonged vasospasm: experimental and clinical studies. Neurol Med Chir (Tokyo) 16: 103–114 (Eng Abstr)

46. Mizukami M, Kawase T, Usami T et al (1982) Prevention of vasospasm by early operation with removal of subarachnoid blood. Neurosurgery 10: 301–307

47. Niizuma H, Kwak R, Otabe K et al (1979) Angiographic study of cerebral vasospasm following rupture of intracranial aneurysms. Part II: Relation between the site of aneurysm and the occurrence of the vasospasm. Surg Neurol 11: 263–267

48. Ohmoto T, Yoshioka J, Morooka H et al (1979) Effect of ascorbic acid on cerebral vasospasm. In: Wilkins RH (ed) Cerebral arterial spasm. Williams and Wilkins, Baltimore, pp 619–624

49. Ohta T, Kawamura J, Osaka K et al (1969) Angiographic classification of so-called cerebral vasospasm. Correlation between existence of vasospasm and postoperative prognosis in subarachnoid hemorrhage. No To Shinkei 21: 1019–1027 (Eng Abstr)

50. Okada H, Endo S, Kamiyama K et al (1980) Oxyhemoglobin-induced cerebral vasospasm and sequential changes of vascular ultrastructure. Neurol Med Chir (Tokyo) 20: 573–582 (Eng Abstr)

51. Owada K, Hori S, Suzuki J (1979) Results of cervical sympathectomy for cerebral vasospasm following aneurysmal rupture. In: Suzuki J (ed) Cerebral aneurysms. Neuron Pub, Sendai, pp 435–441

52. Owada K, Hori S (1977) Cervical sympathectomy for cerebral ischemic lesions; a follow-up study. Tohoku Med J (Sendai) 90: 183–204

53. Pennink M, White RP, Crockarell JR et al (1972) Role of prostaglandin $F_{2\alpha}$ in the genesis of experimental cerebral vasospasm. Angiographic study in dogs. J Neurosurg 37: 398–406

54. Pool JL, Jacobson S, Fletcher TA (1958) Cerebral vasospasm, clinical and experimental evidence. JAMA 167: 1599–1601

55. Pool JL (1958) Cerebral vasospasm. New Engl J Med 259: 1259–1264

56. Pritz MB, Gianotta SL, Kindt GW et al (1978) Treatment of patients with neurological deficits associated with cerebral vasospasm by intravascular volume expansion. Neurosurgery 3: 364–368

57. Quintana L, Konda R, Ishibashi Y et al (1982) The effect of prostacyclin on cerebral vasospasm. An experimental study. Acta Neurochir (Wien) 62: 187–193

58. Raynor RB, McMurtry JG (1963) Prevention of serotonin-produced cerebral vasospasm. An evaluation of blocking agents. J Neurosurg 20: 94–96

59. Robertson EG (1949) Cerebral lesions due to intracranial aneurysms. Brain 72: 150–187

60. Sasaki T, Wakai S, Asano T *et al* (1982) Prevention of cerebral vasospasm after SAH with a thromboxane synthetase inhibitor, OKY-1581. J Neurosurg 57: 74–82

61. Sato S, Suzuki J (1975) Anatomical mapping of the cerebral nervi vasorum in the human brain. J Neurosurg 43: 559–568

62. Sato T, Sato S, Suzuki J (1980) Correlation with superior cervial sympathetic ganglion and sympathetic nerve innervation of intracranial artery—electron microscopic studies. Brain Res 188: 33–41

63. Someda K, Morita Y, Matsumura H (1979) Intimal change following subarachnoid hemorrhage resulting in prolonged arterial luminal narrowing. Neurol Med Chir (Tokyo) 19: 83–93 (Eng Abstr)

64. Sonobe M, Suzuki J (1978) Vasospasmogenic substance produced following subarachnoid hemorrhage, and its fate. Acta Neurochir (Wien) 44: 97–106

65. Sonobe M, Takahashi S, Otsuki T *et al* (1981) Preventive effect on intracranial arterial spasm using combined ventriculocisternal and cisternal drainage. No Shinkei Geka 9: 1393–1397 (Eng Abstr)

66. Stein BM (1975) Modification of cerebrovascular spasm. In: Smith RR, Robertson JT (eds) Subarachnoid hemorrhage and cerebrovascular spasm. ChC Thomas, Springfield, Illinois, pp 236–244

67. Suzuki J (ed) (1979) Cerebral aneurysms. Neuron Pub, Tokyo, pp 13–66

68. Suzuki J, Iwabuchi T, Hori S (1975) Cervical sympathectomy for cerebral vasospasm after aneurysm rupture. Neurol Med Chir (Tokyo) 15: 41–50 (Eng Abstr)

69. Suzuki J, Onuma T, Yoshimoto T (1979) Results of early operation on cerebral aneurysms. Surg Neurol 11: 407–412

70. Suzuki J, Yoshimoto T, Hori S (1974) Continous ventricular drainage to lessen surgical risk in ruptured intracranial aneurysm. Surg Neurol 2: 87–90

71. Suzuki S, Sobata E, Ando A *et al* (1981) Anaerobic change of bloody CSF in subarachnoid hemorrhage. Its relation to cerebral vasospasm. Acta Neurochir (Wien) 58: 15–26

72. Symon L (1978) Disordered cerebrovascular physiology in aneurysmal subarachnoid hemorrhage. Acta Neurochir (Wien) 41: 7–22

73. Tani E, Maeda Y, Fukumori T *et al* (1984) Effect of selective inhibitor of thrombozane A$_2$ synthetase on cerebral vasospasm after early surgery. J Neurosurg 61: 24–29

74. Tani E, Yamagata S, Ito Y (1978) Intercellular granules and vesicles in prolonged cerebral vasospasm. J Neurosurg 48: 179–189

75. Tanishima T (1980) Cerebral vasospasm: contractile activity of hemoglobin in isolated canine basilar arteries. J Neurosurg 53: 787–793

76. Toda N, Shimizu K, Ohta T (1980) Mechanism of cerebral arterial contraction induced by blood constituents. J Neurosurg 53: 312–322

77. Tomlinson BE (1966) Ischemic lesions of cerebral hemispheres. Following rupture of intracranial aneurysms. Part 1. Newcastle Med J 29: 81–94

78. Wellum GR, Irvine TW Jr, Zervas NT (1980) Dose responses of cerebral arteries of the dog, rabbit and man to human hemoglobin *in vitro*. J Neurosurg 53: 486–490

79. White RP, Cunningham MP, Robertson JT (1982) Effect of the calcium antagonist nimodipine on contractile responses of isolated canine basilar arteries induced by serotonin, prostaglandin F$_{2\alpha}$, thrombin, and whole blood. Neurosurgery 10: 344–348

80. White RP, Hung SP, Hagen AH *et al* (1979) Experimental assessment of phenoxybenzamine in cerebral vasospasm. J Neurosurg 50: 158–163

81. Wilkins RH (ed) (1979) Cerebral arterial spasm. Williams and Wilkins, Baltimore

82. Wilkins RH, Levitt P (1970) Intracranial arterial spasm in the dog. A chronic experimental model. J Neurosurg 33: 260–269

83. Wilkins RH, Levitt P (1971) Potassium and the pathogenesis of cerebral arterial spasm in dog and man. J Neurosurg 35: 45–50

84. Wilkins RH (1979) Attempted prevention or treatment of intracranial arterial spasm: a survey. In: Wilkins RH (ed) Cerebral arte-

rial spasm. Williams and Wilkins, Baltimore, pp 542–555

85. Yaşargil MG, Fox JL (1975) The microsurgical approach to intracranial aneurysms. Surg Neurol 3: 7–14

86. Zervas NT, Hori H, Rosoff CB (1974) Experimental inhibition of serotonin by antibiotic: prevention of cerebral vasospasm. J Neurosurg 41: 59–62

CHAPTER 12

1. Amine ARC, Moody RA, Meeks W (1977) Bilateral temporal-middle cerebral artery anastomosis for moyamoya syndrome. Surg Neurol 8: 3–6

2. Ausman JI, Moore J, Chou SN (1976) Spontaneous cerebral revascularization in a patient with STA-MCA anastomosis. Case report. J Neurosurg 44: 84–87

3. Boone SC, Sampson DS (1978) Observations on moyamoya disease: a case treated with superficial temporal middle cerebral artery anastomosis. Surg Neurol 9: 189–193

4. Brodner RA (1981) An unusual moyamoya syndrome treated with superficial temporal-middle cerebral artery anastomosis. Case report. Milit Medicine 146: 52–54

5. Huber P (1982) Cerebral angiography. Thieme, Stuttgart New York, pp 295–300

6. Isono M, Yonemitsu T, Fujiwara S et al (1981) Epidemiological study on Moyamoya desease.—From the experience of 100 cases. Proc 10th Jap Conf Surg Cereb Stroke: 3–7

7. Kameyama M, Shirane R, Tsurumi Y et al (1986) Evaluation of cerebral blood flow and metabolism in childhood moyamoya disease: an investigation into "re-build-up" on EEG by positron CT. Child's Nerv Syst 2: 130–133

8. Karasawa J, Kikuchi H, Furuse S et al (1977) A surgical treatment of "moyamoya" disease, "Encephalomyosynangiosis". Neurol Med Chir (Tokyo) 17: 29–37 (Eng Abstr)

9. Karasawa J, Kikuchi H, Furuse S et al (1978) Treatment of moyamoya disease with STA-MCA anastomosis. J Neurosurg 49: 679–688

10. Karasawa J, Kikuchi H, Kawamura J et al (1980) Intracranial transplantation of the omentum for cerebrovascular moyamoya desease. A two-year follow-up study. Surg Neurol 14: 444–449

11. Karasawa J, Kikuchi H, Kobayashi K et al (1981) Evaluation of angiographical changes after ST-MC anastomoses in "moyamoya" disease. Proc 10th Jap Conf Surg Cereb Stroke: 313–317

12. Karasawa J, Kikuchi H, Matsumoto A et al (1981) Surgical treatment of "Moyamoya" desease.—Follow up study during 3.5 years and over. Proc 10th Jap Conf Surg Cereb Stroke: 306–312

13. Kasai N, Fujiwara S, Kodama N et al (1982) The experimental study on causal genesis of moyamoya disease—correlation with immunological reaction and sympathetic nerve influence for vascular changes. No Shinkei Geka 10: 251–261 (Eng Abstr)

14. Kikuchi H, Karasawa J (1976) Extra-intracranial arterial anastomosis in ten patients with moyamoya syndrome (occlusion of the circle of Willis). In: Schmiedek P (ed) Microsurgery for stroke. Springer, Berlin Heidelberg New York, pp 260–263

15. Kobayashi K, Takeuchi S, Tsuchida T et al (1981) Encephalo-myosynagiosis (EMS) in moyamoya disease—with special reference to postoperative angiography. Neurol Med Chir (Tokyo) 21: 1229–1238 (Eng Abstr)

16. Kodama N, Aoki Y, Hiraga H et al (1979) Electroencephalographic findings in children with moyamoya disease. Arch Neurol 36: 16–19

17. Kodama N (1971) The study on the aging of the perforating branches and its possibility of collateral pathway: concerning with cerebrovascular "moyamoya" disease. No To Shinkei 23: 1389–1402 (Eng Abstr)

18. Kodama N, Mineura K, Suzuki J et al (1976) Ventricular hemorrhage due to chronic cerebral ischemia. No To Shinkei 28: 823–831 (Eng Abstr)

19. Kodama N, Suzuki J (1975) Cerebrovascular moyamoya disease. IIIrd Report. The study on the aging of the perforating branches and the possibility of collateral pathway. Neurol Med Chir (Tokyo) 15: 55–67 (Eng Abstr)

20. Kodama N, Suzuki J (1978) Moyamoya disease associated with aneurysm. J Neurosurg 48: 565–569

21. Kodama N, Fujiwara S, Horie Y et al (1980) Transdural anastomosis in Moyamoya disease—Vault Moyamoya. No Shinkei Geka 8: 729–737 (Eng Abstr)

22. Krayenbühl HA (1975) The moyamoya syndrome and neurosurgeon. Surg Neurol 4: 353–360

23. Kuru M, Karasawa J, Kuriyama Y et al (1981) Anesthetic management of "Moyamoya" disease in children. Proc Jap Conf Surg Cereb Stroke: 207–211

24. Lesoin F, Ijomin M, Viaud C et al (1983) Encephalo-arteriosynangiosis in the treatment of chronic cerebral ischemia. Preliminary report based on 30 cases. Surg Neurol 20: 318–322

25. Mathew NT, Abraham C, Chandy J (1970) Cerebral angiographic features in tuberculous meningitis. Neurology 20: 1015–1023

26. Matsushima Y, Fukai N, Tanaka K et al (1980) A new surgical treatment of moyamoya disease in children. A preliminary report. Surg Neurol 15: 313–320

27. Matsushima Y, Tomita H, Takei H et al (1985) Changes in symptoms after encephaloduroarteriosynangiosis (EDAS) in pediatric moyamoya disease. In: Spetzler RF, Carter LP, Selman WR et al (eds) Cerebral revascularization for stroke. Thieme-Stratton, New York, pp 578–583

28. Nagamine Y, Takahashi S, Sonobe M (1981) Multiple intracranial aneurysms associated with moyamoya disease. Case report. J Neurosurg 54: 673–676

29. Nakagawa Y, Ikota T, Ohtsuka K et al (1981) Indication of reconstructive operation for moyamoya disease and ideal operative methods. Proc 10th Jap Conf Surg Cereb Stroke: 230–235

30. Nakagawa Y, Tsuru M, Gotoh S et al (1984) Reconstructive surgery in moyamoya disease with hemorrhagic attack: Does surgical intervention reduce the risk for repeated bleeding? In: Handa H et al (eds) Microsurgical anastomoses for cerebral ischemia. Igakushoin, Tokyo, pp 233–239

31. Nakagawa Y, Tsuru M, Mabuchi S et al (1985) Reconstructive surgery in 28 patients of moyamoya disease. Operative methods, outcome and postoperative angiography. In: Spetzler RF et al (eds) Cerebral revascularization for stroke. Thieme-Stratton, New York, pp 308–317

32. Ohyama H, Niizuma H, Fujiwara S et al (1985) EEG findings in moyamoya disease in children.—Concerning with the causal genesis of re-build up. No Shinkei Geka 13: 727–733 (Eng Abstr)

33. Owada K, Hori S, Suzuki J (1979) Result of cervical sympathectomy for cerebral vasospasm following aneurysm rupture. In: Suzuki J (ed) Cerebral aneurysm. Neuron Pub, Tokyo, pp 435–441

34. Poor GY, Gacs GY (1974) The so-called "Moyamoya disease". J Neurol Neurosurg Psychiatry 37: 370–377

35. Rajakulasingam K, Cerullo J, Raimondi AJ (1979) Childhood Moyamoya syndrome. Postirradiation pathogenesis. Child Brain 5: 467

36. Richman DP, Watts HG, Parsons D et al (1977) Familial Moyamoya associated with biochemical abnormalities of connective tissue. Neurology 27: 382

37. Sato S, Suzuki J (1975) Anatomical mapping of the cerebral nervi vasorum in the human brain. J Neurosurg 43: 559–568

38. Sato T, Sato S, Suzuki J (1979) Correlation with superior cervical sympathetic ganglion and sympathetic nerve innervation of intracranial artery—electron microscopical studies. No To Shinkei 31: 375–384 (Eng Abstr)

39. Sobata E (1980) Efficacy of the cervical sympathectomy in cerebral ischemic patients.—Measurement of blood flow with a square wave electomagnetic flowmeter and clinical study. No Shinkei Geka 8: 739–748 (Eng Abstr)

40. Sonobe M, Takahashi S, Kubota Y et al (1982) Chronic subdural hematoma developing after EMS for moyamoya disease. No Shinkei Geka 10: 857–859 (Eng Abstr)

41. Spetzler RF, Roski RA, Kopaniky DR (1980) Alternative superficial temporal

artery to middle cerebral artery revascularization procedure. Neurosurgery 7: 484–487

42. Stockman JA, Nigro MA, Mishkin MM et al (1972) Occlusion of large cerebral vessels in sickle cell anemia. N Engl J Med 287: 846–849

43. Suzuki J, Takaku A, Asahi M et al (1965) Diseases showing the "fibrille" like vessels at the base of brain (frequently found in Japan). No To Shinkei 17: 767–776 (Eng Abstr)

44. Suzuki J, Takaku A, Asahi M et al (1968) The disease showing the abnormal vascular net-work at the base of brain, particularly found in Japan. II. A follow-up study. No To Shinkei 18: 897–908 (Eng Abstr)

45. Suzuki J, Takaku A (1968) The disease showing the abnormal vascular net-work at the base of brain, particularly found in Japan. III. Our opinion. Shindan To Chiryo 56: 469–472 (Eng Abstr)

46. Suzuki J, Takaku A (1968) The disease showing the abnormal vascular net-work at the base of brain, particularly found in Japan. IV.—our opinion of the dynamic change of these vascular net-work. No To Shinkei 20: 35–40 (Eng Abstr)

47. Suzuki J, Takaku A (1969) Cerebrovascular "moyamoya" disease. Disease showing abnormal net-like vessels in base of brain. Arch Neurol 20: 288–299

48. Suzuki J, Kodama N, Takaku A (1970) Collateral pathway via "ethmoidal moyamoya" in cerebrovascular "moyamoya" disease: disease showing abnormal net-like vessels in base of brain. No To Shinkei 22: 417–424 (Eng Abstr)

49. Suzuki J, Kodama N (1971) Cerebrovascular "Moyamoya" disease—second report. Collateral routes to forebrain via ethmoid sinus and superior nasal meatus. Angiology 22: 223–236

50. Suzuki J, Takaku A, Kodama N et al (1975) An attempt to treat cerebrovascular moyamoya disease in children. Child Brain 1: 193–206

51. Suzuki J, Kodama N, Mineura K (1976) Mechanism of symptomatic occurrence in cerebrovascular moyamoya disease. No To Shinkei 28: 459–470 (Eng Abstr)

52. Suzuki J, Kodama N (1980) Correlation between ventricular hemorrhage and chronic cerebral ischemia in adult moyamoya. In: Pia HW, Langmaid C, Zierski J (eds) Spontaneous intracerebral hematoma. Springer, Berlin Heidelberg New York, pp 145–152

53. Suzuki J, Kodama N (1983) Moyamoya disease.—A review. Stroke 14: 104–109

54. Suzuki J (ed) (1983) Moyamoya disease. Igaku-shoin, Tokyo

55. Takahashi A, Fujiwara S, Suzuki J (1985) Cerebral angiography following hyperventilation in moyamoya disease.—In relation to the "re-build up" phenomenon on EEG. No Shinkei Geka 13: 255–264 (Eng Abstr)

56. Takahashi A, Fujiwara S, Suzuki J (1986) Long-term follow-up angiography of moyamoya disease.—Cases followed up from childhood to adolescence. No Shinkei Geka 13: 23–29 (Eng Abstr)

57. Takeuchi S, Kobayashi K, Tsuchida T et al (1981) Effect of encephalo-myo-synangiosis in moyamoya disease. Proc 10th Jap Conf Surg Cereb Stroke: 281–285

58. Tomsick TA, Lukin RR, Chambers AA et al (1976) Neurofibromatosis and intracranial arterial occlusive disease. Neuroradiology 11: 229–234

59. Tsubokawa T, Kikuchi M, Asano S et al (1964) Surgical treatment for intracranial thrombosis. Case report of "Durapexia". Neurol Med Chir (Tokyo) 6: 428–429 (Eng Abstr)

60. Ximin L, Xuzhong R, Zhuan C et al (1980) Moyamoya disease caused by leptospiral cerebral arteritis. Chin Med J 93: 599–604

61. Yaşargil MG, Yonekawa Y, Denton I et al (1974) Experimental intracranial transplantation of autogenic omentum majus. J Neurosurg 40: 213–217

62. Yonekawa Y, Yaşargil MG (1977) Brain vascularization by transplanted omentum. A possible treatment of cerebral ischemia. Neurosurgery 1: 256–259

63. Yonemitsu T, Fujiwara S, Kodama N et al (1984) The experimental study on causal genesis of moyamoya disease—immunohistochemical investigation. Angiology (Tokyo) 24: 537–547 (Eng Abstr)

SUBJECT INDEX

B. George / C. Laurian

The Vertebral Artery

Surgery and Pathology

1987. 97 figures. Approx. 300 pages.
Cloth approx. DM 170,–, approx. öS 1200,–
ISBN 3-211-81968-1
Prices are subject to change without notice

In considering vascular surgery in the neck, the carotid artery has usually been regarded as the only vascular axis amenable to surgical approach; surgery on the vertebral artery was contemplated with the idea of mastering a challenge.

The authors demonstrate that the surgical approach, exposure and control of the VA on any part of its cervical course has become a reliable technique at whatever level. Precise technique permits exposure of the VA with a sufficiently large field to allow any surgical procedure identical to that on any other vessel to be performed.

The authors have now performed more than 150 VA operations for various indications with excellent results, no mortality and a very limited morbidity. But the choice of the procedure remains a difficult and controversial problem. Profound knowledge of the anatomy and the frequent variations and anomalies of each VA are of utmost importance.

Therefore the book includes: anatomy of the VA and its variations as far as relevant for the surgeon; extensive description of the pathologies involving the VA and consequences upon the cerebral blood supply; detailed description of surgical approaches to any part of the cervical VA including an original approach with as much safety as carotid artery surgery; report on the authors' personal experience with successes, pitfalls and failures.

These possibilities should be known by every specialist having to cope with deep lesions in the neck or managing cerebral ischemia.

Springer-Verlag Wien New York

Moelkerbastei 5, P.O. Box 367, A-1011 Wien
175 Fifth Avenue, New York, NY 10010, USA
Heidelberger Platz 3, D-1000 Berlin 33
37-3, Hongo 3-chome, Bunkyo-ku, Tokyo 113, Japan

Transcranial Doppler Sonography

Edited by R. Aaslid

1986. 94 figures. XI, 177 pages. ISBN 3-211-81935-5
Soft cover DM 68,–, öS 476,–

Prices are subject to change without notice

From the Foreword by M. P. Spencer, M. D., Director of the Institute of Applied Physiology and Medicine, Seattle, Washington, U.S.A.:
"Every few years a dissertation comes to the area of clinical application of medical technology which carries us forward as on a magic carpet into new regions of understanding and patient care. This book is such a magic carpet. It brings together, in a clear and incisive fashion, important hemodynamic principles with a simple non-invasive method of application to a part of the cerebral vasculature which has been relatively inaccessible. To the lucky and perceptive person who reads this book, a feeling of excitement and hope for progress is engendered. The diligent application of the potentials of transcranial Doppler ultrasound brings new power to our efforts in understanding the cerebral circulation and the causes, treatment and prevention of cerebrovascular disorders."

Springer-Verlag Wien New York

Moelkerbastei 5, A-1010 Wien;
175 Fifth Avenue, New York, NY 10010, USA;
Heidelberger Platz 3, D-1000 Berlin 33;
37-3, Hongo 3-chome, Bunkyo-ku, Tokyo, Japan

A. Harders

Neurosurgical Applications of Transcranial Doppler Sonography

1986. 109 figures. X, 134 pages. ISBN 3-211-81938-X
Soft cover DM 58,–, öS 406,–
Prices are subject to change without notice

In 1981 Dr. Rune Aaslid developed a transcranial Doppler device with a pulsed sound emission of 2 MHz, which enabled blood flow velocities to be measured in the large branches of the circle of Willis. With this innovation, it has become possible to record atraumatically and repeatedly the intracranial hemodynamic changes in neuro-vascular diseases.

The book describes the hemodynamic principles in cerebral vascular circulation and the factors which can effect the blood flow velocities (such as collateral circulation, diameter of the vessels, vascular resistance, arterial partial CO_2 pressure, autoregu-latory factors, and position of the body). Normal values of blood flow velocities and the changes under physiological deviations are measured by transcranial Doppler technique. For patients suffering from subarachnoid hemorrhage, individual time courses of velocity changes are evaluated and the application in clinical routines is stressed: Better defined timing of angiography, surgery and postoperative hyper-tension therapy has significantly reduced the incidence of delayed ischemic deficits. Patients indicating for extracranial-intracranial bypass surgery, as well as the post-operative changed hemodynamics are also investigated. The contribution of the bypass to the brain circulation can be tested by compression tests. The "activity" of an angioma and the influence of superselective embolization procedures for arterio-venous malformations are described.

Furthermore, cerebro-vascular blood flow arrest in brain death patients, can clearly be seen without angiography by evaluating a reverberating flow pattern. The book gives an account of the role of a still very young but exciting technique in diagnostic and therapeutic procedures of cerebral vascular disease based upon three years of experience at the Neurosurgical Department of the University of Freiburg.

Springer-Verlag Wien New York

Springer-Verlag, Moelkerbastei 5, A-1010 Wien;
175 Fifth Avenue, New York, NY 10010, USA;
Heidelberger Platz 3, D-1000 Berlin 33;
37-3, Hongo 3-chome, Bunkyo-ku, Tokyo, Japan